LASER SPECTROSCOPY IX

Proceedings of the
Ninth International Conference on
Laser Spectroscopy

Bretton Woods, New Hampshire
June 18–23, 1989

LASER SPECTROSCOPY IX

Proceedings of the
Ninth International Conference on
Laser Spectroscopy

Bretton Woods, New Hampshire
June 18–23, 1989

Edited by

Michael S. Feld

Department of Physics
Massachusetts Institute of Technology
Cambridge, Massachusetts

John E. Thomas

Department of Physics
Duke University
Durham, North Carolina

Aram Mooradian

MIT-Lincoln Laboratory
Lexington, Massachusetts

ACADEMIC PRESS, INC.

Harcourt Brace Jovanovich, Publishers

Boston San Diego New York
Berkeley London Sydney
Tokyo Toronto

ACADEMIC PRESS, INC.
1250 Sixth Avenue, San Diego, CA 92101

United Kingdom Edition published by
ACADEMIC PRESS INC. (LONDON) LTD.
24-28 Oval Road, London NW1 7DX

Library of Congress Cataloging-in-Publication Data

International Conference on Laser Spectroscopy (9th : 1989 : Bretton
  Woods, N.H.)
  Laser spectroscopy IX : proceedings of the Ninth International
  Conference on Laser Spectroscopy, Bretton Woods, New Hampshire, June
  18-23, 1989 / edited by Michael S. Feld, John E. Thomas, Aram
  Mooradian.
    p.   cm.
  Includes bibliographical references.
  ISBN 0-12-251930-2
  1. Laser spectroscopy — Congresses. I. Feld, Michael S., 1940-
  II. Thomas, John E. III. Mooradian, Aram, 1937-   IV. Title.
  V. Title: Laser spectroscopy 9. VI. Title: Laser spectroscopy nine.
  QC454.L3157   1989
  621.36'6 — dc20   89-28991
                                                              CIP

Printed in the United States of America
89 90 91 92   9 8 7 6 5 4 3 2 1

PREFACE

The Ninth International Conference on Laser Spectroscopy, NICOLS'89, was held in the United States at the Mount Washington Hotel in Bretton Woods, New Hampshire on June 18-23, 1989. In the tradition of its predecessors at Vail, Megeve, Jackson Lake, Rottach-Egern, Jasper Park, Interlaken, Maui and Åre, NICOLS'89 was held in a scenic and remote location, where scientists active in the field of laser spectroscopy discussed recent developments in a relaxed and informal atmosphere. Evening poster sessions, which were spread throughout the historic lobby meeting rooms of the Mount Washington Hotel, greatly facilitated a lively and stimulating exchange of ideas. Substantial free time during the afternoons enabled participants and their families to enjoy the natural beauty of this mountainous location.

The meeting was attended by 201 scientists from 22 countries, including Australia, Austria, Brazil, Canada, China, Denmark, Finland, France, Germany, Iraq, Ireland, Israel, Italy, Japan, The Netherlands, the Republic of China, the Soviet Union, Spain, Sweden, Switzerland, the United Kingdom and the United States.

The scientific program consisted of oral and poster presentations. There were 52 invited talks organized into 14 topical sessions, some with panel discussions. About 60 additional invited contributions were presented in three evening poster sessions. Also included were 15 post deadline oral and poster presentations. These proceedings contain summaries of essentially all of these contributions.

We thank the participants for making NICOLS'89 a stimulating conference, particularly the presenters for the high quality of their contributions and for preparing the manuscripts which comprise this volume. We thank the members of the International Steering Committee for their thoughtful suggestions and advice. We are particularly indebted to the Program Committee for composing a stimulating, broad and well balanced scientific program.

NICOLS'89 was held under the auspices of the International Union of Pure and Applied Physics. We gratefully acknowledge the financial support of IUPAP, the National Science Foundation, the Army Research Office, and the Office of Naval Research. We are also indebted to our corporate sponsors, Academic Press, Coherent, Inc., Lumonics Inc., Newport Research Corporation, Spectra Physics, SPEX, and Quantel. They provided funds which helped defray the cost of travel to the United States for many of our colleagues, and enabled us to create a relaxed and intellectually stimulating atmosphere during our stay in Bretton Woods.

Many people worked behind the scenes to make NICOLS'89 a success. We thank our colleagues of the MIT Graphic Arts Service for their diligent efforts. We thank Scott Anderson, our initial Conference Coordinator, for his help during the Program Committee meetings. We thank John Faherty, the Conference Coordinator, whose hard work, enthusiasm, keen intelligence, and sense of humor in dealing with our many colleagues, contributed greatly to the conference and proceedings. We are indebted to Ramachandra R. Dasari, whose advice and selfless efforts were essential in planning the conference. We thank Wenyue Hsu, Claudio Cesar, Jim Childs and Mike Donovan, whose logistical support made the conference run smoothly and on time. We thank Firooz Partovi for his help with the manuscript. We thank our secretary, Liz Shergill, for cheerfully working many long hours in the final days before the meeting, and also Jennifer Whipple, for careful attention to many details. We also thank Norman Bolter and his fellow musicians of the Boston Symphonic Brass for a delightful concert. Finally, we acknowledge the outstanding facilities of the Mount Washington Hotel and thank its excellent staff, especially Nicola Broderick and Bea Dorsey, for their expert and cheerful help.

August 1989   M.S. Feld
                                                    A. Mooradian
                                                    J.E. Thomas

A Laser Commercial; NICOLS After-Dinner Talk

N. Bloembergen
Division of Applied Sciences, Pierce Hall, Harvard University,
Cambridge, MA 02138 USA

Ladies and gentlemen,

We are approaching the end of a busy scientific conference and an equally successful and equally exhausting banquet. You and I deserve a rest. We should simply thank the chairman, Mike Feld, and his staff for all the arrangements and call it a day. However, Mike Feld wanted to fill a much-needed void in the opulent meeting program and appointed me as after-dinner speaker. So here I am, filling the needed void.

There are several reasons and excuses to put me in this predicament. My first name, Nico, resembles the name of the meeting: Nicols. In fact, several colleagues have addressed me in the past as Nicole. Others continue to call me Nikko. This version of my first name caused some hilarity at an International Conference in Japan held in 1953. The Japanese hosts always have a polite smile when a guest is introduced, but I could not help noticing that the smile was a little more pronounced when they heard my first name. When visiting Nikko National Park, 3 hours north of Tokyo, I found out that the Japanese have a saying, "Do not call anything Kekko until you have seen Nikko," and "Kekko" means "beautiful."

A second reason why I am up here to speak after dinner is that I had nothing of new scientific interest to present at the regular sessions. When I so informed Mike Feld, he said "Fine, then you can be the after-dinner speaker." I am rather inexperienced in this new role into which my waning productivity has put me. To prepare myself for this occasion, I went to a bookstore in Harvard Square and found an interesting title, "The ABC and XYZ of After-Dinner Speaking," I learned that ABC stands for "Always Be Calm," and XYZ stands for "[e] Xamine Your Zipper."

NICOLS is the ninth in the series of International Conferences on Laser Spectroscopy. The first one was held in 1973 in Vail, Colorado. The originators of the series were Dick Brewer and Aram Mooradian. Aram as well as many other young and not-quite-so-young faithful are again present at this meeting. The second through eighth in this series of meetings were all held in odd-numbered years, in Megeve, France (1975), in the Grand Tetons at Jackson Lake, Wyoming (1977), Rottach-Egern in the Bavarian Alps (1979), at Jasper in the Canadian Rockies (1981), Interlaken, Switzerland (1983), Maui, Hawaii (1985), Are, Sweden (1987), and now we have settled down at the foot of Mt. Washington in New Hampshire, we have discussed at length where we are going to meet two years from now. It will be in France, but the difficult choice between the Alps and the Pyrenees requires further research.

The ICOLS meetings have been quite successful because they have been restricted in size, as well as in subject matter. They have brought together leaders in the field of laser spectroscopy, which is concerned with the fine features, or fingerprints, of matter, revealing the structure of atoms, molecules and electromagnetic radiation. The interactions and discussions at a truly international level during the

ICOLS series have undoubtedly provided a stimulus for the continued growth in this active field of scientific endeavor. At every meeting we have had the opportunity to hear about new developments, admire new blooms and offshoots, but also to give and receive critique, to define areas of disagreement which could only be resolved by further theoretical and experimental investigations.

The restricted nature of these meetings has been criticized by outsiders. This critisism would be justified if ICOLS were the only meetings in the field. There are, however, numerous unrestricted meetings which also cover the field of laser spectroscopy. In fact, there are so many of these, that it would be most undesirable to add another unrestricted meeting to the group that already includes the International Conference on Quantum Electrons, the Conference on Lasers and Electro-Optics, the Interdisciplinary Laser Science Conference, among others. Most of these convene at least once a year. They are big and sometimes brassy, *i.e.*, with large hardware exhibitions. They have a dozen or more simultaneous sessions, and attendance at some of them exceeds five thousand. They certainly fulfill an important function, and their very existence permits the ICOLS series to retain its well-protected niche in the larger spectrum (no pun intended) of activity of Quantum Optics, Opto-Electronics, Photonics, or whatever other nomenclature may be used to denote the field of interest.

We are indeed fortunate that the laser field has so many practical ramifications. Personally, I have always enjoyed my contacts with industry. The direct link between fundamental physics and important practical applications enhances the vitality of our field. It attracts many younger students, scientists and engineers. While the conference program emphasizes issues of basic physics and precision measurements, it may be appropriate in this after-dinner talk to call attention to the enormous industrial development which has taken place concomitant with the ICOLS series during the past two decades. Undoubtedly we derive enormous indirect benefits from these commercial developments.

During the 1960's a large variety of lasers, employing hundreds of different substances were demonstrated. This period may be considered as the birth of a new technology. At that time, lasers were described as solutions looking for problems. The decade of the 1970's may be called the period of laser engineering development. The physics of lasers was well understood, possible fields of applications were well recognized, but it proved difficult to turn lasers into economically competitive devices. During the 1980's sizeable comerical markets for laser products have been established. The businesslike question is now which laser — or other electro-optic device — has the best characteristics at a price that the market will bear.

The total commercial laser industry world market in 1984 amounted to near 3 billion dollars, with an annual growth rate of about 40 percent. So in 1987, the total market exceeded 6 billion U.S. dollars. The main categories of commerical applications are: Materials Working, Communication, Medical-Surgical, Metrology and Inspection, Data Capture, Alignment (Construction Industry), and Printing and Graphics.

In 1984 the last category had the largest commercial importance. While the application of lasers in this field and in construction alignment is reaching maturity, laser applications in communication and medicine are still growing fast, and are about to become the most important segments of the market. It is important to note that the commercial laser market exceeded the military market by a margin of more than two-to-one in '84 and this difference is even more pronounced at present.

Last year, in 1988, over 200 million semiconductor lasers have been produced. Mass production techniques, similar to those developed for microelectronic chips, are used. In dollar value, the semiconductor lasers now comprise 30 percent of all lasers. They are by far the cheapest and most durable of all lasers. Unit prices are in the 1–10 dollar range. Laser diode production has become a commodity process with commodity-pricing strategies. Small diode lasers (100 μW output) are used in every compact-disk player. Somewhat more powerful semiconductor lasers are used for low-cost laser printing, and for optical communications. More powerful solid state arrays are under development. Integrated opto-electronic, all solid-state devices, useful in optical data processing, are in the prototype design phase. Semiconductor lasers will become serious competitors for the He-Ne lasers. These are the next most numerous lasers with an annual production of one-quarter million in 1988. More powerful lasers, such as argon-ion lasers, CO_2 lasers and solid state lasers are produced in much smaller quantities, but their total dollar value is comparable to that of all semiconductor lasers.

Most applications of lasers are based on the concentration of light in time and space. The increase in intensity due to focusing the light was already known and used in Greek antiquity. Lactinius observed and wrote in 303 B.C. that a glass globe filled with water is "sufficiently good to light a fire, even in the coldest of weather," Archimedes (212 B.C.) proposed to set fire to a hostile fleet of Phoenicians in the harbor of Syracuse by reflecting the sun's rays from the metal shields of his soldiers. I remember as a young boy using a looking glass to burn shoelaces under the pale sun of the Netherlands.

Laser light can be focused on an area λ^2, which is a few hundred times smaller than the cross-section of a human hair. By concentration in space and time, power flux densities of 10^{12} watts/cm^2 equal to those prevalent in the interior of stars are readily obtained. New physics is reported at power levels of 10^{18}watts/cm^2. Thus laser beams can readily drill holes through textiles, metals, through diamond. They can be used for cutting and welding. They can equally well coagulate the vaporize human tissue, or cut into teeth or bone. More importantly, laser light can be concentrated into and guided by optical fibers, with a dimension of a human hair. Bundles of fibers can be inserted into blood vessels or other channels inside the human body, and used for interior viewing, and for removal of diseased tissue or blockages.

Most of the surgical applications are based on localized absorption of laser light. The use of lasers in the printing industry is also based on the vaporization of thin metal film on a transparent substrate, or on the absorption of light in a photoresist material, with subsequent preferential etching and removal of unexposed material. Such processes are used also in recording compact disks, etc.

Optical information may be stored on magnetic tapes, which are magnetized in a direction opposite to that of a magnetic bias field. As a laser spot hits an area λ^2, localized heating reduces the magnetic anisotropy or remanence, and the magnetization is flipped locally by the bias field. The magnetically stored information can again be read out optically by using reflection of polarized light.

The process of heating by light adsorption is not very exciting from a fundamental physics point of view, but many large-scale commercial applications are based on it. Similarly, the propagation of laser beams along straight lines through the atmosphere or in a vacuum, or in single-mode optical fibers over long distances, is, scientifically-speaking, "old hat," but the widespread use of laser beams for alignment in the construction industry or for optical communications depends on it.

Communication by optical fibers is in an advanced stage of engineering development. An optical cable under the Atlantic is now in operation, and connects a point on the coast of New Jersey with points in England and France. It has the capacity to take care of all present-day telephone communications between two continents. A cable under the Pacific connecting the U.S. West Coast via Hawaii with Japan is scheduled to be inaugurated this year.

Future developments in optical communications will utilize more sophisticated physics and nonlinear optics. Soliton pulses will travel without distortion, the inevitable dispersion in group velocity being compensated by an intensity-dependent index of refraction. The very small losses (< 0.5 db/km) could in principle be compensated by Raman gain, so that the information could be transmitted over 5000 km without the use of repeater stations, involving detection and recoding.

Of course the scientific applications of lasers to spectroscopy, to metrology, to astrophysics, chemistry, biology and physics, etc., are numerous and often based on quite advanced physical concepts. They are appropriate subjects for the scientific sessions of this conference, but not for this after-dinner talk. The point I wanted to make is that large-scale commerical application, the spin-off of advanced laser research, has taken place, and that this is very beneficial to our endeavor. The history of laser development is a textbook case of the give-and-take between pure science and technology.

The general public is the ultimate beneficiary as people can communicate through optical fibers with the whole world, as surgery is revolutionized by less traumatic laser procedures, and as compact disk players become standard equipment in households of industrialized societies. Commercial enterprises benefit as they produce laser devices for mass markets. We as scientists probably benefit the most. We reap all the same advantages as the general public but we can, in addition, serve as consultants to the commercial enterprises, while we retain the fun of carrying out interesting experiments in our laboratories using technologically advanced equipment. An on top of all this, we report our results at nice places such as this conference.

Enjoy it while you can, and thank you for your attention.

INTERNATIONAL STEERING COMMITTEE

F.T. Arecchi, National Institute of Optics, Italy
N. Bloembergen, Harvard University, USA
Ch. J. Borde, University Paris-Nord, France
D.J. Bradley, Trinity College, Ireland
R.G. Brewer, IBM Almaden Research Center, USA
V.P. Chebotayev, Institute for Thermophysics, USSR
W. Demtroder, University Kaiserslautern, FRG
M.S. Feld, Massachusetts Institute of Technology, USA
A.L. Ferguson, University of Southampton, UK
J.L. Hall, Joint Institute for Laboratory Astrophysics, USA
T.W. Hansch, University of Munich, FRG
S. Haroche, University of Paris VI, France
S.E. Harris, Stanford University, USA
T. Jaeger, Norsk Elektro Optikk, Norway
W. Kaiser, Technical University of Munich, FRG
V.S. Letokhov, Institute of Spectroscopy, USSR
S. Liu, South China Normal University, China
A. Mooradian, Massachusetts Institute of Technology, USA
F.P. Schafer, MPI Biophysiks, FRG
Y.R. Shen, University of California, USA
T. Shimizu, University of Toyko, Japan
K. Shimoda, Keio University, Japan
B.P. Stoicheff, University of Toronto, Canada
S. Svanberg, Lund Institute of Technology, Sweden
J.E. Thomas, Duke University, USA
H. Walther, University of Munich, FRG
H.P. Weber, University of Berne, Switzerland
Z.M. Zhang, Fudan University, China

PROGRAM COMMITTEE

J. Bjorkholm, AT&T Bell Laboratories
M. Ducloy, University of Paris-Nord
M.S. Feld, Massachusetts Institute of Technology, Chairman
J.L. Hall, JILA Boulder
T.W. Hansch, University of Munich
E.P. Ippen, Massachusetts Institute of Technology
V.S. Letokhov, Institute of Spectroscopy, USSR
A. Mooradian, MIT Lincoln Laboratory, Cochairman
B.P. Stoicheff, University of Toronto
J.E. Thomas, Duke University, Cochairman

SPONSORS

This volume is the written record of NICOLS'89, the Ninth International Conference on Laser Spectroscopy, and would not have been possible without the generous support of a number of companies and sponsoring agencies. The participants, authors and editors express our sincere gratitude to them.

Corporate Sponsors

Academic Press
Coherent, Inc.
Lumonics Inc.
Newport Research Corporation
Spectra Physics
SPEX
Quantel

Agency Sponsors

Army Research Office
National Science Foundation
Office of Naval Research

NICOLS'89 was conducted under the auspices of IUPAP, the International Union of Pure and Applied Physics, which also provided financial support.

CONTENTS

xiv

Part III	CAVITY QED

Part IV   NOISE AND COHERENCE

Part VIII	TRAPPED ION SPECTROSCOPY

Part XI   MOLECULAR SPECTROSCOPY AND DYNAMICS

Part XII   APPLICATIONS IN RADIATION FORCES

xxi

Part XIII   HIGHLY EXCITED STATES AND DYNAMICS

Part XIV   LASER SPECTROSCOPY FOR BIOMEDICINE

Part I

New Cooling Mechanisms

New Physical Mechanisms in Laser Cooling

Y. Castin, K. Mølmer[*], J. Dalibard and C. Cohen-Tannoudji
College de France et Laboratoire de Spectroscopie Hertzienne de l'Ecole
Normale Supérieure, 24 rue Lhomond, F-75231 Paris Cedex 05, France[**]

1. Introduction

Since the last Laser Spectroscopy Conference, there have been major advances in laser cooling. Two important limits have been overcome. First, precise measurements have shown that the temperatures in optical molasses were about one order of magnitude lower than the expected Doppler limit T_D given by $k_B T_D = \hbar\Gamma/2$, where Γ is the natural width of the excited state e (1-3). Simultaneously, it was shown that it is even possible to go below the recoil limit T_R given by $k_B T_R = \hbar^2 k^2/2M$, where $\hbar^2 k^2/2M$ is the recoil energy associated with the absorption or the emission of a single photon (4).

In this paper, we will focus on the new physical mechanisms allowing one to beat the Doppler limit. A first qualitative explanation of these mechanisms in terms of non-adiabatic effects for a multilevel atom moving in a gradient of laser polarization has been given independently by the Paris and Stanford groups at the last International Conference on Atomic Physics (2,3). We present here a few new theoretical results concerning polarization gradient cooling. After some general considerations on the connection between friction and non-adiabaticity (§2), we show in §3 that, in the simple case of 1-d molasses, there are two different cooling mechanisms which occur respectively when the two counterpropagating waves have orthogonal linear polarizations (lin ⊥ lin case) or orthogonal circular polarizations (σ^+-σ^- case). A more detailed study of these two cases can be found in (5). We present here new velocity distribution profiles deduced from a full quantum treatment of atomic motion in the σ^+-σ^- configuration and we interpret the structures appearing in these profiles in terms of a large step momentum diffusion (§4).

2. General Considerations

An atom at rest in $\vec{r}$ has in general an internal steady state $\sigma_{st}(\vec{r})$ which depends on the characteristics of the laser electric field in $\vec{r}$ (amplitude, phase, polarization). Suppose now that the atom is moving with a velocity $\vec{v}$. Because of the finite internal response time τ_{int}, the internal state $\sigma(\vec{r}, \vec{v})$ for an atom passing in $\vec{r}$ with a velocity $\vec{v}$ lags behind $\sigma_{st}(\vec{r})$. These non-adiabatic effects are characterized by a dimensionless parameter $\epsilon = v\tau_{int}/\lambda = kv\tau_{int}$, which is the ratio between the distance travelled by the atom during τ_{int} and the laser wavelength $\lambda = 1/k$ which is the typical length scale for the laser field. The term linear in ϵ, in the perturbative expansion of $\sigma(\vec{r}, v)$ in powers of ϵ, gives rise to a mean radiative force linear in v, $-\alpha v$, which can be interpreted as a friction force, α being the __friction coefficient__ (6). On the other hand, general arguments using the connection between fluctuations and dissipation (7), allow one to relate the __momentum diffusion coefficient__ D to α, and to show that the __equilibrium temperature__ T of the atoms is given by

----

(*)   Institute of Physics, University of Aarhus
      DK-8000 Aarhus C. Denmark
(**) Laboratoire associé au C. N. R. S. et à l'Université Pierre et Marie Curie

$k_BT \sim D/M\alpha \gtrsim \hbar/\tau_{int}$. When applied to a 2-level atom e-g, having a single internal time $\tau_{int} = \tau_R = 1/\Gamma$, which is the radiative lifetime of the excited state e, these general considerations lead to an equilibrium temperature $k_BT \sim \hbar/\tau_R = \hbar\Gamma$ which is nothing but the Doppler limit T_D.

Consider now an atom, such as an alkali atom, having several Zeeman sublevels in the ground state g. There is, in this case, new internal times τ_p, which are the mean optical pumping times between the Zeeman sublevels of g. We can write $\tau_p = 1/\Gamma'$, where Γ' is the mean scattering rate of the incident photons and can be considered as the radiative width of g due to the optical excitation. Note that such an optical excitation introduces also a light-shift $\hbar\Delta'$ of the Zeeman sublevels of g. At low laser intensity I_L, Γ' and Δ' are both proportional to I_L and can be much smaller than the natural width Γ of e. It follows that the new non-adiabaticity parameter $\epsilon' = kv\tau_p = kv/\Gamma'$ associated with τ_p can be much larger than the previous one $\epsilon = kv/\Gamma$. Non-adiabatic effects can thus appear at much lower velocities, giving rise to higher friction coefficients. The same general considerations predict also an equilibrium temperature $k_BT \gtrsim \hbar/\tau_p = \hbar\Gamma'$, much smaller than T_D, since $\Gamma' \ll \Gamma$ at low I_L.

Finally, let us discuss the importance of gradients of laser polarization. Since we want $\Gamma' \ll \Gamma$, the laser intensity must be small, so that the atom is mainly in g. On the other hand, non-adiabatic effects are important only if $\sigma_{st}(\vec{r})$ varies rapidly with $\vec{r}$. Since the total population in g is nearly equal to 1, and therefore independent of $\vec{r}$, the only possibility leading to large spatial gradients of $\sigma_{st}(\vec{r})$ is to have an anisotropy in g, usually described in terms of orientation or alignment, which depends strongly on $\vec{r}$. This can be achieved simply if the laser polarization varies rapidly in space.

3. The Two Types of Polarization Gradient Cooling at 1-Dimension

We summarize here the results derived in (5) in the simple case of 1-d molasses.

A first type of polarization gradient corresponds to a gradient of ellipticity with fixed polarization axis. It occurs for example when the two waves have equal amplitudes and orthogonal linear polarizations (lin ⊥ lin configuration). In such a case, the polarization of the total field changes every $\lambda/8$ from σ^+, to linear, to σ^-, to linear, to σ^+... One can then show that the light-shifts and populations of the Zeeman sublevels of g are spatially modulated. For a moving atom, optical pumping between these sublevels gives rise to a "Sisyphus effect" analogous to the one occuring in stimulated molasses (8,9) : because of the finite optical pumping time, the atom can climb a potential hill and reach the top of this hill before being optically pumped to a potential valley. During the climbing, part of the kinetic energy is transformed into potential energy, the decrease of atomic momentum being due to the dipole forces, or, equivalently, to a redistribution of photons between the two counterpropagating waves. On the other hand, the energy dissipation, which is essential for the cooling process, is achieved by spontaneous anti-Stokes Raman processes, which carry away the potential energy gained during the climbing.

The second type of polarization gradient corresponds to a pure rotation of the polarization axis, with a fixed ellipticity. It occurs for example when the two counterpropagating waves have equal amplitudes and orthogonal circular polarizations (σ^+-σ^- configuration). In such a case, the resulting polarization is always linear and rotates around the propagation axis Oz of the two waves, forming an helix with a pitch λ. Since the laser intensity is independent of z, the light-shifts of the Zeeman sublevels in g remain constant in space, so that no dipole forces can exist in this case. A new

3

cooling mechanism occurs if there are more than two Zeeman sublevels in g ($J_g \geqslant 1$) and is due to the fact that the wave functions of the light-shifted sublevels in g vary in space. More precisely, one can show that, even at very low velocity, atomic motion in a rotating linear polarization produces a population difference between the Zeeman sublevels in g defined along Oz, giving rise to a large imbalance between the radiation pressures exerted by the two counterpropagating waves. This imbalance varies as kv/Δ', where $\hbar\Delta'$ is the light-shift splitting in g, and is due to a contamination of the wave functions in g induced by atomic motion. It is much larger than the corresponding imbalance occuring for Doppler cooling and which varies as kv/Γ. Let us mention finally that, in the $\sigma^+-\sigma^-$ configuration, the energy dissipation is not due to anti-Stokes Raman processes. In steady state, there are as many Stokes as anti-Stokes processes. As in Doppler cooling, the energy dissipation is due to the fact that, on the average, the fluorescence photons have a blue Doppler shift.

All the previous results are derived from a semi-classical theory, where the position $\vec{r}$ of the atom is treated as a classical variable, and are summarized (for $|\delta| \geqslant \Gamma$) in the table 1. The temperature achievable by polarization gradient cooling is predicted to be proportional to the laser power (5). This could suggest that one can reach arbitrarily low temperature by using sufficiently low laser power. Actually, this is not true : the previous semi-classical approach is valid only if the atomic de Broglie wavelength remains small compared to the laser wavelength λ, or, in other words, if the atomic kinetic energy remains large compared to the recoil energy. This imposes a lower limit on the laser power range that can be studied semi-classically. Below this limit, a full quantum approach has to be developped. The outline of such an approach is presented in the next section.

	2-level	Lin $\perp$ Lin	$\sigma_+-\sigma_-$
Capture range	$kv \simeq -\delta$	$kv \simeq \Gamma'$	$kv \simeq -\Delta'$
Friction α	$-\hbar k^2 \dfrac{\Gamma'}{\delta}$	$-\hbar k^2 \dfrac{\Delta'}{\Gamma'}$	$-\hbar k^2 \dfrac{\Gamma'}{\Delta'}$
Diffusion D	$\hbar^2 k^2 \Gamma'$	$\hbar^2 k^2 \Gamma' \dfrac{\Delta'^2}{\Gamma'^2}$	$\hbar^2 k^2 \Gamma'$
Temperature $k_B T = \dfrac{D}{\alpha}$	$-\hbar\delta$	$-\hbar\Delta'$	$-\hbar\Delta'$

Table 1. Predictions of the semi-classical theory for the Doppler cooling of a 2-level atom, and for the two polarization gradient coolings discussed here. The detuning $\delta = \omega_L - \omega_0$ is negative and large compared to Γ, so that the light shift Δ' is also negative and large compared to Γ'. For simplicity, numerical factors have been omitted.

4

4. Quantum treatment of atomic motion in the $\sigma_+-\sigma_-$ case

Such a treatment uses a quantum description of both internal and external degrees of freedom. This amounts to consider the following density matrix elements

$$< e \text{ or } g, \, m, \, p_x, \, p_y, \, p_z \, |\rho| \, e \text{ or } g, \, m', \, p'_x, \, p'_y, \, p'_z > \qquad (1)$$

where m and m' stand for the Zeeman sublevels indices and p_x, p'_x, ... for the cartesian components of the atomic momentum.

The evolution of ρ is given by the generalized optical Bloch equations including recoil (10,11). Consequently, once the laser field configuration is known, the problem is in principle solvable at least numerically. However, it requires a very large size of computer memory ((8×10^6) x (8×10^6) array for a $J_g = 1 \leftrightarrow J_e = 2$ transition with 100 steps in p along each direction) and we have not performed such a 3-dimensional quantum analysis. On the other hand, the 1-dimensional case is much less memory consuming especially for the $\sigma_+-\sigma_-$ case as we show now.

In this latter case indeed, conservation of angular momentum implies that many non-diagonal matrix elements of the type of (1) are zero in the stationary state. Consider for example the following family of states :

$$\mathcal{F}(p) = \left\{ \, |g, \, m, \, p + m\hbar k >; \, |e, \, m', \, p + m'\hbar k > \, \right\}$$

$$\text{with } |m| \leq J_g \qquad |m'| \leq J_e = J_g + 1$$

The laser atom interaction only couples the states of this family to states belonging to the same family. For instance, the atom in the state $|g, \, m = -1, \, p - \hbar k >$ can absorb a σ_+ photon (momentum $\hbar k$) and go to the state $|e, \, m = 0, \, p >$. There, it can emit a σ_+ stimulated photon and jumps into the state $|g, \, m = 1, \, p + \hbar k >$ and so on... On the other hand, since there is a perfect correlation between the direction $\pm \hbar k$ of the photons and their angular momentum, two states belonging to two different families cannot be connected by absorption-stimulated emission processes. Only spontaneous emission can cause population transfers between families, but it cannot build any off-diagonal density matrix element between two states belonging to two different families.

This important remark considerably reduces the number of non-zero density matrix elements to be considered. For example, for a 1 $\leftrightarrow$ 2 transition with 100 steps in p, this number is only 800 x 8 = 6400 which is easy to deal with on a reasonable size computer.

Two ways can be used to find the stationary state of the atomic density operator. The first one is to calculate the time evolution of the density matrix elements and to wait for a long enough time. The second way is to solve directly the equation $\dot{\rho} = 0$ by a matrix inversion. We have checked that the two methods lead to similar results, the latter being the fastest one.

Examples of results are given in fig. 1, representing four stationary momentum distributions. These curves have been calculated for a Cesium type atom (atomic mass 133, resonant wavelength 852 nm, resonant transition linewidth $\Gamma/2\pi = 5$ MHz). The parameters of the atom laser interaction are the same for the four curves (Rabi frequency 0.5 Γ, detuning $\delta = -3\Gamma$). The only change between the four calculations concerns the angular momenta of the atomic levels involved in the cooling process, from $J_g = 1 \rightarrow J_e = 2$ (curve a) to $J_g = 4 \rightarrow J_e = 5$ (curve d, corresponding to the real situation for Cesium).

These four curves exhibit qualitatively the same types of features : narrow peak around p = 0 with a width of a few recoil, superimposed on a much broader pedestal. The relative size of the narrow peak with respect to the pedestal depends strongly of the angular momenta of the transition, the pedestal being more important when J_g and J_e increase.

Let us now discuss briefly the origin of these two features. First, the narrow peak appears to have characteristics very close to the Gaussian predicted by the semi-classical model, at least if the power is not too low so that the width of the semi-classical Gaussian remains much larger than the recoil momentum.

We think that the large pedestal originates from an anomalous momentum diffusion appearing at high velocities and due to correlations between the directions of two successively absorbed photons. Indeed, because of optical pumping, after the absorption of a σ_+ photon, the atom has a high probability to be in a state $|g, m \rangle$ with a high value m of J_z. Because of the Clebsch–Gordan coefficients, it has then a higher probability to absorb another σ_+ photon than a σ_- one, the difference between these probabilities increasing with J_g. It follows that the atom absorbs in sequence several σ_+ photons, then several σ_- ones, and so on..., thus performing a large step random walk in p-space, the size of the steps increasing with J_g (5, 12, 13).

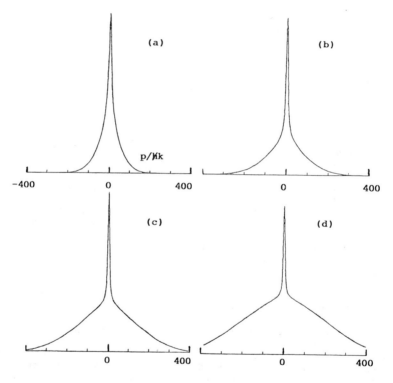

Fig. 1. Stationary momentum distributions in the $\sigma_+ - \sigma_-$ configuration, calculated from a full quantum theory of atomic motion. The parameters are the same for the 4 curves (see text) except for the angular momenta J_g and $J_e = J_g + 1$: (a) : $J_g = 1$, (b) : $J_g = 2$, (c) : $J_g = 3$, (d) : $J_g = 4$.

Such an enhancement of the momentum diffusion coefficient D occurs only if the eigenstates of J_z can be considered as stationary between two fluorescence cycles, i.e. on a time scale $\tau_p = 1/\Gamma'$. This is achieved either for a moving atom with $kv \gg |\Delta'|$ or for a slow atom if the detuning is small ($kv \leqslant |\Delta'|$, $|\delta| \leqslant \Gamma$). Otherwise, for example if $kv \simeq 0$ and $|\delta| \gg \Gamma$, the correlations between successive absorptions is destroyed by a complete redistribution of populations among the various Zeeman sublevels. Such a redistribution is due to absorption–stimulated emission processes which, for an atom at rest, have a rate $|\delta|/\Gamma$ larger than spontaneous processes.

Using these results, it is then possible to understand, at least qualitatively, the momentum profiles of fig. 1. Around $v = 0$, the anomalous diffusion is small and the friction force is important, giving rise to a narrow structure. For larger velocities, the diffusion coefficient D increases (this increase being more important for large J_g and J_e). This gives rise to the large pedestals of fig. 1.

We are now working on the extension of this quantum calculation to more general laser configurations.

References

(1)   P. Lett, R. Watts, C. Westbrook, W. D. Phillips, P. Gould and H. Metcalf, Phys. Rev. Lett. 61, 169 (1988)

(2)   J. Dalibard, C. Salomon, A. Aspect, E. Arimondo, R. Kaiser, N. Vansteenkiste and C. Cohen-Tannoudji, in Atomic Physics 11, S. Haroche, J. C. Gay and G. Grynberg eds, World Scientific 1989, p. 199

(3) a. S. Chu, D. S. Weiss, Y. Shevy and P. J. Ungar, in Atomic Physics 11, same reference as in (2), p. 636

    b. Y. Shevy, D. S. Weiss, P. J. Ungar and S. Chu, Phys. Rev. Lett. 62, 1118 (1989)

(4) a. A. Aspect, E. Arimondo, R. Kaiser, N. Vansteenkiste and C. Cohen-Tannoudji, Phys. Rev. Lett. 61, 826 (1988)

    b. Same authors, to appear in J.O.S.A. B, 1989, Special Issue on Laser Cooling and Trapping

(5)   J. Dalibard and C. Cohen-Tannoudji, to appear in J.O.S.A. B, 1989, Special Issue on Laser Cooling and Trapping

(6)   J. P. Gordon and A. Ashkin, Phys. Rev. A21, 1606 (1980)

(7)   J. Dalibard and C. Cohen-Tannoudji, J. Phys. B18, 1661 (1985)

(8)   J. Dalibard and C. Cohen-Tannoudji, J.O.S.A. B2, 1707 (1985)

(9)   A. Aspect, J. Dalibard, A. Heidmann, C. Salomon and C. Cohen-Tannoudji, Phys. Rev. Lett. 57, 1688 (1986)

(10)   S. Stenholm, Appl. Phys. 16, 159 (1978)

(11)   C. Bordé in Advances in Laser Spectroscopy, F. T. Arecchi, F. Strumia and H. Walther eds, Plenum, 1983

(12)   Y. Castin and K. Mølmer, to be published

(13)   P. J. Ungar, D. S. Weiss, E. Riis and S. Chu, to appear in J.O.S.A. B, 1989, Special Issue on Laser Cooling and Trapping.

A Heterodyne Measurement of the Fluorescence Spectrum of Optical Molasses

W. D. Phillips, C. I. Westbrook, R. N. Watts, S. L. Rolston,
C. E. Tanner, P. D. Lett, and P. L. Gould*

National Institute of Standards and Technology
Phys. B-160, Gaithersburg, MD 20899

Optical molasses has proven to be an exciting subject for both experimental and theoretical investigation since it's demonstration in 1985 by a group at AT&T Bell Labs[1]. One of the most important features of optical molasses is the extremely low temperatures that have been achieved. The limiting temperature of optical molasses has recently been a topic of intense experimental and theoretical investigation. The discovery last year of temperatures well below the accepted cooling limit[2] has prompted the proposal of entirely new mechanisms by which light can damp atomic motion[3,4]. These mechanisms predict much larger damping coefficients and can account for measured temperatures much lower than previously thought possible. To date, optical molasses temperatures have been measured by ballistic methods. We report here a new technique for measuring the atomic velocity distribution by a direct measurement of the Doppler-broadened fluorescence spectrum. In addition, the observed spectrum exhibits an unexpected narrow feature, which we attribute to the confinement of atoms in optical potential wells, producing motional (Dicke) narrowing.

Previous temperature measurements were performed by turning off the molasses laser beams, releasing the atoms to travel ballistically. The temperatures were then obtained by observing some aspect of the atomic trajectory. These methods have the disadvantage of measuring the velocities only after release of the atoms rather than while they are interacting with the laser beams. Effects on the measured temperature due to the release process have already been reported.[5,6] For example if the beams are turned off slowly compared to the molasses damping time (which we have measured to be about 3 μs for typical conditions in Na[6]), the measured temperature will be significantly lower than what it was in steady state. In addition, other difficulties with ballistic techniques are currently limiting the accuracy with which the temperatures can be measured.[6] Finally, several groups have raised the possibility that the velocity distribution of the atoms cannot be described by a simple Maxwell-Boltzmann distribution.[4,6,7,8] This possibility presents additional difficulties for the ballistic techniques because their analysis requires the assumption of an initial velocity distribution.

Motivated by these problems we have developed a heterodyne technique to observe the spectrum of the atomic fluorescence while the atoms are in the molasses. The Doppler-broadening of the elastically (Rayleigh) scattered light directly measures the atomic velocities, without making any assumption about the initial velocity distribution.

The optical heterodyne technique is well known in other contexts, for example laser-Doppler velocimetry.[9] Although this technique has been suggested[10] for the measurement of the spectrum of atomic fluorescence, it has not, to our knowledge, ever been successfully used. Figure 1 shows a diagram of the appartus.

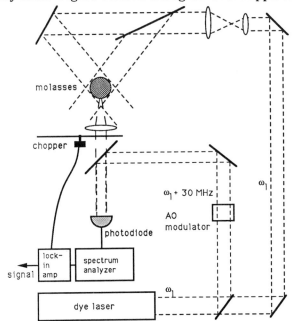

Figure 1.   Sketch of experimental apparatus.

The fluorescence imaged from a small region in the molasses (containing about 10^3 atoms) is combined with a strong local oscillator laser beam on a photodiode. The local oscillator is derived from the same laser that produces the molasses, but is shifted by 30 MHz with an acousto-optical modulator. The beat signal is detected with an RF spectrum analyzer. The intensity of each of the molasses laser beams is about 26 mW/cm^2 on axis and they are tuned below resonance (10 to 30 MHz red of the Na D2 F=2 $\rightarrow$ F=3 resonance). The local oscillator power is 5 mW, enough to operate the detector in the shot-noise limit.

Under these conditions two level atoms would scatter approximately one half of the light into the Mollow triplet[11]. The central peak will have the natural linewidth, Γ (10 MHz for this transition). The remaining light will be elastically scattered, forming a δ-function superimposed on the three-peaked Mollow spectrum. This δ-function will be Doppler-broadened to a profile reflecting the distribution of velocities of the atoms in the molasses. Atoms with a temperature of 100 µK would produce a Gaussian profile with a FWHM of 750 kHz. The spectral density of the Mollow triplet is too small to be observed with our present experimental sensitivity.

Figure 2 shows an example of the heterodyne signal obtained from molasses fluorescence. The most striking feature of the data is that, in addition to the relatively broad peak at the expected width, there is also a very narrow peak located at the center of the broad line. The fit to the data shown is the sum of a narrow Lorentzian and a Gaussian, plus a constant background. The full width at half maximum (FWHM) of the Lorentzian is 90 kHz and the FWHM of the Gaussian is 690 kHz, which corresponds to a temperature of 83 μK. At this detuning (Δ = -20 MHz), ballistic measurements of the sort previously reported[2] give results for the temperature that are within experimental error of these measurements as long as a fast shut-off is used in turning off the molasses laser beams. The ratio of the areas of the Lorentzian to the Gaussian in the fit is 0.4. Note that the width of the narrow peak is less than that which would result if the atoms were at the recoil limit (2.4 μK).

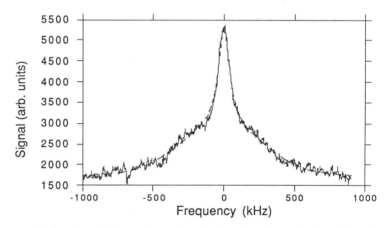

Figure 2.   Heterodyne signal showing Doppler-broadened width (690 kHz FWHM) and narrow feature (90 kHz FWHM).

We have investigated the behavior of the narrow feature as a function of several parameters. The ratio of the areas of the Lorentzian to the Gaussian decreases with detuning, with essentially no narrow feature at Δ = -10 MHz. The ratio is relatively insensitive to intensity changes over the range of 8 to 26 mW/cm[2] per beam on axis.

A tentative explanation of this narrow feature is that it is due to motional (Dicke) narrowing.[12] Motional narrowing induced by the viscous damping forces in molasses (analogous to collision-induced narrowing in a gas) would tend to smoothly change the Gaussian profile into a Lorenztian profile, in contrast to our observation of a narrow Lorentzian on top of the Gaussian distribution. We ascribe the observed results to confinement of at least some of the atoms in optically-induced potential wells. The standing wave optical field is strong to produce local dipole-force potential wells of a depth on the order of 100 μK or more, with a size of λ/2. Such wells can trap the atoms for a significant amount of time, producing a motionally-narrowed Lorentzian peak on top of the free atom Gaussian profile. A one dimensional simulation of the atomic motion produces spectra similar to those observed. Additional mechanisms such as the interruption of the scattering due to optical pumping may reduce the strength of the motionally-narrowed peak, and add a broad peak onto the Gaussian profile, possibly complicating the extraction of a velocity distribution and temperature from the broad peak.

The heterodyne technique offers a new way to measure the properties of optical molasses. It not only provides information about the velocity distribution but also provides information not available from the ballistic techniques. If our explanation for the narrow peak is correct, then it offers us an opportunity to study the details of the atomic motion. It will be interesting to learn if the trapping and scattering of the atoms by the dipole potential wells is partly responsible for properties of optical molasses not explained by atomic diffusion in the presence of simple viscous damping. Finally, the method can be spatially selective and polarization sensitive and questions about the effects of having non-uniform beams (temperature gradients, etc.) and asymmetric loading can potentially be addressed.

This work was partially supported by the Office of Naval Research. Two of us (CET and RNW) would like to acknowledge the support of NRC/NIST post-doctoral fellowships

*   Dept. of Physics, Univ. of Connecticut, Storrs, CT 06268
1.   S. Chu, L. Hollberg, J. Bjorkholm, A. Ashkin, and A. Cable, Phys. Rev. Lett. **55**, 49 (1985).
2.   P. Lett, R. Watts, C. Westbrook, W. Phillips, P. Gould, and H. Metcalf, Phys. Rev. Lett. **61**, 169 (1988).
3.   P. Ungar, D. Weiss, E. Riis, and S. Chu, J. Opt. Soc. Am. B, (to be published in the special issue on laser cooling, S. Chu and C. Wieman, eds., 1989).
4.   J. Dalibard and C. Cohen-Tannoudji, J. Opt. Soc. Am. B, (to be published in the special issue on laser cooling, S. Chu and C. Wieman, eds., 1989).
5.   Y. Shevy, D. Weiss, and S. Chu, Proceedings of the Conference on Spin Polarized Systems, Torino, June 1988.
6.   P. Lett, W. Phillips, S. Rolston, C. Tanner, R. Watts, and C. Westbrook, J. Opt. Soc. Am. B, (to be published in the special issue on laser cooling, S. Chu and C. Wieman, eds., 1989).
7.   D. Weiss, E. Riis, Y. Shevy, P. Ungar, and S. Chu, J. Opt. Soc. Am. B, (to be published in the special issue on laser cooling, S. Chu and C. Wieman, eds., 1989).
8.   Y. Shevy, D. Weiss, P. Ungar, and S. Chu, Phys. Rev. Lett. **62**, 1118 (1989).
9.   L. E. Drain, *The Laser Doppler Technique*, (Wiley, New York, 1980).
10.   J. F. Lam and P. R. Berman, Phys. Rev. A **14**, 1683 (1976).
11.   B. R. Mollow, Phys. Rev. **188**, 1969 (1969).
12.   R. Dicke, Phys. Rev. **89**, 472 (1953).

Optical Molasses with a New Twist

Steven Chu, Erling Riis, P. Jeffery Ungar, David Weiss
Physics Dept., Stanford Univ., Stanford, CA, 94305

When the three dimensional cooling and confinement of atoms was first demonstrated in 1985[1], the so-called optical molasses was consistent with the predictions of the two-level theory. As more experiments were done, it became clear that many of the features of molasses worked better than predicted. The most startling discovery was that atoms could be cooled to temperatures well below the Doppler limit $k_BT=\hbar\Gamma/2$ predicted for a two-level system, and that the lowest temperatures were found at detunings larger than the optimum detuning $\Delta\omega=\Gamma/2$ for a two level system.[2] Initial explanations of the new cooling mechanism were first proposed by J. Dalibard, et al.[3] and S. Chu, et al.,[4] at the ELICAP conference in Paris in 1988. These ideas have been expanded and refined and will be published in a special issue of JOSA B devoted to laser cooling and trapping.[5,6] In this paper we outline the essential features of the present understanding of optical molasses for multilevel atoms, and mention several measurements we have made in order to test the new ideas.

I. Enhanced Cooling of Multilevel Atoms

The new cooling mechanism depends on the presence of Zeeman sublevels of an atom and the interaction of these levels in a light field with polarization gradients. The basic ideas can be see by considering one dimensional molasses (two counterpropagating laser beams) with various polarization configurations. First consider counterpropagating beams with crossed linear polarizations. The laser field polarization varies between right circular, linear, left circular, and linear over a distance of $\lambda/2$. An atom at rest in a right circularly polarized light field will optically pump into the $m_F=+2$ ground state. For $F{\to}F+1$ transitions, this state is most strongly coupled to the excited state manifold so the light shift of the ground state will be the largest. Thus, the atom will optically pump into the lowest energy state. If the atom moves into a region of space with different polarization in a time short compared to the optical pumping time, the atom remains primarily in the original $m_F=+2$ state. This state is no longer the lowest energy state, and the increase in the internal energy of the atom is extracted from its kinetic energy. The internal energy is dissipated by the optical pumping process in the form of spontaneously Raman scattered photons that are blue shifted with respect to the laser photons.

The cooling process is effective as long as the motion of the atom in the light field is non-adiabatic; i.e. as long as the population of the Zeeman levels differs from the steady state population. Even at low velocities, the motion can be non-adiabatic if the intensity and detuning of the cooling laser is adjusted to have a long optical pumping time. For example, the optical pumping time for sodium atoms for the conditions in Fig.1 is on the order of 10 μs, while the time an atom with v=10 cm/s (T=mv²/k_B=28 μK) needs to go a distance $\lambda/2$ is 1.5 μs. This cooling process is similar to stimulated molasses, where the laser is tuned to the high frequency side of the resonance curve for a two-level atom.[7] A major difference is that non-adiabatic motion of atoms in two-level stimulated molasses is governed by the excited state relaxation time of 16 ns, while the multilevel molasses has a relaxation time that can be several orders of magnitude longer. Thus, atoms with velocities on the order of the recoil velocity v=ℏk/m can still move non-adiabatically in the laser field.

A second form of cooling in a light field with polarization gradients, pointed out by J. Dalibard and C. Cohen-Tannoudji[5], occurs when the two opposing laser beams have opposite helicity ($\sigma^+\sigma^-$). The combined light field is linearly polarized at every point in space, but twists by 2π in a distance λ along the direction of the laser beams. The motion of an atom in this "corkscrew" polarization establishes an asymmetric population of the Zeeman sublevels quantized along the beam axis. Once there is an unbalanced occupancy of the ground state sublevels, the unequal oscillator strengths for σ^+ and σ^- light will cause the atom to scatter more photons from the beam opposing the motion of the atom. In our treatment of the problem using the full optical Bloch equations for the case of sodium[6], we found that the atoms moving through the $\sigma^+\sigma^-$ field at 20 cm/s are in a superposition containing 97% of the m_F=+2 state while moving toward the σ^+ beam. An atom in the m_F=+2 ground state scatters 15 times more σ^+ photons than σ^- photons. Thus, the drag force is due to radiation pressure analogous to recoil cooling in classical two-level molasses, except that the net scattering imbalance is due to a population imbalance in the atomic sublevels caused by the non-adiabatic motion of the atom in the light field rather than by the Doppler shift of the two laser beams. Again, because of the long optical pumping times, this new cooling mechanism can be especially effective at low velocities.

In one dimensional molasses, enhanced cooling can be obtained even in the absence of polarization gradients if a weak magnetic field is applied perpendicular to the laser quantization axis. Consider the case of $\sigma^+\sigma^+$ polarization and a magnetic field along an axis normal to the beam axis. The laser field is circularly polarized everywhere and the amplitude of the field is spatially modulated by cos(2kz). In regions of high laser intensity, the atoms will be optically pumped towards the m_F=+2 state, but when they move in regions of low laser intensity, the magnetic field will cause the atom to Larmor precess into a mixed set of states with an average light shifted energy that is higher than in the optically pumped state. Since the relevant relaxation time is still the optical pumping time and the energy level shifts are the same, we expect this magnetic field induced cooling will be about as effective as cooling in a polarization gradient if the strength of the magnetic field is chosen properly. Instead of having the alignment of the ground state population lag behind a changing laser polarization, the alignment lags behind the redistribution caused by the magnetic field. Our measurements of magnetic orientational cooling are reported in the JOSA B article.[8]

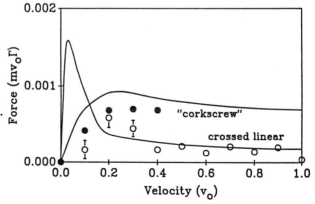

Fig. 1. Steady-state drag forces and averaged Monte-Carlo forces for $\sigma^+\sigma^-$ and crossed linear polarizations. The conditions are I=7.9 mW/cm² and $\Delta\nu$=-2.73Γ. v_o=41 cm/s.

13

II. Calculation of the Cooling Forces

Quantitative calculations of the cooling forces for the case of sodium $F=2 \rightarrow F=3$ transitions were done using a generalization of the optical Bloch equations. The twelve Zeeman sublevels require the solution of a set of 144 coupled differential equations which can be solved numerically.[6] The solid lines in Fig. 1 shows the results of the drag force on a sodium atom. Because of the long optical pumping times, the population of the ground states will not reach their steady state values before fluctuations in the light forces on the atom will cause significant changes in the velocity of the atom. Thus, the steady-state forces will be modified since the actual cooling force the atom feels will be a response based on the past history of the atomic trajectory as well as its instantaneous velocity. We have also performed Monte-Carlo trajectory calculations based on solutions to the optical Bloch equations for atoms in optical molasses fields with polarization gradients.[6] The average force derived from the Monte-Carlo simulation is shown in Fig. 1, where it is apparent that the long time lag in reaching the steady state population smoothes out any sharp features in the force vs. velocity curve.

III. Momentum Diffusion in Multilevel Atoms and Bimodal Velocity Distributions

The Monte-Carlo simulation also allows us to quantitatively compare our measurements in 1D molasses[8] to theory for the specific case of sodium[6]. The calculated velocity distributions give a very cold "temperature" (v_{rms}=6.9cm/s for crossed linear pol.) with less than 5% of the atoms in a warmer tail. The observed distributions have ≈25-30% of the atoms in a cold group (v_{rms}=7.2cm/s), and the remainder in a warmer group (v_{rms}≈30cm/s). We believe that the primary reason for this discrepancy is due to an enhanced momentum diffusion for multilevel atoms in 1D molasses. A large momentum diffusion means that there are large fluctuations in the cooling imposed on the average value. For example for $\sigma^+\sigma^-$ polarization with $I=I_{sat}$, $\Delta\omega$=-2.73Γ, and $v=v_0$=41cm/s, the net force is $\langle f \rangle$=-6.87 x10^{-4}m$v_0\Gamma$, while the diagonalized steady-state density matrix shows that 58% of the atoms are subject to a force 2.03$\langle f \rangle$, 16% see an opposite force -1.62$\langle f \rangle$, 12% feel a force 1.22$\langle f \rangle$, etc. These large fluctuations occur with a correlation time which can be larger than the optical pumping time and can significantly enhance the momentum diffusion. Also, since the ground state populations are a sensitive function of velocity, the diffusion is also expected to be velocity sensitive. The enhanced diffusion is discussed in more detail in our theory paper[6], and the Dalibard and Cohen-Tannoudji paper[5]. Experimentally, we have reduced the enhanced diffusion by introducing additional molasses beams in a second dimension and have seen the velocity distribution get colder.[8]

The shape of the force vs. velocity curve and the velocity distribution of the atoms cooled in 1D molasses can also be understood in a qualitative manner. For a given laser field, there is a critical velocity where the cooling due to the non-adiabatic motion in a polarization gradient is a maximum. For example, in the case of crossed linear polarization, the maximum cooling will occur at a velocity $v_c = \lambda/4\,\tau_{op}$ when an atom moves a distance $\lambda/4$ (from a region of right circularly polarized light to the region of left circularly polarized light) during the optical pumping time τ_{op}. For $v < v_c$, the cooling force will vary linearly with the velocity, and for $v \gg v_c$, one would expect that the cooling would then be dominated by the usual recoil cooling due to Doppler shifts. In that case the cooling force again varies linearly with velocity, but increases far less rapidly. Since the atom temperature is inversely proportional to the slope of the force vs. velocity curve through zero, the force at low velocity will lead to a small velocity spread. The decrease in the damping force at intermediate velocities ($v_c < v < \Gamma/k$) will separate the atoms into a group with a small velocity spread and a group with a velocity distribution more characteristic of a force due to Doppler cooling. We have observed bimodal velocity distributions for both cooling in polarization gradients and cooling in magnetic orientational cooling.[8] The ratio of the number of atoms in the "cold" peak to the number in the "hot" peak depends critically on the size of the momentum diffusion. In our simulations, doubling the size of the diffusion increased the fraction of atoms in the hot tail from less than 5% to ≈70%. Thus, the ratio of hot to cold atoms is a sensitive measure of the momentum diffusion.

IV. Other Experimental Verifications of the New Cooling Theory

Other experimental tests we have done include (i), the verification of the analytic two-level theory when the sodium atom is optically pumped into the $m_F=+2$ ground state, (ii) an observed linear dependence on temperature vs. laser intensity for the cold atoms in the polarization gradient configurations, and (iii), the measured sensitivity of temperature vs. detuning and the fraction of atoms in the cold distribution for the $\sigma^+\sigma^-$ polarization case relative to the crossed linear case. Measurements (ii) and (iii) are in qualitative agreement with the analytic predictions of the simple model systems studied in the Dalibard-Cohen-Tannoudji paper.[5]

There are several caveats in the comparison of our calculations with the experimental data. The Monte-Carlo simulations predict a temperature of $13.4\,\mu K$ for the case of crossed linear polarization using our experimental parameters. The measured temperature of the cold peak for identical experimental parameters is ≈30 μK. A temperature of $13.4\,\mu K$ corresponds to a $v_{rms}=6.9$ cm/s and a deBroglie wavelength of 249nm. At these low velocities, the treatment of the atomic coordinate as a classical parameter begins to break down. Dalibard and Cohen-Tannoudji[4] have outlined a full quantum treatment of the problem (so far applicable only to the $\sigma^+\sigma^-$ configuration) where both the atom's internal degrees of freedom and its momentum states $p\pm n\hbar k$ are written into the eigenstates. Since the change in velocity of an atom due to a single photon momentum kick is $\Delta v=\hbar k/m=3$cm/s for sodium, we expect that the full quantum treatment will broaden any sharp feature by an amount $v\approx\sqrt{((v_{class})^2+(v_{recoil})^2)}$. Comparisons of the full quantum treatment with a semiclassical calculation show this approximate relationship.[9]

Even after correcting for the errors introduced in our semi-classical calculation, the measured temperature of the cold peak is still too hot relative to the predicted value. The excess temperature may be due to effects that result from excitations to other hyperfine levels, particularly the $F=2\to2$ transitions. By tuning the rf sideband frequency of the laser (used to avoid optical pumping) 8 MHz below the $F=1\to2$ transition, we reduced the temperature by a factor of 2 relative to the usual 38 MHz detuning that we typically used. Effects of other hyperfine levels have not yet been included in the new treament of molasses cooling.

We wish to thank C. Cohen-Tannoudji and C. Wieman for helpful discussions. This work was supported in part by grants from the NSF and the AFOSR.

References

1. S. Chu, et al., Phys. Rev. Lett. 55,48 (1985).
2. P.D. Lett, et al., Phys. Rev. Lett. 61,169 (1988).
3. J. Dalibard and C. Cohen-Tannoudji, Atomic Physics 11, ed. S. Haroche, (World Scientific, 1989).
4. S. Chu, Y. Shevy, and D.S. Weiss, op cit.
5. J. Dalibard and C. Cohen-Tannoudji, to be published in JOSA B, special issue on "Laser Cooling", Nov. 1989, eds. S. Chu and C. Wieman.
6. P.J. Ungar, et al., op cit.
7. J. Dalibard and C. Cohen-Tannoudji, JOSA B 2, 1707 (1985).
8. D.S. Weiss, et al., to be published in the JOSA B special issue.
9. Cohen-Tannoudji, private communication.

Trapping and Cooling of Neutral Atoms with the Dipole Force of a Laser Beam

Koichi Shimoda
 Department of Physics, Keio University,
 3-14-1 Hiyoshi, Kohokuku, Yokohama 223, Japan

It is generally believed that the dipole force is ineffective for cooling, because it is a conservative force and not dissipative. This is not the case, however, if the potential of the force varies as the particle moves. Assuming an elastic interaction between a ball and a racket, for example, the ball can be speeded up by bouncing with the moving racket.

The author proposed a novel method of trapping and cooling neutral atoms with the dipole force of a laser beam [1,2]. This method allows us to cool neutral atoms to a temperature of much lower than the Doppler limit. Further cooling even beyond the recoil limit is feasible, because cooling with the dipole force is free from fluctuations of the photon recoil.

1. Outline of the Proposed Trap

The proposed trap is schematically shown in Fig.1(a), where the laser beam is concave upwards. Pre-cooled atoms fall freely in the gravitational field and are reflected by the dipole force exerted by the gradient of the laser beam. Then atoms will be repeatedly rebounding on the laser beam.

The potential of the dipole force is given by $-\alpha' U$, where α' is the real part of the atomic polarizability at the frequency of the laser, and U is the electric energy density of the laser beam. The value of α' is negative at positive detuning, so that the dipole-force potential increases with the light intensity. On the other hand, the gravitational potential for the atom of mass M is Mgz, where g is the gravitational acceleration, and z the vertical coordinate. Then the resulting potential has a valley as shown in

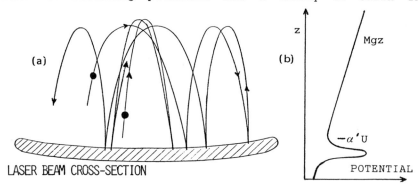

(a)

(b)

z

Mgz

$-\alpha' U$

POTENTIAL

LASER BEAM CROSS-SECTION

Fig.1. (a)Low velocity atoms rebounding on a concave laser beam and (b)the effective potential

Fig.1(b). The falling atoms are repelled by the potential hill as if elastically reflected by the laser beam. Thus these atoms are trapped around the valley between the hill and the gravitational slope. In order to prevent atoms spilling over at either end of the valley, the three-dimensional profile of the laser beam may be formed in shape of a shallow tray.

The power density of the laser beam that is necessary to reflect the falling atoms with the kinetic energy equal to kT is calculated by using

$$P = 2Uc, \qquad -\alpha' U > kT, \qquad \alpha' = \gamma \lambda^3 / 8\pi^2 (\omega_0 - \omega)$$

when $|\omega_0 - \omega| \gg \gamma$. Here γ is the rate of spontaneous emission, $\omega = 2\pi c / \lambda$ is the frequency of the laser, and ω_0 is the resonant frequency of the atom. In order that the dipole force dominates the scattering force which is proportional to the imaginary part of the atomic polarizability, the detuning must be large enough to be

$$\omega - \omega_0 > \gamma w \omega / c$$

where w is the half width of the laser beam that defines the gradient of the potential. The required power at this detuning is therefore

$$P > 32\pi^3 c\, kTw / \lambda^4, \qquad \lambda = 2\pi c / \omega.$$

Numerical values for kT=0.5mK and w=0.2mm are 0.8W/mm^2 at 852nm with a detuning of 7.8GHz for cesium and 3.4W/mm^2 at 589nm with a detuning of 34GHz for sodium. The maximum altitude of the atomic trajectory above the laser beam is calculated to be $z_0 = kT/Mg = 3.2mm$, and 18.4mm for cesium and sodium atoms respectively at 0.5mK.

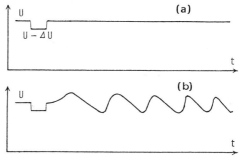

Fig.2. (a)Laser intensity modulation for atomic velocity selection

(b)Laser intensity modulation for cooling

(c)Motion of the atom under cooling

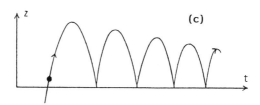

An atom entering the trap at any position and velocity will escape the trap sooner or later, if the laser beam is stationary. Therefore the laser beam must be chopped so as to transmit pre-cooled atoms that have been prepared beneath the laser beam. Furthermore, if the laser beam is chopped by a small amount of its intensity as shown in Fig.2(a), only those atoms having the kinetic energy in a narrow range can be trapped. Then a velocity-selective trap is realized.

2. Cooling by a Modulated Laser Beam

Neutral atoms trapped with the gravity and the dipole force of the laser beam can be cooled by modulating the laser beam intensity as shown in Fig.2(b) in synchronism with the rebounding motion of atoms as shown in Fig.2(c). Alternatively the laser beam may be wagged up and down, but the intensity modulation is more convenient. The rate of intensity modulation dU/dt is equivalent to a vertical motion of the laser beam at the position of the rebounding atom. The effective velocity of the laser reflector is

$$v_{eff} = (dU/dt)/grad U.$$

Non-sinusoidal modulation may well be used, but sinusoidal modulation is assumed here for simplicity:

$$U(t) = U_0[1 - m \sin(2\pi t / t_0)]$$

where $t_0 = 2v/g$ is the time of flight for the initial upward velocity v. Then it gives $v_{eff} = -m\pi g w / v$.

Since the laser reflector is moving effectively downwards, the falling atoms having the velocity v are slowed down by reflection. The change in atomic velocity becomes $\Delta v = 2 v_{eff}$.

The rate of cooling, as expressed by the change in kinetic energy, is given for m=1 by

$$\frac{dW}{dt}\bigg|_C = -\frac{Mv\Delta v}{t_0} = -\frac{\pi Mg^2 w}{v}.$$

The rms velocity for cesium atom at T=0.5mK is 25cm/s, since M=2.21×10^{-25} kg. Thus the cooling rate is calculated to be 5.3×10^{-26} J/s or 3.9mK/s, when w=0.2mm. The cooling rate for sodium atom with v=1.0m/s is calculated to be 2.3×10^{-27} J/s or 0.17mK/s. Smaller modulation, m<1, may be used, because this process of cooling becomes more and more efficient as the atomic velocity is reduced.

Heating of atoms by absorption of the laser power can not be ignored, if the detuning is not sufficiently large. The fractional population of the upper state, when the two-level atom is irradiated by a detuned laser, is given by

$$f = \omega_R^2 / 2 (\omega - \omega_0)^2$$

where ω_R is the Rabi frequency. The field intensity seen by the rebounding atom changes with time and it takes a maximum value at the turning point, where the maximum value of the Rabi frequency is given by

$$\omega_{R,\,max}^2 = 2\,(\omega - \omega_0)\,kT/\hbar.$$

The average Rabi frequency for the moving atom is roughly evaluated by multiplying $w/2z_0$, where z_0 is the altitude of the atomic trajectory, to become

$$< \omega_R^2 > = wkT\,(\omega - \omega_0)\,/z_0\,\hbar = Mgw\,(\omega - \omega_0)\,/\hbar.$$

Thus the heating rate is written as

$$\left.\frac{dW}{dt}\right|_H = f\,\gamma\,\frac{(\hbar\omega/c)^2}{2M} = \frac{\pi^2\,\hbar gw\gamma}{\lambda^2(\omega - \omega_0)}.$$

When $w = 0.2$mm, it gives 1.9×10^{-27} J/s or 0.14mK/s for cesium, and 1.7×10^{-27} J/s or 0.13mK/s for sodium.

It is concluded therefore that the heating rate is smaller than the cooling rate of slow atoms when

$$v < Mg\lambda^2(\omega - \omega_0)/2\pi\hbar\gamma.$$

As the modulation frequency of the laser is gradually increased, while the amplitude of modulation is reduced, some fraction of atoms can be cooled to submicrokelvin.

3. Limit Temperature

The amplitude of atomic motion in the gravity-laser potential becomes much smaller than the beam width, as the atoms are cooled down. Then the trapped atoms oscillate nearly harmonically. The cooling limit by using a stable laser of enough detuning is given by the zero-point vibration of the atom. This quantum limit of temperature in a potential expressed by

$$\phi(z) = Mgw\,(e/2)\,\exp(-z^2/w^2) + Mgz$$

is given by

$$T_{min} = \hbar\sqrt{g/w}/2k.$$

This limit temperature is independent of either the atomic mass or the laser frequency. It is 0.85nK, when $w = 0.2$mm.

Any intensity fluctuation of the laser will heat the atoms, as described by a Langevin equation, and the random noise of a stable laser is estimated to heat the atoms by a smaller amount. The stable laser wothout excessive noise can even cool the ultracold atoms from microkelvin to nanokelvin or subnanokelvin temperatures. The ultimate limit will be achieved by tailoring the laser beam and stabilizing the laser.

References

1. K. Shimoda, 2nd Symposium on Atomic Frequency Standards and Applications: Tsukuba, Japan, Nov. 27-28, 1987.

2. K. Shimoda, IEEE Trans. Instrum. Meas., IM-38, 150(1989).

The Effect of Detuning on the Median Velocity of an Atomic Beam Slowed and Cooled by an Intense Optical Standing Wave

N. Bigelow, M. Prentiss and A. Cable

A.T. & T. Bell Laboratories Crawfords Corner Road, Holmdel, NJ 07733

We describe the effect of detuning on the most probable velocity of a beam of sodium atoms cooled in an intense standing wave laser field colinear with the atomic beam. Two distinct regimes are observed. For the highest intensities, the median velocity is observed to increase with increased detuning for detunings between 300 and 1500 MHz to the red of the $3S_{1/2} \rightarrow 3P_{3/2}$ transition. For more moderate intensities, however, the median velocity is observed to decrease as the detuning increases. These results are first compared to a qualitative model which focuses on the crossover between stimulated heating of slow velocity atoms and spontaneous ('molasses') damping of fast atoms. The observations are then compared directly to numerical calculation of the radiation forces and shown to be in good agreement.

Background

Recent work has shown that it is possible to produce a continuous supply of slow Na atoms using an intense cw standing wave laser field aligned colinear with an atomic Na beam.[1] To understand these results it is necessary to consider both the stimulated and spontaneous processes for the atoms in an intense laser field. The stimulated processes involve the atom mediated exchange of photons between the counter propagating laser fields which form the standing wave. For a detuning of the standing wave frequency Δ below the atomic transition ω_o ('red' detuning), the stimulated process can be shown to cause an atom to accelerate along its direction of motion, or in effect to cause 'stimulated heating'.[2] This process is most important for slow moving atoms. In particular, stimulated heating is significant for atoms whose transit time metween nodes in the standing wave field is long as compared to the excited state lifetime (i.e. $|kv| < \Gamma$).

For high velocity atoms, where the transit time across the standing wave period is much shorter than the excited state lifetime ($|kv| > \Gamma$), spontaneous processes dominate the atom's interaction with the laser field. In this limit the atom can be treated as interacting independently with each of the counter propagating laser fields which form the standing wave. For red detuning, the atom absorbs photons preferentially from the light wave propagating opposite to its direction of motion (for these atoms the Doppler shift brings the atom into resonance with the laser frequency). The atom then reradiates the photon in a random direction. On average the force on the atom is directed opposite to its direction of motion, and the motion is damped. This is exactly the condition for the 'optical molasses' which has been demonstrated to successfully cool and confine atoms in three dimensions.[3]

Consider a system of atoms with a thermal distribution of velocities placed in an intense standing wave. The slow atoms will be accelerated and the fast atoms will be slowed. The final result is the increase in the number of atoms around some intermediate velocity v_f. The value of v_f will be determined by crossover between the two competing effects; the stimulated heating and spontaneous cooling and damping. In this paper we focus on the effect of the detuning Δ on the measured value of v_f.

Theory and Model

Some insight into the dependence of v_f on Δ can be gained by considering the effective potential[4] 'seen' by an atom with velocity $\mathbf{v}$ in the field gradient created by the standing wave $U = -\frac{1}{2}\hbar(\Delta - \mathbf{k} \cdot \mathbf{v})\ln(1 + p)$ where $p = g(\frac{\Gamma}{2})^2/\{(\frac{\Gamma}{2})^2 + (\Delta - \mathbf{k} \cdot \mathbf{v})^2\}$, $g = I/I_{sat}$, $I_{sat} \approx 6.25 mW$ and the natural linewidth $\Gamma = 10$ MHz. For small detunings $|\Delta - \mathbf{k} \cdot \mathbf{v}| < \frac{\Gamma}{2}$, $|U| \sim |\Delta|$ increasing linearly with detuning. As $|\Delta|$ is increased, $|U|$ reaches a maximum, and then decreases as $|U| \sim |\Delta|^{-1}$. As the height of the standing wave potential increases, the energy gained by the atom 'falling' down the potential increases and so the stimulated force also increases (i.e. $F \sim \frac{\partial U}{\partial x}$). We can therefore expect then that v_f should initially increase with increased Δ. At large $|\Delta|$, v_f should reach a maximum and thereafter decrease with further increases in $|\Delta|$. Although it is relatively straight forward to find the extrema in $U(\Delta)$, such a simplified analysis is not sufficient to predict the detailed dependence of v_f on Δ.

The intensity of the standing wave plays an important role in determining the value of v_f as well as the dependence of v_f on Δ. For fixed detuning, the depth and the position of the extreema

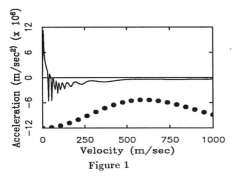

Figure 1

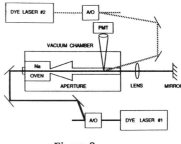

Figure 2

in the standing wave potential as a function of velocity will increase with increasing intensity. This, in turn, will increase the stimulated force at a given velocity and hence increase v_f. It will also cause the turn over in $v_f(\Delta)$ to occur at larger values of Δ. Furthermore, it has been predicted that for high intensities, multi-photon processes will become important in the interaction of the atom with the field, and that these processes can give rise to additional structure in the force versus velocity curve for the atom. In comparison, for sufficiently low intensity ($I \ll I_{sat}$), $\Gamma_P \sim \Gamma$ and the spontaneous molasses effects will dominate the force for all atomic velocities.

Clearly the final value of v_f depends not only on the processes which cause stimulated heating, but on the spontaneous forces which give rise to the molasses cooling and slowing. In order to gain a detailed understanding of the crossover in the spontaneous to stimulated forces, as well as to accurately include the possibility of multi-photon processes, we have solved for the acceleration as a function of velocity numerically using a continued fraction procedure.[5] The spatially averaged longitudinal acceleration (the acceleration along the standing wave averaged over one wavelength of the laser light) calculated using this method is plotted in Figure 1 as a function of atom velocity for fixed values of intensity and detuning. This curve shows a number of important features which are characteristic of the numerical solutions. It can be seen that the size of the stimulated acceleration (heating for low velocity atoms) greatly exceeds the maximum spontaneous deceleration ($a_{max}^{spont} \sim 10^6$ m/sec for this Na transition). There is also significant spontaneous cooling at higher atomic velocities. The cooling covers most of the Maxwell-Boltzman velocity distribution of the atoms in a Na beam at 475 K (solid circles in Figure 1). The acceleration vanishes at a number of finite velocities, which are the stable accumulation points for the atoms (the velocities at which, in the absence of collisions, an atom's velocity will remain fixed). The multiple fixed points arise because of the multi-photon 'doppleron' resonances.[6]

To derive an expected final velocity distribution for a thermal beam of atoms acted on by the standing wave field, we have simulated the effect of the force on a collection of atoms. The atoms are started with a Maxwell-Boltzman distribution of velocities and the velocities are evolved using the accelerations calculated with the continued fraction method. The simulation neglects collisions, but accounts for finite standing wave beam size and the effect of forces transverse to the standing wave. For beam intensities and widths corresponding to the experimental conditions, the final velocity distribution is characterized by a series of peaks, centered about the accumulation velocities observed in the acceleration versus velocity curves. The most prominent velocity peak is centered about the lowest velocity fixed point. For this work we have focused on the center velocity of this peak, which corresponds to the dominant component of the fluorescence signal observed. Features which correspond to the higher velocity accumulation points are observed in the simulated velocity distribution and have been observed experimentally. These observations will be treated in detail in a forthcoming publication.

Experiment

The experimental apparatus, Figure 2, is almost identical to that described in reference 1. For brevity, we will only highlight modifications in the apparatus relevent to these results. In earlier work, the standing wave laser frequency was detuned approximately 1200 MHz below the $3S_{1/2}(F = 2) \rightarrow 3P_{3/2}(F = 3)$ transition ($\Delta = -1200$MHz). For most of that work, the detuning was held relatively constant, with the tuning range limited by the acousto-optic (A/O) modulator to small variations about this Δ. The ability to examine the dependence of v_f on detuning was

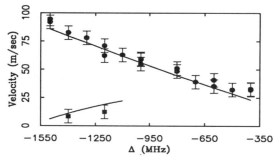

Figure 3. Most probable velocity v_f as a function of detuning Δ. Circles represent more recent results; beam waist $w_o \approx 140\mu$m and 200 mW power in beam ($g \approx 45000$). Squares are earilier results; $w_o \geq 200\mu$m ($g \approx 10000$).

therefore limited. In this work, a different A/O configuration was used on the cooling and slowing laser which allowed detuning between -300 and -1500 MHz. Furthermore the standing wave used in the earlier work was less well collimated than the more recent experiments. For this reason, although the power in the standing wave did not differ radically between the two experiments, the standing wave was on average a factor of about four more intense in the more recent experiments and the region of maximum intensity extended over more of the atomic beam's length.

The fluorescence technique for determining the velocity distribution of atoms in the beam was unchanged; the slowing beam was chopped at a few Hz, the atomic beam was probed for 30 μsec during the chop and the resulting fluorescence signal was measured as a function of probe frequency. In the data described here, the *power* in the standing wave was held constant at about 200 mW to isolate the dependence of v_f on Δ. Throughout the experiment, the probe frequency was simultaneously calibrated against a Doppler-free sodium absorption cell. The measured center frequency of the flourescence peak was later converted to a velocity using the known Doppler shift. The power in the probe beam was chosen to minimize the force on the atoms during probing.

Results

The results of this work are summarized in Figure 3. Two distinct regimes for the change of v_f with Δ have been observed. In the earlier experiments, for detunings near 1200 MHz and lower average standing wave intensity, the peak in the velocity distribution was observed to *decrease* with increased $|\Delta|$ as shown be the circles. However, in the more recent results shown as squares, the value of v_f was observed to *increase* with increased $|\Delta|$. Both observations are consistent with the behavior predicted by the variation in the standing wave potential as discussed above.

For a more quantitative theoretical comparison, the position of the lowest velocity fixed point as determined by our numerical analysis was calculated as a function of detuning and is shown as a solid line in figure 3. The values of standing wave intensity and beam waist size used in these calculations were independently measured. The experimental results are found to be in excellent agreement with the numerical calculations.

In conclusion, we find strong agreement between the experimental results and a numerical solution based on the theory. We find that observed effects can be interpreted in terms of a crossover between stimulated and spontaneous processes which determine the force on an atom in an intense optical standing wave.

References

[1]M. G. Prentiss and A. E. Cable, Phys. Rev. Lett. **62**, 1354 (1989).

[2]J. Dalibard and C. Cohen-Tannoudji, J. Opt. Soc. Am., **11**, 1707 (1985).

[3]S. Chu, L. Hollberg, J. E. Bjorkholm, Alex Cable and A. Ashkin,Phys. Rev. Lett. **55**, 48 (1985).

[4]A. Ashkin, Phys. Rev. Lett. **40**, 729 (1978).

[5]V. G. Minogin and O. T. Serimaa, Opt. Comm. **30**, 373 (1979).

[6]E. Kyrölä and S. Stenholm, Opt. Comm. **22**, 123 (1977).

Coherent Population Trapping: Reversal of the Light-Induced Drift Velocity?

Eric R. Eliel and Marc C. de Lignie

Huygens Laboratory, University of Leiden,

P.O. Box 9504, 2300 RA Leiden, The Netherlands

When two laser fields interact resonantly with a three-level Λ type atomic system coherent population trapping can occur. Then a stationary linear superposition of the ground-state levels exists, that is immune to excitation by the combined laser fields.[1] Spontaneous emission provides a channel to populate this nonabsorbing state in which the atoms remain trapped. Thus no particles can reach the excited state; the fluorescence will vanish. Clearly, any phenomenon, which, in some way, measures the excited state population, will be sensitive to the effects of coherent population trapping. In order to bring about such trapping, the beat frequency between the two laser fields has to be resonant with the Raman transition in the three-level system. As a consequence of the vanishing fluorescence this coherent Raman resonance is known as the black resonance. It has been studied in the Na absorption spectrum in both atomic beams and in vapor cells[1-4], in all cases on the D_1–line.

In Light-Induced Drift (LID), a well-studied transport phenomenon in a gas mixture, the drift velocity is directly proportional to the fraction of particles in the excited state and thus will be sensitive to coherent population trapping. Much work has been done on LID of Na diluted in various buffer gases and drift velocities of the order of 10–20 m/s are easily achieved.[6,7] At present two single-frequency lasers are most often employed in these studies to avoid hyperfine pumping.[5,6] In a simple picture one would then expect to see effects of coherent population trapping, i.e. vanishing fluorescence and a vanishing drift velocity at the black resonance. No such effects were observed in experiments on the D_2–line.[5]

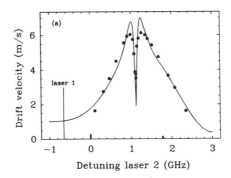

Figure 1: *The drift velocity on the D_1-line of Na as a function of the frequency of laser 2. Laser intensities are 3.9 W/cm^2 and 2.8 W/cm^2. 1.5 Torr xenon acts as a buffer gas. The F=1 and F=2 resonance frequencies are indicated by bars. The solid line represents the result of a model calculation.*

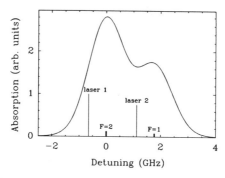

Figure 2: *Selected laser frequencies in the Na absorption spectrum for observing effects of coherent population trapping in LID. The centers of the Doppler-broadened hyperfine transitions are indicated by bars.*

Recently we have observed effects of coherent population trapping on LID by carefully measuring the drift velocity on the D_1- line in a Na/Xe mixture.[8] Two single-mode lasers were used and a sharp decrease was observed in the drift velocity at the Raman resonance. Fig. 1 shows the results of this experiment, obtained in a standard setup for measuring drift velocities.[5]

Co-propagating lasers were used to obtain a Doppler-free black resonance. One laser was tuned 650 MHz in the red Doppler wing of the $^2S_{1/2}(F{=}2)-^2P_{1/2}$ transition and the other laser was scanned through the red wing of the (partially overlapping) $^2S_{1/2}(F{=}1)-^2P_{1/2}$ transition. In the experiment of fig. 1 the drift velocity does not vanish at the expected frequency but remains finite. This is a result of the fact that two separate single-frequency lasers were employed; the finite coherence time of the beat frequency of the two laser fields (fluctuation bandwidth 3 MHz) limits the buildup of the ground-state coherence and provides an escape channel for atoms from the nonabsorbing state.

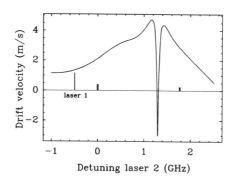

Figure 3: *Calculated drift velocity as a function of the detuning of laser 2 on the D_1 line in the absence of fluctuations in the beat frequency of the lasers. Intensities are $I_1{=}2.0$ W/cm² and $I_2{=}6.0$ W/cm².*

For fully correlated laser fields population trapping can become complete. However, contrary to the naive picture of a vanishing drift velocity, we expect the drift velocity of Na to *reverse sign* at the black resonance. This is a direct consequence of the overlap of the Doppler-broadened transitions from the two lower hyperfine states. Referring to fig. 2, the two lasers cause the drift velocity to vanish for the velocity classes in the red wing of the two Doppler-broadened transitions. However, laser 2 also excites atoms in the blue wing of the $^2S_{1/2}(F=2)-^2P_{1/2}$ transition, leading to a finite drift velocity *with a sign opposite to the normally dominant red wing contribution* (fig. 3). This dramatic feature of coherent population trapping in LID requires overlap of the Doppler broadened transitions from both ground-state hyperfine levels, a condition fulfilled in the case of Na. Note that this discussion implies that the fluorescence will not be vanishing either and thus that the black resonance in a vapor cell experiment on Na is not truly black. An experiment to verify these predictions for the drift velocity is presently being pursued.

References

1. G. Alzetta, A. Gozzini, L. Moi and G. Orriols, Nuovo Cim. B 36, 5 (1976).

2. H.R. Gray, R.M. Whitley and C.R. Stroud Jr., Optics Lett. 3, 218 (1978).

3. M.S. Feld, M.M. Burns, P.G. Pappas and D.E. Murnick, Optics Lett. 5, 79 (1980).

4. J.E. Thomas, P.R. Hemmer and S. Ezekiel, Phys. Rev. Lett. 48, 867 (1982).

5. H.G.C. Werij and J.P. Woerdman, Phys. Rep. 169, 145 (1988).

6. M.C. de Lignie and J.P. Woerdman, submitted for publication.

7. C. Gabbanini, J.H. Xu, S. Gozzini and L. Moi, Europhys. Lett. 7, 505 (1988).

8. M.C. de Lignie and E.R. Eliel, Opt. Commun., in press.

PREPARATION OF A MONOENERGETIC SODIUM BEAM
BY LASER COOLING AND DEFLECTION

J.Nellessen, K.Sengstock, J.H.Müller and W.Ertmer

and

BROAD BAND LASER COOLING ON NARROW TRANSITIONS

H.Wallis and W.Ertmer

Institut für Angewandte Physik, Wegelerstrasse 8, D-5300 Bonn 1, FRG

1. Atomic beam deflection

A sodium atomic beam with a density of approx. 10^5 at/cm³ within a velocity interval of less than 3 m/s and with a mean velocity of typically 50-160 m/s has been produced by laser deflection of a laser cooled atomic beam. Laser cooling with the frequency chirp method decelerates and cools a considerable part of an atomic beam into a narrow velocity group with a temperature of approx. 30 mK as a part of the resulting atomic beam.

This velocity group has been selectively deflected up to 30°-40° using a light field with k vektors always perpendicular to the atomic trajectory (Fig.1). If the light field is prepared by use of a cylindrical lens, the angle of deflection is nearly independent from the actual orbit radius.

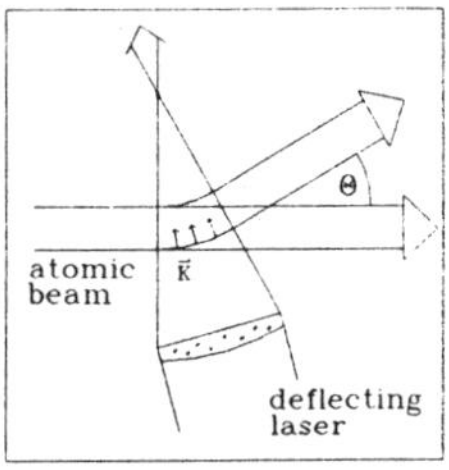

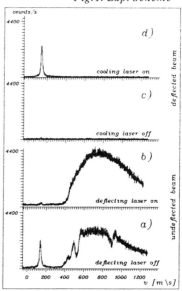

Fig.1: Exp. Scheme

For a laser frequency detuning of about one natural linewidth to the red, the strong frequency dependence of the light pressure force leads to a beam collimation via "detuning-locking" of the atomic trajectory [1]. To avoid optical pumping we used a frequency modulated laser beam with a sideband spacing matched to the hyperfine splitting of the ground state. As the cooling was performed by the frequency chirp method, one can use a part of the cooling laser beam as deflecting laser beam. Typical velocity distributions in the deflected and undeflected atomic beam, measured 22cm downstream the deflection zone, are shown in Fig. 2 . It shows the perfect transfer of the cooled velocity group from the laser cooled beam into the deflected beam; curve c) shows as comparison the result for the deflection of the initial thermal atomic beam.

Fig.2: Velocity distributions

2. Broad Band Laser Cooling

Laser cooling on transitions with a linewidth Γ much narrower than the recoil shift $\hbar k^2/2m$ is studied numerically according to a full quantum mechanical treatment of the photon recoil [2]. We analyzed the quantum limit of a momentum space optical pumping configuration [3] realized by a set of sidebands of a very narrow laser [4] with a frequency spacing ω_{sp} and a blue cut-off frequency ω_c. Atoms with a $(J = 0 \Rightarrow J = 1)$ transition (e.g. alkaline earths) interacting with counter-propagating σ^+-σ^- laser waves represent the ideal scattering force experiment, because induced radiation pressure due to redistribution of momentum between the counterpropagating laser waves is excluded. The steady state is obtained by solving the density matrix equation $d\rho/dt = 0$ for both the internal and external degrees of freedom simultaneously [2,5], i.e. by solving

$$0 = \Gamma[-\pi_+(p)-\pi_-(p) + \int_{p-\hbar k}^{p+\hbar k} (\pi_+(p'-\hbar k) + \pi_-(p'+\hbar k))\Phi(p-p')dp'] \tag{1}$$

where $\pi_+(p) = <e^+, p+\hbar k \mid \rho \mid e^+, p+\hbar k>$ denotes the population which is excited from a ground state with momentum p, whereas the integral represents the probability transfer from other momentum classes. Broad band cooling is now investigated by calculating the excited state populations from a simple rate equation ansatz as a sum over Lorentzians giving an excitation profile as in figure 3. The cut off detuning of the spectral distribution $\delta_c' = \omega_c - \omega_A - \hbar k^2/2m$ (including recoil shift) determines the achievable momentum compression. Solving equ.(1) numerically, we found a minimum 85%-momentum spread of $\Delta p^2/2m \approx \hbar\Gamma$ much smaller than $(\hbar k)^2/2m$, when choosing an optimum cut-off detuning of $\delta_c' = -1.8(\Gamma\hbar k^2/m)^{1/2}$.

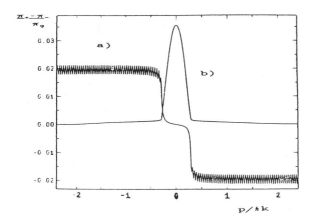

Fig. 3 a) Difference of excitation probabilities for $\hbar k^2/m = 30\Gamma$,
$\omega_{sp} = \Gamma$, $\delta_c' = -1,5(\hbar k^2\Gamma/m)^{1/2}$, b) momentum distribution (arbitrary units)

For a two step cooling scheme for alkaline earths an initial ensemble may be chosen that has been cooled down to the Doppler limit of the strong ($^1S_0$-$^1P_1$) transition with linewidth Γ_{ST}. During the second cooling stage using the narrow transition, the first cooling laser is switched off. The time T required to reduce an initial momentum $p_i \approx (\hbar\Gamma_{ST}m/2)^{1/2}$ down to 0 in a single-line molasses would be estimated [5] as roughly $T \approx (1/\Gamma)(\Gamma_{ST}^2/\Gamma)^2$. The quotient of the broad over the narrow linewidth (Γ_{ST}/Γ) may easily exceed 10^5 (see Table 1) in which case cooling would not be observable. A broad band (multi-line) scheme on the contrary is designed for a maximum deceleration rate $\hbar k/2\tau$ during the whole cooling process, from the molasses momentum of the first strong transition down to the recoil momentum of the second slow transition. In this case the required time would be only $T \approx 2\tau(\hbar\Gamma_{ST}m/2)^{1/2}/\hbar k$. This estimate does not include the time of the final stage of cooling from the recoil momentum down to the quantum limit momentum. We checked within our Monte-Carlo calculations that the average number of photons scattered in this final stage is of the order of the reciprocal probability to hit the interval $(2m\hbar\Gamma)^{1/2}$ within an interval $\hbar k$. An experimental constraint arises from the fact that even for maximum deceleration $\hbar k/2\tau$ a free flight during deceleration over a length $l \approx \tau(p_i)^2/(4m\hbar k) = \tau \cdot (\hbar\Gamma_{ST}/8\hbar k)$ will limit the achievable interaction time.

Table 1 : Broad band cooling parameters for alkaline earths

Wavelength λ, lifetime $\tau = 1/\Gamma$, recoil velocity $\hbar k/m$ and ratio recoil shift over linewidth refer to the ($^1S_0$-$^3P_1$) intercombination line. Γ_{ST} is the linewidth of the strong ($^1S_0$-$^1P_1$) cooling transition. The value of the limit temperature is calculated according to $E_{kin} = \hbar\Gamma$.

	Mg	Ca	Sr	unit
λ	457	657	689	nm
τ	4.6×10^{-3}	0.5×10^{-3}	21×10^{-6}	s
$\hbar k/m$	3.64	1.52	0.66	cm/s
$\hbar k^2/m\Gamma$	2.3×10^3	73	1.26	
Γ_{ST}/Γ	2.3×10^6	1.1×10^5	4.2×10^3	
$(m\hbar\Gamma_{ST}/2)^{1/2}/\hbar k$	22.2	27.3	40.15	
T	204	27.4	1.7	ms
l	234	2.83	0.2	mm
$2\hbar\Gamma/k_B$	3.3	30	727	nK

References

[1]   J.Nellessen, J.H.Müller, K.Sengstock and W.Ertmer, 1989, JOSA B, in press
[2]   Y.Castin, H.Wallis and J.Dalibard, 1989, JOSA B, in press
[3]   D.E.Pritchard et.al., Laser Spectroscopy VIII, ed.W.Persson and S.Svanberg, p.68
       (Springer Verlag, Berlin, 1987)
[4]   H.Wallis, W.Ertmer, Poster at 11thConference on Atomic Physics, Paris, July 1988
[5]   H.Wallis, W.Ertmer, 1988, JOSA B, in press

Cooling Atoms with Extraresonant Stimulated Emission Below the Doppler Limit.

Yaakov Shevy

California Institute of Technology 12-33
Pasadena, CA 91125

The process of cooling atoms with "radiation pressure" is well understood in terms of absorption and spontaneous emission of fluorescence photons. This process imposes a lower limit on the minimum equilibrium temperature of laser cooled two level atoms of $K_bT = \hbar\Gamma_{21}/2$ (the Doppler limit), where Γ_{21} is the excited state decay rate to the ground state. At high laser intensity, it has been demonstrated that the stimulated emission process changes the sign of the force to a heating force at the red side of the atomic resonance and to a cooling force at blue detunings[1]. Although this stimulated force is more efficient than the radiation pressure force, it has been generally accepted that this force cannot lead to lower equilibrium temperatures due to the large heating caused by diffusion of momentum at high intensity.

These conclusions are valid only when the sole damping mechanism is the excited state decay to the ground state by spontaneous emission . However, when the atomic system is "opened", i.e., is allowed to decay to other levels, or the dipole decay rate is altered by dephasing events, the stimulated force is dramatically modified. Under this conditions the stimulated force can occur at lower laser intensity and can even reverse sign to provide damping at the red side of resonance. These phenomena originate from extraresonances in the stimulated emission process between the two counterpropagating waves. These resonances appear as a dispersive feature in pump probe spectra (Two Wave Mixing) and are closely related to the extraresonances in four wave mixing studied originally by Bloembergen and co-workers. This paper establishes this connection and the potential of these phenomena for laser cooling. The implications of these results to the recently observed ultra-cold Na and Cs atoms are also discussed.

In order to demonstrate these effects, a simple model consisting of a two level system (figure 1) which is allowed to interact with a reservoir of levels, is used. This enables the introduction of different decay rate of the ground and excited states Γ_1 and Γ_2 respectively. In addition, the dipole decay rate Γ_2' can also be altered by the introduction of phase interrupting events $\Gamma_2' = 0.5(\Gamma_1 + \Gamma_2) + \Gamma_\phi$ where Γ_ϕ is the rate of these events (in a closed system $\Gamma_2'/\Gamma_{21} = 0.5$). Thus, the Optical Bloch Equations(OBE) are given by:

$$\dot{\rho}_{11} = -\Gamma_1 (\rho_{11} - \rho^0{}_{11}) + \Gamma_{21}\rho_{22} - (g\rho_{12}* + g*\rho_{12})$$
$$\dot{\rho}_{22} = -\Gamma_2\rho_{22} + (g\rho_{12}* + g*\rho_{12})$$
$$\dot{\rho}_{12} = -\gamma\rho_{12} - g(\rho_{22} - \rho_{11})$$

Eq.1

Where $\gamma = \Gamma_2' - i\Omega$, $g = i\mu_{21}E/\hbar$, Ω is the laser detuning from resonance and $\Gamma_1\rho^0{}_{11}$ is a phenomenological repopulation term of the ground state. The OBE are solved to all orders of the field and to the first order of the atomic velocity giving the following expression for the force[2-3].

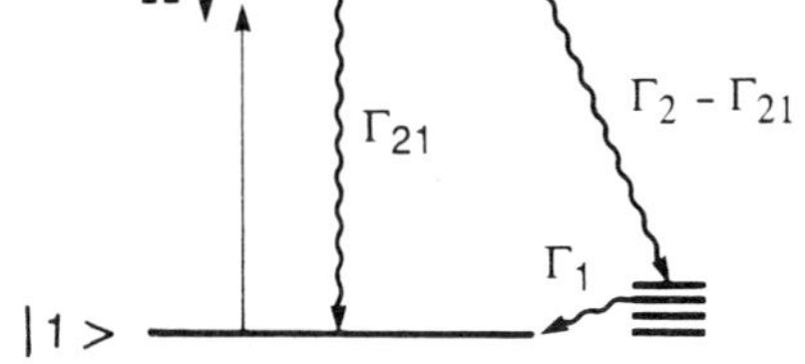

Figure 1. The model used in the calculation, two levels which are in contact with reservoir of levels.

$$F = -\alpha\hbar\Omega\rho^0{}_{11}\frac{p}{N}\left[1-\frac{-A|\gamma|^2p-\frac{\Gamma_2'}{\Gamma_2}(1+\Lambda)^2|\gamma|^2p^2+2\left(1-\frac{\Gamma_2'}{\Gamma_2}(1+\Lambda)p\right)\Gamma_2\Gamma_2'}{\Gamma_2|\gamma|^2N^2}\alpha v\right] \qquad \text{Eq.2}$$

Where $\Lambda=(\Gamma_2-\Gamma_{21})/\Gamma_1$, $N=1+\Gamma_2'(1+\Lambda)p/\Gamma_2$, $p=2|g|^2/|\gamma|^2$ and A is defined by:

$$A = 2\frac{\Gamma_2'}{\Gamma_2}\left[1+\frac{\Gamma_2}{\Gamma_1}\Lambda-\frac{\Gamma_{21}}{\Gamma_1}\right]-(1+\Lambda) \qquad \text{Eq.3}$$

In the limit of a closed two-level system $\Gamma_2-\Gamma_{21}=\Gamma_1=0$, and $\Gamma_\phi=0$ the first term in the numerator is zero while the other terms are reduced to the well-known expression of the force (eq. 18 ref.2).

$$F= -\alpha\hbar\Omega\frac{p}{1+p}\left[1-\frac{-2|\gamma|^2p^2+\Gamma_{21}{}^2(1-p)}{\Gamma_{21}|\gamma|^2(1+p)^2}\alpha v\right] \qquad \text{Eq.4}$$

The appearance of the additional term is due to two classes of extraresonances characterized by the excited and ground state decay rates. In a case of a system which conserves population the extra term reduces to $-2|\gamma|^2$ $(2\Gamma_2'/\Gamma_{21}-1)p$.This term enhances the stimulated force which usually has a cubic intensity dependence whenever $\Gamma_2'/\Gamma_{21}\neq0.5$. The origin of this phenomena is the removal of destructive interference between quantum amplitudes with dephasing events. In a closed system, this resonance occurs only at high intensity due to higher order radiative processes(the $-2|\gamma|^2p^2$ term).

In the case of different decay rates of the excited and ground states to other levels $\Lambda\neq1$ but with $\Gamma_\phi=0$, $\Gamma_2'/\Gamma_2\cong0.5$, another class of extraresonance contributes to the force . In this case, the direction of the stimulated force depends on the value of Λ ; for $\Lambda>1$ the stimulated force opposes the radiation pressure force as usual. However, when $\Lambda<1$, the extra term gives rise to a stimulated cooling force at red detunings. The physical significance of this condition is that if the ground state repopulation rate is larger from the excited state loss to other levels, the stimulated force reverse sign and occur at lower intensity. This phenomenon is due to the pertubation in the coherent population pulsation which is induced whenever $\Lambda\neq1$.

In the analytic solution presented above, the atomic motion is treated only to first order in velocity . In figure 2, however, a numerical solution of the OBE shows the full velocity dependence of the force for the limiting cases discussed above . Trace a shows the tail of the Doppler cooling force in the case of a closed system. In agreement with the analytic solution, trace b shows the appearance of the stimulated force (characterized by $kv\cong\Gamma_{21}$) at the same low intensity of trace a when phase interrupting events are included. Trace c shows the extra cooling force which is maximized at $kv=\Gamma_1/2$,which in this example is chosen as $\Gamma_1=p\Gamma_{21}/3$.

Figure 2:The spatially averaged force as a function of velocity for the case of:
a) a closed two level system, $\Gamma_1=0$, $\Lambda=1$ and no dephasing $\Gamma_\phi=0$.
b)dephasing $\Gamma_1=0$, $\Lambda=1$, $\Gamma_\phi=0.5\Gamma_{21}$.
c)different decay rates $\Gamma_1=0.06\Gamma_{21}$, $\Lambda=0.5$ and no dephasing.

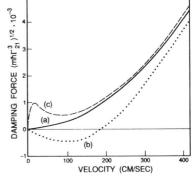

The remarkable appearance of the additional cooling force in trace C implies a profound modification of the minimum equilibrium temperature of laser cooled atoms. For example, with the parameters chosen in curve c, a temperature of $K_b T = \hbar \Gamma_{21}/30$ can be reached provided that the diffusion of momentum remains the the same as in a closed two level system. Other important characteristics of the force in this example are: 1) the temperature continues to decrease at larger detunings ; 2) since Γ_1 depends on p , at low p the temperature is independent on laser power; 3) the velocity distribution of the laser cooled atoms at equilibrium is bimodal.

These new results may explain the recent surprising experiments of cooling atoms below the Doppler limit., The simple model used ,however , does not contain the details of the multi level 3-D cooling used in the experiments, in particular no Zeeman coherence is considered. Nevertheless , numerical solution of the OBE for the Na F=2-->3 transition gives a force very similar to the force shown in trace C. Moreover a recent study of the FWM response in Na[4] shows a narrow dip with a width of the ground state when the lasers are tuned near the Na F=2-->3 transition while a spike is found near the D1 transition. The origin of the dip is ascribed to the optical pumping process which tends to increase the $m_F=0$ population this is due to the fact that the decay rate to this ground state level is larger by a factor of two from the excited state loss. These results are analogous to the laser cooling experiments if one realizes the close connection between the extraresonances in multiwave mixing to the stimulated cooling process (due to the dispersive line shape of two wave mixing, a dip in FWM corresponds to a change of sign of the stimulated force).

In conclusion, the forces exerted on a slowly moving atom in a standing wave are profoundly modified if the atomic system is opened. This phenomenon is closely related to the extraresonances in multi wave mixing . The results shown here may provide an explanation for the recent experiments of cooling atoms below the Doppler limit as well as more exciting possibilities for cooling atoms with stimulated emission.

Acknowledgment: I wish to acknowledge usefull discussions with S. Chu , P. Berman and H. Kimble. I also thank J. Ungar for his help with the computer solutions.

REFERENCES

1)A. Aspect, J. Dalibard, A Heidman, C. Salomon and C. Cohen-Tanoudji, Phys. Rev. Lett., 14, 1688 (1986).
2) J. P. Gordon and A. Ashkin, Phys. Rev., A21, 1606 (1980).
3) Y. Shevy , submitted for publication.
4) P. R. Berman, D. G. Steel , G. Khitrova and J. Liu, Phys. Rev., A38, 252 (1988).

Stimulated Laser Cooling of a Fast Particle Beam

Poul Jessen, Peixiong Shi and Ove Poulsen
Institute of Physics, University of Aarhus
DK-8000 Aarhus C, Denmark

Fast ion-beam laser spectroscopy constitutes a highly specialized tool in atomic physics and in fundamental studies of eg. QED and Special Relativity[1]. One limiting factor in many applications is the final velocity distribution of the accelerated particles. Thus methods to reduce phase-space is of utmost importance also in this field of physics[2] as it certainly is in "low energy" atomic physics, where laser cooling and trapping has been highly developed[3].

The ligth induced forces may be split into the scattering force and the dipole force, the former associated with a traveling plane wave and the latter with a standing plane wave[4]. In general, laser cooling of gases and atomic beams exhibit rich structures, both experimentally and theoretically in one, two or three dimensions and in two and three level atoms. Dealing laser cooling of fast beams, a topic still in its infancy simpler considerations suffice at the present stage of its development[5,6].

1. Dipole Force in a Standing Plane Wave

In contrast to the scattering force which saturate at high laser intensities the dipole force requires the atomic transitions to be saturated to observe deceleration/acceleration for a blue/ red shifted detuning. In this way a dipole force which is much larger than the scattering force can be realized. The conditions for the dipole force to be optimum is that the velocity of the particles must be low relative to the standing wave. This can conviniently be satisfied in a transverse geometry in atomic beams[7], but at sufficiently high Rabi-frequencies cooling is even observed in a longitudinal geometry[8].

Fast beams, produced in ion-sources and extracted by ion-optics have typical relative velocity spreads $\Delta v/v \sim 10^{-3}$ or better, corresponding to velocity spreads $\Delta v \sim 20$ m/sec. This makes stimulated laser cooling an attractive possibility for cooling of fast beams, with cooling time limited by the transit time.

Dalibard and Cohen-Tannoudji[4] have demonstrated the utility of the dressed-atom picture in the description of cooling by a strong standing plane wave. They have calculated the force $F_i = dE_i/dz$ for a two level atom in levels i=1,2 in a standing plane wave, where $E_i(r)$ is the energy of the dressed states

$$E_i(r) = (n+1)\hbar\omega - \tfrac{1}{2}\hbar\delta \pm \tfrac{1}{2}\hbar\Omega(r) \tag{1}$$

$$\Omega(r) = \sqrt{\kappa^2(r)+\delta^2} \quad ; \quad \kappa^2(r) = \kappa^2\cos^2 kz \tag{2}$$

and κ is the Rabi-frequency and δ the laser detuning.

At vanishing particle velocity the dipole force averages to zero over a wavelength. A low velocity introduces a lag in the populations, destroying the symmetry of the transition rate around field nodes and antinodes. This results in a net force, linear in v. At high velocities the populations are nearly independent of position and the average force approaches zero as v^{-1}. At intermediate velocities, a numerical calculation of the populations is required to bridge the asymptotic results. However, in the entire velocity range, the force is a cooling force for blue detunings.

In the intense fields used, the steady-state velocity distribution have a width corresponding to a kinetic energy $\sim \hbar\kappa \gg \hbar\Gamma$ where Γ is the natural linewidth. This is easily derived by considering an atom which has come to rest in one of the vallies of the potential $E_i(r)$. When decaying to the other dressed state it arrives at a potential maximum and is subsequently accelerated to a kinetic energy

$$\tfrac{1}{2}Mv_m^2 = \hbar\{\sqrt{\kappa^2+\delta^2} -\delta\}. \tag{3}$$

Finally fluctuations, due to spontaneous processes, must be included in the force. Monte-Carlo simulations have been performed[5] to study the resulting cooling force, with parame-

32

ters typical of the experiment to be discussed next. The results of such a simulation, also taking into account the degeneracy due to the magnetic sublevels of a J=2 to J=3 transition, is shown in Fig.1.

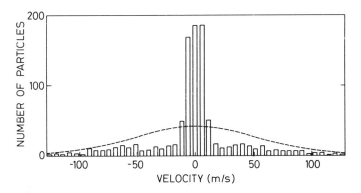

Fig.1: Velocity distribution before and after cooling in a standing plane wave, with $\delta=200\Gamma$ and $\kappa=1100\Gamma$, during the interaction time 900Γ (8 μsec). The dashed curve shows the initial Gaussian velocity distribution with a relative velocity spread $\delta v/v \sim 2\cdot10^{-4}$. The final distribution is from a Monte Carlo calculation with 1000 particles.

2. Experimental setup

The dipole force on an atom in a strong standing plane wave is substantial only when the particle velocity with respect to the wave does not exceed a few times Γ/k which typically is in the order meters per second. The wave must therefore be standing in a frame moving with a velocity βc close to the velocities of the particles in the fast beam.

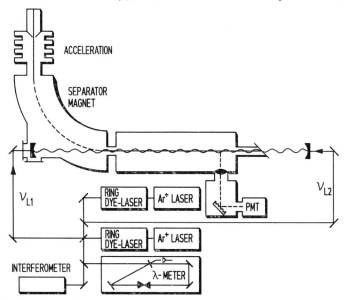

Fig.2: Experimental test setup for the study of the dipole force in a fast atom beam. The probe laser is not shown.

Such a wave is obtained as the superposition of two travelling waves, moving in opposite directions with different frequencies ω_{L1} and ω_{L2} in the laboratory. To obtain sufficiently high intensities, these travelling waves are taken as counterpropagating components of two different standing waves in a high finesse resonator of length L, i.e., $L_i = n_i\pi c/L$, where n_i is the number of nodes. In a frame moving at velocity βc with respect to the laboratory, where $\beta = (n_2-n_1)/(n_2+n_1)$ the Doppler shifted frequencies become equal, i.e., a standing wave is obtained. The dipole force will thus compress the longitudinal velocity distribution around βc, where β is a controlled parameter by locking the "build up" resonator to one dye laser and in turn lock another dye laser to the resonator. A third laser is used to probe the velocity distribution.

3. Results

To test the apparatus, experiments without the laser enhancement resonator has been performed. At a beam energy of 1 kV the interaction time in the apparatus is 8 μsec. This time is sufficiently long to observe cooling due to the scattering force for small red detunings. The resulting velocity distribution has been reduced 10% in agreement with calculations. However, to observe cooling due to the dipole force strong Rabi-frequencies and blue detunings are required. As to the detuning it must be larger enough to fulfil the adiabatic condition $v < v_{cr}$, where the critical velocity v_{cr} is derived by Dalibard and Cohen-Tannoudji[4]. The parameters given in Fig.1 fulfil this criterium as well as representing the design parameters of the experimental equipment in Fig.2. The experimental work is in progress.

4. References

1. Ove Poulsen,in:"Electronic and Atomic Collisions", eds, H.B.Gilbody et. al., Elsvier Science Publ.BV,1988,p.579
2. J.Javanainen, M. Kaivola, U. Nielsen, O.Poulsen and E.Riis, J.opt.Soc.Am.B 2,1768(1985)
3. See JOSA B,topical issue on Laser cooling and trapping of atoms(1989)
4. J.Dalibard and C. Cohen-Tannoudji, J.Opt.Soc.Am.B 2,1707(1985)
5. E.Bonderup and K. Mølmer, J.Opt.Soc.Am.B, accepted
6. E.Bonderup, P. Jessen, K. Mølmer and O. Poulsen in: Proceedings of the Workshop on Crystalline Ion Beams, GSI-89-10(1988)
7. C.Salomon, J.Dalibard, A.Aspect, H.Metcalf and C.Cohen-Tannoudji, Phys. Rev. Lett.59,1659(1989)
8. M.Prentiss and A.Cable, Phys. Rev. Lett.62,1354(1989)

Towards laser cooling of fast Be$^+$ ions in the storage ring TSR

A.Faulstich, W.Petrich, V.Balykin [1] , D. Habs, R.Neumann
D.Schwalm, R.Stokstad [2] , B.Wanner, A.Wolf
Physikalisches Institut, University of Heidelberg,
and Max-Planck-Institut für Kernphysik,
D-6900 Heidelberg, Fed.Rep. of Germany
M.Gerhard, G.Huber, R.Klein, S.Schröder
Institut für Physik, University of Mainz, D-6500 Mainz, Fed.Rep. of Germany
T.Kühl
Gesellschaft für Schwerionenforschung, D-6100 Darmstadt, Fed.Rep. of Germany

^{9}Be$^+$ ions stored in a heavy-ion storage ring seem to be a promising species for laser cooling [1] down to temperatures several orders of magnitude less than those reached for protons by electron cooling at the Novosibirsk ring NAP-M [2]. Short cooling times and microkelvin temperatures can be envisaged, where the structure of the ion beam is dominated by Coulomb repulsion. However, while laser cooling of ion clouds confined in traps has been extensively studied we are confronted with the novel situation of very fast ion beams stored in a ring involving complex dynamical effects (e.g. betatron oszillations, non-adiabatic passages of the ions through magnetic fields, etc.). The present studies are aimed at a clear understanding of laser-ion interaction under storage ring conditions to prepare the basis of laser cooling of fast stored ion beams. In addition, the method of laser-induced fluorescence provides precise data for beam properties like, e.g., absolute velocity, momentum spread, and lifetime.

We used the strong resonant transition at $\lambda= 313$ nm between the $2^2S_{\frac{1}{2}}$ ground state and the $2^2P_{\frac{3}{2}}$ excited states, having a natural linewidth $\Delta\nu = 18.3$ MHz according to the 2^2P lifetime $\tau = 8.7$ nsec (see Fig.1). Due to the spin I=$\frac{3}{2}$ of the ^{9}Be nucleus, both $2^2S_{\frac{1}{2}}$ and $2^2P_{\frac{3}{2}}$ states exhibit a hyperfine structure splitting, having a size of 1.25 GHz and 9 MHz, respectively (see Fig.1).

A schematic view of the experimental configuration is shown in the lower part of Fig.1. The light beam from a homemade frequency-doubled free-jet ring dye laser with typically 30 mW power in the ultraviolet propagates parallel to the ions. Two cylindrical high-voltage (HV) electrodes have been installed in a 1.3 m long zone in the center of the straight ring section. The laser-ion interaction is monitored via the fluorescence emanating from inside the first HV cylinder, using a cooled photomultiplier operated in the single-photon counting mode. If ions populating e.g. the $2^2S_{\frac{1}{2}}$, F=1 substate are excited to the $2^2P_{\frac{3}{2}}$ state by monochromatic laser light, they will end up in the F=2 substate after a few absorption and reemission cycles due to the $\Delta F = 0, \pm 1$ selection rule for electric-dipole transitions. In contrast to a single-pass configuration, this optical pumping mechanism is very effective in our case, since the ions pass the laser beam very often ($\Delta t = 4\mu$sec) during their lifetime in the ring. The HV tubes provide a refilling mechanism for the initial sublevel by a local change of the ion beam velocity. The HV has to be set such that the laser excites e.g. the $2^2S_{\frac{1}{2}}, (F = 2) - 2^2P_{\frac{3}{2}}(F = 1, 2, 3)$ transition outside the HV range and, due to the different Doppler shift, the (F=1-F=0,1,2) transition within the HV region.

---

[1] on leave from Academy of Sciences, Troizk, Moscow Region, USSR
[2] on leave from Lawrence Berkeley Laboratory, Berkeley, California

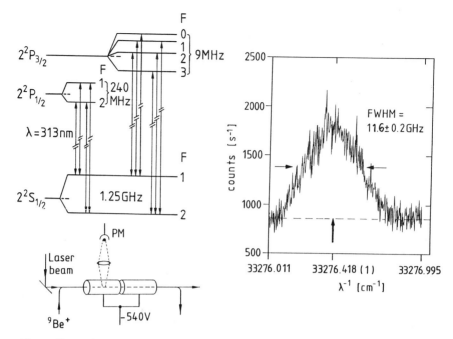

Fig.1. Ground-state and first excited-state energy levels of $^9Be^+$. The lower part shows a schematic view of the experimental configuration.

Fig.2. Doppler profile of $2^2S_{\frac{1}{2}} - 2^2P_{\frac{3}{2}}$ transition in $^9Be^+$ obtained by laser frequency scanning.

Fig.2 shows the ion Doppler profile obtained by laser frequency scanning with the tube voltage kept fixed at a suitable value, from which the beam energy as well as the longitudinal momentum spread can be extracted.

A HV-tuned fluorescence spectrum observed for a fixed laser frequency is displayed in Fig.3. The two peaks represent two separate velocity classes of ions that were transferred by optical pumping from the F=1 to the F=2 ground state sublevel (peak at positive voltage) and vice versa (peak at negative voltage) outside the HV tubes, and are shifted into resonance again inside the tubes. Thus the fluorescence peaks originate from a pumping back from one sublevel to the other. The broad distribution underneath the two peaks in Fig.3 is not yet fully understood, but may arise from a refilling mechanism provided by the storage ring, such as nonadiabatic spin flips occurring when the ions enter or leave the strong fields of the beam bending magnets of the ring. The fluorescence intensity as a function of time is depicted in Fig.4 for HV settings of -450V (black circles) and 0V (open circles). At -450V the ion-velocity class involved in the left peak of Fig.3 is monitored. Its decay time ($\tau \simeq 3$ sec for 30 mW cm^{-2} laser intensity) is much shorter than the lifetime of the ions in the ring ($\tau \simeq 15$ sec). This suggests that these ions are shifted out of laser resonance because their velocity has changed due to multiple photon momentum transfer. This transfer can take place at this tube voltage setting since pumping back from one sublevel to another occurs. With less than 10 mW cm^{-2} of laser intensity the momentum transfer is significantly reduced and the decay time is observed to be correspondingly longer.

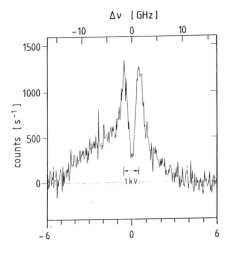

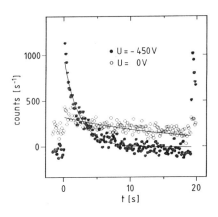

Fig.3. High-voltage tuned $^9Be^+$
resonances for fixed laser frequency.

Fig.4. Fluorescence decay curves for
two different HV-values at fixed
laser frequency.(A new beam
is stored every 19 sec.)

At 0 voltage, the only pumping back that can occur is due to substate-changing processes induced by the ring itself. Thus, the fluorescence signal is weak, and only little photon momentum transfer can occur. Once an equilibrium between laser substate depopulation and ring induced repopulation is established, the decay time should be and is observed to be about equal the ion beam lifetime.

In summary, the experimental results indicate the production of a Bennett hole in the ion velocity distribution, thus providing an important step towards laser cooling of fast Be^+ ions stored in a ring.

References

[1] J.Javanainen, M.Kaivola, U.Nielsen, O.Poulsen,E.Riis,
     J.Opt.Soc.Am.B11(1985)1768

[2] E.N.Dementev, N.S.Dikanskii, A.S.Medvedko, V.V.Parkhomchuk,
     D.V.Pestrikov, Sov. Phys. Tech. Phys. 25(1980)1001

[3] A.Faulstich, W.Petrich, D.Habs, R.Neumann, D.Schwalm, A.Wolf,
     S.Schröder, G.Huber, T.Kühl, D.Marx, Proc. Workshop on Crystalline Ion
     Beams, Wertheim, Germany, Oct.4-7,1988 (ed. by R.W.Hasse,
     I.Hofmann, D.Liesen), GSI-89-10 Report, April 1989
     (ISSN 0171-4546), p.289

Acknowledgement

We would like to thank in particular the TSR group around E. Jaeschke; without their work and their enthusiasm these experiments would not be possible.

Laser Spectroscopy

Observation of Light-Pressure-Induced Line-Shape Asymmetries of Saturated Absorption and Dispersion Resonances

Jürgen Mlynek and Rudolf Grimm
  Institute of Quantum Electronics
  Swiss Federal Institute of Technology (ETH) Zürich
  CH-8093 Zürich, Switzerland

The study of resonant light forces on the motion of free atoms is of strong current interest. So far, most of the experiments on light forces have been performed on atomic beams: It has been shown that atoms in a beam can be strongly deflected, decelerated, and even trapped by laser light. Although, quite obviously, resonant light forces can also act on the atoms in a gas /1/, very few experimental work has been performed on *light pressure effects in gases*; here the total modification of the velocity distribution that can be achieved by resonant light pressure under normal experimental conditions seems to be insignificantly small. Recent investigations, however, showed that even very small light-pressure-induced modifications of the atomic velocity distribution can have substantial effects on the *optical response* of a gas to laser light; here the effect of light pressure shows up in the transmitted light and is manifested in *modifications of dispersion /2,3/ and absorption /3-5/ profiles*.

In this contribution, we want to discuss the observation of the effect of resonant light pressure in saturation spectroscopy /4,5/: If this well-known technique of high-resolution laser spectroscopy is applied on a low-pressure gas, substantial *asymmetries* of the Doppler-free absorption and dispersion resonances can occur as a result of light pressure.

Basic Idea

Let us start with a simple qualitative explanation of the light-pressure-induced line-shape asymmeries in saturation spectroscopy. As is well-known, this method is based on the velocity-selective excitation of a certain velocity subgroup in a gas by a relatively strong optical pump field. The induced hole occuring at the resonant velocity v_0 in the population difference of ground and excited state, known as Bennett hole, is probed by a weak tunable test field: As a consequence, Doppler-free signals occur in both the absorption and the dispersion curve of the test field. Usually, these Doppler-free signals are considered as being completely attributed to a perturbation of the *internal* atomic degrees of freedom, namely the *saturation* of the optical transition. But, in addition to this, also the *external* atomic degrees of freedom can experience a perturbation: the *redistribution of atomic velocities* caused by the spontaneous light pressure of the saturating field (see Fig. 1a). The corresponding distortion of the atomic velocity distribution $N(v)$ leads to an asymmetry of the Bennett hole (see Fig. 1b) in the population difference of ground and excited state $n_g(v)-n_e(v) = N(v) \times [1-2p(v)]$; here $p(v)$ denotes the steady-state excitation probability. Thus, obviously, also asymmetries of the Doppler-free contributions to absorption

40

and dispersion of the test field can occur as a result of resonant light pressure.

A detailed theoretical description of these light-pressure-induced line-shape asymmetries is given in Ref. /5/. It is shown that, in the case of low laser intensities, the Doppler-free signals can be written as a sum of an ordinary contribution that is due to the saturation of the optical transition and a light-pressure-induced contribution resulting from the modification of the velocity distribution. Because of the opposite symmetry of these contributions, the total Doppler-free signals display an asymmetric line shape. The strength of the light-pressure-induced contribution and therewith also the strength of the asymmetry is proportional to an effective atom-field interaction time; in the collision-free regime of a low-pressure gas, this time is determined by the transit times of the atoms through the laser beam and thus is closely related to the laser beam diameter.

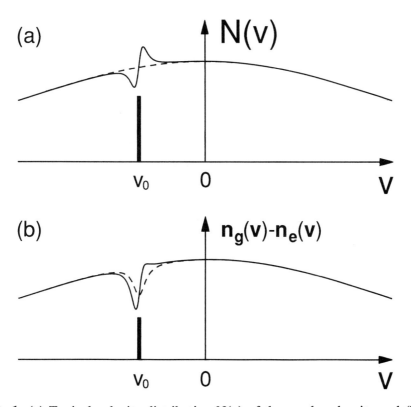

Fig.1: (a) Typical velocity distribution $N(v)$ of the number density and (b) corresponding velocity distribution $n_g(v)-n_e(v)$ of the population difference of an ensemble of two-level atoms interacting with a monochromatic laser field at velocity v_0. Both curves are shown with (solid lines) and without (dashed lines) modifications due to resonant light pressure.

<u>Experiment</u>

Our experiments to demonstrate the existence of the light-pressure-induced line-shape asymmetries of the Doppler-free saturation resonances were performed on the λ=555.65nm J=0-J'=1 transition in atomic ytterbium vapor. We applied frequency-modulated saturation spectroscopy /6/, which allows to observe both the absorption and the dispersion-related signals. In the experiments, we used copropagating laser beams, which were derived from the same laser source; here a variable optical frequency shift $\Delta\omega = \omega_1 - \omega_0$ of the probe field (ω_1) with respect to the saturating field (ω_0) was

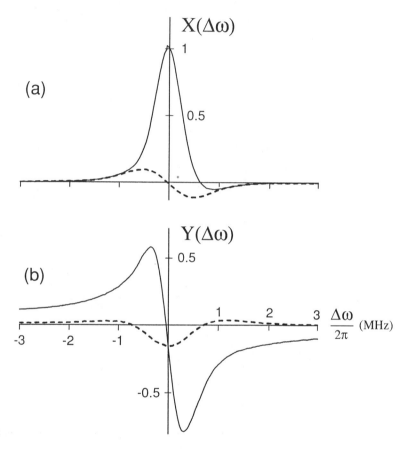

Fig.2: (a) and (b), typical measured line shapes $X(\Delta\omega)$ and $Y(\Delta\omega)$ of the Doppler-free absorption and dispersion signal, respectively. The intensity of the saturating field was 20μW/cm^2, and the probe field intensity was about 10μW/cm^2. The solid curves show the total Doppler-free signals, observed with pump and test field spatially overlapping in the sample. The dashed curves, which were obtained with spatially separated beams, show the light-pressure-induced contributions only.

introduced by acousto-optical modulation techniques /5/. This experimental scheme turned out to be much less sensitive to the residual frequency jitter of our dye laser than an experiment with counterpropagating beams.

Typical measured saturated absorption and dispersion resonances are shown in Fig. 2(a) and (b), respectively. The probe beam of diameter 1.5mm was located in the center of the saturating beam; the latter was expanded to a diameter of about 1.5cm in order to provide atomic transit times being sufficiently long for a strong occurence of the light-pressure-induced signal contributions. As a result of light pressure, the Doppler-free signals observed in this way clearly display an *asymmetric line shape* in accordance with theoretical predictions. Our experiments also confirm that the asymmetric shape of the saturation signals is independent of the laser intensity if saturating and probe field are weak compared with the saturation intensity of the optical transition.

A variation of our experiment allows to give an *unambigous proof* that the observed asymmetries are, in fact, a result of resonant light pressure: With the use of *spatially separated beams*, the light-pressure-induced signal contributions can be observed separately from the ordinary saturation contributions /5/. For a sufficiently large distance between saturating and probing beam, all effects related to the optical excitation cannot play any role in the signal formation since all perturbations of the internal atomic degrees of freedom are completely relaxed when the atoms enter the probe zone. Thus, the signals that we observed with spatially separated beams (dashed curves in Fig. 2) are clearly due to a perturbation of the *external* atomic degrees of freedom, i.e., the light-pressure-induced redistribution of velocities.

Conclusion

Our experiments have demonstrated that resonant light pressure can lead to substantial modifications of the line shapes of Doppler-free absorption and dispersion resonances obtained by saturation spectroscopy in low-pressure gases. Our experimental results are in satisfying agreement with theory /5/. Besides containing basic information on the interaction of light and matter, light-pressure-induced asymmetries and shifts of spectral lines may be of importance in various applications of saturation spectroscopy, e.g., for high-precision frequency measurements using absorption cells.

References

/1/ T. W. Hänsch and A. L. Schawlow, Opt. Commun. **13**, 68 (1975).
/2/ R. Grimm and J. Mlynek, Phys. Rev. Lett. **61**, 2308 (1988).
/3/ A. P. Kazantsev, G. I. Surdutovich and V. P. Yakovlev, JETP Lett. **43**, 281 (1986).
/4/ R. Grimm and J. Mlynek, Phys. Rev. Lett., in press.
/5/ R. Grimm and J. Mlynek, Appl. Phys. B, in press.
/6/ J. L. Hall, L. Hollberg, T. Baer, and H. G. Robinson, Appl. Phys. Lett. **39**, 680 (1981).

Light Induced Drift in a spherical cell.

S.Gozzini, D.Zuppini, C.Gabbanini and L.Moi
Istituto di Fisica Atomica e Molecolare del CNR
Via Del Giardino, 7 - 56100 PISA - Italy

The Light Induced Drift (LID) is an effect proposed few years ago by Gel'mukhanov and Shalagin[1] and observed for the first time by Antsygin et al [2]. It is due to the combined actions of resonant laser excitation and collisions with a buffer gas. Both the difference between the diffusion coefficients of the ground and excited state atoms and the velocity selective excitation produce an asymmetry in the Maxwellian distribution of the atomic velocities and hence a macroscopic flux of atoms inside the cell. LID can both pull or push the vapor depending on the laser detuning inside the Doppler linewidth.

Until now all LID experiments, as well as all the theoretical calculations, have been performed in one-dimension geometry, i.e. in capillary cells whose diameter d « l, where l is the cell length. Under this condition only the drift along the major cell axis is considered while the transverse diffusion is neglected.

We started new experiments in which the capillary cell is exchanged with a spherical one and the laser beam is split in six beams directed along orthogonal directions crossing the center of the cell.

Aim of these experiments is to study the vapor induced diffusion under this geometry and in particular to realize a situation in which the vapor is confined and compressed at the center of the cell. Under these conditions the vapor is isolated from any direct contact with the cell walls, while the thermal equilibrium with the environment is realized by the buffer gas.

This is a new situation in which the vapor is dry and cannot condensate at the cell surface. As a consequence, high soprasaturation regime can be achieved when the temperature is suddenly dropped down. Pure gas phase condensation phenomena might be observable, like, for example, cluster formation.

We have obtained preliminary results by using a 1.5cm diameter spherical cell connected through a long capillary (few cm) to the Na reservoir. The cell is filled with few torr of Kr as buffer gas and it is coated with a hether solution of dimethylpolysiloxane in order to minimize the surface effects[3]. The laser is a c.w. broad-band dye laser with a long cavity configuration which has a better efficiency with respect to the single mode one[4]. The laser beam is sent to a bunch of six optical fibers which are then suitably positioned around the cell. Two other fibers, connected to two photomultipliers, are used to monitor the fluorescence coming from the external region of the cell. If the vapor is optically thick the counterpropagating laser beams do not compensate each other and a drift to the center is obtained.

Preliminary results are shown in Fig.1).

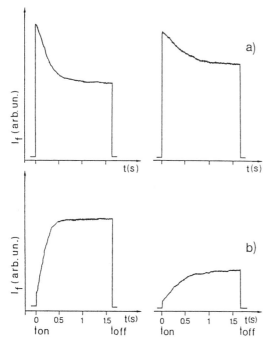

Fig.1) Fluorescence signals obtained with all laser beams (left) and only with two counterpropagating beams (right); a) at the cell wall; b) in the cell center. $I_L = 280mW$, $T = 200°C$

REFERENCES

1.   F.Kh.Gel'mukhanov and A.M.Shalagin, JEPT Lett. 29 (1979) 711.
2.   V.D.Antsygin, S.N.Atutov, F.Kh.Gel'mukhanov, G.G.Telegin and A.M.Shalagin, JEPT Lett. 30 (1979) 243
3.   J.H.Xu, M.Allegrini, S.Gozzini, E.Mariotti and L.Moi, Opt.Commun. 63 (1987) 43; E.Mariotti, J.H.Xu, M.Allegrini, G.Alzetta, S.Gozzini and L.Moi, Phys.Rev. A 38 (1988) 1327.
4.   C.Gabbanini, J.H.Xu, S.Gozzini and L.Moi, Europhysics Lett. 7 (1988) 505

M. DUCLOY, M. ORIA[(*)], A. KLEIN, S. LE BOITEUX, O. GORCEIX,
J.R.R. LEITE[(*)], M. FICHET and D. BLOCH.
Laboratoire de Physique des Lasers, U.A. 282 du C.N.R.S.
Université Paris-Nord
F-93430 Villetaneuse FRANCE

I.   INTRODUCTION

Optical Phase Conjugation (OPC) has been a promising technique for more than a decade and persists in stimulating numerous novel fundamental contributions, from various points of view [1]. Beyond the well-known possibility of image restoration, phase conjugation (PC) has widened its scope of interest to image processing, associative memory , laser frequency stabilization through optical phase locking, oscillation in various types of new, possibly self-aligned or self-pumped, laser cavities. The most widely used technique to generate phase conjugation is degenerate Four-Wave Mixing (FWM) (see Fig.1) [2] on a non linear medium owing a third-order susceptibility and it is the very nature of the material in which wave- mixing takes place which governs the performances of the phase conjugator device. The microscopic properties of the nonlinear material is of such a crucial importance that FWM is currently used for laser spectroscopy (phase conjugate spectroscopy) and is notably a powerful tool in the time-resolved domain.

Effective applications of OPC requires one to determine which materials are the most convenient for specific purposes. The criteria to be considered notably include : efficiency of the PC mirror (measured either as a PC reflectivity R_{PC} , or relatively to the overall input power), time response, required mode of operation (pulsed or c.w.), wavelength range of operation and availability of the corresponding laser sources, price and reproducibility of nonlinear samples, fidelity of wavefront reversal and angular resolution. To this basic list of wishes , one can add extra requirements like the need of multicolor phase conjugation, or of an added versatility by a complementary control parameter (e.g. for purposes of electronic modulation) like incident light polarization, or the voltage of an applied external field.

OPC through FWM has been demonstrated in a large variety of nonlinear materials. In the pulsed regime, non resonant <u>dense</u> media can provide reasonable efficiencies. The wavelength range is large and essentially limited by absorption while the maximum acceptable input power is limited by the damage threshold. In the c.w. regime, materials exhibiting strong

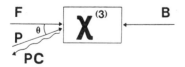

Figure 1: Scheme of FWM geometry (<u>i.e.</u> identical frequencies for F,B and P beams), PC emission is <u>automatically</u> phase-matched, whatever is θ.

[(*)]Work supported by CNPq (Brazil)

enough nonlinearity are rarer. Resonant dense media, like rare earth doped materials, are reasonably fast but with a low reflectivity (typically $R_{PC} \lesssim 10^{-2}$). Photorefractive materials have remarkable efficiency ($R_{PC} \simeq 100$ [3]) and can be used at any available wavelength within approximately the whole visible range. Their main drawback is their very slow response time, of the order of seconds. At last, resonant low density vapors can be used with low power irradiation due to their low saturation intensity. They have been extensively studied but more often from a fundamental point of view than for applications, because they generally require to use tunable sources. They present however important advantages: samples (for which various designs can be chosen) are more reproducible than for doped crystals, and not sensitive to optical damage with intense pulses; typical response times are short ($\sim 10^{-6} - 10^{-9}$ s) ; besides, the universality of resonant processes, with respect to the wavelength, can make gas media still useable in wavelength ranges unpracticable with dense media (absorption window in the IR, VUV...). One can add also that the isotropic nature of gas media contribute to the large versatility of the experiments with respect to the incident polarization, and that multicolor OPC can be operated through multiresonant wave-mixing[4].

It is the aim of this paper to discuss the wide possibilities offered by resonant gas media and how they can compete with non resonant dense media used for devices based upon OPC. Experimental possibilities are illustrated by two recent extensions: (i) to low-power diode laser irradiation on Cs vapor; and (ii) to the UV range.

II. GENERAL PROPERTIES OF FWM IN RESONANT GAS MEDIA

When the third order susceptibility is based upon resonant process, DFWM can be viewed as real time-holography: the forward pump beam (F) and the probe (P) resonantly induce a volumic grating in the $\chi^{(3)}$ medium, which is read out by the backward pump beam (B) (see Fig. 1). From the first studies dealing with the specific case of resonant gas media, one knows the crucial importance of the atomic thermal motion which in most cases restricts the effective angular aperture θ in FWM [5]. Indeed, the induced grating is thermally washed-out in a typical time Λ/u, where u is the atomic velocity and $\Lambda = \lambda/2\sin(\theta/2)$ is the grating spacing. Hence, in the typical case of a predominant Doppler broadening, efficient FWM occurs only for $\theta \ll 1$ (θ is the (F,P) angle , see Fig. 1). Moreover in this case the pump beams axis naturally defines a preferential axis, along which a velocity selection is operated, like in saturated absorption . This implies that the FWM resonance linewidth is governed by the homogeneous width and is only residually sensitive to the Doppler broadening. With a narrow linewidth laser, this contributes to the sharp FWM amplitude dependence with the angular separation θ and sets a major restriction to the angular capability of PC devices based on resonant gas media. However, this limitation can be overcome by homogeneous broadening (collisional or power broadening) or adequate choice of gas media (small Doppler width). Laser cooling of atomic motion, or Dicke narrowing thanks to a buffer gas, could be alternate solutions to increase the maximum acceptable aperture angle.

The analogy between FWM and (spatially modulated) saturated absorption (SA), which helps to interpret FWM angular dependence, is also very fruitful to deal with saturation effects [6]. However, the thermal washout of the induced population grating has no equivalent in SA. While in an open two-level system, population transfer _via_ optical pumping increases SA sensitivity, it dramatically alters PC efficiency by making the medium transparent: indeed, one has to compare the grating thermal lifetime (possibly lengthened by Dicke narrowing) and the optical pumping characteristic time, which is generally much longer. This difference between SA and PC is spectrally revealed by Fig.2.

The magnitude of the PC signal, and the probe beam reflectivity in FWM in gas media have been intensively studied. It has been theoretically predicted and experimentally observed that in a Doppler-broadened optically thin gas medium, there exists an upper limit (in the 1% range) to PC reflectivity [7]. Such an absolute limit is a typical signature of saturation effects. One notes that even when R_{PC} remains low, intense enough pump beams can help bleaching the gas medium, so that FWM induced nonlinear probe amplification can overcome linear probe absorption. In these conditions, with a 1W dye laser, one can observe oscillation in a two-mirror cavity where the pumped vapor is the gain medium [8]. Moreover, one knows since the very first experiments on resonant gas media, that largely exceeding unity reflectivity can be obtained in the pulsed regime (R_{PC} =40 with Na vapor is reported in [9]). This can occur for high enough intensities, off-resonance irradiation (i.e. in Doppler wings) and high atomic density (i.e. the medium is optically thick for on-resonance irradiation). If these requirements are fulfilled in the c.w. regime, R_{PC} is high, and a c.w. PC oscillation has even been observed in a self - aligned empty cavity [10] ended by an ordinary mirror at one end and strongly pumped Na vapor at the other end. A new important step for the use of resonant gas medium in phase conjugation has been the recent achievement of a self-pumped Na vapor phase conjugator [11] where the pump beams, initiated by scattered light are sustained by two-wave-mixing gain. Hence, with these most recent developments, various types of c.w. PC oscillators, previously demonstrated only with slow photorefractive media , can actually be operated with much faster resonant vapors. However, all such experimental demonstrations remain strongly dependent on the limited output power available from c.w. tunable sources, and this is why they have been essentially performed on the Na vapor resonance line (λ = 589 nm) with rather powerful (1W), but not very convenient, dye lasers. Recent experimental work shows that FWM in resonant gas media can still provide promising results well beyond the limits imposed by unhandy dye lasers.

III. RESONANT PHASE CONJUGATION WITH LOW-POWER DIODE LASERS

The recent development of compact and cheap low-threshold temperature-operated diode laser is obviously a major breakthrough for numerous laser applications. However, the limited power available from c.w. single mode

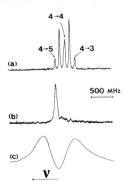

Figure 2: Spectra recorded in the vicinity of the Cs D_2 line $6S_{1/2}$ (F=4) - $6P_{3/2}$ {F'=2,3,4,5} (see[13]):(a) saturated absorption spectrum; not labelled resonances are Doppler cross-overs; (b) typical PC linashape in an optically thin medium; emission occurs only for F=4 - F=5 transition for which optical pumping is forbidden; (c) PC lineshape is an optically thick medium (PC reflectivity is $\sim$ 20%).

diode laser is a severe practical limitation to their use in nonlinear optics. Moreover, their wavelength range located in the near IR is not the most favorable for efficient photorefractive crystals, whose response time considerably slows down with increasing wavelength [12]. Fortunately, the resonance lines of alkali vapors (Rb, Cs) fall down in the emission range of AlGaAs diode laser, for which frequency tunability across the resonance is ensured by control of the injected current.

We have recently demonstrated that diode laser-pumped Cs vapor is a high performance medium for c.w. phase conjugation [13]. With a 30 mW diode laser (λ = 852 nm), up to 50% PC reflectivity was observed in a 1 mm short Cs vapor cell, as well as oscillation in a two-mirror cavity with pumped Cs-vapor as the gain medium (circulating power inside the cavity $\sim$ 100 μW). To get these high PC reflectivities, the incident beams are focused in order to increase saturation, and bleach the optically thick vapor (T$\sim$100-130°C) . The cell length is chosen as short as possible (this is compensated for by an increased vapor pressure in order to deal with an optimized total number of atoms) to ensure satisfactory beam overlap under reasonable angular separation ($\sim$50 mrad). Size parameters (beam waist, cell length, and atomic density) have been optimized relatively to the available output power, and with respect to the main limitations: self-focusing and collisinal effects. Higher level of performances, notably in order to operate other types of PC oscillators already demonstrated with Na vapor, essentially rely on the quick progress in the diode laser state-of-the-art. It should be outlined that because of strong power broadening, the PC reflectivity lineshape, which presents maxima off resonance, is broader ($\sim$ 1 GHz) than the Doppler width ($\sim$ 400 MHz) of the D_2 resonance line [see Fig. 2] This has at least two important consequences : (i) efficient resonant phase conjugation simply requires an easy-to-realize stabilization of the injected current ($\Delta i/i \leqslant 10^{-4}$), and a very small drive current should permit to switch on or off the oscillation of a PC cavity in a typical time governed by the 30 ns lifetime of the $6P_{3/2}$ level; (ii) limitations on the FWM aperture angle solely originate from beam overlap conditions and not from atomic thermal motion.

IV.   EXTENSION TO THE UV RANGE

Extending nonlinear optics from the IR and visible domain to the ultraviolet (UV) range is a task of importance justified by various interests. It lowers the diffraction limit, which can be important for image processing. Shorter wavelengths also considerably increase the efficiency of photochemical processes or photoablation in material processing. In laser induced fusion, shorter optical wavelengths are also of considerable importance, and the intense, but aberrated laser pulses can be in principle optically restored thanks to OPC.However, extension to the UV range is often made difficult by the lack of convenient UV sources, and by absorption bands of dense nonlinear materials. In the deep UV range, it is well-known that only gas media, which are transparent but in the vicinity of an atomic absorption line, can be used for frequency up-conversion to the VUV region.

In a recent work [14] we have operated the first c.w. phase conjugator in the UV region. FWM emission has been observed from Mg vapor irradiated with a laser tuned to the $^1S_0 - {}^1P_1$ resonance line (λ=285 nm). The UV source is an intracavity frequency-doubled single-mode actively stabilized tunable dye laser delivering up to 8 mW. The Mg resonance line was chosen because Mg vapor requires moderate heating [T = 280°C for 5.10^{10} atoms/cm^3

] and because of the absence of hyperfine optical pumping. Up to now, the PC reflectivity has remained low ($R_{PC} \leqslant 10^{-6}$) essentially because it is hard to saturate a broad resonance line ($\gamma=85$ MHz) with a 8 mW source. Indeed, the estimated saturation parameter does not exceed 5.10^{-3}. Along the same principles for optimization as developped for Cs vapor, higher Mg density and a re-designed and shorter cell would be necessary, and stronger beam focusing could help to reach a better efficiency. However, even for comparable intensities, the expected efficiency on Mg remains lower than on Na (and a fortiori, than on Cs) as shown by the γ/λ^3 efficiency factor. This illustrates the specific difficulty of nonlinear processes in the UV originating from the λ^{-3} dependence of stimulated emission processes.

Although the efficiency remains weak, this first demonstration is very encouraging because the main limit is due to the low power of the source, and because the wavelength is already in the absorption band of most nonlinear crystals. In the pulsed regime (frequency-doubled pulsed dye lasers or new solid state sources [15]), remarkable reflectivities should be expected. An exciting development, which seems attainable in the near future, would be to perform OPC in the VUV region [16]. Rare gases or H whose resonance lines are reachable with tunable coherent sources (Xe: 147 nm; Kr: 124 nm; H: 121,5nm) would be good candidates for such demonstrations.

REFERENCES

[1] For reviews, see e. g. R. A. FISHER ed. , Optical Phase Conjugation (Academic, New York, 1983); Special issues of IEEE J. Quant. Electr. QE-22 (8), (1986) and QE-25 (3) (1989).
[2] R. W. HELLWARTH, J. Opt. Soc. Am. 67, 1 (1977).
[3] J. FEINBERG, Opt. Lett. 7, 486 (1982).
[4] J. W. R. TABOSA, S. LE BOITEUX, P. SIMONEAU, D. BLOCH and M. DUCLOY, Europhys. Lett. 4, 433 (1987); M. DUCLOY, Appl. Phys. Lett. 46, 1020 (1985).
[5] M. DUCLOY and D. BLOCH, J. Phys. (Paris) 42, 711 (1981); M. DUCLOY and
D. BLOCH, Phys. Rev A 30, 3107 (1984); C. L. CESAR, J. W. R. TABOSA, P. C. de OLIVEIRA, M. DUCLOY and J. R. R. LEITE, Opt. Lett. 13, 1108 (1988).
[6] M. DUCLOY and D. BLOCH, Opt. Comm. 47, 351 (1983).
[7] S. LE BOITEUX, P. SIMONEAU, D. BLOCH, F. A. M. de OLIVEIRA and M. DUCLOY, I.E.E.E. J. Quant Electr.QE-22, 1229 (1986).
[8] M.PINARD, D. GRANCLEMENT and G. GRYNBERG, Europhys. Lett. 2, 755 (1987); D. GRANCLEMENT, G. GRYNBERG and M. PINARD, Phys. Rev. Lett. 59, 44 (1987).
[9] D. M. BLOOM, P. F. LIAO and N. P. ECONOMOU, Opt. Lett. 2, 58 (1978).
[10] J. R. R. LEITE, P.SIMONEAU, D. BLOCH, S. LE BOITEUX, and M. DUCLOY, Europhys. Lett. 2, 747 (1986).
[11] C. J. GAETA, J. F. LAM and R. C. LIND, Opt. Lett. 14, 245 (1989).
[12] P. H. BECKWITH and W. R. CHRISTIAN, Opt. Lett. 14, 642 (1989).
[13] M. ORIA, D. BLOCH, M. FICHET and M. DUCLOY, Opt. Lett. (Oct. 1989).
[14] A. KLEIN, S. LE BOITEUX and M.DUCLOY, Opt. Lett. 14, 60 (1989).
[15] A. I. KATZ, J. FELD and V. A. APKARIAN, Opt. Lett. 14, 441 (1989).
[16] see the contribution of B.P. STOICHEFF in these proceedings.

DOPPLER-FREE SPECTROSCOPY BY LINEAR SELECTIVE REFLECTION AT GLASS/Cs-VAPOR INTERFACE

M. Oria[*], *D. Bloch and M. Ducloy*
Laboratoire de Physique des Lasers, UA 282 du CNRS
Université Paris-Nord
Ave J.-B. Clément - 93430 VILLETANEUSE - FRANCE

Doppler-free spectroscopy in vapors is restricted to nonlinear techniques, requiring sufficiently high incident intensities. We discuss here the possibility of effective Doppler-free <u>linear</u> spectroscopy through frequency modulated (FM) linear selective reflection (SR) at a dielectric vapor interface, and its applications to <u>atom-surface interaction</u>.

Let us recall that in SR spectroscopy [1], one monitors the spectral variations of the Fresnel reflection coefficient of a resonant beam incident onto the interface, with angle θ with the normal. Due to an asymmetry between the optical response of atoms arriving onto the surface - in a permanent regime - and of atoms departing from the surface - experiencing a transient regime - a sub-Doppler structure [2, 3], singling out the contribution of atoms with null normal velocity (v_z) , is superimposed to the expected Doppler-broadened linear dispersion curve. However, this structure, maximum at normal incidence $\theta=0$, remains sensitive to the Doppler broadening, with a width $\sim\sqrt{\gamma\,\Gamma_D}$ (γ : homogeneous width, Γ_D : Doppler width) and a shape $\sim \ln\left[\gamma^2/(\gamma^2 + \Delta^2)\right]$ (Δ : frequency detuning).

With an extra FM applied to the incident beam, the amplitude modulation (AM) induced on the reflected beam is solely sensitive to the derivative of the overall SR lineshape. At normal incidence, the diverging subDoppler structure is now turned into a regular Doppler-free dispersive Lorentzian $\sim\!\Delta\,/(\gamma^2 + \Delta^2)$, in the infinite Doppler width approximation [4]. An example of the achieved resolution is provided by fig. 1, which compares for the Cs D_2 resonance line (λ = 852 nm, γ = 2.6 MHz, $\Gamma_D \geqslant 200$ MHz) simultaneous recordings of (a) FM saturated absorption (SA) spectrum, and (b) FM linear SR spectrum at a glass/Cs vapor interface near normal incidence. As expected, one notices that the crossover resonances between hyperfine components are observable only in the nonlinear SA technique as they correspond to selection of a $v_z \neq 0$ velocity group. Let us note also that in order to reach an effective high resolution, the c.w. diode laser is passively stabilized by optical feedback of a Fabry-Perot [5] and this is a major improvement relatively to a rather similar previous experiment [4].

It should be also outlined that in Fig. 1 a very good sensitivity (atomic density $\sim 10^{13}$ at/cm³) is achieved for such a type of technique in which only the <u>atoms located within one wavelength</u> from the surface contribute to the signal. Although <u>linear</u> Doppler-free techniques are interesting by themselves (notably for high resolution spectroscopy of hardly saturable weak transitions) high resolution SR is essentially an elegant tool for studying various effects related with atom-wall interaction :

[*] Work supported by CNPq (Brazil)

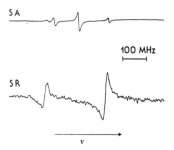

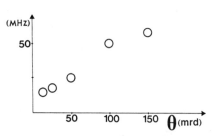

Figure 1: SA and SR spectra of Cs
D_2 line F = 4 → {F'=4,5} are simultaneously recorded with the same FM
tunable diode laser. Atomic densities are respectively ~ 10^{11}
at/cm³ (SA) and ~ 10^{13} at/cm³ (SR).
Incident intensities are not
saturating.

Figure 2: Frequency linewidth
(FWHM) of the Doppler-free SR
signal as a function of the in-
dence angle θ.

(i) short range interaction is essentially related with
atom-wall collision :its signature is the sub-Doppler structure typical of
SR spectroscopy. This structure has been recently predicted to broaden and
vanish with increasing incident angle θ [3]. For instance, in the Doppler
limit, and for θ ≪ 1, the AM reflection signal can be shown to follow the
imaginary part of a Voigt-type lineshape:

$$\int_{-\infty}^{+\infty} dv_x \; \frac{W(v_x)}{\gamma - i(\Delta - kv_x \, \theta)}$$

($W(v_x)$ velocity distribution for the component v_x parallel to the interface; k,wavevector). Thanks to our high resolution, the predicted line-
broadening is easily observable experimentally (Fig. 2). This angular dependence of the SR lineshape gives direct access to the velocity distribution -along the surface- of atoms arriving onto or leaving the surface.
(ii) long-range atom-wall interaction (i.e. typical range ~ λ) essentially
affects the frequency and width of the transition of a freely-propagating
atom.
    Long-range forces may be adequately described by the interaction bet-
ween the atomic dipole and its electric image induced by the partially
reflecting surface, which results in a wall-induced change of the atomic
wavefunction. First, the presence of the induced dipole image alters the
spontaneous radiation rate by interference between the emission diagrams.
The radiation rate is reduced in the case of a dipole polarization parallel to the surface, at short distances ("subradiant" state formed of two
opposite dipoles). On the other hand, the Van der Waals forces between the
oscillating e.m. dipole and its image shifts its resonance frequency.
Therefore both width and frequency of the SR resonance should contain a
spectroscopic signature of the atom-surface long-range interaction (not
yet taken into account in the present theory of SR spectroscopy [3]), and
yield information strongly dependent on the specific nature of the surface
(in contra-distinction to the effect of short range interactions, which
mainly consists of an instantaneous surface-independent erasure of the optical excitation).

In preliminary experiments on Cs vapor, SR signals seem to appear slightly red-shifted (by $\sim 1 - 3$ MHz) relatively to the equivalent SA signal. This should indicate that the atoms close to the surface are sensitive to an attractive Van der Waals potential. However, a residual asymmetry in the SR lineshape could hinder the effective signal linecenter. Enhanced effects can be expected from a metallic coating deposited at the interface in order to increase the surface reflection coefficient. Such an experiment is under progress.This preliminary reports shows that resonant selective reflection at a solid-vapor interface allows one to perform Doppler-free spectroscopy of atoms located within one wavelength from the wall,i.e. in a kind of partly "open" micro-cavity.

REFERENCES

[1] J.P. WOERDMAN and M.F.H. SCHUURMANS, Opt. Comm. 14, 248 (1975) ;
    A.L. BURGMANS and J.P. WOERDMAN, J. Phys. (Paris) 37, 677 (1976).
[2] M.F.H. SCHUURMANS, J. Phys. (Paris) 37, 469 (1976).
[3] G. NIENHUIS, F. SCHULLER and M. DUCLOY, Phys. Rev. A38, 5197 (1988).
[4] A.M. AKUL'SHIN, V.L. VELICHANSKII, A.S. ZIBROV, V.V. NIKITIN, V.V.
    SAUTENKOV, E.K. YURKIN and N.V. SENKOV, JETP Lett. 36, 304 (1982).
[5] B. DAHMANI, L. HOLLBERG and R. DRULLINGER, Opt. Lett. 12, 876 (1987)

Time-domain Far-Infrared Spectroscopy of Water Vapor and
Direct Measurement of Collisional Relaxation Times.

Martin van Exter, Ch. Fattinger and D. Grischkowsky
IBM Watson Research Center, P.O. Box 218, Yorktown Heights, NY 10598

We describe the application of a new sub-picosecond teraHz-beam system (1-2) to time-domain spectroscopy. The spectral content of our source extends from low frequencies up to the teraHz frequency range. By analyzing the propagation of the electromagnetic pulses through water vapor, we have made the most accurate measurements to date of the absorption cross-sections of the water molecule for the 9 strongest lines in the frequency range from 0.2 to 1.45 THz.

The teraHz radiation source (2) is illustrated in Fig.1a and consists of a coplanar transmission line, composed of two parallel 10 μm wide aluminum lines separated from each other by 30 μm, terminating an imbedded ultrafast dipole antenna. The resulting structure was fabricated on an ion-implanted, silicon-on-sapphire (SOS) wafer. The antenna was driven by photoconductively shorting the 5 μm antenna gap with 70 fsec pulses coming at a 100 MHz rate from a colliding-pulse, mode-locked dye laser. The resulting transient current in the antenna generates a short burst of radiation. This radiation is collimated and directed onto the receiver with the teraHz optics illustrated in Fig.1b. The optics consist of two matched crystalline magnesium oxide (MgO) spherical lenses contacted to the sapphire side of the SOS chips located near the foci of two identical paraboloidal mirrors. The resulting teraHz beam propagates the 88 cm distance from transmitter to receiver through an airtight enclosure in which the water vapor content could be controlled.

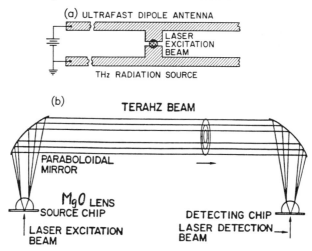

Fig. 1a,b Ultrafast dipolar antenna and teraHz optics.

The THz radiation detector uses the same ultrafast antenna and terminating transmission line as the transmitter. During operation the antenna is driven by the electric field of the incoming teraHz radiation pulse. The induced time-dependent voltage across the antenna gap is measured by shorting the gap with the 70 fsec optical pulses in the detection beam and monitoring the collected charge (current) vs the time delay between the excitation and detection laser pulses.

Figure 2 displays the detected teraHz radiation pulses after propagating through an atmosphere containing 1.5 Torr of water vapor, corresponding to 8% humidity at 20.5° C. This high signal-to-noise measurement with millivolts of signal was made in a single 10 minute scan of the 200 psec relative time delay between the excitation and detection pulses. The fast oscillations behind the main pulse are absent in a pure nitrogen atmosphere and are caused by the combined action of the dispersion and absorption of the water vapor lines. As the described pump-probe experiment directly measures the electric field, a Fourier analysis reveals the spectral content of the teraHz pulses. The result of such a spectral analysis is shown in Fig. 3a for pulses propagated over 88 cm in an atmosphere with and without water. The additional structure on the broad spectra is not related to noise, but results from reflections of the main pulse and is completely reproducible.

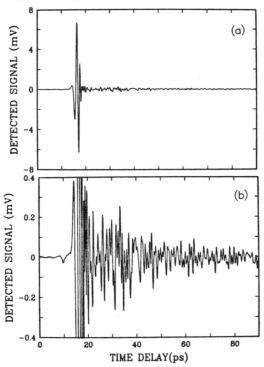

Fig. 2a,b Measured electrical pulse with 1.5 Torr of water vapor in the enclosure.

The absorption of water vapor can be derived by division of the amplitude spectra as shown in Fig. 3b. The erratic behaviour of this ratio below 0.2 THz and above 1.5 Thz is due to noise and is a result of the low spectral power in these frequency ranges. We could identify some 20 lines, between 0.2 and 2 THz and have determined the absolute absorption cross-section of nine. For the strongest line in this region (at 1163 GHz) we found a peak intensity absorption coefficent $\alpha = 38(1)$ m-1 at 100% humidity (18 Torr at 20.5° C) in one atmosphere of nitrogen. With this value the absolute absorption coefficients of the other lines can be derived from Fig. 3b. The values obtained are in good agreement with theory and other experiments (3).

The decay of the fast oscillations trailing the main pulse, as shown in Fig.2, is related to the average transverse relaxation time of the excited rotational modes of water. The relaxation times for the different excited modes vary only about 10%, making the average a good typical value (4). We have extracted this relaxation time from our measurements for various pressures of the nitrogen buffer gas. The results of these measurements are shown in Fig.4. For these measurements it is crucial to work with a low concentration of water vapor to avoid both self-broading due to water-water interactions and the formation of 0π pulses, which distorts the decay. For the shortest relaxation times it proved to

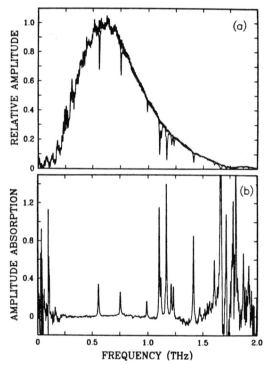

Fig. 3a,b Spectral amplitude of pulses propagated through an atmosphere with and without water and the derived absorption.

56

be easier to use the average frequency width of the absorption lines as a measure for the relaxation time. We have found that the average relaxation time scales inversely with the pressure of the buffer gas. When we choose oxygen as buffer gas the relaxation time increased, while it decreased for carbon-dioxide. The results of these measurements are shown in Fig.4. The relaxation seems to be based on the interaction between the water dipole and the quadrupole of the buffer gas (4). When helium was used as buffer gas the relaxation was indeed very slow, but still observable (about 250 ps.atm).

The described time-domain spectroscopy method has some powerful advantages in producing results which appear to be equivalent to those of traditional c.w. spectroscopy. Firstly, the detection is extremely sensitive. Although the energy per THz pulse is very low (1 attoJoule), the 100 MHz repetition rate and the coherent detection allow us to determine the electric field of the propagated pulse with a signal-to-noise ratio of about 3000. In terms of average power the sensitivity exceeds that of liquid helium cooled bolometers (5), by more than 1000 times. Secondly, because of the gated and coherent detection, the thermal background, which plagues traditional measurements in this frequency range (5), is observationally absent. This research was partially supported by the U.S. Office of Naval Research.

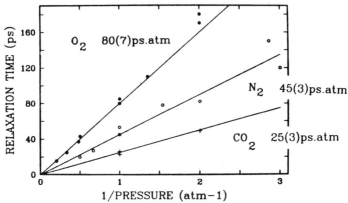

Fig. 4   The average transverse relaxation for the excited rotational water modes as a function of the pressure of the buffer gas.

1. Ch. Fattinger and D. Grischkowsky, Appl. Phys. Lett., Vol.53, 1480 (1988); Vol.54, 490 (1989).
2. M. van Exter, Ch. Fattinger and D. Grischkowsky, Appl. Phys. Lett., Vol.55, 337 (1989).
3. J.M. Flaud, C. Camy-Peyret and R.A. Toth, "Water vapour Line Parameters from Microwave to Medium Infrared", Pergamon Press (Oxford, 1981).
4. S.D. Gasster, C.H. Townes, D. Goorvitch and F.P.J. Valero, J. Opt. Soc. Am. B., Vol.5, 593 (1988).
5. C. Johnson, F.J. Low and A.W. Davidson, Optical Engr., Vol. 19, 255 (1980).

Time Dependent Optical Phase and Imaginary Collision Kernels

J.E. Thomas, P.J. Laverty, and K.D. Stokes
Physics Department, Duke University
Durham, North Carolina 27706

Abstract

New techniques are demonstrated which measure collision induced *phase* in macroscopic optical coherence. The shape of the experimental phase versus time delay curves yields the first information on *imaginary* velocity changing collision kernels for optical radiators. The measured velocity changes are much larger than the diffractive values obtained previously for the corresponding real kernels.

1. Introduction

Since the early 1970's, it has been appreciated that a complete description of velocity changing collisions for optical coherence in vapors requires a quantum mechanical treatment for both the internal and center of mass degrees of freedom [1,2,3]. In general, the collision induced velocity change distribution (i.e. the velocity changing kernel) and the total collision cross section will have both a *real* and an *imaginary* component [1,2,3,4]. This is due to the fact that optical coherence requires a superposition of ground and excited electronic states, each of which is shifted and deflected differently in a collision with a perturber. Figure 1 shows classical deflection angles as a function of impact parameter for the $^1S_0$ and $^3P_1$ states of ^{174}Yb, which is studied in this paper. A van der Waals' potential is assumed and ground and excited state C_6 constants for argon perturbers are estimated from the total cross sections measured for the excited state, $\sigma_P = 834 \ \overset{o}{A}{}^2$, [5] and for the optical radiator, $\sigma_{SP} = (\sigma_S + \sigma_P)/2 = 703 \ \overset{o}{A}{}^2$ [6]. The diffraction angle for scattering of either state from the perturber, which is roughly the atom De broglie wavelength divided by the range of the collision potential is ~ 5 mrad [6]. Since this angle is approximately the uncertainty in the scattering angle, qualitatively, the classical trajectories may be said to "separate" when the difference in the scattering angles is larger than the diffraction angle [4]. For classical scattering to be a valid approximation, the scattering angles must be large compared to the diffraction angle.

Imaginary collision kernels appear to be relatively unexplored, although they are a unique feature of the scattering of superpositions of dissimilar states. Unlike the real kernel, the imaginary kernel requires for its existence a nonzero phase shift. It is therefore particularly well suited to investigating large phase shift effects such as the "trajectory separation" described above. Prior to the work presented here, only indirect information on imaginary kernels has been obtained from the nonlinear pressure dependence of the line shift measured for infrared radiators [7].

2. Experiment

For optical phase measurement, Figure 2, a Ramsey fringe is induced in the population velocity distribution of a two level atomic vapor (^{174}Yb) by exciting the $^1S_0 \rightarrow ^3P_1$

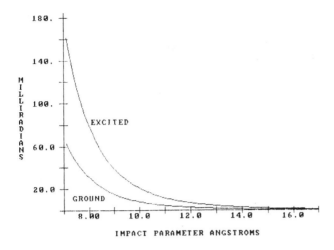

Figure 1: Classical Deflection Angle versus Impact Parameter.

transition at 556 nm using two optical pulses ($\sim 10mW/mm^2$) separated by a time delay, T [8]. Pulses are generated by acoustooptic modulation of stable c. w. dye laser radiation [6]. Absorption, α, of a weak counterpropagating c. w. probe wave ($\sim .2\mu W$) originating from the same laser, is measured as a function of laser frequency detuning, Δ. The absorption change signal peaks and is measured at a time just after $2T$ relative to the first input pulse. Due to the population fringe, the probe absorption signal contains a fringe–like component, which for short pulses takes the form

$$\alpha(\Delta) = A(\Delta)\cos[2\Delta T + \varphi(2T)] \qquad (1)$$

where $\Delta = \omega - \omega_o + \omega_{AO}/2$ and ω is the laser frequency, ω_o the atomic resonance frequency, and ω_{AO} the frequency shift of the acoustooptic modulator. Collision induced *phase* at time $2T$, $\varphi(2T)$, is determined by comparing signals from the two cells (Figure 2) at different perturber pressure as a function of Δ.

The phase $\varphi(2T)$ is determined for a number of time delays T in order to obtain a phase versus time delay curve. As discussed in reference [8], $\varphi(2T)$ is a nonlinear function of T from which the imaginary part of the velocity changing kernel can be determined by Fourier transformation.

3. Results

An estimate of the width of the imaginary collision kernel was obtained by assuming a Gaussian velocity change distribution of $1/e$ width δv. The width, δv, pressure shift δ_s, and the imaginary part of the total collision rate, $Im\,\gamma_{tot}$ are determined by requiring that the calculated phase fit the experimental phase versus time data [8].

The results of the measurements, Table 1, show that the imaginary part of the total collision rate is smaller than the line shift, in contrast to the real part, which is larger than the broadening rate [6]. This is reasonable, since from an experimental standpoint, the imaginary part of the total collision rate is the line shift excluding the contribution of

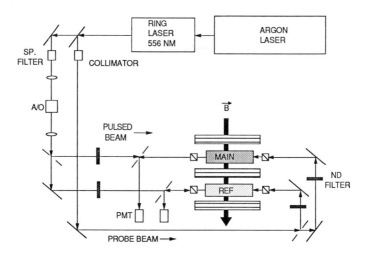

Figure 2: Experimental Scheme.

the coherence velocity changing kernel which reduces the amplitude of the fringe in the population inversion when T is large. Hence, the largest velocity changes which cause the largest collision induced phase shifts (for a monotonic potential) do not contribute. The ratio $r = Im\,\gamma_{tot}/\delta_s$ can be estimated from the total collision cross sections for each state assuming a Van der Waals' potential. In this case,

$$r = \frac{1 - \sigma_S/\sigma_P}{\left(1 - (\sigma_S/\sigma_P)^{\frac{5}{2}}\right)^{\frac{2}{5}}} \tag{2}$$

Using the cross sections for argon perturbers cited in the introduction, one obtains $r = 0.38$ in excellent agreement with the experimental result of 0.4. However, the measured value of the shifts is about a factor 2 smaller than one would predict using simple monatomic potential results. For helium perturbers, one does not expect quite as good an agreement since the potential is unlikely to be monatomic. Assuming an n=12 repulsive potential and cross sections taken from references [5,6] yields r=0.31 compared to the experimental result of 0.45.

A most important feature of this work is that the width of the imaginary kernel appears to be much larger than that of the real kernel, where a Gaussian distribution also was assumed [6]. For argon perturbers, the width falls in the nearly classical region, ~ 43 mr in the center of mass frame (Figure 1). This is about 8 times the diffractive width observed for the real kernel [6]. In addition, the width for helium perturbers falls below that for argon. By contrast, for the real kernel, the width for helium perturbers is twice that obtained for argon, as expected for diffractive collisions [6].

4. Conclusions

Imaginary collision kernels for optical radiators have been studied by exploiting the Fourier transform relationship between the shape of the measured phase versus time curve and the kernel. The results of the experiments suggest that Yb optical radiators

Perturber	δ_s(MHz/Torr)	Imγ_{tot}(MHz/Torr)	δv(cm/s)
He	+0.58(.06)	$+0.26\left(^{+.04}_{-.04}\right)$	$350\left(^{+175}_{-120}\right)$
Ar	−1.50(.15)	$-0.60\left(^{+.04}_{-.05}\right)$	$479\left(^{+122}_{-76}\right)$

Table 1: Imaginary Kernel Collision Parameters

survive much larger velocity changes than previously measured [6] or expected on the basis of qualitative trajectory separation arguments for dissimilar states (see Figure 1) [4]. Further, there is a curious lack of diffractive scattering for the imaginary kernel compared to the real kernel where diffraction was dominant. Further work will be required to understand these interesting features.

Acknowledgement

It is a pleasure to acknowledge stimulating discussions with S.Cameron, S.Zilio, M.S.Feld, and P.R.Berman. This work is supported by the U.S. National Science Foundation through Grant No. PHY-8703664.

References

[1] E. W. Smith, J. Cooper, W. R. Chappell, and T. Dillon, J. Quant. Spectrosc. Radiat. Transfer **11**, 1547(1971); **11**, 1567(1971).

[2] P. R. Berman,Phys.Rev. **A5**, 927(1972).

[3] See V. A. Alekseev, T. L. Andreeva, and I. I. Sobelman,Zh. Eksp. Teor. Fiz. **62**, 614(1972)[Sov. Phys.– **JETP 35**,325(1972)], and references therein.

[4] For a recent review see P. R. Berman, T. W. Mossberg, and S. R. Hartmann,Phys. Rev. **A25**,2550(1982).

[5] A.G.Yodh,J.Golub, and T.W.Mossberg, Phys.Rev.**A32**,844(1985).

[6] R. A. Forber, L. Spinelli, J. E. Thomas, and M. S. Feld, Phys. Rev. Lett. **50**,331(1983).

[7] See for example, S.N.Bagaev,E.V.Baklanov,and V.P. Chebotaev, ZhETF Pis. Red. **16**,344(1972)[JETP Lett. **16**,243(1972)].

[8] P.J.Laverty, K.D.Stokes, and J.E.Thomas, Phys.Rev.Lett. **62**,1611(1989); J.E.Thomas, P.J.Laverty, and K.D.Stokes, "Time Dependent Optical Phase and Imaginary Collision Kernels",Conference on Quantum Electronics and Laser Science 1989 Technical Digest Series, Vol. 12 (Optical Society of America, Washington,D.C.1989) pp.8; J.E.Thomas, P.Laverty, and K.Stokes,Bull.Amer.Phys. Soc.**33**,1011(1988);J.E.Thomas,P.J.Laverty, and K.D.Stokes, to be published in Phys.Rev.**A40**(1989).

Experiments in Cold and Ultracold Collisions

J. Weiner
Department of Chemistry and Biochemistry
University of Maryland
College Park, Maryland 20742

Two-particle and light-field interactions characterize collision physics in the cold and ultracold kinetic energy regimes. Cross sections increasing by orders of magnitude and acute sensitivity to light field intensity and polarization distinguish this new domain from conventional thermal environments. This contribution reports results from collisions within optical traps and atomic beams.

1. Introduction

As kinetic energy decreases below 1 K, many conventional ideas concerning heavy particle collisions are turned upside down. The deBroglie wavelength becomes orders of magnitude longer than the range of the chemical bonding force, the translational Maxwell-Boltzmann distribution assumes a width comparable to the natural width of excited atomic states, radiative lifetimes become long compared to collision times; and weak, long-range interactions control the probability of inelastic events. The present paper reports on the first few exploratory experiments in this largely uncharted territory. We define collisions between 1K and 1mK as "cold" and below 1 mK as "ultracold". In the cold regime alignment and orientation of the weakly interacting atomic populations control the collision probability, while in the ultracold regime optical field state dressing plays an indispensable role. At this writing ultracold collisions have been studied in optical traps and cold collisions within an atomic beam.

2. Associative Ionization in an Optical Trap

Associative ionization (AI) between two resonantly excited Na atoms in an optical trap presented itself as a likely candidate for a first experiment since the process was known to occur[1] and the fraction of excited states in a dipole trap is always near saturation (about 30%). In collaboration with the group of W.D. Phillips at N.I.S.T., therefore, we searched for the dimer ions resulting from

$$Na(3p) + Na(3p) \longrightarrow Na_2^+ + e \qquad (1)$$

using some of the atomic cooling and trapping techniques developed by them and others over the last several years[2]. The optical trap design specific to this experiment has already been described in some detail elsewhere[3], and we will only summarize it briefly here. A sodium atomic beam is first laser cooled by using a tapered magnetic field technique to compensate for the decreasing Doppler shift as the atoms decelerate. At the end of the tapered solenoid some fraction of the cooled atoms drift up to the zone where six orthogonal laser beams intersect to form a volume of "laser molasses" from which the dipole trap is loaded. Trapping the atoms from molasses serves to concentrate the density to about 10^{10} atoms cm^{-3} thereby increasing the collision frequency to an observable level. The trap consists of two counterpropagating, circularly polarized laser beams focused to a waist of about 100 μm with about 40 mW power in each beam. The laser beams are chopped in time to avoid standing-wave heating, and the foci are spatially separated by about 5 cm. Trapping and cooling cycles alternate to prevent the atoms from boiling out of the trap, and both the molasses and trap are situated between field plates which accelerate charged particles onto a multiplier detector mounted vertically above. A time-of-flight measurement determines ion mass, confirming that only Na_2^+ is produced in the trap. Variation of the ion signal intensity with density confirms a quadratic dependence characteristic of a bimolecular collision. Using this experimental setup we determined (see ref. 3) the absolute rate constant (and the corresponding cross section) for process (1) at 0.75 mK. The result is $K=1.5 \times 10^{-11}$ $cm^3 sec^{-1}$. This value is not too different from those measured at conventional temperatures, but the corresponding cross section $\sigma=8.6 \times 10^{-14}$ cm^2 is about three orders of magnitude greater. Increasing cross sections at very low kinetic energy are to be generally expected because the deBroglie wavelength of the reduced mass increases inversely as the velocity.

Another curious consequence of this regime is the long duration of a collision compared to spontaneous emission. Julienne[4] has pointed out that the characteristic time for collision between $^2S_{1/2}$ and $^2P_{3/2}$ Na atoms at 1 mK is about 30 nsec -- four radiative lifetimes of the corresponding Na_2 molecule. Thus associative ionization between two excited Na atoms can only take place if stimulated absorption can maintain population in the excited state over a sufficient fraction of the total collision time. As the two atoms approach, long-range interaction will shift their levels out of resonance with the laser field, and the excited state population will begin to deplete at the spontaneous emission rate. Relative strength of atomic coupling to the laser field compared to collisional interaction will determine the fraction of excited-state population available for associative ionization. In ref. 4 Julienne has worked out a four-state model for this coupling competition for the conditions of the molasses (weak field) and trap (strong field) appropriate to the experiment described above. The model predicts that the ratio of the AI rate constants with the trap on and off will be about 10^4. In order to test this prediction ion production was measured synchronously with the trapping and cooling cycle, and the results shown in fig (1). Julienne's qualitative prediction of marked enhancement in the strong-field AI rate appears to be correct, but experimental data are not yet precise enough to test the model quantitatively.

3 Associative Ionization in a Single Beam

Although optical traps achieve temperatures at which kT approaches the natural width of an atomic line, as containers for collision experiments they are essentially cells of very

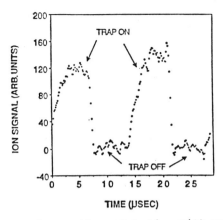

Fig 1.   Ion intensity modulation with optical trap
          switching.

cold gas in which the distribution of collision directions is
isotropic.   Atomic beams, however, provide a principal
laboratory axis and are therefore useful for studying the
effects of orientation and alignment in collision problems.
Cold collisions can even be studied in a beam without recourse
to "standard" cooling or trapping techniques.   Consider a
highly collimated atomic beam crossed by a cw, monomode laser
propagating in the opposite direction and tuned to the red of
the atomic line rest frequency.   The Doppler shift of a narrow
velocity class will just compensate for the laser detuning,
producing a narrow velocity distribution of excited atoms.
Collisions between these excited atoms can result in
associative ionization just as it does in the optical trap.
Analysis[5] shows that the average velocity is 974 cm sec^{-1},
corresponding to a "temperature" (T=2E/3k) of 44 mK. Although
this collision energy cannot be considered "ultracold", it is
cold enough so that normally negligible long-range
interactions dominate the reaction probability, and the effect
of circular and linear polarization as well as laser field
intensity on collision cross section can be studied
systematically.   We have just begun to exploit the
possibilities of cold collisions in beams, and the results
reported here must be regarded as preliminary.   Nevertheless,
they serve to illustrate possibilities of the technique.
Figure (2) is schematic of the atomic beam apparatus which can
be used in either crossed- or single-beam mode.   For the
experiments reported here one of the beams was flagged off.
The obvious first experiment was to measure the cross section
for Na(3p) + Na (3p) AI at a temperature intermediate between
the ultracold value (T=0.75 mK, σ=8.6x10^{-14} cm^2) and that
measured[6] under conventional alkali vapor conditions (T=650K,
σ=1.0x10^{16} cm^2).   Two counterpropagating laser beams are tuned
such that the laser frequency is resonant with the Doppler
blue-shifted γ_{23} transition in one direction while resonant
with the Doppler red-shifted γ_{12} transition in the other.   As
pointed out almost 15 years ago[7], this two-level, single-
frequency technique maximizes excited-state population and

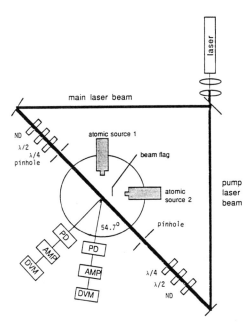

Fig 2.   Crossed-beam apparatus used to measure
           cold collision properties.

avoids intensity-dependent optical pumping effects even when
the laser light is linearly polarized.   The rate constant
using circular polarization was measured to be 5.8×10^{-13} cm^3
sec^{-1} corresponding to a cross section of 5.1×10^{-16} cm^2.   Light
polarized linearly in the collision plane yields a rate
constant greater by a factor of 2.0.   Figure (3) shows a plot
of absolute measurements for three temperature regimes:
ultracold, cold and conventional, together with a theoretical
curve calculated from a semiclassical model recently published
by Geltman[8].   The cross section decreasing with decreasing
temperature to about 75 K is due to the closing of those
entrance   channels   with   repulsive   quadrupole-quadrupole
interaction   terms.   Only   channels   with   a   quadrupole
orientation   leading   to π states   remain   open.   As   the
temperature further decreases, the cross section rises again
as the effects of a decreasing angular momentum barrier and
increasing deBroglie wavelength begin to dominate. The theory
does   not   take   account   of   state-dressing   effects   that   are
essential for understanding collision dynamics at ultracold
temperatures;   and,   therefore   lack   of   agreement   with   the
optical trap results at 0.75 mK is not surprising.
     The single-beam technique can also be used to study the
effects of optical-field state dressing, but a narrow velocity
class must be isolated in one of the Na ground state hyperfine

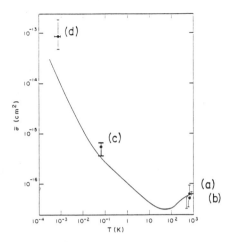

Fig 3.   Cross section for process (1) as a
          function of collision "temperature" over
          seven orders of magnitude--theory and
          experiment. Points a,b ref 6; c, ref 5;
          d, ref 3.

levels so as to avoid an increased velocity dispersion by
power broadening. We have used a setup originally suggested
earlier[9] in which the atomic beam is crossed by the laser beam
three times: the first pass optically pumps all the
population into F=1 of the ground state Na atom, the second
pass selects a narrow velocity group by excitation to
Na($3p^2P_{3/2}$) followed by decay to F=2 in the ground state. At
this point only the optically selected velocity class is
present in the F=2.   The third pass again excites this
velocity class to the upper level from which AI takes place.
In this third step, however, the intensity and polarization
of the laser field can be varied to observe the effect on the
rate constant.   The transition is power broadened but the
velocity class is not. Figure (4) shows some recent results
employing this technique. Polarization-dependent intensity
effects on the AI cross section are clearly evident, but to
date a definitive interpretation has not been carried out.
The observed behavior may arise either from competition
between optical and collisional coupling, leading to increased
"survival"   of   the   excited-state   population   (against
spontaneous emission) during the incoming branch of the
collision, or to variation in the atomic excited-state density
matrix (due to optical pumping) as laser intensity increases.
The hope is that these new results will inspire a redoubled
theoretical effort, and that a detailed analysis of intensity
and polarization effects in the very-cold collision regime
will soon be forthcoming.

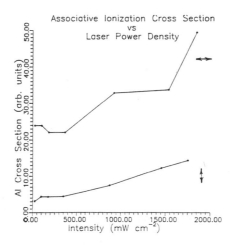

Fig 4.   AI cross section as a function of
intensity in single-beam experiments.
Note that for polarization parallel and
perpendicular to the atomic beam axis,
cross sections increase markedly with
laser intensity.

References

1.   J. Weiner, F. Masnou-Seeuws, and A. Giusti-Suzor, Adv.
      At. Mol. Opt. Phys. **26** (in press)

2.   For an early review of "The Mechanical Effects of Light"
      see JOSA B **2** 1705 (1985) and for a more recent account,
      JOSA B **6** No. 11 (to be published)

3.   P.L. Gould, P.D. Lett, P.S. Julienne, W.D. Phillips, H.R.
      Thorsheim, and J. Weiner, Phys. Rev. Lett. **60**, 788 (1988)

4.   P.S. Julienne, Phys. Rev. Lett. **61**, 698 (1988)

5.   H. A. Thorsheim, Y. Wang, and J. Weiner (to be published)

6.   J. Huennekens and A. Gallagher, Phys. Rev. A **28**, 1276
      (1983) and R. Bonanno, J. Boulmer, and J. Weiner,
      Comments At Mol Phys **16**, 109 (1985)

7.   G. M. Carter, D. E. Pritchard, and T. W. Ducas, Appl.
      Phys. Lett. **27**, 498 (1975)

8.   S. Geltman, J. Phys B: At. Mol. Opt. Phys. **21**, L375
      (1988)

9.   F. Schuda and C. R. Stroud, Optics Comm. **9**, 14 (1973) and
      D.G. Steel and R. A. McFarlane, Optics Lett. **8**, 33 (1983)

Dressed–Atom Approach to Collision–Induced Resonances

P. R. Berman
  Physics Department, New York University, 4 Washington Place,
  New York, NY 10003, USA
G. Grynberg
  Laboratoire de Spectroscopie Hertzienne de l'Ecole Normale
  Supérieure, Université Pierre et Marie Curie,
  75252 Paris CEDEX 05, France

Motivated in large part by the work of Bloembergen and coworkers, there has been considerable interest over the past ten years in the study of pressure–induced extra resonances, a class of resonant structures that appear in spectroscopic line shapes only in the presence of collisions (1). What is particularly intriguing about the pressure–induced resonances is that collisions, which are often thought to broaden and destroy coherent structures, are essential for producing these resonances. Moreover, the pressure–induced resonances can be very narrow, in some cases having widths equal to the inverse lifetime of the ground states of the atoms which are interacting with the laser fields. It is relatively simple to obtain theoretical expressions for the pressure–induced resonances. What has been more elusive, however, is a physical explanation of their origin.

We present an interpretation of pressure–induced resonances based on a dressed–atom picture of the atom–field interaction (2)–(4). Both semiclassical (classical fields – quantum–mechanical atoms) and fully–quantized (quantized fields – quantum–mechanical atoms) dressed–state theories are employed. Using this dressed–atom approach, we are able to show that the vanishing of the pressure–induced resonances in the absence of collisions is a direct consequence of the conservation of energy. The positions and widths of the resonances can be attributed either to a modulated dressed–state population or to a level–crossing of the dressed states.

In order to illustrate the physical concepts, we consider a pressure–induced resonance that is produced in fluorescence beats (4),(5). A two–level atom is subjected to two copropagating laser fields (Fig. 1). The first field has frequency Ω and is detuned by $\Delta = \omega - \Omega$ from the atomic resonance, while the second field has frequency $\Omega + \delta$. It is assumed that $|\delta| \leq \gamma_2 \ll |\Delta|$ (γ_2 is the decay rate of the excited state) and that $|\Delta|$ is much greater than the Doppler width associated with the 1–2 transition. If the incident fields are relatively coherent, it is found that part of the fluorescence from level 2 is modulated at frequency δ. To lowest order in the applied fields, this component of the fluorescence, denoted by $I(\delta)$, is given by

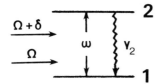

Fig. 1. Copropagating laser fields having frequencies Ω and $\Omega + \delta$ are incident on on a two–level atom. Modulated fluorescence from level 2 is monitored as a function of δ.

$$I(\delta) = \frac{\chi_1\chi_2^*}{\Delta^2} e^{i\delta t}\left[1 + \frac{2\Gamma}{\gamma_2 + i\delta}\right] + c.c.,$$

where χ_i is a Rabi frequency associated with field i and Γ is a collision rate associated with the atomic coherence ρ_{12}. The modulated fluorescence consists of a background term (present even in the absence of collisions) and a pressure–induced term which exhibits a resonant structure centered at $\delta=0$, having width γ_2.

To explain the physical origin of this resonance, we use a fully–quantized dressed–atom approach. To lowest order in the applied fields, the appropriate dressed states are given by

$$|A;n_1,n_2\rangle = |1;n_1,n_2\rangle + \theta(n_1)|2;n_1-1,n_2\rangle$$
$$+ \theta(n_2)|2;n_1,n_2-1\rangle$$

$$|B;n_1,n_2\rangle = |2;n_1,n_2\rangle - \theta^*(n_1+1)|1;n_1+1,n_2\rangle$$
$$- \theta^*(n_2+1)|1;n_1,n_2+1\rangle,$$

where the n's label the photon number states of the fields,

$$\theta(n) = ig\sqrt{n}/\Delta,$$

and g is a coupling constant. Some of the dressed energy levels are shown in Fig. 2

The structure of the fluorescence beats can now be understood as follows: In the absence of collisions, all population remains in the "A" dressed states, since there is no physical mechanism to provide the energy difference $\hbar\Delta$ to transfer population from states A to B. Since states $|A;n_1+1,n_2\rangle$ and $|A;n_1,n_2+1\rangle$ each contain an admixture of state $|2;n_1,n_2\rangle$, they can undergo a radiative decay to state $|A;n_1,n_2\rangle$. The component of the modulated fluorescence associated with this collision–free contribution is

$$I_{cf}(\delta) = \sum_{n_1 n_2} \theta(n_1+1)\theta^*(n_2+1)\rho^F_{n_1+1,n_2;n_1,n_2+1}(t) + c.c.,$$

where $\rho^F(t)$ is the free–field density matrix. There is no resonant structure centered at $\delta=0$. As can be seen in Fig. 2, the collision–free contribution to the fluorescence occurs at the laser frequencies.

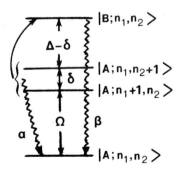

Fig. 2. Dressed states of the atom–field system. Collisions couple states A and B within a given manifold (curved arrow). A coherence between states $|A;n_1,n_2+1\rangle$ and $|A;n_1+1,n_2\rangle$ produced by the incident fields leads to background (α) and pressure–induced (β) modulated fluorescence.

With collisions present, states $|A\rangle$ and $|B\rangle$ within a given dressed-state manifold are coupled since the collisional interaction can provide the energy mismatch $\hbar\Delta$ between these states. Radiative decay from level $|B;n_1,n_2\rangle$ to $|A;n_1,n_2\rangle$ results in fluorescence centered at the *atomic* frequency (see Fig. 2 – recall that $\Delta+\Omega = \omega$). The modulated component of this fluorescence is given by

$$I_c(\delta) = \theta(n_1+1)\,\theta^*(n_2+1)\,[2\Gamma/(\gamma_2+i\delta)]$$
$$\times\ \rho^F_{n_1+1,n_2;n_1,n_2+1}(t)\ +\ c.c.$$

The amplitude of this collision-induced component vanishes in the absence of collisions. Owing to the modulation of ρ_{BB}, the fluorescence exhibits a resonant structure centered at $\delta=0$, having width γ_2. The modulated fluorescence can be traced to a combined collisional-radiative process that couples the initial coherence $\rho_{A,n_1+1,n_2;A,n_1,n_2+1}$ to the dressed-state population $\rho_{B,n_1,n_2;B,n_1,n_2}$.

It is seen that the fluorescence vanishes unless $\rho_{n+1,n} \neq 0$ for each of the incident fields. In other words, the incident fields must be phase coherent to produce the fluorescence beats. The same conclusion would have been reached had we used a semiclassical dressed atom approach (4). In that case, the collision-induced modulated signal is proportional to $\chi_1\chi_2^*$ which vanishes, on average, for uncorrelated fields. In a manner analogous to that presented for fluorescence beats, one can use a dressed-atom approach to explain the pressure-induced resonances that can be produced when (a) a three-level atom is excited by four incident fields (3); (b) a four-level atom is excited by four incident fields (6); an atom is ionized by four fields (6); and (d) four-wave mixing signals are generated in active media of two or three-level atoms. In contrast to cases (a)–(c), in case (d) the signal depends only on the average number of photons in each incident field; the incident fields need only be spatially coherent to generate the appropriate phase-matched emission.

Acknowledgments. This research is supported in part by NSF Grants INT–8413300 and INT–8815036. The research of PRB is supported by the U. S. Office of Naval Research and NSF Grant PHY–8814423. The Laboratoire de Spectroscopie Hertzienne de l'Ecole Normale Supérieure is "associé au Centre National de la Recherche Scientifique."

References

1. For recent reviews of pressure-induced resonances, see L. Rothberg, in *Progress in Optics*, edited by E. Wolf (Elsevier Scientific, Amsterdam,1987), pp. 39–101; G. Grynberg, in *Spectral Line Shapes*, edited by R. Exton (A. Deepak, Hampton, VA, 1987), Vol. 4, pp. 503–521.
2. G. Grynberg, J. Phys. B14, 2089 (1981).
3. P. R. Berman and G. Grynberg, Phys. Rev. A39, 570 (1989).
4. G. Grynberg and P. R. Berman, Phys. Rev. A39, 4016 (1989).
5. G. S. Agarwal, Opt. Comm. 57, 129 (1986).
6. G. Grynberg and P. R. Berman, submitted to Phys. Rev. A.

Speed-Dependent Inhomogeneities in Collision-Broadened H_2 Vibrational Transitions*

R. L. Farrow, L. A. Rahn, and G. O. Sitz[†]
   Combustion Research Facility, Sandia National Laboratories, Livermore, CA 94550, USA
G. J. Rosasco
   Center for Chemical Technology, National Institute of Standards and Technology, Gaithersburg, MD 20899, USA

Introduction

The collisional broadening of an isolated rotation-vibration transition of a gas is considered to be in the impact regime[1] when the the collisional width is small compared to the inverse of the collision duration and large compared to the Doppler width, and the collisional width and shift have no significant dependence on relative molecular speeds. Such conditions have been shown theoretically[2] to produce Lorentz line profiles.

We report observations of inhomogeneous broadening in the vibrational line profiles of H_2 perturbed by a heavy collision partner, at densities normally considered to be in the impact regime. At densities up to 27 amagat (where the spectra are clearly dominated by collision broadening), non-Lorentzian, asymmetric features are observed in Raman Q-branch transitions of H_2 diluted in a heavy perturber gas. We compare these measurements with an inhomogeneous line-profile model based on strong speed dependences in the collisional shift cross sections. Quantitative agreement is obtained only when spectral line narrowing resulting from speed-changing collisions is included.

Experimental Observations

We used a high-resolution, quasi-cw inverse Raman (IRS) system[3] to measure Q-branch spectra and transition frequencies of pure H_2 and H_2 diluted in Ar. Figure 1 shows experimental spectra (data points) of the $Q(1)$ transition of 2 mole percent H_2 in Ar, at 295 K and 1000 K, and at a total density of ~14 amagat. In (a) and (c) the data are compared to best-fit Lorentzian profiles (curves), illustrating pronounced asymmetries and Gaussian-like spectral cores which increase with temperature. The effects are only present for perturber molecules having masses much greater than that of H_2 (Ar, Kr, and N_2), and vanish for perturbers with comparable masses (He and H_2). Consequently, the anomalies decrease rapidly as the percentage of H_2 increases in a given mixture. This rapid decrease in linewidth is actually faster than linear (with respect to H_2 mole fraction), leading to an apparent violation of the additivity rule for perturber broadening. The spectral shapes of the anomalous profiles were not sensitive to total density (in the range from 3.6 to 14 amagat at room temperature) although the widths increased linearly.

Speed-Dependent Collisional-Shift Model

In considering Doppler-broadened profiles, Berman[4] and Ward et al.[5] and others have shown that systems with small radiator/perturber mass ratios can exhibit spectral asymmetries and inhomogeneities as a result of speed-dependent collisional shift and

---

*This research was supported by the U. S. Department of Energy, Office of Basic Energy Sciences, Chemical Sciences Division and, in part, by the U. S. Army Research Office.
†Sandia Postdoctoral Fellow

broadening coefficients. Berman and Pickett[6] pointed out that such effects should persist to high densities. We observe strong temperature dependences in the line-shift density coefficients for H_2-H_2 and H_2-Ar, indicating that strong variations in the collisional line-shifts with relative speed are indeed present. Thus, radiators with different speeds are associated with different transition frequencies, giving rise to a new inhomogeneity in the thermally averaged line profile. The observed temperature dependence of the asymmetry further supports the speed-dependent mechanism.

An inhomogeneous line profile is computed using a speed-dependent shift coefficient $\delta(\upsilon_{rad}, T)$ based on our line-shift measurements vs. temperature and the expression for a speed-dependent Lorentzian profile:[4]

$$I_{inhom}(\omega) = \frac{1}{2\pi} \int \frac{f_M(\mathbf{v}_{rad}, T)}{\gamma(T)\rho - i[\omega + \delta(\upsilon_{rad}, T)\rho]} d\mathbf{v}_{rad} + c.c.$$

Here, γ is the coefficient for the total dephasing width (HWHM), ω is the detuning from the zero-density Raman frequency, ρ is the density in amagat, $f_M(\mathbf{v}_{rad}, T)$ is the Boltzmann distribution of radiator velocities, and υ_{rad} is the radiator speed. This expression alone is generally in poor agreement with experimental asymmetric spectra in that it overestimates linewidth and asymmetry. However, if *speed*-changing collisions occur more frequently than dephasing collisions, scattering amplitudes associated with different speed-groups can coherently interfere, resulting in spectral narrowing. The narrowing and symmetrization are analogous to those produced in Doppler profiles by *velocity*-changing collisions, and can significantly alter the shape of the speed-dependent profile. We included speed-changing collisions (frequency = v) by using I_{inhom} and a hard-collision narrowing model analogous to that for Dicke narrowing.[7]

Fig. 1b and 1d show a comparison of the speed-dependent, line-narrowed model (curves) with the same data seen earlier. The agreement is essentially within the experimental noise. Remarkably, at 1000 K we find that ~75% of the linewidth arises

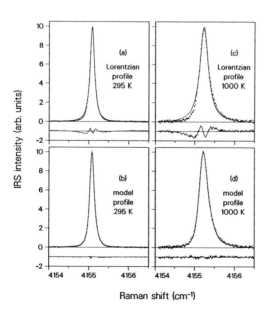

Fig 1. Experimental IRS spectra (data points) of the $Q(1)$ transition of H_2 broadened by Ar at 295 K and 1000 K. The H_2 mole fraction is 2% and the total densities are 13.9 and 13.7 amagat, respectively. The data are compared to best-fit Lorentzian profiles and to the model profile described in the text. The model includes inhomogeneous broadening arising from speed-dependent collisional shifts.

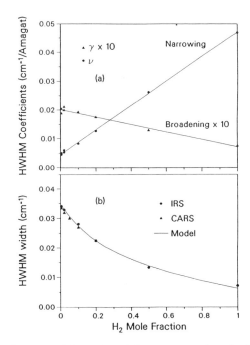

Fig 2. (a) Best-fit collision-broadening (γ) and -narrowing (ν) coefficients, indicated by symbols, obtained from IRS spectra of the $Q(1)$ transition of H_2/Ar mixtures at 295 K and 7.3 amagat total density. (These coefficients are used by the proposed line-profile model.) Also plotted are best-fit straight lines through the data. (b) Spectral linewidths determined from fitting Lorentzian profiles to the data (symbols) and to the theoretical profiles (curve) calculated using the corresponding parameters in (a). The CARS measurements are to be reported fully elsewhere.

from speed-dependent, inhomogeneous broadening. Moreover, the best-fit speed-changing rate was found to be only ~8% of the rate of *velocity*-changing collisions,[8] implying that H_2-Ar collisions tend to change the direction much more than the magnitude of the H_2 velocity.

We also investigated the ability of the model to describe observations obtained at one temperature but with varying relative concentrations of H_2 and Ar. At 295 K, we find that the data are consistent with linearly varying dephasing (γ) and speed-changing (ν) coefficients (see Fig. 2a). The values for ν are found to increase with the mole fraction of H_2, implying that H_2-H_2 collisions with are more likely to change the speed of H_2 than are H_2-Ar collisions. As shown in Fig. 2b, increasing ν causes the line profile to narrow rapidly, thus predicting the nonlinear variation in linewidth with H_2 mole fraction described previously.

References
1. A. Ben-Reuven, Adv. Chem. Phys. **33**, 235 (1975).
2. See, for example, G. Peach, Contemp. Phys. **16**, 17 (1975).
3. L. A. Rahn and R. E. Palmer, J. Opt. Soc. Am. B **3**, 1166 (1986).
4. P. R. Berman, J. Quant. Spectrosc. Radiat. Transfer **12**, 1331 (1972).
5. J. Ward, J. Cooper, and E. W. Smith, J. Quant. Radiat. Transfer **14**, 555 (1974).
6. H. M. Pickett, J. Chem. Phys. **73**, 6090 (1980).
7. S. G. Rautian and I. I. Sobel'man, Sov. Phys. Usp. **9**, 701 (1967).
8. We calculated β using $\beta = kT/mD$ as discussed in Ref. 7, where the diffusion coefficient, D, was computed per J. O. Hirschfelder, C. F. Curtiss, and R. B. Bird, *Molecular Theory of Gases and Liquids* (Wiley, New York, 1964), p. 539.

Collisionally Induced Four Wave Mixing: Breakdown of the Impact Approximation

Nissim Asida and Michael Rosenbluh
Department of Physics, Bar Ilan University, Ramat Gan, 52100, ISRAEL

The idea that incoherent perturbations, such as dephasing collisions, can lead to the formation of coherent processes, has been demonstrated by numerous experiments (1) since the pioneering work of Bloembergen and co-workers (2). The generally presented explanation for this counter-intuitive phenomenon is that the collisional dephasing removes the destructive interference between quantum mechanical inter- action pathways possessing time reversal symmetry. Such an explanation was deemed necessary for the experiments thus far performed, in all of which the laser was near resonant to an atomic transition. In these cases the collision can be well treated within the impact approximation, i. e. the duration of the collision is much less than $(\Delta)^{-1}$, where Δ is the detuning of the laser from resonance.

We have performed four wave mixing (FWM) experiments at very large detunings from atomic resonance, where the impact approximation is no longer applicable. The observation of the emission of a coherent fourth wave in these experiments can be understood using a very simple and intuitive model of the collision process. Thus the generation of collision-induced coherent phenomenon in this regime is a natural consequence of the quasi-static model of the collision (3-4).

The quasi-static model is presented schematically in Fig. 1, which shows some of the molecular-state potentials for the Na-He system as a function of distance between the Na and the He atom. At very large distances, the resonance transition frequencies of the Na-He system are just the unperturbed Na resonances, corresponding to the D lines. As the two atoms approach each other, the resonance frequencies of the pertur- bed Na atom shift; to lower frequencies for the $A^2\pi$ molecular states and higher frequencies for the $B^2\Sigma$ state. Thus FWM at a frequency which is significantly detuned from one of the

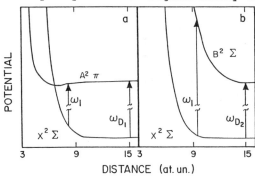

Fig. 1. Molecular-state potentials for the Na-He sytem (from Ref. 6) as a function of interatomic distance. In order to expand the scale an energy close to the unperturbed resonance energy was subtracted from the excited state potentials.

unperturbed Na D lines will be a "resonant process" for exactly those collisional atomic pairs that have the right interatomic distance. In this model, therefore, the role of the collision is to bring a certain fraction of the atoms into resonance with an arbitrary laser frequency, and it is these atoms that contribute to the coherent fourth wave emission.

We have derived an equation for the intensity of the fourth wave for the case of non-impact degenerate FWM in a two-level system with lasers of arbitrary strength and propogation direction (5). In the case of the Na-He system the two-level approximation is a very good one as long as the laser is not tuned to a frequency between the D lines. As can be seen from Fig. 1, for blue detunings of the laser only the $B^2\Sigma$ molecular state can contribute while for red detunings only the $A^2\pi$ states contribute. Our expression for the fourth-wave intensity predicts the FWM signal dependence on laser detuning, laser intensity, buffer gas pressure and Na density, and has only one adjustable parameter: the non-impact collisional lineshape. Therefore, a measurment of the FWM signal over a large range of all the experimental parameters can be used to determine the non-impact lineshape. We have performed such measurments on the Na-He system for both laser detuning to the red of the D_1 line and to the blue of the D_2 line. From these measurments we have determined the absolute non-impact collisional lineshape of both of the Na transitions.

Our experiments were performed using a pulsed dye laser, pumped by a doubled Nd:Yag laser. The three collinearly polarized incident beams were obtained from a single laser and entered the Na cell in a folded boxcar geometry. The 40 mm long Na cell was designed to enable us to vary either the Na density or He buffer gas pressure while maintaining a uniform atomic density through the optical path. The energy of each laser pulse was independently measured and the fourth wave signal was averaged for each incident laser energy within a 5% range. This feature of the experiment was of critical importance due to the nonlinear and strong dependence of the FWM signal on laser intensity.

Representative data, showing the detuning dependence of the FWM signal for red detunings from D_1 line is shown in Fig. 2a. The solid lines are from calcuation. The calculation was performed using the expression for the FWM signal derived by us. The non-impact collisional lineshape of the Na D_1 line was obtained by consistently fitting a large set of such data with many different experimental parameters, to a single form given by $L = P\, C/\Delta^x$, where P is the buffer gas pressure and C and x are determined from the fitting. We found in this case that x = 0.8. If the potential difference of the states involved in the transition was given by a power law interaction, $V \propto R^{-n}$, x would have to be greater than one (4). Thus our lineshape determination indicates that the potential difference in this case is not a simple power law, but could for example be a sum of power laws. A similiar conclusion can be reached from the shape of the $A^2\pi$ potential curve (Fig. 1a) obtained from the ab-initio calculation of Baylis (6).

A similiar proceedure was used to determine the lineshape for the $X^2\Sigma - B^2\Sigma$ transition, and data for detunings to the blue of the D_2 line are shown in Fig. 2b. In this case we found that $x = 1.1$, which is also qualitatively in agreement with the shape of the potentials obtained from the ab-initio calculations (6). (See Fig. 1b.)

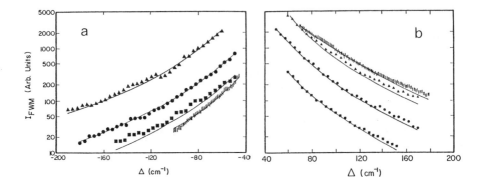

Fig. 2. FWM signal as a function of detuning (a) from the D_1 resonance. The data indicated by ▲, ▮, •, ▪, are for Na densities of 2.2, 0.4, 1.7, 1.3 $\times[10^{13}]_2$ cm^{-3}, incident laser intensities of 3.7, 2.2, 2.8, 1.2 MW/cm^2, and He pressures of 169, 347, 347, 722 torr, respectively: (b) from the D_2 resonance. The data indicated by ▲, ▮, •, ▪, are for Na densities of 1.7, 1.5, 2.2, 1.3 $\times[10^{13}]_2$ cm^{-3}, incident laser intensities of 2.9, 2.4, 1.6, 0.8 MW/cm^2, and He pressures of 169, 347, 347, 722 torr, respectively.

In conclusion, we have observed collisionally induced FWM at very large detunings from resonance, in the non-impact collisional regime. Using a theory derived for arbitrarily strong laser intensities, we have determined the absolute non-impact collisional lineshape for the Na-He system. The derivation of the potential difference from the experimentally determined lineshape is in progress.

References

1. See recent review by L. Rothberg, Progress in Optics XXIV, E. Wolf ed., page 40, Elsevier Science Publishers, 1987.
2. N. Bloembergen, H. Lotem, R. Lynch, Ind. J. of Pure and Appl. Phys., 16, 151 (1978).
3. B. Sayer, Acta Physica Polonica, A61, 531 (1982).
4. N. Allard, J. Kielkopf, Rev. of Mod. Phys., 54, 1103(1982).
5. N. Asida, M. Rosenbluh, S. Mukamel, submitted for publication, and N. Asida, M. Rosenbluh, to be published.
6. W. Baylis, J. Chem. Phys., 51, 2665 (1969), and private communication.

Perturbative Theory of Higher-Order Collision-Enhanced Wave Mixing*

Rick Trebino and Larry A. Rahn
Combustion Research Facility
Sandia National Laboratories
Livermore, California 94551

Collision-enhanced resonances represent an interesting class of non-linear-optical processes. They occur because collisional dephasing can re-phase quantum-mechanical amplitudes that ordinarily cancel out exactly, thereby allowing otherwise unobservable wave-mixing resonances to be seen. This is an especially interesting phenomenon because these resonances are coherent effects that are induced by an incoherent process (collisional dephasing). First predicted[1] in the late 1970s and eventually observed[2] in 1981, these novel effects have now been seen in a wide variety of four-wave-mixing experiments, ranging from self-focusing to coherent anti-Stokes Raman spectroscopy. Recently, we have extended these observations to higher order, where we have shown both experimentally[3] and theoretically[4] that higher-order, collision-enhanced effects exist in nonlinear optics, ap-pearing as subharmonics of two-photon resonances. Indeed, we have found that collision-enhanced processes are ideal systems for studying higher-order, nonlinear-optical effects because very high orders can be made to con-tribute with little or no saturation broadening. Experiments on sodium in a flame using six- and eight-wave-mixing geometries have revealed still higher-order effects (at least as high-order as $\chi^{(13)}$). In addition, we have de-

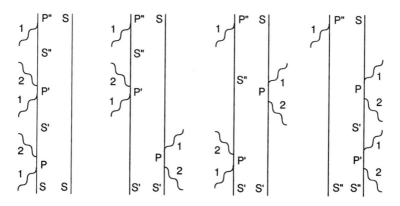

Figure 1. The four most important Feynman-like diagrams describing collision-enhanced, ground-state resonances in six-wave mixing. The states, S, S', and S", are electronic-ground-state levels, while P, P', and P" are electronically excited states. The numbers, 1 and 2, represent the laser frequencies ω_1 and ω_2. The number of such diagrams for colli-sion-enhanced N-wave mixing is $2^{(N/2-1)}$. When state, frequency, and polarization combi-nations and permutations are included, the number of terms that must be summed is much larger. (Figure reprinted from Ref. 6)

veloped a computer model of these effects that computes nonlinear-optical spectra as high-order as $\chi^{(11)}$. It uses diagrammatic perturbation theory[5] and includes all sixteen hyperfine and Zeeman levels of the sodium $3S_{1/2}$ and $3P_{1/2}$ states. The use of two-photon matrix elements eliminates sums over excited states, but sums over permutations of the eight ground-state levels remain. While perturbation theory clearly cannot explain all features of collision-enhanced phenomena (e.g., power-broadening and saturation effects), it is better able to handle the large number of states in this problem than other high-intensity methods, such as dressed-state or continued-fraction techniques.[4]

We have determined the most important perturbation-theory diagrams for collision-enhanced, ground-state resonances in all orders and have derived analytical expressions[6] for $\chi^{(N)}$ (see Figure 1). Our computer program sums these expressions over all combinations and permutations of the eight Zeeman

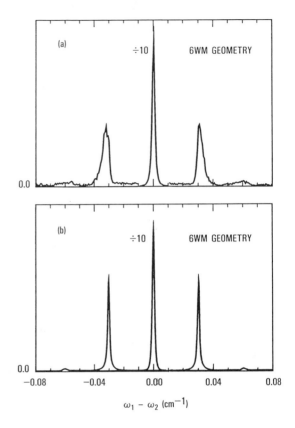

Figure 2. Experimental (a) and theoretical (b) spectra for experiments on sodium in a flame using a six-wave-mixing $3\omega_1-2\omega_2$ geometry. The theoretical spectrum has been calculated by adding all six- and eight-wave-mixing processes that satisfy the phase-matching condition for the $3\omega_1-2\omega_2$ beam geometry. Resonances at 0 cm^{-1} and 0.06 cm^{-1} correspond to two-photon Zeeman and hyperfine resonances, respectively, and are also seen in low-intensity four-wave-mixing experiments. Higher-order features in this spectrum are the subharmonics—resonances at ±1/2 of the hyperfine splitting—which are due to four-photon resonances.

and hyperfine electronic-ground-state levels and all laser frequencies and polarizations. As a result, the program generates theoretical spectra within the framework of perturbation theory with few additional assumptions.

Figures 2 and 3 show experimental and theoretical spectra for six- and eight-wave-mixing-geometry experiments. Agreement between these spectra is good. Slight discrepancies occur because we have not convolved the theoretical spectra with the laser line shape (a much more difficult problem in higher order). Instead, we have artificially increased the dephasing rates to approximate the line widths. In addition, the Zeeman line in the experimental eight-wave spectrum is a factor of five stronger than in the theoretical spectrum, due probably to ten- and twelve-wave contributions, where we have found that the Zeeman line is stronger.

We have also found that higher-order theoretical spectra display some unexpected effects. For example, contributions due to processes at least four orders higher than that of the geometry [e.g., $2\omega_1-\omega_2+(\omega_1-\omega_1)+(\omega_2-\omega_2)$ in a $2\omega_1-\omega_2$ geometry] show deep holes in spectral lines. This is due to cancellations between singly and triply resonant terms, the latter yielding narrower lines and having opposite sign.

Computed spectra also predict new high-order selection rules. Specifically, the twelve-wave-mixing contribution in a twelve-wave-mixing $(6\omega_1-5\omega_2)$ geometry exhibits an eight-photon selection rule against the subharmonic at $\pm 1/4$ of the hyperfine splitting of sodium. This selection rule has not previously been derived analytically.

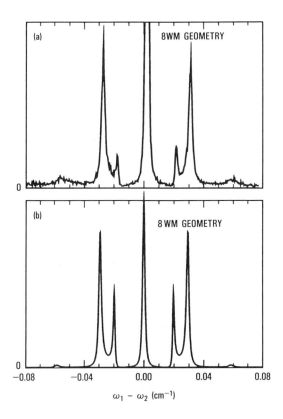

Figure 3. Experimental (a) and theoretical (b) spectra for experiments on sodium in a flame using an eight-wave-mixing $4\omega_1-3\omega_2$ geometry. This spectrum contains only the eight-wave-mixing processes that satisfy the phase-matching condition for the $4\omega_1-3\omega_2$ beam geometry. Resonances at 0 cm^{-1} and 0.06 cm^{-1} correspond to two-photon Zeeman and hyperfine resonances, respectively, and are seen in low-intensity four-wave-mixing experiments. Higher-order features are the subharmonics—resonances at $\pm 1/2$ and $\pm 1/3$ of the hyperfine splitting—which are due to four- and six-photon resonances, respectively. (Figure reprinted from Ref. 6)

References
* This work was supported by the U.S. Department of Energy, Office of Basic Energy Sciences, Division of Chemical Sciences.
1. N. Bloembergen, H. Lotem, and R.T. Lynch, Indian. J. Pure Appl. Phys. 16, 151 (1978).
2. Y. Prior, A.R. Bogdan, M. Dagenais, and N. Bloembergen, Phys. Rev. Lett. 46, 111 (1981).
3. R. Trebino and L.A. Rahn, Opt. Lett. 12, 912 (1987).
4. R. Trebino, Phys. Rev. A 38, 2921 (1988); see also G.S. Agarwal, Opt. Lett. 13, 482 (1988).
5. T.K. Yee and T.K. Gustafson, Phys. Rev. A 18, 1597 (1978).
6. R. Trebino and L.A. Rahn, submitted to Opt. Lett.

VELOCITY-CHANGING COLLISIONS IN THE Sm I 570.7 nm ($J_{lower}=1$ to $J_{upper}=0$) SATURATED ABSORPTION SPECTRUM

A.P. Willis
  Monash University, Clayton, Victoria 3168, Australia.
P. Hannaford and R.J. McLean
  CSIRO Division of Materials Science and Technology,
  Clayton, Victoria 3168, Australia.
H.-A. Bachor and R.J. Sandeman
  Department of Physics and Theoretical Physics,
  Australian National University, Canberra 2601, Australia

When laser saturation spectra are obtained from vapour cells containing buffer gas, the lineshapes are modified by the occurrence of velocity-changing collisions (VCC) (e.g.(1,2)). The VCC can be separated phenomenologically into two types: those that produce a large velocity change (strong VCC), giving a broad background pedestal to laser saturation lineshapes, and those that change the velocity by a small amount (weak VCC), leading to broadening in the region of the Doppler-free peak. Recent investigations have shown that the effect of VCC can be significantly reduced in saturated absorption spectra by using a high chopping frequency of the pump beam (3). Here we report a systematic study into the effect of VCC on saturated absorption lineshapes using a range of pump chopping frequencies (1 to 100 kHz) and rare-gas pressures (0.1 to 0.9 Torr). The 570.7 nm ($J_{lower}=1$ to $J_{upper}=0$) transition in samarium was chosen for the study. This system is theoretically tractable, its isotopic structure allows a single isotope to be easily studied, and it permits the investigation of metal atoms in a rare gas environment, a situation important in experiments involving cathodic sputtering.

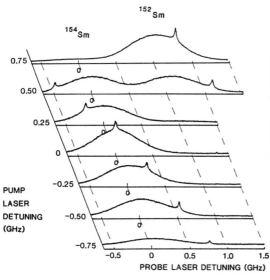

Figure 1: Two-laser saturated absorption spectra in the region of the ^{154}Sm and ^{152}Sm components of the Sm I 570.7 nm transition, for various pump laser detunings.

The standard saturated absorption arrangement was modified so that the pump and probe laser beams originated from different single-frequency cw ring dye lasers. The pump laser frequency was fixed at a particular detuning from the atomic resonance while the frequency of the probe laser was scanned through 2 to 3 GHz, making it possible to displace the Doppler-free component from the VCC component (see for example (4,5)). The absorption cell consisted of a demountable glass envelope containing a heated filament used to produce the samarium vapour.

Figure 1 shows a series of two-laser saturated absorption spectra taken with different detunings of the pump laser from resonance with the ^{154}Sm component of the 570.7 nm transition. When the pump laser is detuned to the low frequency side (bottom two traces of Fig. 1), the contribution from other isotopes is negligible. Figure 2 shows both the in-phase and quadrature components of two-laser saturated absorption signals obtained with various chopping frequencies and argon pressures and with a pump laser detuning of -0.5 GHz. Figure 3 shows an expanded spectrum recorded at a pump detuning of -0.42 GHz, a pump chopping frequency of 100 kHz, and an argon pressure of 0.9 Torr. The profiles in Figs. 2 and 3 are identified by the parameter p/f (where p is the pressure of rare-gas in Torr and f is the pump chopping frequency in kHz), which is proportional to the number of VCC occurring during a pump chopping cycle. This parameter varies by almost three orders of magnitude between Figs. 2a and 2e. When p/f = 0.9 (Fig. 2a) the pedestal is large and is clearly distinguished

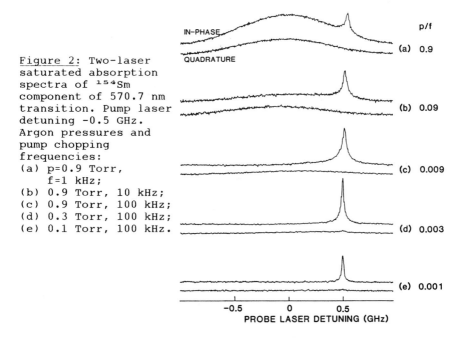

Figure 2: Two-laser saturated absorption spectra of ^{154}Sm component of 570.7 nm transition. Pump laser detuning -0.5 GHz. Argon pressures and pump chopping frequencies:
(a) p=0.9 Torr, f=1 kHz;
(b) 0.9 Torr, 10 kHz;
(c) 0.9 Torr, 100 kHz;
(d) 0.3 Torr, 100 kHz;
(e) 0.1 Torr, 100 kHz.

IN-PHASE

QUADRATURE

p/f

(a) 0.9
(b) 0.09
(c) 0.009
(d) 0.003
(e) 0.001

-0.5   0   0.5
PROBE LASER DETUNING (GHz)

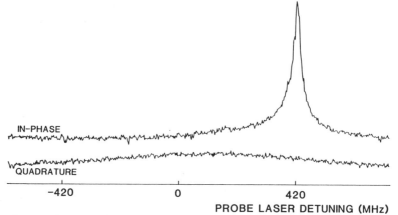

IN-PHASE

QUADRATURE

-420   0   420

PROBE LASER DETUNING (MHz)

Figure 3: Two-laser saturated absorption spectrum
showing asymmetric Doppler-free peak. Pump laser
detuning -0.42 GHz. Pump laser chopping frequency
100 kHz. Argon pressure 0.9 Torr.

from the Doppler-free peak. The quadrature signal, which lags
the in-phase signal by 90°, shows only the pedestal component.
For p/f = 0.009 (Fig. 2c, Fig. 3), the in-phase signal
exhibits almost no pedestal, though it does show an asymmetry
in the wings of the Doppler-free peak due to weaker VCC. The
quadrature signal shows a broad pedestal centered not quite at
the atomic transition frequency. When p/f is reduced to 0.001
(Fig. 2e) then these remaining VCC effects do not appear in
the spectra and a narrow, symmetric Doppler-free peak is
observed.

As expected, reducing the number of collisions within a
chopping cycle (by using high frequency chopping or low rare-
gas pressure) reduces the effects of VCC on saturated
absorption spectra. This leads to high quality saturated
absorption spectra that allow the effects of strong and weak
VCC to be separated and should ultimately allow the effects of
phase- and velocity-changing collisions to be distinguished
and collision kernels to be determined.

1. C. Brechignac, R. Vetter and P.R. Berman, J. Phys. B 10,
   3443 (1977).
2. Ph. Cahuzac, E. Marie, O. Robaux, R. Vetter and
   P.R. Berman, J. Phys. B 11, 645 (1978).
3. D.S. Gough and P. Hannaford, Opt. Comm. 67, 209 (1988).
4. C. Brechignac, R. Vetter and P.R. Berman, J. Physique Lett.
   39, L-231 (1978).
5. M. Gorlicki, A. Peuriot, M. Dumont, J. Physique Lett. 41,
   L-275 (1980).

HIGH QUALITY SATURATED ABSORPTION SPECTROSCOPY IN A SPUTTERED VAPOUR: APPLICATION TO ISOTOPE SHIFTS IN Zr I.

P. Hannaford and D.S. Gough
CSIRO Division of Materials Science and Technology,
Clayton, Victoria 3168, Australia.

Recent investigations in this Laboratory (1) have shown that narrow, pedestal-free saturated absorption spectra may be readily obtained in a sputtered atomic vapour by employing a planar-cathode sputtering discharge operated at low pressure (e.g., 0.1 Torr Xe) and using a high chopping frequency (e.g., 100 kHz) of the laser pump beam. Under these conditions the number of velocity-changing collisions occurring during one chopping cycle can be kept sufficiently low that the saturated absorption signals are essentially free from background pedestals. Furthermore, the low pressures at which the planar-cathode discharge can be operated permit very narrow Doppler-free resonances to be obtained (~ 5 MHz FWHM for transitions involving long-lived levels) and the high chopping frequency allows excellent signal-to-noise. Reduction of the pedestals by use of high modulation frequencies has previously been observed in optical heterodyne saturation spectroscopy (2) and velocity-selective optical pumping experiments (3).

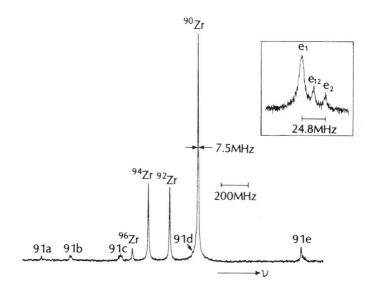

Fig. 1. Saturated absorption spectrum for the Zr I 613.46 nm ground-state transition in a sputtered vapour at 0.1 Torr of Xe. Single 100 s scan; time constant of lock-in 30 ms. Inset: expanded scan of the 91e peak, comprising the ^{91}Zr hyperfine doublet e_1, e_2 and cross-over resonance e_{12}.

83

Use of the sputtering vaporization technique should allow high-resolution saturation spectroscopy to be readily extended to almost any element, including the highly refractory elements. This paper reports the application of saturated absorption in a sputtered vapour to the determination of accurate isotope shifts in zirconium, a highly refractory element for which there are few reliable isotope shift data and for which the isotope shifts are small compared with the Doppler broadening. Isotope shifts in zirconium are of special interest since they permit investigation of the effect on the nuclear charge distribution of adding a single neutron and also pairs of neutrons to a closed shell of fifty neutrons (^{90}Zr).

Figure 1 illustrates the quality of the saturated absorption spectra obtained in a sputtered vapour of zirconium using the above technique. The small ^{91}Zr peaks each consist of two or three hyperfine components and cross-over resonances separated by the excited-state hyperfine splittings (see inset to Fig. 1). The ^{91}Zr isotope shifts were measured by first determining the hyperfine structures for the upper and lower levels (1) and then finding the centroid of the hyperfine structure pattern. The isotope shifts found for twenty Zr I $4d^25s^2-4d^25s5p$ transitions are summarised in Table 1. The precision is typically ± 0.5 MHz, which represents an improvement of about an order of magnitude (or greater) over both the earlier Doppler-limited data obtained with enriched isotope samples (4,5) and the previous Doppler-free data (6-9).

Table 1 Isotope shifts for $4d^25s^2-4d^25s5p$ transitions in Zr I.

λ (nm)	Transition	90-91 (MHz)	90-92 (MHz)	92-94 (MHz)	94-96 (MHz)
588.56	$a^3F_3-z^3F_4$		-224.6(3)	-165.2(6)	-120.8(6)
593.52	$a^3F_2-z^3F_3$		-220.3(7)	-163.7(6)	-120.9(6)
612.74	$a^3F_4-z^3F_4$	-95.5(5)	-230.1(6)	-170.5(4)	-124.2(3)
613.46	$a^3F_2-z^3F_2$	-91.5(8)	-210.7(6)	-163.4(7)	-127.0(3)
614.32	$a^3F_3-z^3F_3$	-94.0(6)	-226.3(4)	-167.6(5)	-123.4(4)
612.08	$a^3P_2-z^3P_1$	-88(2)	-195.8(4)	-159.1(5)	-131.5(3)
612.48	$a^3P_0-z^3P_1$	-91.7(5)	-198.7(4)	-168.7(7)	-145.1(6)
614.05	$a^3P_2-z^3P_2$	-88.8(1.)	-200.5(8)	-160.6(5)	-129.9(8)
619.30	$a^3P_1-z^3P_1$	-92.3(9)	-199.8(4)	-170.6(4)	-147.7(8)
621.31	$a^3P_1-z^3P_2$	-92(2)	-205.1(3)	-171.6(3)	-145.1(9)
630.43	$a^3P_1-z^3P_0$		-312.5(4)	-215.2(1.)	-142.2(1.)
573.57	$a^3F_2-z^3D_1$		-241.7(8)	-168.5(6)	-112.0(7)
579.77	$a^3F_3-z^3D_2$		-219.1(5)	-167.9(7)	-126.1(6)
587.98	$a^3F_4-z^3D_3$		-239.8(8)	-173.2(7)	-122.9(6)
595.54	$a^3F_2-z^5F_1$	-99.6(1.)	-263.4(9)	-167.8(8)	- 96.9(1.)
606.28	$a^3F_3-z^5F_2$		-272.7(4)	-170.8(7)	- 93.3(1.)
566.45	$a^1D_2-y^1D_2$		-360.0(6)	-226.0(4)	-124.9(1.)
592.51	$a^1D_2-z^3S_1$		-336.7(1.)	-218.4(8)	-127.7(1.)
568.09	$a^3P_1-z^3S_1$		-334.2(3)	-220.2(7)	-135.9(1.)
612.19	$a^1G_4-y^1F_3$		-308.2(6)	-200.0(3)	-119.1(6)

Multi-dimensional King plots were performed on the isotope shift data for the twenty Zr I transitions and the shifts were found to be consistent within the quoted uncertainties. Field and mass shifts were separated using values of $\delta\langle r^2\rangle^{AA'}$ deduced from the combined analysis of muonic atom isotope shifts and electron scattering data (10,11). The $\delta\langle r^2\rangle^{AA'}/\delta\langle r^2\rangle^{90,92}$ values, determined from the ratios of the field shifts, and the absolute $\delta\langle r^2\rangle^{AA'}$ values, obtained by calibration against the muonic atom-electron scattering data, are summarised in Table 2, along with earlier results deduced from the Doppler-limited isotope shift data (4,5). The new $\delta\langle r^2\rangle^{AA'}$ values obtained from the combined analysis of optical isotope shift, muonic atom and electron scattering data represent a significant improvement in precision and allow reliable information to be deduced about the systematics of the nuclear charge distribution in the region of the magic number N=50. Contrary to previous experimental results (Table 2) and theory (4,5), there is definite evidence of significant odd-even staggering ($\delta\langle r^2\rangle^{90,91}/\delta\langle r^2\rangle^{90,92}=0.427$) when a single neutron is added to the magic number closed shell of fifty neutrons. In addition, these results together with the isotope shift data for strontium (5) reveal the expected sharp jump in $\delta\langle r^2\rangle^{AA'}$ when a pair of neutrons is added to the closed neutron shell.

Table 2   Change in mean-square nuclear charge radii for Zr.

AA'	$\delta\langle r^2\rangle^{AA'}/\delta\langle r^2\rangle^{90,92}$		$\delta\langle r^2\rangle^{AA'}$ (fm^2)	
	This work	Doppler-limited (4,5)	This work	Doppler-limited (4,5)
90-91	0.427(7)	0.500(13)	0.128(2)	0.123(24)
90-92	1.0	1.0	0.301(4)	0.246(49)
92-94	0.766(12)	0.707(35)	0.230(4)	0.174(43)
94-96	0.586(12)	0.530(47)	0.176(4)	0.130(37)

References
1. D.S. Gough and P. Hannaford, Opt. Comm. 67, 209 (1988).
2. R.K. Raj, D. Bloch, J.J. Snyder, G. Camy and M. Ducloy, Phys. Rev. Lett. 44, 1251 (1980).
3. C.G. Aminoff and M. Pinard, J. Physique 43, 263 (1982).
4. P. Aufmuth and M. Haunert, Physica 123C, 109 (1983).
5. P. Aufmuth, K. Heilig and A. Steudel, Atomic Data and Nuclear Data Tables 37, 455 (1987).
6. P. Hannaford and D.S. Gough, Laser Spectroscopy VII, Springer Series in Optical Sciences 49, 379 (1985).
7. G. Chevalier and J.M. Gagne, Opt. Comm. 57, 327 (1986).
8. G. Chevalier, J.M. Gagne and P. Pianarosa, Opt. Comm., 64, 127 (1987); JOSA B 5, 1492 (1988).
9. Ch. Bourauel, W. Rupprecht and S. Buttgenbach, Z. Phys. D 7, 129 (1987).
10. T.Q. Phan, P. Bergem, A. Ruetschi, L.A. Schaller and L. Schellenberg, Phys. Rev. C 32, 609 (1985).
11. H.J. Emrich et al, Proc. Fourth Int. Conf. on Nuclei far from Stability (Helsingor, Denmark), Vol 1, p 33 (1981).

High-resolution Pulsed Laser Spectroscopy in the UV/VUV Spectral Region

J. Bengtsson, J. Larsson, S. Svanberg and C.-G. Wahlström

Department of Physics, Lund Institute of Technology
P.O. Box 118, S-221 00 Lund, Sweden

In the present paper we address the problem of achieving high-resolution spectroscopy at short UV and at VUV wavelengths where narrow-band lasers are not readily available. In the early days of dye lasers when CW systems could only be operated broadbanded, high-resolution spectroscopy could still be achieved by combining broadband CW excitation with resonance spectroscopy, observing rf- or level-crossing signals (see, e.g. (1)) or by combining pulsed laser excitation with quantum-beat spectroscopy (see, e.g. (2)). With the great advances in single-mode laser technology and the development of numerous powerful Doppler-free laser spectroscopic techniques, high-resolution spectroscopy has become straightforward at UV, visible and infrared wavelengths. However, at short UV and VUV wavelengths only pulsed lasers of considerable bandwidth can readily be used, even if low-power single-mode radiation has been generated and successfully applied (see, e.g. (3)). Again, optical double-resonance (ODR), level-crossing (LC) or quantum-beat spectroscopy (QBS) can readily provide a spectroscopic resolution only limited by the Heisenberg uncertainty relation. Such techniques can thus be put to work to enable studies of e.g. light atoms, Rydberg states and ions for which frequently very short excitation wavelengths are needed. For measurements of small splittings the quantum-beat method is very convenient, while pulsed double-resonance and level-crossing spectroscopy have a more general applicability. Different aspects of pulsed resonance spectroscopy are reviewed in (4).

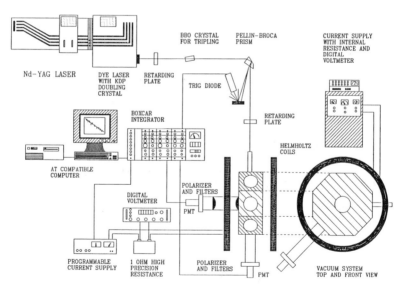

Fig. 1. Experimental set-up for pulsed level-crossing spectroscopy [9]

With the introduction of the β-bariumborate (BBO) crystal, the limit for direct frequency doubling has been moved to about 205 nm. Using non-linear frequency conversion in vapours and gases, VUV radiation of generally low intensity can be generated (5,6). We have found frequency tripling of dye laser radiation in BBO (7) very useful and have generated about 10 mJ of tunable radiation around 200 nm using a Nd:YAG pumped system. Such high-power pulses can be efficiently anti-Stokes Raman shifted in H2 into the upper VUV region, gaining 0.5 eV in photon energy per shift. Evacuation or N2 flushing of the optical paths outside the scattering chamber can be utilized to prevent the onset of absorption due to the O2 Schumann-Runge bands. In this way, high-resolution laser spectroscopy can be readily extended to shorter wavelengths. A dual PMT detection scheme using boxcar integration with signal linearization (8) is very effective for low-noise pulsed signal recording.

We will illustrate high-resolution experiments along these lines by work performed on Cu and Ag atoms. These atoms have alkali-like spectra but because of the configuration interaction with broken d-shells, strong perturbations of hyperfine structure and radiative properties occur. An experimental set-up for pulsed LC spectroscopy is shown in Fig. 1. An LC illustration for the $5p \, ^2P_{3/2}$ level of Cu is given in Fig. 2, where an experimental recording and a theoretically calculated Breit formula curve are shown. The zero-field Hanle signal and high-field level-crossing signals can be seen. In Fig. 3 Pashen-Back ODR signals for the $6p \, ^2P_{3/2}$ level of Ag are given, also providing a "school-book" example of the resolution improvement obtained by restricting the detection to "old" atoms. Two signals, corresponding to a nuclear spin $I = 1/2$ are obtained. The splitting yields the hyperfine structure shile the Landé g_j factor determines the center of gravity for the signals. Finally, an illustration of determinations of short lifetimes by the pulsed Hanle method is given in Fig. 4, where a recording for the $3s4p \, ^1P_1$ state of Mg is given. This technique should be valuable for measuring lifetimes down to 1 ns, also for laser-produced ions.

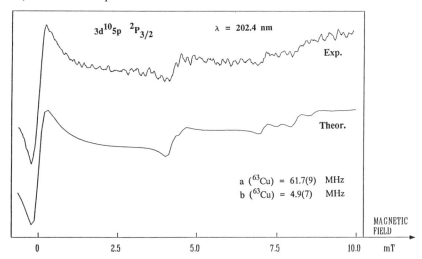

Fig. 2. *Experimental and theoretical level-crossing curves for Cu atoms. The evaluated magnetic dipole and the electric quadrupole interaction constants a and b, respectively, are given for the dominant ^{63}Cu isotope [9]*

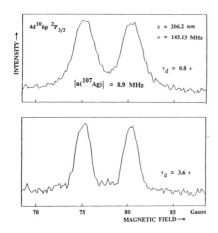

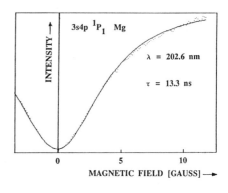

Fig. 3. ODR recordings for Ag [10]   *Fig. 4. Mg Hanle signal [10]*

This work was supported by the Swedish Natural Science Research Council.

References

1. S. Svanberg, in *Laser Spectroscopy*, ed. by R.G. Brewer and A. Mooradian (Plenum, New York 1974), p. 205; S. Svanberg, in *Laser Spectroscopy III*, ed. by J.L. Hall and J.L. Carlsten, Springer Series in Optical Sciences, Vol. 7 (Springer, Berlin, Heidelberg 1977), p. 187

2. S. Haroche and J.A. Paisner, in *Laser Spectroscopy*, ed. by R.G. Brewer and A. Mooradian (Plenum, New York 1974), p. 445

3. H. Hemmati, J.C. Bergquist and W.M. Itano, Opt. Lett. **8**, 73 (1983)

4. J. Larsson, L. Sturesson and S. Svanberg, Phys. Scr., in press

5. W. Jamroz and B.P. Stoicheff, Progress in Optics **XX**, 325 (1983)

6. R. Hilbig, G. Hilber, A. Lago, B. Wolf and R. Wallenstein, Comments At. Mol. Phys. **18**, 157 (1986)

7. W.L. Glab and J.P. Hessler, Appl. Opt. **26**, 3181 (1987)

8. U. Wolf and E. Tiemann, Appl. Phys. **B39**, 35 (1986)

9. J. Bengtsson, J. Larsson, S. Svanberg and C.-G. Wahlström, to appear

10. J. Bengtsson, J. Larsson and S. Svanberg, to appear

Laser Spectroscopy of Relativistic Beams of H- and H: Observation of e-Detachment from H- by Multiphoton Absorption

W.W. Smith
   Department of Physics, U-46, The University of Connecticut,
   Storrs, CT   06268   USA
C.R. Quick, Jr., J.B. Donahue, and S. Cohen
   CLS and MP Divisions, Los Alamos National Laboratory,
   Los Alamos, NM 87545   USA
C.Y. Tang, P.G. Harris, A.H. Mohagheghi, H.C. Bryant, R.A. Reeder, and H. Toutounchi
   Department of Physics and Astronomy, The University of New Mexico,
   Albuquerque, NM 87131   USA
J.E. Stewart
   Department of Physics, University of Western Washington,
   Bellingham, WA 98227   USA
H. Sharifian
   Department of Physics, California State University,
   Long Beach, CA   90840   USA

Laser spectroscopy on near-light velocity H- ions and H atoms has been carried out at the Los Alamos Meson Physics Facility using a variety of fixed frequency lasers intersecting accelerated beams at variable angles. Beam energies up to 800 MeV $(v/c) = 0.84$ make possible an unusually wide tuning range at modestly high resolution. A dedicated beam line, the High Resolution Atomic Beam (HIRAB), also makes possible Stark effect and field ionization studies in the multi-megavolt/cm range. Preliminary results on multiphoton detachment of fast H-ions using a pulsed CO_2 laser focussed to ~10^{11} W/cm^2 over a factor 10 photon energy range (CM frame) are presented here.

1. Introduction and Review

A collaboration from several universities and from Los Alamos National Labovatory initiated by H.C. Bryant et al of the University of New Mexico has been studying UV photodetachment resonances in H- ions (1) in a crossed-beam configuration by using a fixed frequency laser intersecting, at a variable angle α, a relativistic ion beam from the LAMPF proton linac. LAMPF produces beams of H- ions, as well as H atoms and H+, from a few hundred MeV up to 800 MeV, with a momentum purity $\Delta p/p \sim 1/2000$ or better. The Doppler tuning range in the center-of-mass frame of the fast ions is from $\gamma(1+\beta)$ for colliding beams to $\gamma(1-\beta)$ for merged beams, i.e. a factor of ~10. H- photodetachment resonances have been observed at high motional electric fields ($\beta\gamma cB$) up to the MV/cm range. (2) One can use the 3rd or 4th harmonic of a YAG laser (1.064μ fundamental) to provide a broadly-tunable CM-frame Lyman-α range laser. High-field Stark effects in v~c H atoms have been studied using this technique. (3) The Bohr levels of H atoms in the fast beam provide convenient optical markers to measure the absolute beam velocity via the relativistic Doppler formula, which has been tested "to all orders" with a precision of ~1/2000 at β=0.84. (4)

2. Multiphoton Detachment of H-

We recently reported the first observation of multiphoton electron detachment of H-. (5) The relativistic Doppler shift of the H- beam at Los Alamos was exploited to provide a large tuning range as in (1)-(4). References 1-5 of (5) give an indication of the strong interest of theorists in this problem: H- has a low electron affinity ("ionization potential") of 0.754 eV and only one bound state. Multiphoton detachment (MPD) can take place with lower energy photons than typical multiphoton ionization (MPI) with neutrals. The lack of intermediate-states simplifies the calculations as does the fact that the final-state interaction (in single e- MPD) is of short range rather than Coulomb: H- may thus provide a particularly clean test of MPI/MPD theory. Large changes in the MPD rate can be expected when a large static electric field is applied in addition to a laser field. (6)

The apparatus [Figure 1] uses a CO_2 TEA laser (oscillator-amplifier combination) at up to ~1J/pulse at 0.5 Hz. After passing through collimating optics, a half-wave plate and a ZnSe meniscus lens (f=25.4 cm), the 10.6μ beam is directed up the rotation axis, reflected by ~99% reflecting copper mirrors and made to intersect the fast H- beam from LAMPF at a variable angle α. Detached e- are detected with high efficiency and low background by the electron spectrometer; fast H atoms are detected by a downstream scintillator. The waist diameter (~350μ focus) is << the ~2mm H- beam size but the Rayleigh range z is large (~1cm). Maximum average intensity at the focus is ~2x10^{10} W/cm^2. The free-electron quiver energy (ponderomotive potential) $W_p=e^2E^2/4m\omega^2$ = 9.34 x 10$^{-14}\lambda^2$I eV (for λ in μm) = 0.21 eV, exceeding the photon energy of 0.117 eV (lab frame) and comparable to the electron affinity. Thus MPD is expected to be a relatively efficient process.

The angle α ranges from 21^0 - 159^0, implying a continuous CM-frame tuning of photon energy at β=0.84 between 0.387 and 0.046 eV. Ignoring the ~0.2 eV ponderomotive shift, the photon energy is thus above threshold for MPD by 2-17 photons, or 3-21 photons when W_p is added to the electron affinity. The photon flux F=I/hω, in photons/cm^2-s, transforms under a Lorentz transformation to the CM frame the same way as the frequency $\omega'=\gamma(1+\beta\cos\alpha)\omega$, so I/ω^2 = I'/ω'^2 is a relativistic invariant. (7) Thus the ponderomotive potential in the CM frame is a constant, independent of α, simplifying the scaling of the data.

During our first beam time (in 1988), we saw strong MPD signals above background with, successively, a minimum of 3-5 photons before the signal was lost in the noise. By turning up the phototube gain, signals could then be obtained down to ~0.1 eV CM photon energy (min. 7-photon process). In our 1989 run (still underway), a low-pressure gain cell (smoothing tube) was added to the multimode CO_2 oscillator to provide mode selection and eliminate most pulse spiking. (Unfortunately we discovered that our more intense H- micropulses produced radiation along the beam line that led to irregular firing of the laser!) MPD signals were still easily seen despite the lower peak intensity with the smoothing tube. We confirm the generally monotonic decrease of the MPD e- yield with decreasing photon energy seen in the 1988 run. (5) There is an apparent intensity-dependent threshold and a hint of some structure also (with linewidth > the ~6 meV instrumental width), particularly near $h\omega' = 0.32\text{-}0.34$ eV, which may be due to a transition from the 3 to 4-photon threshold region. Observed MPD rates with the smoothing tube were somewhat smaller than without it, as expected, but at the highest intensities we were still able to observe partial depletion saturation of at least the 3-photon threshold detachment on the ~1 psec scale of the H- transit time through the laser focus.

This work was performed under the U.S. Dept. of Energy, in part under Grant No. DE-AS04-77ER03998, and was also supported in part by the U.S. Army SDC.

References
1. H.C. Bryant, B.D. Dieterle, J. Donahue, H. Sharifian, H. Tountounchi, D.M. Wolfe, P.A.M. Gram, M.A. Yates-Williams, Phys. Rev. Lett. 38, 228 (1977).
2. For a review of some earlier experiments, see H.C. Bryant et al, Atomic Physics 7, D. Kleppner and F.M. Pipkin, eds., Plenum Press, NY 1981, pp. 29-64.
3. T. Bergeman, et al, Phys. Rev. Lett. 53, 775 (1984).

4. D.W. MacArthur, et al, Phys. Rev. Lett. 56, 282 (1986).

5. C.Y. Tang, et al, Phys. Rev. A 39, 6068 (1989).

6. T. Mercouris and C.A. Nicolaides, J. Phys. B 21, L285 (1988); and preprint (1989).
7. L.D., Landau and E.M. Lifshitz, The Classical Theory of Fields, 1st ed. (Addison-Wesley, Reading, MA, 1951), p. 122.

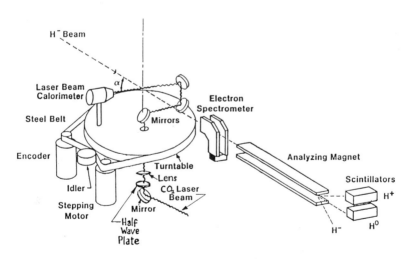

FIGURE 1. Schematic diagram of the 2-beam interaction region.

LINE SHAPES IN FORWARD SCATTERING

P. Jungner, B. Ståhlberg and T. Fellman

Accelerator Laboratory, Department of Physics,
University of Helsinki, Hämeentie 100,
SF-00550 Helsinki, Finland

M. Lindberg

Optical Science Center, University of Arizona,
Tuscon, Arizona, U.S.A.

The magneto-optic effects refered to as forward scattering are experimentally easy to observe. A correct interpretation of the measured signals is, however, a problem. Nonlinearities induced by moderate laser intensities in simple level configurations show lineshapes that are hard to understand intuitively. An appropriate theoretical model and a detailed mathematical analysis are required to explain the recorded lineshapes. As pointed out by Lange et al. [1], only recently have these effects been studied systematically with coherent light fields.

In this paper we report forward scattering lineshape studies in excited Ne20 systems. Our experimental arrangement is very simple. A cw single mode dye laser beam traverses a neon gas absorber. The frequency of light is tuned to the center of a selected Doppler profile. The beam is linearly polarized and it is focused into a 4 cm long DC discharge. A variable magnetic field is introduced parallel (or anti-parallel) to the laser beam. The forward scattered light intensity is detected with a photo-diode after it has passed an analyzer prism which is crossed with respect to the beam polarization.

The curves in Fig. 1 are obtained with the laser wavelength 607.4 nm and the neon transition $1s_4$ (J = 1, g = 1.464) - $2p_3$ (J = 0) (Paschen notation). Around zero magnetic field the upper curve shows a structure which derives from the nonlinear laser-atom interaction. When the beam power is attenuated to the μW level the nonlinear contribution can not be distinguished from the noise. This is shown by the lower curve in Fig. 1. Lineshapes like those in Fig. 1 are obtained with transitions where the levels have approximately the same linewidths.

In Fig. 2 we show recordings obtained with the laser wavelength 594.5 nm and the $1s_5$ (J = 2, g = 1.503) - $2p_4$ (J = 2, g = 1.298) system. The lower level $1s_5$ is metastable. Therefore the linewidths associated with it are narrow. With relatively high laser powers and with low neon pressures the lineshapes show interesting structures to be explained.

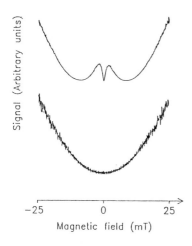

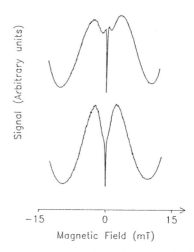

Fig.1 Experimental lineshapes obtained with $1s_4$ ($J = 1$)– $2p_3$ ($J = 0$) neon systems. Lower curve: $P_{laser} = 5$ μW. Upper curve: $P_{laser} = 20$ mW. Both curves: $p_{Ne} = 75$ Pa.

Fig.2 Experimental lineshapes obtained with $1s_5$ ($J = 2$) – $2p_4$ ($J = 2$) neon systems. Lower curve: $P_{laser} = 20$ mW, $p_{Ne} = 40$ Pa. Upper curve: $P_{laser} = 50$ mW, $p_{Ne} = 15$ Pa.

Our theoretical treatment is based on the semiclassical laser theory. We consider $J = 1$ to $J = 0$ and $J = 1$ to $J = 1$ systems. The calculations are performed up to the second order in the laser intensity. Thus, the model can not allow occurence of higher order nonlinear mixing of the levels than those corresponding to the selection rule $|\Delta m| = 1$. The expressions for the signal contain a nonlinear part and a linear part. The nonlinear part consists of the dispersive responses (D) of the resonances involved. In addition to the $\Delta m = 2$ coherences (D_a and D_b) associated with the coherence relaxation parameters (γ_a and γ_b) there is a broader contribution (D_{ab}) associated with the homogenous linewidth parameter (γ_{ab}). For a system with $J_b = 1$ to $J_a = 0$, the signal becomes

$$S \propto \left[I^2 \left[D_b(B) + D_{ab}(B) \right] - IB \right]^2, \tag{1}$$

where I is the laser intensity and B is the magnetic field. For a system with $J_b = 1$ to $J_a = 1$, we obtain

$$S \propto \left[I^2 \left[D_a(B) + D_b(B) + D_{ab}(B) \right] - IB \right]^2. \tag{2}$$

Our calculated results agree well with those presented by Giraud-Cotton et al. [2].

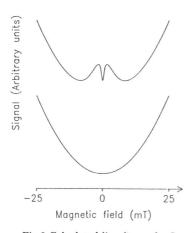

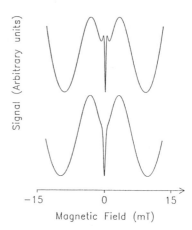

Fig.3 Calculated lineshapes for J = 1 to J = 0 to be compared with Fig.1.

Fig.4 Calculated lineshapes for J = 1 to J = 1 to be compared with Fig.2.

In Fig. 3 we show the calculated curves corresponding to the $J_b = 1$ to $J_a = 0$ case of Fig. 1. As can be seen, for low enough laser intensities the lineshape ·becomes a parabolic function of the magnetic field. The agreement between the experimental lineshapes (Fig. 1) and the calculated lineshapes (Fig. 3) is excellent.

To explain the experimental lineshapes of Fig. 2 we have used equation (2) for the calculations. For simplicity we use the same Landé g-factor values for both levels $J_b = 1$ and $J_a = 1$. Letting $(\gamma_a/ku) \approx (\gamma_{ab}/ku) \approx 0.1$ and $(\gamma_b/ku) \approx 0.01$ (ku = Doppler width) we end up with lineshapes shown in Fig. 4. We emphasize that the additional structure around zero magnetic field derives from the different linewidths for the $\Delta m = 2$ coherences only. Already for the case $J_b = 1$ to $J_a = 0$, this structure can be produced provided that $\gamma_b \ll \gamma_a \approx \gamma_{ab}$. In our neon experiments only the 1s₅ (J = 2) - 2p₄ (J = 2) transition is found to fulfil $\gamma_b \ll \gamma_a \approx \gamma_{ab}$. In principle, in the case of a J = 2 to J = 2 transition also higher order (hexadecapole $\Delta m = 4$) couplings may occur. Our model does not include these couplings and since even a J = 1 to J = 0 system can produce the observed lineshapes we do not expect higher order coherences to contribute significantly to the observed signals.

This work has been supported by the Academy of Finland.

References

1. W. Lange, K.-H. Drake and J. Mlynek: In Laser Spectroscopy VIII, Eds. W. Persson and S. Svanberg (Springer, Berlin 1987) p.300

2. S. Giraud-Cotton, V.P. Kaftandjian and L. Klein, Phys. Rev. **A32**, 2223 (1985)

Novel Lutetium Spectroscopic Interactions via cw RIMS

B. L. Fearey and C. M. Miller
Isotope Geochemistry Group, INC-7
Isotope and Nuclear Chemistry Division
Los Alamos National Laboratory,
Los Alamos, New Mexico   87545, USA

Novel spectroscopic interactions of argon-ion laser enhanced resonance ionization of lutetium are observed and discussed; these include line-narrowing, non-linear power dependences and anomalous optical pumping effects of the hyperfine transitions. In addition, isotopically saturation dip spectra are observed and presented, allowing for precise determination of hyperfine constants of rare isotopes.

1.   Introduction

Interest in the technique of Resonance Ionization Mass Spectrometry (RIMS) developed over recent years due to: 1) the selective manner in which isobaric interferences are discriminated against,[1,2] and 2) the large dynamic range available for measuring isotopic ratios.[3] Previous work[3] has shown the ability of cw RIMS to measure lutetium isotopic ratios down to the 0.4 ppm level on very small samples ($\sim$60 ng); however, the ultimate isotope ratio dynamic range was limited by low signal. Recently, we have succeeded in using a second, high-power, non-resonant laser for the ionization step to dramatically increase ionization efficiency.[4-6] Several effects are observed in this configuration, including spectral line-narrowing and superlinear power dependences.[4,5] To significantly increase the dynamic range further, it will be necessary to perform isotopically selective resonance ionization. In the present work, secondary enhanced ionization and Doppler-free saturation spectroscopy via RIMS are utilized for this goal, *i.e.*, to precisely determine the position of the various HF components for the lutetium isotopes as well as increase overall ionization efficiency. Full details of the processes and phenomena which occur in these experiments will be discussed in forthcoming papers.[6,7]

2.   Experimental

Briefly, a ring dye laser was tuned to the one-photon transition of lutetium ($^2D^0_{3/2} \leftarrow {}^2D_{3/2}$) at $\sim$22125 cm^{-1} (see Figure 1. for a simplified energy level diagram). To increase the ionization efficiency[4-6], an Ar$^+$ laser tuned to the 457.9 nm line ionized the excited atom. Both laser beams were propagated parallel to and $\sim$2 mm above the sample filament, with typical beam diameters of $\sim$100 μm. A spherical mirror was inserted to retroreflect the laser beams collinearly. A magnetic sector mass spectrometer was used for detection of the lutetium ions.

3.   Results and Discussion

Additional experiments have addressed the use of resonance ionization for isotopically selective Doppler-free spectroscopy. Bennet[8] and Lamb[9] recognized that the narrow resonances that appeared in the center of inhomogeneously broadened gain profiles interacting with counterpropagating laser beams resulted from "holes" burned into the Maxwell-Boltzmann velocity distribution. The advantages this phenomenon provide are a simultaneous means for determining the center frequency of a transition and for removing the inhomogeneous line-broadening. Extensive saturation spectroscopy has been performed in a pump-probe scheme using relatively high pressure static cells, with inherent difficulty in examining rare isotopes. In contrast, for the present RIMS experiments, the pump and probe are the same laser, and a low-pressure mass spectrometer is used as the detector, which removes possible pressure broadening effects and permits isotopically selective saturation spectroscopy of unenriched samples.

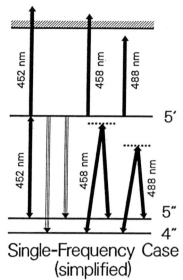

Single-Frequency Case
(simplified)

Fig. 1. Simplified lutetium energy level diagram showing the Raman pumping interaction mechanism (note: only part of the hyperfine transitions are shown). The arrows labelled "452 nm" correspond to the ring dye laser, while those labelled "458 nm" and "488 nm" correspond to high power Ar$^+$ ion laser lines.

Rather unusual spectroscopic interactions have been observed for lutetium upon the addition of a non-resonant argon ion laser to the interaction region. Observations include line-narrowing of spectral features and super-linear power dependencies of the ionization signal on Ar$^+$ laser power.[5,6] A Raman pumping mechanism is invoked to explain these observations. Basically, the high power of the Ar$^+$ laser remixes the hyperfine populations that would otherwise be lost through optical pumping, via a non-resonant process (see Figure 1.). This Raman process accounts for 1) the non-linear power dependence (i.e., the Raman pumping process is proportional to the square of the Ar$^+$ laser power), and the line-narrowing seen in the hyperfine spectra when the Ar$^+$ laser is added (i.e., remixing reduces optical pumping and hence, saturation). This process is verified by recent experiments where an increase in ionization efficiency occurs even when the secondary laser photon is of insufficient energy to ionize the excited atom. Since ionization the only occurs via the resonant laser (see Figure 1.), the increased ionization must be due to the remixing of the optically pumped hyperfine states.

Two spectral features are generally observed: (1) dips which reflect line centers approaching their natural linewidth, and (2) crossover peaks which occur at the mean frequency of two hyperfine lines whose Doppler profiles overlap. A typical expanded experimental spectrum for ^{175}Lu is shown in Figure 2., which illustrates the signal-to-noise and resolution attained using this technique. The dips and peaks correspond to the line centers and crossovers for the atomic hyperfine components for lutetium as indicated in the figure. Similar ^{176}Lu spectra were taken with comparable quality.

Included in the Figure 2. spectrum are the transmission peaks of a calibrated ~300 MHz confocal etalon used for frequency difference determination. With this and the near-natural linewidth characterizing the dips and peaks, the hyperfine splitting constants for the ^{175}Lu and ^{176}Lu ground and excited states were measured with high precision and accuracy. These determinations are discussed and compared to previously values[10,11] in a forthcoming detailed paper.[7]

4. Conclusions

Several rather novel effects have been observed and are understood in terms of a non-resonant Raman pumping mechanism[7]. Additionally, a technique utilizing RIMS to obtain very high resolution atomic spectra with isotopic selectivity has been demonstrated. Advantages of this technique are increased spectral resolution, isotope-specific spectral information from unenriched samples and relative ease of operation. Importantly, a combination of modulated photoionization and

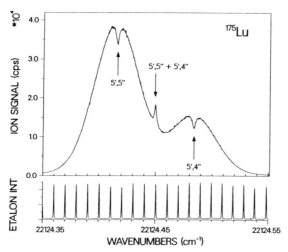

Fig. 2. Expanded, unsmoothed saturation spectrum of the $(^2D^0_{3/2} \leftarrow {}^2D_{3/2})$ transition at $\sim$22125 cm^{-1} for the first two hyperfine components of ^{175}Lu. The arrows point out the dips and crossover for the indicated hyperfine transitions. Also included are the transmission peaks of a calibrated $\sim$300 MHz confocal etalon.

phase-locked detections may lead to significantly increased isotopic selectivity in mass spectral analysis. Future work includes spatial-modulation of the counter-propagating laser beam to remove Doppler pedestals,[12] spectral analysis of very rare lutetium isotopes, and demonstration of increased selectivity in isotopic analysis.

References

1. L. J. Moore, J. D. Fassett, and J. C. Travis, Anal. Chem. 56, 2770 (1984).
2. C. M. Miller, N. S. Nogar, E. C. Apel, and S. W. Downey, Resonance Ionization Spectroscopy 1986, G. S. Hurst and C. G. Morgan, Eds. (Bristol: Inst. of Physics, 1986), p. 109.
3. N. S. Nogar, S. W. Downey, and C. M. Miller, Resonance Ionization Spectroscopy 1984, G. S. Hurst and M. G. Payne, Eds. (Bristol: Inst. of Physics, 1984), p. 91.
4. D. C. Parent, B. L. Fearey, C. M. Miller, and R. A. Keller, 35th ASMS Conference Proceedings, p. 1006 (1987).
5. B. L. Fearey, D. C. Parent, R. A. Keller, and C. M. Miller Resonance Ionization Spectroscopy 1988, T. B. Lucatorto and J. E. Parks, Eds. (Bristol: Inst. of Physics, 1989), p. 263.
6. D. C. Parent, B. L. Fearey, R. A. Keller, and C. M. Miller, (in preparation) (1989).
7. B. L. Fearey, D. C. Parent, R. A. Keller, and C. M. Miller, (submitted, JOSA B) (1989).
8. W. R. Bennet, Jr., Phys. Rev. 126, 580 (1962).
9. W. E. Lamb, Phys. Rev. 134A, 1429 (1964).
10. C. M. Miller, R. Engleman, Jr., and R. A. Keller, J. Opt. Soc. Am. B 2, 1503 (1985).
11. R. Engleman, Jr., R. A. Keller, and C. M. Miller, J. Opt. Soc. Am. B 2, 897 (1985).
12. T. P. Duffey, D. Kammen, A. L. Schawlow, S. Svanberg, H. R. Xia, G. G. Xiao, and G-Y Yan, Opt. Lett. 10, 597 (1985).

Optical echoes without phase reversal

A V Durrant and T B Smith

Physics Department, The Open University, Milton Keynes MK7 6AA, U.K.

Over the past 15 years or so a variety of optical echo effects have been exploited to measure relaxation mechanisms, especially in gaseous samples (see reviews [1,2]). In all of this work it is assumed that the echo-generating mechanism is the "phase-reversal" produced by the second or by following pulses, as in the original spin-echo discovered by Hahn [3]. Other echo-generating mechanisms are possible, however, and have been studied experimentally, though not as yet in the optical regime. During the period 1964-1970 a number of groups in the U.S. investigated echoes observed from free electrons in weakly-ionised plasmas situated in slightly-inhomogeneous magnetic fields and excited by rf pulses tuned to the mean electron cyclotron resonance. Electrons excited in cyclotron orbits by a sequence of two or three rf-pulses, emitted sequences of rf-echo pulses. (See reviews [4,5].) The mechanisms responsible for plasma cyclotron echoes do not require phase-reversal by the second or other pulses. They are characterised instead by perturbative mechanisms acting on the radiating electrons after the second pulse has passed, and are described by classical electrodynamics since the cyclotron orbital energies are large (of order eV). Specifically two mechanisms were identified: velocity-dependent collision rates for electrons with neutral atoms or ions in the plasma, and the velocity-dependent relativistic mass increase (see refs [4,5] for details).

While the plasma cyclotron mechanisms are specific to plasma dynamics, the question arises: are there any analogous optical spectroscopic mechanisms at work under the conditions of an optical echo experiment? If so, then the observed optical echo might be the result of superposition of the familiar phase-reversal echo with new echoes generated by these other mechanisms. The question is of some practical importance for experiments designed to measure collisional relaxation rates in gases since echoes generated by different mechanisms may relax at different rates.

In a recent paper [6] we enlarged the familiar two-pulse echo theory for two-level atoms to include the possibility of a class of new optical echo effects generated by perturbations acting selectively on the radiating dipoles or their radiation fields after the second pulse has passed, and have worked out the details for a specific perturbation: the absorption (or amplification) of the atomic dipole radiation as it escapes in the forward direction through the sample. The following outlines a mathematical framework in which these and other possible new echo effects can be described.

In the usual theory of two-pulse echoes from two-level atoms [1,2], the atomic dipole moment after the second pulse has passed is of the form $d(t) = D(t) \exp[i(kz - \omega t)]$ + complex conjugate, where

$$D(t) = A \exp[-i\Delta t] + B \exp[-i\Delta(t-\tau)] + C \exp[-i\Delta(t-2\tau)] \qquad (1)$$

Here ω and k are the laser frequency and wavenumber, Δ the dipole free-evolution frequency in the "rotating frame", and the two short laser pulses are applied at times

$t = 0$ and $t = \tau$. A, B and C depend on the laser pulse areas and the dipole moment matrix element. Spontaneous decay and collisional relaxation are neglected. When this expression is averaged over the inhomogeneous broadening of the sample (i.e. over Δ) the third term gives the familiar phase-reversal polarisation echo at time $t = 2\tau$, while the first two terms give the free induction decay (FID) of polarisation immediately following the two laser pulses. The coherent FID is essentially zero at the echo time provided $\tau \gg T_2^*$, the inhomogeneous dephasing time, but the individual dipoles continue to oscillate incoherently. The optical FID and echo are calculated in the usual way from their polarisation sources. For three-pulse stimulated echoes [1,2] with the third pulse applied at $t = t_3$ there are additional terms to those in Eq(1), two of which,

$$E \ \exp[-i\Delta(t - t_3)] + F \ \exp[-i\Delta(t - t_3 - \tau)] \tag{2}$$

lead respectively to an FID immediately following the third pulse and the usual stimulated echo at $t = t_3 + \tau$.

To see formally how new echo effects might arise in a two-pulse echo experiment we first consider a modified or perturbed dipole moment for $t > \tau$,

$$D'(t) = \exp(-a + ib) \ D(t) \tag{3}$$

where $D(t)$ is given by Eq(1) and the exponential factor is introduced to describe the effects of a perturbing mechanism, unspecified for the moment. The essential requirement is that either the attenuation factor a or the frequenty shift b should be a periodic function in $\Delta\tau$ with period 2π. The modified dipole-moment can then be expressed as a Fourier series

$$D'(t) = \sum_{n=-\infty}^{\infty} G_n \ \exp(in\Delta\tau) \ \{A \ \exp[-i\Delta t] + B \ \exp[-i\Delta(t - \tau)]$$

$$+ C \ \exp[-i\Delta(t - 2\tau)]\} \tag{4}$$

where the G_n are the Fourier amplitudes. Clearly there are three terms in Eq(4) that will lead to an echo at $t = 2\tau$ after the averaging over Δ has been carried out. They are

$$(G_2 A + G_1 B + G_0 C) \ \exp[-i\Delta(t - 2\tau)] \tag{5}$$

while other terms lead to multiple echoes at $t = m\tau$ ($m = 3,4,5 \ldots$). The term in $G_0 C$ in Eq(5) leads to the familiar phase-reversal echo, modified by G_0, while the terms in $G_1 B$ and $G_2 A$ are additional contributions to the echo at $t = 2\tau$. It can be seen that the perturbative mechanisms introduced in Eq(3) generate an echo sequence from each of the two FID terms in Eq(1) as well as from the phase-reversal echo. Extension to the case of three-pulse stimulated echoes is straightforward. In addition to the usual stimulated echo at $t = t_3 + \tau$ from the second term in Eq(2) there are new contributions from the action of the perturbative mechanism on the FID term in Eq(2), and there is also a sequence of echoes at later times $t = t_3 + n\tau$ ($n = 2,3,4 \ldots$).

To appreciate how a perturbing mechanism with a $\Delta\tau$-periodicity might actually arise we note that the dipole moment of Eq(1) is itself periodic in $\Delta\tau$ as is the inversion W which has the form

$$W = W_0 + W_m \cos (\Delta\tau) \tag{6}$$

following a two-pulse sequence [1,2,6] where W_0 and W_m are functions of the laser pulse areas. Any inversion-dependent mechanism acting on the atomic dipoles or on their radiation fields in the way indicated by Eq(3) is a possible echo-generating mechanism. One such mechanism is the self absorption by the sample of the dephased incoherent forward propagating FID radiation. The frequency-dependent absorption coefficient depends on the inversion W and will therefore be periodic in $\Delta\tau$, providing the homogeneous broadening is not large compared with the modulation frequency $\Delta_m = 2\pi/\tau$. Using a simple model based on Beer's law and Eqs(4) and (6) we have worked out some of the details of echoes generated by this mechanism [6]. We estimate that for a sample of about one absorption length the contribution of this mechanism to the primary echo amplitude (at $t = 2\tau$) can be of order 20% of that of the phase-reversal echo. The validity of the model is restricted to small pulse areas.

These predictions of new echoes by absorption have not yet been tested experimentally. Our analysis for two-pulse echoes [6] shows that while the primary echo is expected to occur at the same time ($t = 2\tau$) as the phase-reversal echo, it has distinguishing features that might lead to observable effects. Firstly the collisional decay of the new echo is expected to occur from relaxation of $\Delta\tau$-periodicity of the absorption coefficient as well as from relaxation of optical coherence. Secondly the predicted echo amplitude for small pulse areas [6] is normally negative with respect to that of the phase-reversal echo. A non-exponential pressure-induced decay of the composite echo might therefore be observable in precision experiments. Similar remarks apply to the three-pulse stimulated echo. Finally, the dependence of the echo amplitude on the laser pulse areas is different from that of the phase-reversal echo. In particular, the $\pi/2 - \pi$ pulse area sequence which is optimum for two-pulse phase-reversal echoes produces no modulation of the inversion and so the new echoes are not then formed. This may be one reason why previous workers have not reported anomalies that might be attributable to the new effects.

References

1   R. Shoemaker, in *Laser and coherence spectroscopy,* ed. Steinfield J.I. (Plenum 1978)

2   T.W. Mossberg, R. Kachru, K.P. Leung, E. Whittaker and S.R. Hartmann, in *Spectral lineshapes*, ed. Wende B (W de Gruyter, Berlin, New York 1981)

3   E.L. Hahn, Phys. Rev. **80**, 580 (1950)

4   G.F. Hermann, R.M. Hill, and D.E. Kaplan, Phys. Rev. **156**, 118 (1967)

5   R.W. Gould, Am. J. Phys. **37**, 585 (1969)

6   A.V. Durrant and T.B. Smith, J. Mod. Opt., **36**, 261 (1989)

LASER EVAPORATION AS A SOURCE FOR SMALL FREE RADICALS

Maarten Ebben, Gerard Meijer and J.J. ter Meulen
Department of Molecular and Laser Physics
Katholieke Universiteit Nijmegen
Toernooiveld, 6525 ED Nijmegen, The Netherlands

Introduction

Laser evaporation in combination with supersonic expansion
has been shown to produce rotationally cold beams of species,
normally to be found only in very hot environments.
The cluster source, as described by Smalley [1], is known to
produce large amounts of clusters and cluster-ions. The main
experimental parameters that can be varied to influence the
production of the different products are the color and the
fluence of the evaporation laser and the geometry of the
expansion channels. In the present experiment we investigated
whether this source is suitable for the production of high
densities of small free radicals. For this purpose we aimed
at diatomics in the first place formed from one atom from a
solid target and one atom from a gas added to the seeding
gas, if not the seeding gas itself.

Experimental

The nozzle part of our experimental setup is similar to the
one, described in reference 1. We deliberately designed the
flow channel as short as possible to prevent the products
from clustering and to avoid secundary reactions as much as
possible. A schematic diagram of the flow channel is depicted
in figure 1.

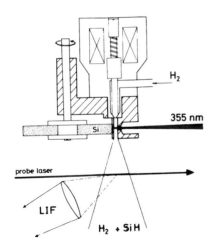

Fig. 1.

Schematic diagram of the flow
channel in which the
evaporation takes place. The
solid target disk rotates and
translates continuously to
achieve a good shot to shot
production stabilty. The pulsed
nozzle assembly is also shown.
The probe laser crosses the jet
perpendicularly, 10 mm
downstream from the orifice.
The LIF is detected by a
photomultiplier and send to a
boxcar averager and a digital
oscilloscope.

Discussion

This experimental setup enables us to monitor in real time all the light involved in the experiment : direct beam emission (mostly from locally produced, highly excited atoms), laser induced fluorescence and black body radiation as well as Mie scattering from macroscopically large particles. These large paricles were present, despite of the fact that we used high energy (355 nm) photons for evaporation and heating up the plasma in combination with a short expansion channel in order to preferentially produce small compounds. The radicals detected in the beams were all found at the time of maximum beam emission.

Different target materials were used (Si, Cu, Fe, C, Al,...) in combination with different gasses (He, H, O, ...). As an example, the spectrum of the first electronic transition in the SiH radical is shown in fig.2.

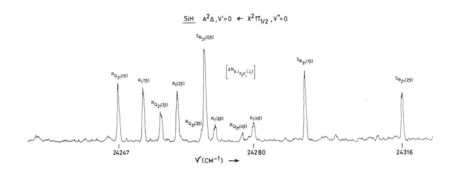

SiH $A^2\Delta, V'=0 \leftarrow X^2\Pi_{1/2}, V''=0$

Fig.2. The SiH spectrum. The rotational temperature obtained was 30 K, however not yet in the final stage of the expansion.

From the time resolved fluorescence signal resulting from the excitation of molecules in a single rotational state, we calculated the absolute number of SiH molecules produced in this way to amount 10^9 molecules/sr in a single pulse within one order of magnitude.

Besides SiH, we detected CuH, CuO, CH, and FeO with LIF. The technique indeed is very promising to yield cold spectra of species like FeH which possess many low lying electronic states with high multiplicity. Also more elusive species with unknowm electronic structure may be produced in this way.

In figure 3 the spectra of CuH and CH are shown. These spectra simultaneously occured when a copper disk, hosting an unknown small concentration of carbon was evaporated in the presence of hydrogen. The spectra were unraveled by simply changing the time delay between the probe laser and the detection gate, thus probing different molecular states with different radiative lifetimes.

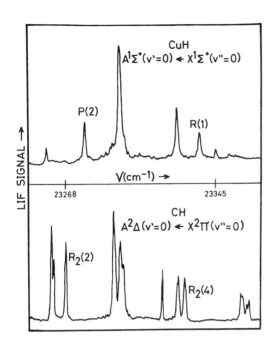

Fig.3.

The excitation spectra of CuH and CH in the same frequency region under identical conditions. A trace of carbon in the copper target disk led to this CH spectrum. This demonstates the vigor of the production method.

In the case of copper as target material, the beam emission was particularly intense in the green as could readily be observed with the naked eye. This emission extended over a range of 50 mm from the channel orifice. Dispersed it showed to originate from cascade decay involving the allowed transitions in the copper atom from states with energies just below the ionisation limit down to the ground state. To investigate this emission, an electric field was applied both parallel and antiparallel to the jet. The parallel emission did not influence the emission at all, while the antiparallel field shortened the range of green emission proportional to the applied field strength. This led us to the conclusion that positive (cluster-) ions are involved in the pocess leading to the highly excited copper atoms, while low energy electrons can be excluded. This rules out that the collisional dissociation of larger clusters induced by such electrons is the mechanism leading to this production of excited copper atoms.

[1] R.E. Smalley
    Laser Chem. 2 (1983) 167

PROGRESS IN THE SPECTROSCOPY OF THE RARE GAS DIMERS

D.J. Kane, S. B. Kim, and J. G. Eden
Everitt Laboratory
University of Illinois
Urbana, IL 61801

Abstract

Rydberg $\longleftarrow$ Rydberg state transitions have been observed in Ne_2, Ar_2 and Kr_2 by laser excitation spectroscopy. Molecular Rydbergs converging to the dimer ion electronic ground state ($A^2\Sigma^+_{1/2u}$) and with principal quantum numbers as large as 26 for Ar_2 and 17 for Ne_2 have been recorded despite the predissociative nature of these states. Several vibrational bandheads have been rotationally resolved and vibrational autoionization (in competition with predissociation) has also been observed in Ne_2.

I. INTRODUCTION

Two of the more conspicuous difficulties that have hindered the spectroscopy of the rare gas dimers in the past are the large ionization potentials of the atomic species (and Ne and Ar, in particular) and the ~ 1 Å difference in the equilibrium internuclear separations (R_e) for the lowest molecular excited level ($ns^3\Sigma^+_u$ ($1_u, 0^-_u$); n = 3-5 for Ne_2, Ar_2 and Kr_2, respectively) and its ground state (van der Waals) counterpart. Laser excitation spectroscopy exploits the fact that the higher-lying Rydberg states of these molecules are predissociated[1] and that all of the rare gas dimer excited states are Rydberg. Consequently, the strongly bound A core Rydberg states are now conveniently examined near R_e and excitation spectra of Kr_2 have been reported previously.[1] Recent results that have been obtained by this experimental approach for Ne_2 and Ar_2 are briefly described here.

II. EXPERIMENTAL RESULTS AND DISCUSSION

In these experiments, the $ns^3\Sigma^+_u$ metastable species is produced in the afterglow of a pulsed corona discharge. Excitation spectra are obtained by monitoring the atomic fluorescence that follows the predissociation of high-lying Rydberg states that are pumped by a time-delayed dye laser pulse having a spectral linewidth of $\sim 0.04-0.2$ cm^{-1}. Figure 1 shows the spectrum that was recorded for Ar_2 and

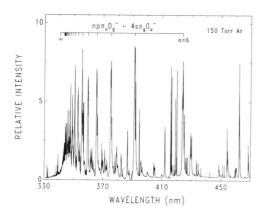

Fig. 1 Laser excitation spectrum for Ar_2 $4s^3\Sigma_u^+(1_u$,0_u^-) in the visible and near ultraviolet ($330 \lesssim \lambda \lesssim 470$ nm) in which one of the four distinct Rydberg series observed in these experiments is identified. These data were acquired with a laser linewidth $\Delta\tilde{v}$ of 0.2 cm^{-1}.

principal quantum numbers as high as 26 have been observed. The limit for this series has been determined to be 29373 ± 2 cm^{-1} and the quantum defect (μ) is a function of energy ($\mu = \mu_0 + \mu_1 E + \mu_2 E^2 + \ldots$) where the constant μ_0 and linear coefficient μ_1 are 1.6366 ± 0.0001 and 0.265 ± 0.096, respectively. Similar constants were measured for the corresponding $np\pi_u 0_g^- \leftarrow 4s\sigma_g 0_u^-$ (v',v" = 0,1; $5 \lesssim n \lesssim 10$) series. The remaining two series were identified with the aid of Edlén plots as $np\sigma_u \leftarrow 4s\sigma_g$ ($6 \lesssim n \lesssim 10$) and $nf\pi_u \leftarrow 4s\sigma_g$ ($4 \lesssim n \lesssim 11$). Only the n = 9 term was not observed in the latter series. Quantum defect calculations[2] of the Ar_2 $4s^3\Sigma_u^+$ ($1_u,0_u^-$) photoionization cross-section, based on the $np\pi_u$ Rydberg series parameters given earlier, yield a maximum value of $2.8 \cdot 10^{-18}$ cm^2 at $\lambda \sim 245$ nm. Furthermore several inter-Rydberg transitions of Ar_2 in the green ($\tilde{v} \sim 19750$ cm^{-1}) have been rotationally-resolved, yielding $R_e"$ ($4s\sigma_g(1_u,0_u^-)$) as 2.23 Å.

Similar experiments have been carried out for Ne_2 and 13 terms of the $np\pi_u {}^3\Pi_g \leftarrow 3s\sigma_g {}^3\Sigma_u^+$ series (n $\leq$ 17) have been observed for which the limit is 34399 ± 3 cm^{-1} and μ_0 and μ_1 are 0.8121 ± 0.0001 and 0.223 ± 0.098, respectively, for $^{20}Ne_2$. The other major series identified to date is $np\sigma \leftarrow 3s\sigma_g$ for which the limit was determined to be ~34420 cm^{-1}. Several Ne_2 vibrational bandheads have also been rotationally resolved and Fig. 2 illustrates the results for the (0,0) band of the Ne_2 $4p^3\Pi_g \leftarrow 3s^3\Sigma_u^+$ transition lying near 417 nm ($\tilde{v} \simeq 23970$ cm^{-1}). Analysis of the combination differences yields $B_{v"} = 0.564 \pm 0.004$ cm^{-1} and $B_{v'} = 0.574 \pm 0.006$ cm^{-1} for $^{20}Ne_2$. Consequently, $R_e(3s^3\Sigma_u^+)$ is measured to be 1.728 ± 0.005 Å which is within 4% of theoretical (<u>ab initio</u>) values.[3]

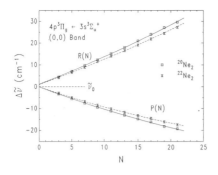

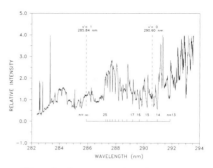

Fig. 2 R and P branch data for the (0,0) band of the $4p^3\Pi_g \leftarrow 3s^3\Sigma_u^+$ transition of Ne_2 ($\lambda \sim 417$ nm) for two stable Ne isotopes.

Fig. 3 Laser excitation spectrum of $^{20}Ne_2$ above the $A^2\Sigma_{\frac{1}{2}u}^+(v' = 0)$ ionization limit, clearly showing autoionizing resonances of $v' = 1$ Rydberg states superimposed onto the predissociation continuum.

Recently, vibrational autoionization has been observed in Ne_2. As depicted in Fig. 3, autoionizing resonances, associated with an $np\pi_u$ Rydberg series converging to $v' = 1$ of the Ne_2^+ $A^2\Sigma_{\frac{1}{2}u}^+$ ion ground state, appear as suppressions superimposed onto a predissociation continuum. The calculated limit for this series is in excellent agreement with the vibrational frequency for the Ne_2^+ ground state.

III. SUMMARY

Detailed information concerning the Rydberg states of Ne_2 and Ar_2 as well as the heavier dimers may be obtained near R_e by laser excitation spectroscopy. Results to date demonstrate that this technique permits one to measure quantum defects and ionization limits, for example, with reasonable precision as well as to investigate the onset of autoionization phenomena in the vicinity of the first ionization limit.

This work was supported by the Air Force Office of Scientific Research under grant 89-0038 and the National Science Foundation under grants ECS 88-07679 and 86-11474.

REFERENCES

1. M. N. Ediger and J. G. Eden, J. Chem. Phys. 85, 1757 (1986).

2. R. A. Sauerbrey, IEEE J. Quant. Electron. QE-23, 5 (1987).

3. For a more complete discussion, see D. J. Kane and J. G. Eden, Phys. Rev. A 39, 4906 (1989).

Detection of Photon Bursts from Single 200eV Mg Ions, Progress in Photon Burst Mass Spectrometry

W. M. Fairbank, Jr. and R. D. LaBelle
  Department of Physics, Colorado State University
  Fort Collins, CO 80523, USA
Richard. A. Keller and E. Philip Chamberlin
  Los Alamos National Laboratory,
  Los Alamos, NM 87545, USA

Modern atom counting methods, based on advances in laser and accelerator technology, provide a valuable complement to traditional decay counting methods for radioisotope dating and tracer work. Tandem Accelerator Mass Spectrometry (TAMS) has already had a large impact on ^{14}C dating and is beginning to provide new opportunities with [10]Be and several other isotopes. We report here on progress in the development of a laser-based technique, Photon Burst Mass Spectrometry,[1,2] which is potentially capable of analyzing many of the elements which are forbidden in TAMS because they do not form negative ions. We are especially interested in the noble gases, which have a variety of potential scientific and environmental applications.[3]

Photon Burst Mass Spectrometry (PBMS) is based on the coupling of a highly selective laser single atom or ion counter, the photon burst detector, with a mass spectrometer, which serves as an on-line prefilter to reduce the abundant isotopes by 10^6 (Fig. 1).In two previous papers, calculations have been presented for several sample isotopes, which indicate that isotope ratios as low as 10^{-15} are measurable in 30 hours or less by the PBMS method.[2,4] In the last year or so, our primary effort has been in demonstrating experimentally the parameters upon which these calculations were based and in designing and testing the important components of the proposed PBMS instrument. Progress on these goals is the major focus of this paper.

The photon burst detector is being developed and tested with a simple Colutron ion source, equipped with an ExB (Wien) filter. It provides mass-selected Mg^+ ions at 200 eV at a low

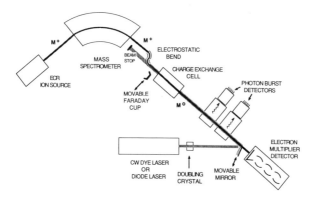

Fig. 1. Photon Burst Mass Spectrometer.

current on the order of picoamps. The ions pass collinearly
through a photon burst detector with a focussed laser beam
tuned to the $^2S_{1/2}- {}^2P_{3/2}$ Mg$^+$ resonance at 279.5 nm. Focussing
was required because the laser power, 40 μW, was insufficient
to saturate the full diameter of the ion beam.

Results obtained with the mass filter off are shown in Fig.
2. In the total fluorescence signal, peaks are seen from all
three isotopes, but a large stray light background (2.5×10^{-9} of
the incident light) limits the signal-to-background ratio to
about 2. Much greater contrast is seen in the burst signal,
with nearly zero background. The steepness of the sides of the
peaks indicates that the contributions from neighboring
isotopes at line center are well below the stray light
background. With measured experimental parameters and an
estimated 7% collection/quantum efficiency, predicted maximum
values for the two ^{25}Mg$^+$ and ^{26}Mg$^+$ peaks from left to right are
165, 1900 and 6600 bursts/sec, respectively. The predicted
stray light background is 2.5 bursts/sec, including measured
effects of double pulsing in the photomultiplier. All these
predictions agree well (within a factor of two or so) with the
observations. The derived average burst size for resonant ^{25}Mg+
and ^{26}Mg$^+$ ions is, respectively 3.7 and 5.2 counts.

This is the first time that photon bursts from single ions,
other than in a trap, have been observed. It is also the first
time that superthermal particles have been detected by the
photon burst method. It is worth noting that the high optical
efficiency and burst sizes demonstrated in these experiments,
as well as the good agreement of the experimental data with
theoretical predictions, represent important milestones in the
development and testing of the PBMS method. The relatively
high stray light level in these experiments, due partly to
light from the ion source, prevented a better demonstration of
the isotopic selectivity of the photon burst method.
Nevertheless, an abundance sensitivity of at least 10^3 is shown
in Fig. 2. The calculated average background from neighboring
isotopes is 10^4 lower than the stray light level, and
contributed negligibly to background bursts.

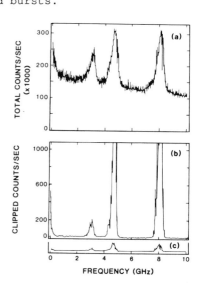

Fig. 2. Fluorescence and
photon burst spectra of Mg
ions obtained with the Wien
filter off: (a) total count
rate, and (b) rate of bursts
with ≥5 counts. Curve (c)
represents the calculated
burst rate if the photon
arrival rate in (a) were
uncorrelated, except for a
0.3% afterpulsing correction.
From the left the peaks are:
^{24}Mg$^+$, ^{25}Mg$^+$ (F"=2→F'=1,2,3),
^{25}Mg$^+$ (F"=3→F'=2,3,4) and
^{26}Mg$^+$.

Recently we have installed an electrostatic bend between the ion source and the photon burst detector, as illustrated in the PBMS diagram (Fig. 1). This effectively eliminates background photons from ion source. A stray light level of 10^{-10} for the whole system has been achieved at 280 nm. In more careful work with the photon burst detector alone, 10^{-11} was measured at 280 nm and 10^{-12} with a clean TEM_{00} beam at 633 nm. It is anticipated that proper spatial filtering of the beam with a pinhole or single-mode optical fiber will allow the latter level to be reached at all wavelengths. Thus we have demonstrated essentially all of the required properties of the proposed photon burst detector.

The planned mass spectrometer for our full PBMS demonstration is a 126° magnetic isotope separator at Los Alamos National Laboratory.[5] A microwave source for this instrument[6] has been installed and tested. It produces a 5 μA current in a 1-2 mm spot at the exit of the spectrometer with 3% overall ionization and transmission efficiency and reasonable divergence. An abundance sensitivity of 3×10^3 at the krypton masses has been measured. Further improvements to 10% overall efficiency and 10^6 abundance sensitivity are expected in the near future. The decelerator, required at the exit of the mass spectrometer has been designed by computer and is under construction.

In conclusion, many of the important features of the Photon Burst Mass Spectrometry method, including selective counting of single fast ions, have now been confirmed experimentally. Further studies are still required on the most important remaining uncertainty in the method, the magnitude of longitudinal velocity changes during charge exchange to metastable atomic states. However, this is not expected to be a severe problem because most charge exchange collisions occur at large impact parameters.

References

1. R. A. Keller, D. S. Bomse and D. A. Cremers, Laser Focus (October 1981) p.75.

2. W. M. Fairbank, Jr., Nucl. Instr. and Methods B29, 407 (1987).

3. B. E. Lehmann and H. H. Loosli, "Use of noble gas radioisotopes for environmental research", in Resonance Ionization Spectroscopy 1984, G. S. Hurst and M. G. Payne, eds. (Institute of Physics, Bristol, 1984) p.219.

4. W. M. Fairbank, Jr. et al., "Prospects for large dynamic range isotope analysis using photon burst mass spectrometry", in Resonance Ionization Spectroscopy 1988, T. Lucatorto and J. E. Parks, eds. (Institute of Physics, 1989) p. 53.

5. E. P. Chamberlin et al., Nucl. Instr. and Methods B26, 21 (1987).

6. E. P. Chamberlin et al., Nucl. Instr. and Methods B26, 227 (1987).

Coherent Multiple Pulse Spectroscopy of the Hydrogen 1S-2S Transition

M.A. Persaud, J.M. Tolchard[*] and A.I. Ferguson
Department of Physics and Applied Physics, University of Strathclyde,
Glasgow, G4 0NG, Scotland.

Coherent multiple pulse spectroscopy using a train of ultrashort light pulses from a tunable laser has many desirable features for the frequency calibration of Doppler-free two-photon resonances (1). One of the most interesting transitions which can be studied using this technique is the 1S-2S transition in atomic hydrogen. This experiment requires tunable radiation in the region of 243nm. Several methods have been developed in recent years for the generation of light at this wavelength using intracavity frequency doubling or sum frequency mixing of single frequency lasers. We have developed a new method for the generation of a picosecond pulse train at 243nm. Average ultraviolet powers of greater than 50mW have been generated by frequency doubling in an external enhancement ring cavity corresponding to an energy conversion efficiency of about 25%. Using this system it will be possible to generate coherent multiple pulse spectra of the hydrogen 1S-2S transition with excellent signal to noise ratios. The signal strength obtained will be the same as that using a single frequency laser with the same average power. The expected linewidth is comparable to that obtained using single frequency lasers.

The main power of the technique will be in providing a frequency calibration. One method of implementing this will be to lock the comb of modes of the mode-locked laser to the two-photon resonance in hydrogen. A second single frequency laser will be locked either to a standard calibration line or to a deuterium resonance. This laser will be heterodyned with the modes of the mode-locked laser to provide an absolute frequency calibration modulo the precisely determined mode spacing.

The system we have developed is one of the most intense ultraviolet sources of picosecond pulses available. There are many applications for such a source outside the field of high resolution spectroscopy. The photon energy is in the region where chemical bonds can easily be broken and hence the system may find application in photo-chemistry and photo-biology.

A schematic diagram of the laser system is shown in figure 1. A krypton laser operating on all lines in the violet is mode-locked using an acousto-optic rhomb mode-locker. The mode-locker is driven at three times the cavity mode spacing to produce three pulses in the cavity at any given time. This virtually triples the average power available to about 1.5W compared to mode-locking at the fundamental frequency. The pulse duration is about 100psec and the repetition

---

[*]Present Address: Shell Thornton Research Centre, Chester, England.

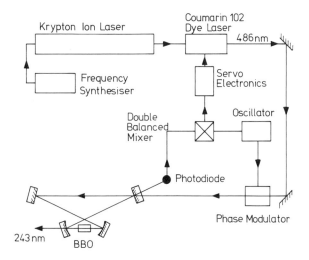

Figure 1: Experimental system for high average power UV generation of tunable picosecond pulses at 243nm.

rate is 235MHz. The dye laser is a modified commercial model (Coherent 699-21) and is matched in length to the krypton laser repetition rate. The dye laser is tuned using a two-plate birefringent filter and produces pulses of about 10psec duration with an average power at 486nm of up to 500mW. This pulse train is injected into an enhancement ring cavity where the mode spacing of the cavity is matched to the dye laser repetition rate. A crystal of beta barium borate is placed at an intracavity waist and angle tuned for harmonic generation. For most of the work the crystal was close to normal incidence and uncoated which represented an intracavity loss of about 12%. Preliminary measurements have been made using a Brewster angled crystal. The laser carrier frequency is locked to the enhancement cavity using a frequency modulation sideband technique (3). The error signal is fed to the commercial servo electronics supplied with the dye laser and drives an intracavity tipping Brewster plate for slow frequency corrections and to a piezo mounted mirror for fast corrections. The reflectivity of the input mirror has been chosen to 'impedance match' the cavity (3,4).

The calculated conversion efficiency from 486nm to 243nm as a function of average pump power for a range of intracavity crystal transmissions is shown in figure 2. The conversion efficiency is strongly dependent on the intracavity loss. The calculation has been carried out for a crystal of 7mm length and the reduction in efficiency due to the limited acceptance bandwidth has not been taken into account. A more careful analysis shows that the crystal length which maximises the conversion is less than 1mm. The majority of the measurements have been taken using a normal incidence beta barium

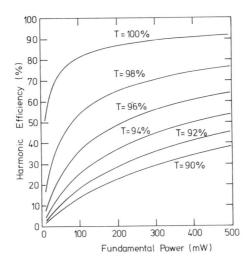

Figure 2: Calculated second harmonic conversion efficiency as a function of fundamental power for a range of crystal transmissions in an impedance matched UV enhancement ring cavity.

borate crystal with Fresnel losses in the region of 12%. In this case we have obtained powers of about 30mW at 243nm for a pump power of 300mW. In more recent experiments preliminary results using a 5mm Brewster angled crystal has produced powers of greater than 50mW with a 200mW pump. This is a substantial amount of ultraviolet radiation and more than adequate for the observation of the hydrogen 1S - 2S transition. We anticipate that with minor improvements to the system, average powers of greater than 100mW will be available at 243nm using this method.

Acknowledgments

This work has been supported by the Science and Engineering Research Council.

References

1. J.N. Eckstein, A.I. Ferguson and T.W. Hänsch, Phys. Rev. Lett. 40, 847 (1978).

2. R.W.P. Drever, J.L. Hall, F.V. Kowalski, J. Hough, G.M. Ford, A.J. Munley and H. Ward, Appl. Phys. B31, 97 (1983).

3. A.Ashkin, G.D. Boyd and J.M. Dziedzic, IEEE J. Quant. Electron. QE-2, 109 (1966).

4. W.J. Kozlovsky, C.D. Nabors and R.L. Byer, IEEE J. Quant. Electron, 24, 913 (1988).

CW Laser Spectroscopy of the Triplet Manifold of Magnesium

J. A. Gelbwachs and Y. C. Chan
Chemistry and Physics Laboratory
The Aerospace Corporation, P. O. Box 92957, Los Angeles, CA 90009

Introduction

There has been considerable interest in the spectroscopy of the triplet manifold of magnesium. For instance, energy-pooling reactions of the metastable Mg $(3p^3P)$ atoms are considered as an attractive means for generating high yields of excited-state Mg population for developing new short wavelength lasers.[1] In addition, metastable Mg is used as the medium in the atomic filter that operates at a strong Fraunhofer line of the solar spectrum.[2] Metastable Mg atoms are frequently prepared by optical excitation from their ground state with powerful pulsed lasers. High intensity excitation is required to overcome the extremely small transitional probability ($f = 4 \times 10^{-6}$) of the intercombination line. Intrinsic characteristics of the pulsed lasers, such as low resolution and short interaction time, are undesirable in certain experiments. A cw single-mode dye laser provides an attractive alternative. It offers advantages such as ultra-high resolution ~ 0.001 cm^{-1}; high reproducibility; linear wavelength scans; well-characterized spatial modes; long integration times; and negligible nonlinear effects. To our knowledge, there has not been any study on an metastable atomic system with cw dye lasers.

We have been conducting a series of spectroscopic studies on the Mg triplet manifold system by using cw single-mode dye lasers. We now summarize our results to date and discuss the aim of our current experiments.

Discussion

A. Lineshifts and line broadening due to collisions with noble gases

A two-laser double-resonance technique was employed to measure, for the first time, the broadening and shift of the Mg I intercombination line $(3p^3P_1 - 3s^1S_0)$ and the first triplet $(4s^3S_1 - 3p^3P_2)$ transition due to the presence of helium, neon, argon, krypton and xenon. Details of the study can be found in Ref. 3. Magnesium vapor and buffer gas were contained in a stainless steel cell heated to 400°C. Two cw single-frequency lasers were used. One laser was tuned to the 457 nm intercombination line and the other was tuned to 518 nm triplet resonance line. Magnesium emission was collected by a lens and focused onto a photomultiplier tube. Lineshapes of the intercombination line were recorded by monitoring the green emission from the $4s^3S_1$ level to the metastable level as the blue laser was electronically scanned over the intercombination line. The resolution of the method is limited by the wavelength stability of the ring dye lasers, and it is nominally 3 MHz. The collisionally-broadened linewidth of the triplet resonance line was obtained in a similar fashion.

The interaction potentials between the two pairs of levels and the noble gases are determined by a comparison of the data with semiclassical impact theory.[4] Our data indicate that van der Waals dispersive forces are responsible for the long range interaction of the heavier noble gases, i.e. argon, krypton and xenon with these magnesium levels. Strong repulsive forces dominate the interactions with He and Ne.

B. Fine-structure splitting

Spin-orbit interactions are examined by studying the energy splittings of the $4p^3P_J$ level. Precise measurement of the three allowed $4s^3S_1$ - $4p^3P_J$ (J = 0, 1, 2) transitions gives the energy separation between the fine structure components. Similar to part (A), two cw ring dye lasers operating at 457 nm and 518 nm were used to excite Mg atoms into the $4s^3S_1$ level. The wavelength of a cw single mode F-center laser operating in the 1.5 μm region was scanned over the $4p^3P_J$ triplet. Coincidences of the IR laser wavelength with the $4s^3S_1$ - $4p^3P_J$ transitions were monitored by detecting 384 nm emission from the $3d^3D$ - $3p^3P$ transition. When the IR laser wavelength matches the energy separations of the $4s^3S_1$ - $4p^3P_J$ transitions, population in the $4s^3S_1$ level is pumped into the $4p^3P_J$ states. The energy separation between the $4p^3P_J$ level and the neighboring $3d^3D$ level is only 109 cm^{-1}, so that the two levels are effectively mixed by collisions with the buffer gas at room temperature and above. Atoms in the $3d^3D$ level decay back to the $3p^3P$ level by emitting 384 nm photons. Figure 1 shows a spectrum of the $4s^3S_1$ - $4p^3P_J$ transitions.

C. Diffusion coefficients of metastable species

A novel technique is employed to extract diffusion coefficients of metastable Mg atoms among buffer gases, such as He and Ne. The method involves monitoring of the laser intensity required to saturate the intercombination Mg transition in the presence of buffer gas at various pressures. Saturation phenomenon can readily be observed by monitoring emission from the excited state (See Figure 2). Application of the technique can be extended to other metastable atomic systems. The method is briefly outlined below.

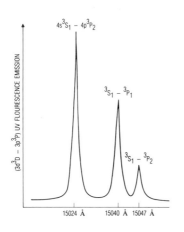

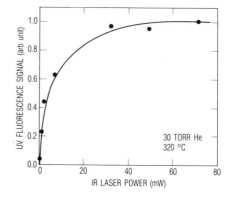

Fig. 1. Fluorescence spectrum from the $3d^3D$ level as an IR laser scans across the $4s^3S_1$ - $4p^3P_J$ level.

Fig. 2. Laser saturation measurement of the $4s^3S_1$ - $4p^3P_2$ transition.

By solving the rate equations of a simple two level atomic system, one finds that the ratio of the excited state population (N_2) to the total population (N_T) is related to the pumping rate (W) by the expression:

$$N_2/N_T = W/(2W + K_T).$$

where K_T is the decay rate of the excited state, and W is proportional to the laser intensity. Assuming that quenching losses are negligible, such that radiative decay and diffusional losses are the two dominant decay channels, K_T can be expressed as:

$$K_T = 1/\tau + k/P$$

where τ is the lifetime of the excited state, k is the diffusion rate normalized for pressure, and P is the buffer gas pressure.

For long-lived species, the laser beam can be focussed to a waist size for which the diffusion rate of metastable atoms out of the beam exceeds the lifetime. Under this condition

$$W_s = 2k/P$$

Thus the diffusion rate can be determined from the slope of the laser saturation intensity vs buffer gas pressure plot.

D. Collisional mixing of $4p^3P$ - $3d^3D$

The collisional mixing rate of the $4p^3P_J$ - $3d^3D_J$ levels by noble gases is a critical parameter for the determination of internal conversion efficiency of the Mg atomic filter.[2] It also provide insight into the nature of atomic interactions. We are conducting experiments to elucidate the collisional mixing rate by monitoring laser saturation intensity of the $4s^3S_1$ -$4p^3P_J$ atomic transition as a function of buffer gas pressure. The measurement scheme can best be described with the aid of a simple three-level atomic system. Let us label the three atomic levels: 1, 2 and 3. Further, we assume that level 2 and 3 are mixed by buffer gas collisions, and 1 and 2 are radiatively connected and pumped at a rate W. Radiative decay of levels 2 and 3 to level 1 and the collisional transfer between levels 2 and 3 are assumed to be the only significant decay channels. Solving the rate equations of a three level system we find that the pumping rate of the laser required to saturate the 1 - 2 transition is:

$$W_s = \alpha\, k_3/(2 + \alpha) \quad \text{and} \quad \alpha = a\, K_m\, P/(K_m\, P + k_3)$$

where a is the Boltzmann factor, K_m is the mixing rate of levels 2 and 3, and k_3 is the radiative decay rate of level 3 that is assumed known. At low pressure, the slope of the W_s vs pressure plot is K_m/k_3. Thus by measuring saturation intensity at various buffer gas pressures, the mixing rate can be obtained.

The triplet Mg system is slightly more complicated than the simple three-level model discussed above. Hence solving the rate equations of the real system is more involved. Nevertheless, the mixing coefficient of the $4p^3P_J$ - $3d^3D_J$ levels can be obtained in a similar manner.

E. Quenching of the excited triplet levels

During the study of the collisional-mixing efficiency of the $4p^3P$ and the $3d^3D$ levels, we observed non-radiative quenching of the excited triplet level populations. As illustrated in Figure 3, the radiative quantum efficiency for the $3d^3D$ decay decreases as the He buffer gas pressure is increased above 100 Torr. While quenching due to collisions with helium atoms is negligible, the observation indicates the presence of unidentified decay channels in $3d^3D$ population removal. Possible depleting channels are collisional energy pooling and collisional enhanced ionization among the excited Mg atoms. We are currently investigating such collision-enhanced quenching processes.

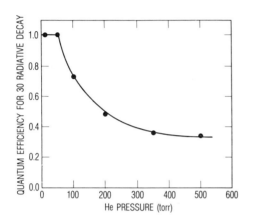

Fig. 3. Radiative quantum efficiency for $3d^3D$ decay as a function of helium buffer gas pressure.

Summary

We have employed cw lasers to conduct spectroscopic studies on the triplet manifold of Mg. Broadening and line-shifting cross sections were obtained for the intercombination line and first triplet transition for the first time. High resolution spectra of the $4P^3P_J$ were acquired. Saturation curves are being used to measure both diffusion coefficients of the metastable species and the $4p^3P$ - $3d^3D$ mixing rate. Further, non-radiative decay of the $3d^3D$ level in the presence of helium has been recorded.

Acknowledgement

This work was supported by the Aerospace Sponsored Research Program.

References

1. W. J. Stevens and M. Krauss, J. Chem. Phys. **67**, 1977 (1977).

2. J. A. Gelbwachs, IEEE J. Quantum Electron. **24**, 1266 (1988).

3. J. A. Gelbwachs, Phys. Rev. A **39**, 3343 (1989).

4. E. L. Lewis, Phys. Reports **58**, 1 (1980).

Two-Photon 205-nm Excitation of Atomic Hydrogen in Flames: Stimulated Emission, Time-Resolved Fluorescence, and Doppler-Free Excitation*

J. E. M. Goldsmith and L. A. Rahn
   Combustion Research Facility, Sandia National Laboratories
   Livermore, California 94551, USA
R. J. M. Anderson and L. R. Williams
   High Temperature Interfaces Division, Sandia National Laboratories
   Livermore, California 94551, USA

1. Introduction

Multiphoton excitation techniques are powerful methods for detecting a variety of species in harsh environments, such as flames, where single-photon vacuum-ultraviolet techniques cannot be used (1). Quantitative application of these techniques requires an understanding of the dynamics of the processes that play roles in the excitation and detetion steps. In this paper, we describe three recent studies that shed new light on these processes in atomic hydrogen. All three studies were conducted using 205-nm excitation of the n=1→n=3 transition, followed by detection of the subsequent 656-nm Balmer-α n=3→n=2 fluorescence decay (2,3). First, we discuss observations of stimulated emission occurring on the same n=3→n=2 transition used for fluorescence detection. Next, we report quenching rates measured in flames using direct time-resolved fluorescence techniques. Finally, we describe the first Doppler-free measurements of atomic hydrogen in flames.

Three laser systems were used to meet the various needs of these studies. In all three systems, 615-nm dye-laser pulses were up-converted to 205 nm using frequency-doubling in KD*P followed by frequency-mixing in beta-barium-borate. We performed most of the stimulated emission studies using a conventional Q-switched Nd:YAG-laser-pumped system (5-ns pulses). Because the time-resolved fluorescence measurements required shorter excitation pulses, we used a dye laser pumped by an actively-modelocked Nd:YAG laser (~50 ps pulses). Narrowband radiation was the main requirement for the Doppler-free study, leading to the use of a pulse-amplified cw dye laser system (~40 MHz bandwidth).

2. Stimulated Emission

Two-photon-excited stimulated emission (SE) has been studied in a variety of atomic and molecular systems for over a decade; this process is also called amplified spontaneous emission. The effect that SE may have on diagnostic applications of multiphoton-excited fluorescence (FL) detection, and the potential use of SE detection as a useful technique in its own right, have received little attention until a recent report describing SE from atomic oxygen in flames (4). We have made similar two-photon excitation measurements to study SE and FL from atomic hydrogen in flames.

One of the most striking characteristics in our atomic hydrogen study was easy observation of the red SE by eye on an ordinary white card. The SE had the same spatial behavior as the 205-nm laser beam, propagating in both directions as a well-collimated beam. This behavior readily identifies the emission as a separate process from fluorescence, which is emitted (with some angular dependence) in all directions.

Intensity dependences of the FL and SE signals provide an excellent means of confirming their different dependences on the density of excited atoms. Fig. 1 displays intensity dependences of FL and SE signals measured simultaneously in a 72-Torr flame. In the lower-energy region of the plot, the data representing the FL measurements lie along the dashed line drawn with a slope of 2, consistent with the I^2 dependence of the two-photon process. The data representing the SE measurements have a stronger intensity dependence in the lower-energy region, becoming very similar to the FL measurement in the higher-energy region. The low-intensity behavior illustrates the exponential-gain regime of SE,

---

* This work was supported by the U.S. Department of Energy, Office of Basic Energy Sciences, Division of Chemical Sciences and Division of Materials Sciences.

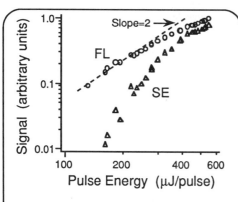

Fig. 1. Intensity dependences of FL and SE signals

followed by saturation and a linear-gain regime (5), where the SE and FL signals have similar intensity dependences.

We also observed SE in studies using the other laser systems mentioned in the introduction. Using the 50-ps pulses, we observed pulses with a scope-limited FWHM of 350 ps. We thus can only put an upper limit of the duration of the SE pulse, but it is much shorter than the duration of the FL pulse (several ns observed in the low-pressure flames, as described in the next section). Using the smooth 9-ns pulses from the single-mode laser system, we observed spiking, as seen in other SE studies.

We believe that the presence of SE in these and other studies has two important consequences for diagnostic applications of two-photon excitation. First, SE competes with other deexcitation pathways in the atomic system, and can affect the quantum yields of the fluorescence and photoionization channels. SE may therefore complicate quantitative interpretation of measurements made using those detection methods. Second, the strength and the coherent nature of SE beam are very attractive features in a diagnostic technique. There are difficulties in accurately modeling this process, but we have had some success measuring flame profiles with SE detection, and we are continuing to investigate this application.

2. Time-Resolved Fluorescence

Quantitative use of laser-induced fluorescence techniques in flames requires knowledge of the variation of collisional quenching rate through the flame in order to convert a measured LIF profile into a relative concentration profile. Direct fluorescence decay measurements of atomic hydrogen quenching rates in flames using conventional (~5-ns pulse length) dye laser systems are generally not practical because of the short natural lifetimes of the excited states and the rapid quenching rates characteristic of combustion environments. In this section, we describe measurements of quenching rates obtained by direct time-resolved fluorescence decay measurements of atomic hydrogen in flames using ~50 ps laser pulses and a fast photomultiplier system.

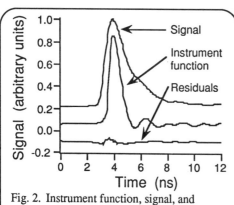

Fig. 2. Instrument function, signal, and residuals from least-squares fit in time-resolved fluorescence study

The response characteristics of our detection system were carefully taken into account using a deconvolution procedure. We recorded an instrument function representing the response of the system by attenuating the laser beam and monitoring it directly with the dtection system; the result is shown in Fig. 2. A fluorescence decay curve measured in a 20-Torr flame is also shown in Fig. 2. We convolved the instrument function with an exponential decay and iteratively compared it to the measured decay curve under the control of a least-squares fitting routine. The fit yielded a 1/e decay rate of 788 MHz; the residuals from the fit are also shown in Fig. 2.

We have measured the variations

in quenching rates with position in a variety of flames. These measurements can be used not only for converting LIF profiles to relative atomic hydrogen concentration profiles, but also for determining absolute concentrations by comparing fluorescence signals to a flame with a known concentration. We have also recorded measurements in flames tailored such that one (or perhaps a few) collision partners dominate the quenching process to determine species-dependent quenching cross sections.

3. Doppler-Free Excitation

Doppler-free 243-nm two-photon excitation of the hydrogen 1S-2S transition has been studied in detail for over a decade. The detection methods used in those studies are not compatible with flame measurements, however. Using 205-nm two-photon excitation with the conventional counter-propagating-beam geometry, we have recorded Doppler-free spectra of atomic hydrogen in a variety of low-pressure flames.

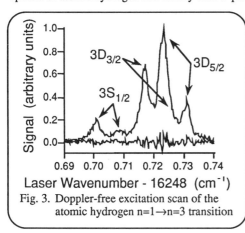

Laser Wavenumber - 16248 (cm^{-1})

Fig. 3. Doppler-free excitation scan of the atomic hydrogen n=1→n=3 transition

The top curve in Fig. 3 shows a spectrum recorded in a 15-Torr acetylene-oxygen-argon flame plotted as a function of the dye laser wavenumber. Each fine-structure transition appears as a doublet because of the hyperfine splitting of the ground $1S_{1/2}$ state (we cannot resolve the much smaller hyperfine structure of the n=3 states). The lower curve shows the residuals from a least-squares fit of six Lorentzian profiles to the spectrum with a single width used to fit all six transitions (the spacings between the transitions were fixed at the known values, and the relative strengths of the two resolved hyperfine structure transitions for each fine-structure transition were fixed at 3:1). Linewidth values from these measurements provide values for the total collisional broadening rates ($1/T_2$) under flame conditions. The combination of these rates with quenching rates ($1/T_1$) from above can be used to unravel the details of the collisional processes in flames.

5. Conclusion

This paper has touched on a few aspects in each of three studies using two-photon 205-nm excitation of atomic hydrogen in flames. The first, two-photon-excited stimulated emission, may have the adverse effect of modifying the quantum yields using fluorescence and photoionization detection, but may also have the advantage of providing a new diagnostic technique. The second, time-resolved fluorescence, provides direct measurements of collisional quenching rates in flames. Finally, studies using Doppler-free excitation provide additional information on the collisional dynamics of the excited state.

References

1. J. E. M. Goldsmith, in *Process Diagnostics: Materials, Combustion, Fusion*, A. K. Hays, A. C. Eckbreth, and G. A. Campbell, eds., Materials Research Society Symposium Proceedings Vol. 117 (Materials Research Society, Pittsburgh, Pa., 1988), pp. 193-201.
2. J. Bokor, R. R. Freeman, J. C. White, and R. H. Storz, Phys. Rev. A **24**, 612 (1981).
3. R. P. Lucht, J. T. Salmon, G. B. King, D. W. Sweeney, and N. M. Laurendeau, Opt. Lett. **8**, 365 (1983).
4. M. Aldén, U. Westblom, and J. E. M. Goldsmith, Opt. Lett. **14**, 305 (1989).
5. L. Allen and G. I. Peters, J. Phys. A **4**, 564 (1971).

Laser Polarization Spectroscopy of Atomic Oxygen

M. Inguscio, L. Gianfrani, A. Sasso, and G.M. Tino
Dipartimento di Scienze Fisiche, Università di Napoli,
Mostra d'Oltremare, 80125 Napoli, Italy

We report on laser polarization spectroscopy of oxygen atoms produced and excited in an O_2-rare gas radiofrequency discharge. Several ^{16}O, ^{17}O, ^{18}O transitions are investigated, possibly evidencing a contribution of the nuclear size change to the isotope effect. Homogeneous narrow linewidths are recorded also for anomalously broadened transitions.

Recently we made atomic oxygen accessible to cw laser sub-Doppler investigations by means of intermodulated optogalvanic spectroscopy [1]. Excited atoms were produced from O_2 traces in a Ne or Ar substained radiofrequency discharge. Fine structure and $^{16}O-^{18}O$ isotope shifts could be measured for the first time in the optical region [1,2,3]. However, collisions affected the sub-Doppler lineshapes with the presence of a strong pedestal caused by velocity-changing collisions. In order to improve sensitivity and resolution and to extend investigations to enriched samples including all the three stable isotopes of oxygen we perform in the present work laser polarization spectroscopy [4] of the same discharge samples. Respect to previous measurements we exploited high sensitivity offered by polarization technique to reduce buffer gas pressure to lower, though still non neglible, values. The measurements reported here have been performed using 1 Torr of argon as buffer gas and 0.05 Torr from a 50% $^{17}O-^{18}O$ enriched sample. The improvement comes not only from the reduction in the homogeneous width but, more important, from the complete elimination of the collisional Doppler background. In Fig.1 we show the experimental results obtained for the 615.8 nm ($3p^5P_3-4d^5D$) transition.

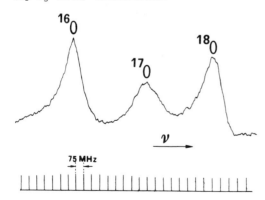

Fig.1. Polarization spectroscopy of the $^5P_3-^5D_4$ oxygen transition using a 50 % $^{17}O-^{18}O$ enriched sample. Atomic oxygen was produced in a radiofrequency excited discharge in presence of .05 Torr of O_2 and 1 Torr of argon.

Table 1

transition	wavelength (nm)	^{16}O-^{18}O IS (MHz)	^{16}O-^{17}O IS (MHz)	^{16}O-^{18}O IS, Ref. [3] (MHz)
$3p^5P_3$-$4d^5D_4$	615.8	1320 (5)	686 (10)	1310 (40)
$3p^5P_3$-$5s^5S_2$	645.7	1163 (9)	646 (10)	1160 (20)
$3p^3P_2$-$6s^3S_1$	604.6	1278 (10)	723 (10)	1300 (40)

An argon pumped, actively stabilized ring dye laser (Coherent mod.699/21) was used with rhodamin 6G or DCM as dyes. Measurements were performed with about 100 mW for the circularly polarized laser pump beam and about 10 mW for the linearly polarized probe beam. The frequency markers were provided by a 1 m long confocal Fabry-Perot interferometer.

Both the levels involved in the transition are radiative with a lifetime of the order of hundred nanoseconds, leading to a natural broadening of tens of MHz. The observed width ($\approx$150 MHz) is then essentially caused by collisional and saturation broadening. These results bring to an improvement of the accuracy in the ^{16}O-^{18}O isotope shift measurements previously reported [3]; in addition, for the first time it is possible to measure ^{17}O isotope shift. It must be noticed that for each isotope only one of the three fine structure components, namely the $^5P_3$-$^5D_4$, is observed. This is due to different relative intensities of the lines but also to the fact that, when using a circularly polarized pump beam, transitions corresponding to $\Delta J=0$ have a smaller cross-section respect to transitions with $\Delta J=\pm1$. This induces a simplification in the spectra avoiding overlapping of fine structure components of different isotopes.

The larger observed linewidth of ^{17}O peak can be ascribed to the unresolved hyperfine structure, nuclear spin being $I=5/2$.

Results obtained for the isotope shift in this transition and in the others we have investigated are reported in Table 1. Analysis of these data, deducing normal and specific mass contributions, are possibly evidencing a volume effect in the isotope shift, which is not completely surprising considering that ^{16}O nucleus is "doubly magic", with a structure different from ^{17}O and ^{18}O ones. Nevertheless, further measurements on other transitions are necessary to confirm this observation.

Another interesting question arises for the mechanism of the O_2-Ar collisions, leading to anomalously broadened oxygen transitions [3] as early evidenced in the O laser emission [5]. For instance, the 604.6 nm ($3p^3P_{1,2,0}$-$6s^3S_1$) transition displays a width of 6 GHz, which would imply a temperature of about 5000°K if interpreted as a maxwellian inhomogeneous broadening process. Nevertheless, another interpretation assuming a very fast collisional decay of O levels was suggested. In this case, however, the line broadening would be of homogeneous type.

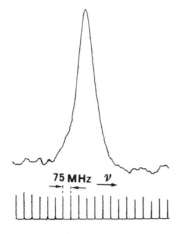

Fig.2. Polarization signal for the $3p^3P_2-6s^3S_1$ oxygen transition at 604.6 nm. Triplet transitions show an anomalous Doppler broadening in presence of argon buffer gas($\approx$ 6 GHz for this transition).

75 MHz ν

On the contrary, the clear observation of the narrow homogeneous linewidth, recorded in Fig.2 by means of polarization spectroscopy, confirms that the origin of the effect is purely kinetic; this observation provides an additional information for a complete modelling of O_2-rare gas discharges [6].

In conclusion, we believe that the powerful means of laser spectroscopy we are applying to oxygen will allow a deeper understanding of the structure of this atom which has a relevant position both in atomic and nuclear physics. Furthermore, the investgation could produce a deeper insight in the complex collisional mechanisms involved in the production of the atoms.

References

1. M. Inguscio, P. Minutolo, A. Sasso and G.M. Tino, Phys. Rev. A 37, 4056 (1988).

2. A. Sasso, P. Minutolo, M. Schisano, G.M. Tino, and M.Inguscio, J. Opt. Soc. Am. B 5, 2417 (1988).

3. K. Ernst, P. Minutolo, A. Sasso, G.M. Tino, and M. Inguscio, Opt. Lett. 14, 11 (1989).

4. C. Wieman and T.W. Hansch, Phys. Rev. Lett. 36, 1170 (1976).

5. M.S. Feld, B.J. Feldman, and A. Javan, Phys. Rev. A 7, 257 (1973).

6. A. Sasso, M. Inguscio, G.M. Tino, and L.R. Zink, in "Non-Equilibrium Processes in Partially Ionized Gases", M.Capitelli and J.H. Bardsley Eds., NATO ASI Series, Plenum, New York (1989).

A NEW APPROACH TO THE KINETIC ENERGY DETERMINATION OF PHOTOFRAGMENTS: Towards Pair-Correlation in Dissociation Dynamics

Reiner Lindner and Klaus Müller-Dethlefs

Institut für Physikalische und Theoretische Chemie der
Technischen Universität München
Lichtenbergstrasse 4
D-8046 Garching, West Germany

A problem of reaction dynamics lies in the determination of the quantum states of correlated products originating from a parent molecule which is prepared in a definite predissociative quantum state. It is well known that this problem can be solved by the measurement of the kinetic energy of one of the two fragments for each of its quantum states. Energy and momentum conservation can then be used to determine the internal energy of the correlated fragment, and with spectroscopic assignments, the state(s) of the correlated fragment. Such information would lead to a complete understanding of the reaction dynamics and be extremely valuable to test theoretical models.

For a resolution down to rotational states, translational energies have to be measured down to the wavenumber range. The most commonly used Doppler-photofragment spectroscopic techniques generally do not offer such resolution potential.
In excellent earlier work by Welge et al.[1] a time-of-flight method was used to measure the kinetic energy of the photofragments. In this experiment H-atoms originating from the process $NH_3(\tilde{A})$ --> $NH_2 + H$ were detected by 1+1 ionization via the 1s --> 2p transition. The resolution of approximately 40 meV in these measurements was sufficient to show resolved structures in the TOF spectra ascribed to rovibrational states of NH_2. However, the conventional TOF methods are inappropriate for the detection of the slowest particles and therefore lead to severe difficulties for lower fragmentation energies.
To this end, we have applied to photofragments the "delayed pulsed field technique" already employed successfully by us for high-resolution (near-) zero kinetic energy photoelectron spectroscopy[2,3,4].The technique is based on the fact that photoions (in a similar manner as photoelectrons) will drift apart under field-free conditions depending on their initial kinetic energy. If a pulsed extraction field is applied after some μs and a suitable time-of-flight detection is used a kinetic energy resolution down to cm^{-1} can be obtained.In first experiments we have carried out kinetic energy measurements on H-atoms originating from the $\tilde{A}$ (v=0) state of NH_3.

In planar structures there is a correlation between the $\tilde{A}$ state of NH_3 and the ground state of NH_2, while the ground state of NH_3 correlates with the $\tilde{A}$ state of NH_2. The conical intersection of the $\tilde{X}$ and $\tilde{A}$ state potential energy surfaces results in two possible dissociation channels, $NH_3(\tilde{A})$ --> $NH_2(\tilde{X}) + H(1s)$ and

$NH_3(\tilde{A})$ --> $NH_2(\tilde{A})$ + $H(1s)$. The first process is dominant. The potential energy function is shown in Fig. 1 (McCarthy et al.[5]).

The experiments reported here were carried out in a skimmed supersonic beam apparatus described earlier[6]. The ammonia $\tilde{A}$ state was excited at 46200 cm^{-1}. For the detection of the H-atoms originating from the process $NH_3(\tilde{A})$ --> NH_2 + H we used 2+1 ionization via the 1s --> 2s two-photon transition. This transition is not the best choice because of electron recoil, and therefore no well re-solved TOF spectra could be seen, but nevertheless it is useful for showing the advantages of the delayed pulsed field method. The following measurements with various field free delays

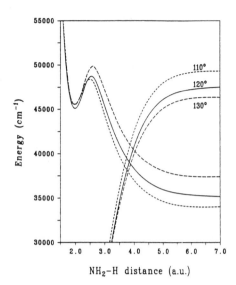

Fig. 1 Potential Energy functions of the $\tilde{X}$ and $\tilde{A}$ states for planar NH_3 for various angles α_{HNH}

and the use of a 1m long time-of-flight mass-spectrometer (LIN-TOF) show the principles of our method and point out that the low energy fragments are not disfavored.

Our preliminary results show that the delayed pulsed field technique should be extremely useful to kinetic energy measurements of photofragments

1. J. Biesner, L. Schnieder, J. Schmeer, G. Ahlers, Xiaoxiang Xie, and K. H. Welge, M.N.R Ashfold and R.N. Dixon, J. Chem. Phys. **88** (6) 3607
2. K. Müller-Dethlefs, M. Sander, and E.W. Schlag, Z. Natur-forsch. Teil A **39** (1984) 1089
3. L.A. Chewter, M. Sander, K. Müller-Dethlefs, and E.W. Schlag, J. Chem. Phys. **86** (9) (1987) 4737
4. M. Sander, L.A. Chewter, K. Müller-Dethlefs, and E.W. Schlag, Phys. Rev. A **36** (9) (1987) 4543
5. M.I. McCarthy, P. Rosmus, H.-J. Werner, P. Botschwina, and V. Vaida, J. Chem. Phys **86** (12) 6693
6. W. Habenicht, R. Baumann, K. Müller-Dethlefs, and E.W. Schlag, Ber. Bunsenges. Phys. Chem. **92** (1988) 414

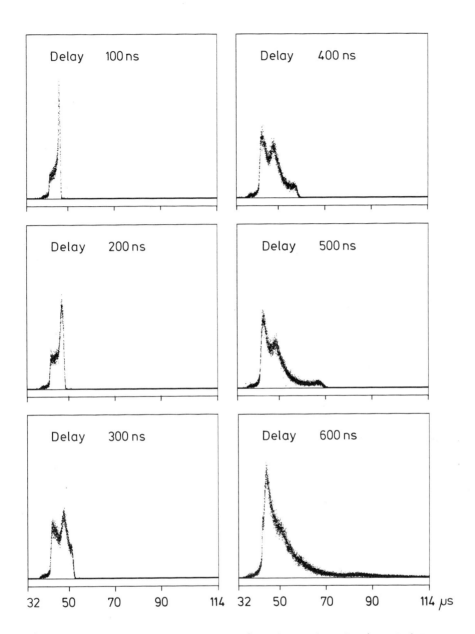

Fig. 2 H⁺-ion TOF spectra resulting from photolysis of jet-cooled NH_3 at 46200 cm^{-1} within the $\tilde{A}$-$\tilde{X}$ 0_0^0 absorption band. Figures (a)-(f) show measurements with various delays of the extraction puls (6 V/cm) vs. light.

INFRARED DIFFERENCE FREQUENCY LASER AND SRS SPECTROMETERS. Q–BRANCH OF CD_3H ν_1 BAND.

D. Bermejo, J.L. Doménech, P. Cancio, J.Santos and R. Escribano

Instituto de Estructura de la Materia (C.S.I.C.)

c/Serrano 123, 28006–Madrid. SPAIN

1.- Introduction

A cw difference frequency infrared laser spectrometer has been constructed in our laboratory, following the design of Pine (1). The output of a frequency stabilized Ar^+ laser and that of a commercial ring dye laser are mixed collinearly and focused into a 5 cm long, a–cut, $LiNbO_3$ crystal, held in a temperature controlled oven (stability better than 0.1 °C) to achieve phase matching. The generated IR radiation is recollimated, to allow long absorption paths. Two InSb detectors are used in a dual beam configuration to obtain transmittance spectrum. Amplitude modulation and lock–in detection is used. Data are acquired, stored and processed with an AT–compatible computer.

The stabilized Ar^+ laser, the ring dye laser (after pulse amplification) and the frequency calibration and data acquisition systems are also part of a q–cw S.R.S. spectrometer, whose performance and early results have been recently reported (2,3). The whole experimental arrangement is schematically depicted in fig. 1. This setup allows consecutive recording of Stimulated Raman and IR spectra, just by the insertion of a couple of mirrors. At present we are working in a setup capable of simultaneously recording both kind of spectra referred to a common IR wavelength standard. In this way we expect to be able to measure Raman frequencies with IR precision.

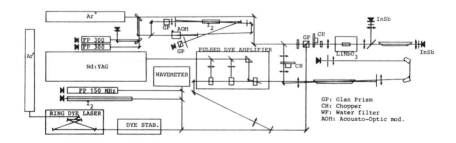

Fig. 1. Schematic arrangement of the IR and SRS spectrometers.

2.- Performance

About 10 μW, 5 MHz linewidth, are generated at 3000 cm^{-1} from 400 mW Ar$^+$ and 20 mW dye, linewidth being mainly limited by our dye laser frequency jitter. The wavelength coverage is from 2.2 μm to 4.2 μm, limited by the crystal. Although the generated IR power is 10^5 to 10^6 times greater than the noise equivalent power of the detectors, the best S/N ratio attainable in the spectrum is about 10^3 because of interference fringes and IR beam wandering.

The performance of the SRS spectrometer has been already described (3) and only characteristics relative to frequency determinations will be mentioned here. The instrumental resolution is estimated to be 90 MHz limited by the temporal width of the pump pulses (10 ns).

The I_2 absorption spectrum is currently used as standard, and limits the absolute accuracy of the frequency scale of both spectrometers, that is 10^{-3} cm^{-1} according to the accuracy of the tabulated I_2 wavenumbers (4). As mentioned above we could improve this figure using IR standards with smaller Doppler width and better defined absorption maxima.

Using the 2990 cm^{-1} band of ethylene, the precision of the IR frequencies has been estimated to be 5x10^{-4} cm^{-1}, from the standard deviation between our observed frequencies and those of ref. (6). Frequencies of Raman lines are 2x10^{-3} cm^{-1} red-shifted from the IR ones. This can be attributed to the AC Stark effect (5) due to the high electric field at the focus in the Raman cell. The standard deviation of the differences between Raman and IR lines leads to a precission in the S.R.S. spectra of 1.2x10^{-3} cm^{-1}.

3.-CD_3H spectrum

The Q–branch of CD_3H ν_1 band recorded with both spectrometers in shown in fig. 2. Raman lines appear broader than the IR ones, which is also a hint of the Stark effect.

We have tried to refine our data with a simple harmonic oscillator–rigid rotor + quartic centrifugal distortion model, but the fit does not progress beyond the first 15 or 20 lines. Possible perturbations may come from the $\nu_3 + 2\nu_6$ (3051 cm^{-1}), $\nu_2 + \nu_6$ (3181 cm^{-1}) or even $\nu_4 + \nu_6$ (3283 cm^{-1}) bands (7).

This work has been partially supported by U.S.-SPAIN Joint committee for scientific and technological cooperation

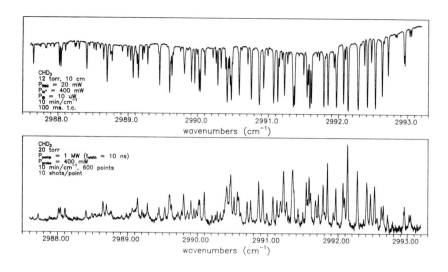

Fig. 2. Q-branch of ν_1 band of CD_3H. a) Infrared. b) Raman.

4.- References

1. A. S. Pine, J. Opt. Soc. Amer., **64**, 1683 (1984)

2. D. Bermejo, J.M. Orza, J. Santos, P. Cancio, J.L. Doménech and C. Domingo, International Conference on Raman Spectroscopy, Calcuta, November 1988.

3. D. Bermejo, J. Santos, P. Cancio, J.L. Doménech, C. Domingo, J.M. Orza, J. Ortigoso and R. Escribano. Submitted to Journal of Raman Spectroscopy.

4. S. Gernsternkorn and P. Luc, Physiqe Apliquée, **14**, 791 (1979)

5. G. Guelachvili and K. Narahari Rao, Handbook of Infrared Standards, Academic Press 1986.

6. L.A. Rahn, R.L. Farrow, M.L. Koszykowsky and P.L. Mattern, Phys. Rev. Lett., **45**, 620 (1980)

7. J.W. Perry, D.J. Moll, A. Kuppermann and A.H. Zewail, J. Chem. Phys., **82**, 1195 (1985)

Phenomenon of Dynamical Rainbow in the Field of Laser Radiation

S.M.Burkitbaev
Institute of Mathematics and Mechanics, Kazakh AS,
Pushkin str. 125, Alma-Ata, 480021, USSR

1. Introduction

Nowadays the phenomenon of combination /Raman/ light scattering from liquid drop-shape-oscillations is very intensively investigated [1,2]. The essence of this effect is in following. The drop-shape distortion exites the drops surface waves which have the discrete frequency spectrum. The light, scattered by oscillating drops, have the spectral components corresponding to droplets oscillations.

The Raman frequencies can give us the information about scatterer size. But, really, the amplitude of oscillations is very small and to measure frequency shifts in scattered light it's necessary to swing the drops by external source [2,3]. In this article we project the alternative approach for observation of the drop-shape-oscillation. It concerns in amplification of the modulated component of the scattered light in the definite direction. This direction is the angle of rainbow. The internal-reflectance angle in the drop, for rainbow generated shaft of ligth, is near the critical angle. Hence the small changes in the drop's shape cause the oscillations in the direction and amplitude of rainbow. This non-stationary [4] /oscillated/ component of the rainbow signal we call the Dynamical Rainbow

2. Numerical model

Rainbow have long been a source of the new mathematical approaches to explain the detailed picture of rainbow phenomenon [5]. For our purposes it is enough the simple approximation of geometrical optics. On the Fig.1

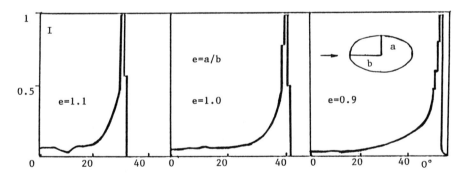

Fig. 1. Rainbow angular position vs drop's shape.

one can see the results of the numerical calculation. The basic model is the model of Descartes [5]. It shows the very interesting picture. The angle of rainbow is very sensitive indicator of the drop's shape.

3. Experimental consideration

In the experiments be reported here, we have attempted to investigate rainbow formation by oscillated water drops. The measurements were obtained with computer-drived goniometer using distilled and undistilled water droplets of 1÷4 mm diameter, which suspended in a linearly polarized Ion Argon Laser beam from the tip of a hypodermic needle. The scattered light angular and temporal behavior is detected by photon counter and data is processed by computer-realized correlator in accordance to equation $G(T)=\langle n(t)\ n(t+T)\rangle$, where T-delay time [6]. The ultra-sound source with the constant power on the wide frequency range was used to excite the drop's oscillations by harmonic and impulsive signals. The experimentally observed angular scattering pattern from water droplets displays fine details in rainbow structure that are not predicted by methods ignoring the phase of scattered light. But in this article we interested to temporal behavior of the rainbow main peak. To the point, the experiment shows that one can observe the drop's oscillations in any components of fine interference structure of rainbow. We measured as well single drops as drops ensemble.

Let's consider the results of light scattering by single drop. The experimental results for drop under acoustical excitation is like to one for ordinarily oscillator (echo-effects for ensemle, too). That's why we'l not consider it here.

4. The self-excitation of drop's oscillation under laser illumination

Once, during the experiments, we used the tea solution for drop creation. The measured correlation function displayed stationary auto-oscillations without any external excitation source. Detailed measurements have been performed with the difference diameter (0.5 mm,..., 4.0 mm) un-distilled water droplets under laser beam of power (0.1÷1.0) W. As the result one can formulate the following conclusion. To complete comparison of experimental results it's convenience to support constant laser-power-density and for decay length in water about 50 m one can measure the specific energy absorbance dy drop Q_{exp} about 5 mW/cm^3. Fig.2 shows temporal correlation functions of rainbow signal for different drop diameters $d_0=1.8$ mm $< d_a= (2.0÷2.5)$ mm $< d_b= (2.5÷3.0)$ mm.

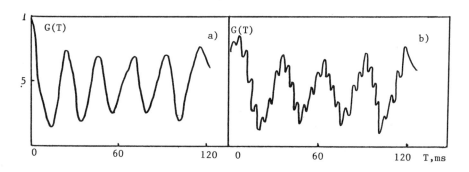

Fig. 2. Rainbow signal generated by self-excited water drop.

For d_0 there is the absence of harmonical excitations. On Fig.2a one can observe the appearance of self-induced oscillations with the period T= 24 ms; and at last on the Fig.2b there is the higher frequency mode with the period T' (T/T' about 7). The same experimental measurements where repeated with double-distilled water drops. There are not any regular oscillations for diameter from 1.5 mm to 4.0 mm.

Let's discuss the experimental results. First of all let's attempt to estimate the correspondence of these results to the article [7], where the thermo-capillary mechanism of self-induced oscillations for droplet under homogeneous heating was predicted. The essence of this effect is in the temperature dependence of the surface tension γ. For example, for water $\partial\gamma/\partial t < 0$. In this case one can receive the eguation for the excitation level Q. For the water droplets suspended in air,and for mode index j=2, radius of drop is 0.1 cm, one can receive (see [7]) Q=3 mW/cm^3. This result is in a good agreement with the experiment. For the case of Q more than Q_{exp} , i.e. for double-distilled water droplet, there are not the self-excitation phenomenon.

In accordance with the theoretical expression for Q[7]:

$$Q \geqslant \frac{j(2j+1)(j-1)^2}{10\ R^2}\ \frac{\chi_i}{\chi_e}\ C_e\ \rho_e\ \nu_i\ (\frac{1}{\gamma}(-\frac{\partial\gamma}{\partial t}))^{-1},$$

where γ - thermal condactivity, C - specific thermic capacity, ρ - density, ν -kinematic viscosity; index "e"-external (air), index "i"- internal (drop).It is obviously that with the growth of drop's size R, one can observe the appearance of j=2,3,... (Fig.2). To the point, the lower frequency modulation of correlation function may be interpreted as suspended drop's pendulum like swinging.

Thus in this article the effect of Dynamical Rainbow from oscillating drop and drop ensemble was predicted and experimentally registered. The self-excitation of drop shape oscillations under laser illumination, due to thermo-capillary mechanism, was registered for the first time too.

The effect of Dynamical Rainbow, to all appearances, should be very sensitive and effective technique to investigate the dynamical properties of drops and drop-like objects (nuclei, for example).

References

1. Yu.A.Bykovskii et.al., J.of Quantum Electronics (Russ.). 2,1803(1975).
2. V.Kalechitz et al., Soviet J.of Technical Physics,Letters,5,485(1979).
3. V.Kalechitz et al., Ibid., 5,1184(1979).
4. A.D.Bishigaev, S.M.Burkitbaev, Soviet Physics, Dokladi. 1989, to be published.
5. H.M.Nussenzweig, Sci.Am.. 236,116(1977).
6. Photon correlation spectroscopy/ ed.by H.Z.Cumminz and E.R.Pike. N.-Y., 1973.
7. V.Kalechitz et.al., J.of Quantum Electronics (Russ.). 9,1274(1982).

FLUORESCENCE SPECTRA

OF SINGLE PHASE SUPERCONDUCTOR $YBa_2Cu_3O_x$

EXCITED BY 280, 295, 308 AND 337.1nm RADIATION

Ji-ye Cai, Chen-xi Wang
Yu-ming Gao, Jun-de Chen and Song-hao Liu
Laboratory of Laser Spectroscopy, Tel.91534
Anhui Institute of Optics and Fine Mechanics,
Academia Sinica, Hefei, P.R.C
Jia-fu Wang, Jin-guang Wu and Zeng-fu Song
Department of Chemistry, Beijing University, P.R.C

Introduction

By spectroscopy method, information of the structure of solid molecules can be obtained[1,2,3]. Up to now a great deal of work has been reported on high Tc superconductor, but there is few work done about fluorescence spectra of superconductor. We are interested in finding the energy bands and the electronic transition properties of superconductor. So we measured the fluorescence of single phase superconductor $YBa_2Cu_3O_x$.

Experiment

The ceramics samples were obtained from Beijing University (Prof. Jin-guang Wu and coworkers). The samples were synthesised from powder mixture of stoichiometric Y_2O_3, Ba_2CO_3 and CuO. Four terminal dc resistance measurements were performed on these samples and these high Tc (Tc≳85K) samples were examined by X-Ray diffraction method with the Philips APD-1700 (Cu, Kα, 40KV, 35mA).

The samples of single phase $YBa_2Cu_3O_x$ superconductor were selected to perform fluorescence measurement. The exciting wavelength are 280, 295, 308 and 337.1nm, respectively. The fluorescence spectra were recorded following excitation.

Result

The fluorescence spectra of single phase $YBa_2Cu_3O_x$ under exciting wavelength of 280, 295, 308 and 337.1nm are shown in Fig.1 and Fig.2, respectively. These spectra are obtained from more than 10 samples. The relative intensity of the emission peaks will change with different samples, but the wavelength of the peaks is unchanged. There are three stronger peaks at 410, 380 and 390nm under exciting wavelengths 280, 295, 308nm respectively. These three peaks are nearly within the bandwidth of fluorescence peak of YBa$_2$Cu$_3$O$_x$ superconductor. We consider these three peaks are likely to come from the emission center of Cu$^+$ ^{1}Eg(3d^{9}4s^1) --→ ^{1}A$_{1g}$(3d^{10}). There is another stronger peak at 560nm in Fig.1(b) which always appears when the pumping wavelengh is 295nm. It is interesting to compare these peaks with reflection-absorption spectra in the Y-Ba-Cu-O superconductor[4]. There are two broad absorption bands at 350 and 500nm. When the wavelength of exciting light changes, the molecules will be pumped to different levels of the exciting band (^{1}A$_{1g}$(3d^{10}) --→ ^{1}Eg(3d^{9}4s^1), in the same time part of the energy transferring to the lattice vibration. And the molecules transfer to certain levels, such as ^{1}Eg and ^{3}Eg , and emit fluorescence accompanied by phonon emission. So the 350nm absorption band is the transition of ^{1}A$_{1g}$ -- ^{1}E$_g$ and the 500nm absorption band is of ^{1}A$_{1g}$ -- ^{3}E$_g$. The 390nm fluorescence is the ^{1}E$_g$ -- ^{1}A$_{1g}$ transition and the

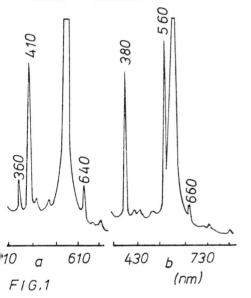

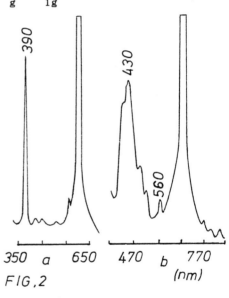

FIG.1

FIG.2

(nm)

Fig.1(a). Fluorescence spectra of YBa$_2$Cu$_3$O$_x$ phase. The wavelength of the exciting light is 280nm. The emission bands are at 360, 410, 490 ,640 and 720nm.
(b). The wavelength of the exciting light is 295nm. The emission bands are at 380, 430, 460, 510, 560, 660, 720 and 840nm.

Fig.2(a). Fluorescence spectra of YBa$_2$Cu$_3$O$_x$ phase. The wavelength of the exciting light is 308nm. The emission bands are at 390, 460, 490 and 560nm.
(b). The wavelength of the exciting light is 337.1nm. The emission bands are at 410, 430, 460, 490, 510, 560 and 840nm.

560nm is of ^3E_g -- $^1A_{1g}$. In the case of free Cu^+ ions the transition between $3d^94s^1$ -- $3d^{10}$ is forbidden. When Cu^+ is in the octahedral and the orthorombic crystal field, because of the unsymmetry of the ligands[5] in the Y-Ba-Cu-O superconductor, the transition probability is changed owing electron--lattice interaction.

The saturation peaks in Fig.1 and Fig.2 are the second order diffraction beams of the exciting light. There are two peaks at 720 and 840nm. When the pumping light is 266nm the intensity of these two peaks is bigger. We consider the 720 and 840nm fluorescence peaks are emitted from Cu^{2+} ions. The ^2D spectral term of Cu^{2+} ion($3d^9$ electrons) will split into ^2E_g (e orbit) and $^2T_{2g}$ (t orbit) levels in the distorted octahedral. These levels may split further in the orthorombic crystal field of the Y-Ba-Cu-O superconductor. The five orbits of 3d electron will split into 5 levels. The detail is to be finished. There are other weaker emission bands, such as 360, 460, 490, 510, 540 and 660nm. These peaks often appear in the spectra. When the wavelength of the exciting light is 337.1nm, many peaks overlap together which are shown in Fig.2(b). Some of these peaks may be emitted from Cu^{3+} ions. We are continuing to study all these spectra.

This work is supported by Chinese National Natural Science Foundation.

References
1. R. G. Greenler. et al., J. Catal, <u>23</u>, 42, (1971).
2. S. A. Francis and A. H. Ellision, J.Opt.Soc.Amer. <u>49</u>, 131, (1958).
3. R. G. Greenler, J.Chem.Phys. <u>50</u>, 1963, (1969).
4. J. Y. Cai et.al., Reflection-absorption Spectra in the Y-Ba-Cu-O Superconductor, 4th National Conference on Solid Optics (Wuxi, China,1988).
5. P. Steiner et.al., Z. Phys. B69, 449(1988).

Part III

Cavity QED

Nonclassical Radiation in the One-Atom Maser

G. Rempe, G. Babst, F. Schmidt-Kaler, and H. Walther
Sektion Physik der Universität München and Max-Planck-Institut für Quantenoptik, D-8046 Garching, Fed. Rep. of Germany

In this paper, we review recent results obtained with the one-atom maser which represents the most fundamental system for studying radiation-matter coupling: a single two-level atom is interacting with a single mode of an almost lossless superconducting cavity [1,2]. It could be demonstrated, that with a quality factor of 3×10^{10} for the cavity, operated at a temperature of 0.5K, on the average only 0.005 atoms in the cavity are sufficient to reach threshold for maser oscillation [2].

The experimental setup was described earlier [1,3]. Rubidium atoms of a velocity-selected atomic beam are excited with frequency doubled light of a cw dye laser to the $63p_{3/2}$ Rydberg level. They are then injected into the cylindrical superconducting cavity operating in the TE_{121} mode. The resonance frequency of the cavity is tuned to the $63p_{3/2}$ <-> $61d_{5/2}$ transition of ^{85}Rb at a frequency of 21.456 GHz by slightly squeezing the almost circular cross section by means of a piezoelectric transducer. After emerging from the cavity, the atoms are monitored using field ionization which can be performed state-selectively by choosing the proper field strength.

With very low atomic beam fluxes, the cavity of the single-atom maser contains only thermal photons, whose number varies randomly obeying Bose-Einstein statistics. In the older setup [3], the cavity was operated at 2.5K corresponding to about 2 thermal photons; the new cryostat with 0.5K gives an average of 0.15 photons. When the flux is increased, the atoms deposit energy in the cavity, and the maser reaches the threshold leading to a change in the photon statistics.

It was demonstrated in earlier experiments that the Rabi oscillation shows a complicated dynamics leading to collapse and revivals [3] as predicted by the Jaynes-Cummings theory [4]. Most of the observed effects are purely quantum features [5].

There are two approaches to the quantum theory of the one-atom maser. Filipowicz et al. [6] use a microscopic approach to describe the device while Lugiato et al. [7] use a modification of the standard macroscopic quantum laser theory both obtaining the same steady-state photon number distribution. The special features of the one-atom maser or micromaser are not emphasized in standard laser theory because the broadening due to spontaneous decay obscures the Rabi cycling of the atoms. When similar averages in the micromaser theory associated with inhomogeneous broadening are performed, equivalent results are obtained.

Both theoretical approaches predict that the photon number distribution in the maser cavity depends mainly on the interaction time of the atoms with the cavity field. It is mostly sub-Poissonian, and using a cavity with a high quality factor $Q > 10^{10}$ without thermal photons, even number states can be generated [8].

Recently, it could be shown theoretically that a further decrease in temperature of the micromaser cavity leads to a steady state photon number distribution with sharp minima which are characterized by fixed numbers of both photons and Rabi nutations. Such "trapping states" [9] may be used to generate a preselected photon number in the micromaser cavity. Since the minima are very sharp, very low cavity temperatures and atoms with a quite homogeneous velocity distribution are required.

The investigation of the micromaser photon statistics gives new insight into the statistical properties of ordinary masers and lasers where, in contrast to the single-atom maser, Poisson statistics are observed. The reason for this is that the resonators of macromasers and lasers are damped much stronger than the Rydberg micromaser cavity. In addition, other atomic and molecular transitions have much shorter lifetimes than Rydberg states. Furthermore, the selected velocity of the atoms used in the micromaser [3] leads to a fixed interaction time with the cavity field helping to reduce the photon number fluctuations since the photon exchange between atom and cavity field can be exactly controlled. The smallest fluctuations are achieved when the atoms leave the cavity again in the upper state. Of course it is necessary that energy is deposited in the cavity in order to maintain maser oscillation. However, since the losses are very small, the probability that the atoms leave the cavity in the upper state can be made very close to unity, which corresponds to a quantum non-demolition situation.

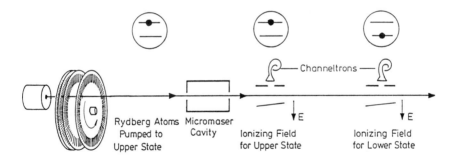

Fig.1: Scheme of the experimental setup. The maser cavity and excitation region are inside a He3 cryostat.

In a new experiment, we succeeded in measuring the sub-Pois-sonian statistics of the one-atom maser field. These results will be reported in the following. The setup for this experi-ment is shown in Fig. 1. Velocity-selected Rydberg atoms in the upper maser level $63p_{3/2}$ are injected into the cavity. They are probed by the first static electric field which io-nizes all atoms in the upper level. All the atoms that are not ionized have emitted a photon in the cavity. When these atoms are counted in the second ionizing field (Fig. 1), the total number of photons in the cavity can be inferred [2,8,10].

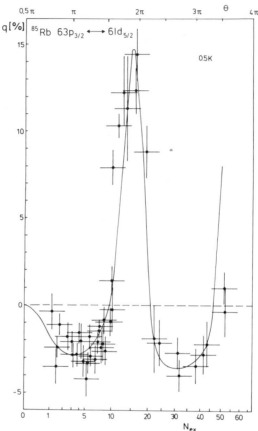

Fig.2: Normalized variance q of atoms in the lower maser level as a function of the atomic flux. N_{ex} is the number of atoms per cavity decay time. The pump parameter $\theta = \Omega t(N_{ex}+1)^{\frac{1}{2}}/2$ with the one-photon Rabi frequency Ω and the interaction time t. The solid line gives the result of the micromaser theory.

One result of our measurements is displayed in Fig. 2 which shows the parameter q versus the atomic flux whereby $q = [<m^2>-<m>^2-<m>]/<m>$, and m is the number of atoms in the lower state. Negative q values correspond to a sub-Poissonian distribution. As a consequence, the photons in the cavity also have sub-Poissonian statistics [2].

Our results show that the fluctuations in the number of maser photons can be 70% below the standard quantum limit. The measurements shown in Fig. 2 reproduce nicely the predicted photon number fluctuations [6,7] in the region with a 2π-rotation of the Bloch vector.

The one-atom maser also allows an intriguing test of the complementarity principle (according to which wave- and particle-like descriptions are complementary). If the atoms are prepared in a coherent superposition of two states, the subsequent population transfer to two other levels in two consecutive micromaser cavities does or does not preserve the coherence depending on the statistics of the photons in the cavities: if there is a Fock state, the coherence is destroyed while for a Poissonian distribution, the coherence is maintained.

Phase arguments which are vital in Bohr's arguments in connection with the Young interference experiment are clearly not applicable here, so the micromaser allows to perform basically a new test of complementarity [11].

References

1. D. Meschede, H. Walther, G. Müller, Phys. Rev. Lett. <u>54</u>, 551 (1985).

2. G. Rempe, G. Babst, F. Schmidt-Kaler, H. Walther to be published.

3. G. Rempe, H. Walther, N. Klein, Phys. Rev. Lett. <u>58</u>, 353 (1987).

4. E.T. Jaynes, F.W. Cummings, Proc. IEEE <u>51</u>, 89 (1963).

5. H.I. Yoo, J.H. Eberly, Phys. Rep. <u>118</u>, 239 (1985) and P.L. Knight, P.M. Radmore, Phys. Lett. <u>90A</u>, 342 (1982)

6. P. Filipowicz, J. Javanainen, P. Meystre, Phys. Rev. <u>A34</u>, 3077 (1986).

7. L.A. Lugiato, M.O. Scully, H. Walther, Phys. Rev. <u>A36</u>, 740 (1987).

8. J. Krause, M.O. Scully, H. Walther, Phys. Rev. <u>A36</u>, 4547 (1987).

9. P. Meystre, G. Rempe, H. Walther, Opt. Lett. <u>13</u>, 1078 (1988).

10. J. Krause, M.O. Scully, T. Walther, H. Walther, Phys. Rev. <u>A39</u>, 1915 (1989).

11. M.O. Scully, H. Walther, Phys. Rev. <u>39</u>, 5229 (1989).

Rydberg atoms in a cavity: a new method to generate photon number states

J.M.RAIMOND, M. BRUNE, J. LEPAPE, S. HAROCHE

Laboratoire de Spectroscopie Hertzienne de l'Ecole Normale Supérieure*

24 Rue Lhomond, 75005 PARIS

(*Associé au CNRS UA18 et à l'université Pierre et Marie Curie)

ABSTRACT: We propose a new method to generate Fock states of the radiation field (pure photon number states). The method uses an adiabatic rapid passage scheme for N excited atoms in a superconducting cavity. An experiment is under way to test this method

It is well known that one can generate light fields with statistical properties very different from the ones of ordinary coherent or thermal fields. Most experiments focus on the generation of squeezed states of light, for which the noise in a phase component of the field is smaller than the usual quantum limit [1]. Among the non-classical fields, those with sub-poissonian photon number statistics are also very interesting [2]. In these states, the reduced photon number variance $V = \sqrt{\Delta N^2 / \bar{N}}$ is smaller than unity. Ultimately, $V=0$ corresponds to a pure Fock state, with a perfectly determined photon number.

Fock states are extremely useful tools to perform high sensitivity optical experiments or to test various aspects of quantum theory of measurement involving light beams [3]. For this reason, generation of field states with very small photon number variance is a very challenging goal. Evidence for sub-poissonian fields has already been reported in the optical [2] as well as in the microwave [4] domain. Single photon states have also been generated in fluorescence cascade experiments [5].

In this communication, we present a new method, very simple in its principle, for the generation of field states which can in principle have arbitrarily small photon number variance. Although we discuss here the method in the context of Rydberg atoms radiating microwave photons in a superconducting cavity, its principle can be extended to the optical domain, resulting in a potentially versatile "Fock states source".

Our Fock state generator is a Rydberg atom maser [6][7] operating in a transient regime. Consider Rydberg atoms prepared in the upper level e of a millimeter wave transition nearly resonant with a low order mode of a high Q microwave cavity (f is the lower level of the transition). Assume that N atoms are prepared in e just before entering the cavity. During the time the atoms cross the cavity, its frequency is swept across the e —>f atomic resonance. The atoms therefore experience an adiabatic rapid passage process which leaves them all in the lower state f by the time they exit the cavity. If cavity relaxation can be neglected during the atom cavity crossing, the field contains exactly N photons at the end of the process.

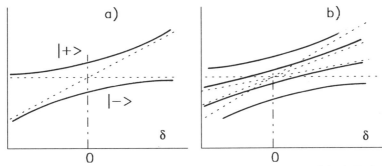

Figure 1: Atom-field energy levels anti-crossings relevant to the adiabatic Fock state generation process. The energies are plotted against cavity-atom detuning δ. The uncoupled states are shown by dotted lines, the "dressed states" by solid lines. a):one atom case. b)Three atoms case.

We consider now in more details the single atom case. The atomic excitation is weak enough so that at most one atom at a time is coupled to the cavity, which is initially empty (temperature is low enough for the thermal field to be negligible). The two relevant atom-field states are then |e,0> and |f,1> where 0 and 1 are the photon numbers in the field (we neglect for the time being cavity relaxation). These two states are eigenstates of the atom field hamiltonian, when the dipole coupling term is omitted. Their energies plotted versus the atom cavity detuning δ are shown on Fig 1a (dotted lines). When atom-field interaction is taken into account, the eigenstates of the full hamiltonian are the "dressed states" |+> and |-> (solid lines on Fig 1a). These levels, which are linear superpositions of the |e,0> and |f,1> states, avoid crossing for δ=0, the minimun distance between the dressed levels being 2Ω, where Ω is the vacuum Rabi precession frequency (typically about 1MHz in a Rydberg atom maser). At the beginning of the cavity frequency sweep, the atom-field system is in state |e,0>, with a negative detuning large compared to Ω. This initial state mainly projects on the |+> dressed state. The cavity is then slowly tuned across resonance, up to a large positive δ value. Provided the cavity frequency variation rate is small compared to Ω^2, transition rates to the other dressed level |-> remain negligible (adiabatic condition) and the system stays throughout the process in the same |+> dressed state which projects mainly for δ>>0 onto the |f,1> uncoupled state: the field thus contains exactly one photon at the end of the process.

The same argument holds for an experiment in which an N atom sample is coupled to a frequency-swept cavity: the system is initially (δ<<0) in the |e,e,...e;0> state representing N excited atoms with zero photon in the field. This state is coupled by the atom-field interaction to the N states |(N-p)e;p> representing N-p atoms excited (with a symmetrical distribution of the excitation among the N atoms) and p photons in the field (p=1...N) [8]. The N+1 uncoupled states are represented as a function of the detuning δ on Fig 1b for N=3 atoms (dotted lines). When atom field interaction is taken into account, the eigenstates become N+1 dressed states (solid lines on Fig 1b). These dressed states anti-cross for δ=0, with a mean spacing between adjacing levels of the order of $\Omega\sqrt{N}$, which is in fact the frequency of the collective Rabi oscillation for N atoms in a resonant cavity [8]. If the adiabatic passage condition is fullfilled, the system follows the uppermost dressed state

141

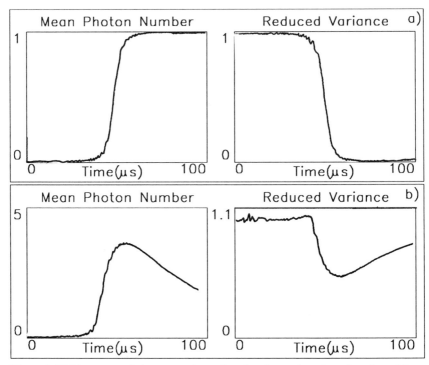

Figure 2: Field energy and photon number reduced variance plotted as a function of time during the 100 µs adiabatic cavity frequency sweep. δ is varied from -3MHz to 3MHz. a): one atom case with no cavity relaxation. b) five atom case with a 50µs cavity relaxation time.

and ends up in the |f,f,f...f;N> state for δ>>0. The field thus contains exactly N photons.

These qualitative arguments are confirmed by direct integration of the system evolution equations. Fig2a presents the field energy and reduced variance V as a function of time in the one atom case and clearly shows the strong sub-poissonian character of the final field state. Fig2b presents the result of the calculation for 5 atoms, in the case of finite cavity damping. The field does not reach a 5 photon value because of competition with relaxation. A strong transient sub-poissonian character is however observed during the cavity damping time.

For atom number N larger than a few tens, the Bloch vector model can be used[8]. It modelizes the atomic system as a classical pendulum driven by a random noise representing the field fluctuations. We have verified by numerical integration of the Maxwell-Bloch equations that the adiabatic passage works for arbitrarily large atom numbers, provided the cavity-frequency excursion is made large enough (compared to $\Omega\sqrt{N}$).

An experiment is under way in our laboratory to demonstrate this method. We prepare by weak diode laser excitation Rydberg atoms in the 39P state of Rubidium. These atoms enter one at a time in the cavity, initially tuned slightly below atomic resonance (39P—39S transition at 68.4GHz). The laser diode is pulsed into atomic resonance at a given time and we thus know when an atom has entered the resonator and when the cavity sweeping must start. The cavity tuning will occur during the 30μs the atom spend in the mode. We will then resume the experiment with another atom, the time between atoms being long enough for the cavity to relax back to vacuum. We expect to detect all the atoms exiting the cavity in the lower 39S state when adiabatic condition is fullfilled. This has to be contrasted with the outcome of the experiment when the cavity is kept exactly at resonance: since we have no atomic velocity selection, and thus a large dispersion of atom-field interaction times, we observe then a random distribution of atoms exiting in the lower and upper states. The main difficulty of the experiment is the cavity tuning. Its frequency has to be modulated on a few MHz interval within 30μs. We change the cavity frequency by pressing on its walls using a stack of piezoelectric crystals. Taking advantage of mechanical resonances of the cavity tuning system, we are actually able to perform the required frequency scans without affecting the cavity high Q factor (10^8). With a stronger pulsed laser excitation, and a somewhat wider frequency scan, the same experiment will allow us to prepare N photon Fock states. The knowledge of the N value will have to be deduced from a count of the number of atoms exiting the cavity after each pulse.

REFERENCES:

[1]   R. LOUDON and P.L. KNIGHT Journal of Modern Optics 34,709 (1987) and references therein.

[2]   M.C. TEICH and B.E.A. SALEH in "Progress in Optics XXVI",p 1, E. WOLF ed. North Holland, Amsterdam (1989).

[3]   M.O SCULLY and H. WALTHER Phys. Rev; A39,5229,(1989)

[4]   G. REMPE and H. WALTHER, postdeadline poster, NICOLS proceedings

[5]   P. GRANGIER, P.G. ROGER, A. ASPECT, A. HEIDMANN and S. REYNAUD, Phys. Rev. Lett, 57,687,(1986)

[6]   D. MESCHEDE,H. WALTHER and G.MÜLLER Phys. Rev. Lett, 54,551,(85)

[7]   L. DAVIDOVICH, J.M. RAIMOND, M. BRUNE, S. HAROCHE Phys. Rev. A36,3771,(1987); M. BRUNE, J.M. RAIMOND, P. GOY, L.DAVIDOVICH, and S. HAROCHE Phys. Rev. Lett 59, 1899 (1987).

[8]   S. HAROCHE and J.M.RAIMOND in "Advances in atomic and molecular Physics XX", D.R. BATES and B. BEDERSON eds. Academic Press,New York(1985)

Dissipative Quantum Dynamics in Cavity Q.E.D.

H.J. Kimble, M.G. Raizen*, R.J. Thompson, R.J. Brecha, H.J. Carmichael[+], and Y. Shevy
  Norman Bridge Laboratory of Physics 12-33, California Institute of Technology, Pasadena, CA 91125

Apart from the obvious interest from the perspective of atomic spectroscopy, the interaction of a small collection of atoms with the modes of a resonant cavity offers exciting opportunities for the investigation of quantum dynamics in a dissipative setting. In a nonperturbative regime, the coupling coefficient g of an individual atom to the cavity field is comparable to the dissipative rates $(\gamma, \gamma_\perp, \kappa)$, where $(\gamma, \gamma_\perp)$ describe spontaneous decay of the atomic inversion and polarization to modes other than those of the resonator and κ accounts for the damping of the cavity field. In terms of dimensionless parameters, the single-atom cooperativity parameter $C_1 \equiv g^2/2\kappa\gamma_\perp$ provides a measure of the significance of the quantum fluctuations of the polarization field from a single atom. Likewise, the saturation photon number $n_s \equiv \gamma\gamma_\perp/4g^2$ offers a qualitative indication of the significance of the addition or deletion of a single photon from the intracavity field and is the number of photons required to produce appreciable excitation of the intracavity medium. As documented in Ref. (1) & (2), the conditions $C_1 > 1$ and $n_s < 1$ are sufficient to ensure that quantum fluctuations play an important role in the dynamics of the atom-cavity system, even for the case of a number of atoms $N \gg 1$. Indeed, in this regime measurements of the photon statistics for the driven damped atom-cavity system offer a realizable avenue toward an investigation of quantum state reduction in a dissipative setting (2).

Underlying these considerations of the dynamical nature of the interaction are questions related to the structure of the composite atom-cavity entity (3). In the limit of weak excitation, we note that the parent Hamiltonian for N two state atoms coupled to a single cavity mode endows the system with a particularly simple eigenvalue structure. The coupling of the atomic polarization and the intracavity field is described by the eigenvalues $\lambda_\pm$, where

$$\lambda_\pm = -1/2(\kappa+\gamma_\perp) \pm [1/4(\kappa-\gamma_\perp)^2 - g^2N]^{1/2} , \qquad (1)$$

while the decay of the atomic inversion is described by the eigenvalue $\lambda_0 = -\gamma$. Equation (1) is derived for coincident atomic transition frequency ω_A and cavity resonant frequency ω_c and is written in a rotating frame at $\omega_A = \omega_c$. The coupling frequency $g = (\mu^2\omega_c/2\hbar\varepsilon V)^{1/2}$, where μ is the transition dipole moment and V the effective cavity volume. These eigenvalues are obtained quite generally from the coupled hierarchy of moment equations derived from the master equation. For weak excitation, a basis set limited to three states (ground state with no excitation and two excited states each with one quantum of energy) is sufficient for describing the coupled dynamics of the amplitudes of the atomic polarization and of the intracavity field. In this basis the amplitude equations decouple from the hierarchy of moment equations and yield the eigenvalues $\lambda_\pm$.

* Present Address: Time & Frequency Division, NIST, Boulder, CO 80303
[+] Present Address: University of Oregon, Dept. of Physics, Eugene, OR 97403

To make contact with other formalisms in cavity Q.E.D., we observe that for large cavity damping and weak coupling ($\gamma \ll g^2 N/\kappa \ll \kappa$), the eigenvalues become

$$\lambda_+ \cong -\gamma_\perp(1 + g^2 N/\kappa\gamma_\perp)$$
$$\lambda_- \cong -\kappa(1 - g^2 N/\kappa^2), \tag{2}$$

where λ_+ describes cavity-enhanced spontaneous emission and λ_- indicates a corresponding reduction in the cavity damping rate. We emphasize that $(\gamma, \gamma_\perp)$ are not the free-space decay rates but rather are defined in terms of the usual (perturbative) sum over modes excluding those of the cavity. In the complementary limit of large atomic damping and weak coupling ($\kappa \ll g^2 N/\gamma_\perp \ll \gamma_\perp$), we find

$$\lambda_+ \cong -\kappa(1 + g^2 N/\kappa\gamma_\perp)$$
$$\lambda_- \cong -\gamma_\perp(1 - g^2 N/\gamma_\perp^2), \tag{3}$$

where now λ_+ describes atom-enhanced cavity decay and λ_- indicates a corresponding reduction in the atomic decay rate. Of course there is an obvious symmetry to the results (2) and (3) as is well known in the context of optical bistability, where $g^2 N/\kappa\gamma_\perp = 2C$ with C as the atomic cooperativity parameter (4).

By contrast to the two limiting cases expressed in Equations (2) and (3), we wish to explore the strongly coupled regime for which $g\sqrt{N} > (\kappa, \gamma_\perp)$ and for which

$$\lambda_\pm = -1/2(\kappa + \gamma_\perp) \pm i[g^2 N - 1/4(\kappa - \gamma_\perp)^2]^{1/2}. \tag{4}$$

The eigenmodes are now formed from equal admixtures of atom and cavity states. Decay is described by Re $(\lambda_\pm)$ and is given neither by the atomic nor cavity rate, but rather by the average of decay rates independent of g. The otherwise degenerate states of weakly excited cavity and ground state atoms and of weakly excited atoms and vacuum cavity mode are now split by the interaction frequency Im$(\lambda_\pm) \simeq g\sqrt{N}$. This splitting has been termed a vacuum-field Rabi splitting (5); we observe simply that it is a normal-mode splitting as is found for any coupled oscillator system and that the above considerations are valid for N=1, 2, 3,...

Somewhat more explicitly, we note that there is no particular notoriety to be attached to the case of a single atom in the cavity with respect to the normal-mode structure in the weak-field limit. In this case, the atom-field system behaves as a pair of coupled oscillators satisfying the following equations for arbitrary N (6-9).

$$\dot{x} = -gy - \gamma_\perp x$$
$$\dot{y} = Ngx - \kappa y, \tag{5}$$

where x corresponds to the amplitude of the atomic polarization and y to that of the intracavity field and where (5) is written for equal cavity and atomic resonance frequencies in a rotating frame at $\omega_A = \omega_c \equiv \omega_0$. The eigenvalues of Eq. (5) are straightforwardly found to be as given by Eq. (1). More generally for nonzero cavity detuning $\theta = (\omega_c - \omega_0)/\kappa$ and atomic detuning $\Delta = (\omega_A - \omega_0)/\gamma_\perp$ relative to the frequency of a rotating frame at ω_0, the eigenvalues $\lambda_\pm$ (scaled by the rate γ) are found from the roots of the secular equation

$$\lambda^2 + \lambda[(1 + i\Delta)/2\Gamma + (1 + i\theta)\mu] + \mu C/\Gamma + \mu(1 + i\theta)(1 + i\Delta)/2\Gamma = 0, \tag{6}$$

where $\Gamma \equiv \gamma/2\gamma_\perp$ accounts for nonradiative relaxation and $\mu \equiv \kappa/\gamma$. For weak excitation (to validate the coupled oscillator approximation), the dependence of line positions and linewidths on detuning (through (Δ, θ)) and on coupling coefficient g (through C) can thus be deduced from Eq. (6) from a unified perspective that treats the atomic and cavity contributions on an equal footing. Cavity enhanced or inhibited atomic processes are deduced as limiting cases from the more general formalism, as illustrated for the particular choice of parameters in Eq. (2). We note that Equations (5) and (6) can be given a rigorous interpretation even for the case of a spatially varying coupling coefficient, as is appropriate for many experiments. Furthermore, in discussing the structure of the system as expressed by $\lambda_{\pm}$, there is no reason to place special emphasis on a particular initial condition such as an initial vacuum state.

In order to investigate experimentally the structure of the composite atom-field system, we have made direct spectroscopic measurements for a collection of two-state atoms in a high finesse cavity (8,9). The particular apparatus employed consists of an atomic beam of optically prepared sodium atoms (3 $S_{1/2}$, F=2, M_F=2 →3 $P_{3/2}$, F=3, M_F=3 transition at 589 nm) intersecting at 90° the axis of a spherical mirror cavity formed by two mirrors of radii of curvature of 1 m separated by either 1.7 or 3.2 mm. With the various mirror configurations employed, the cavity finesse ranged from 18,000 to 26,000. A more detailed discussion of the experiment will be presented elsewhere (9).

For weak excitation of the cavity by an external field E_{in} (Ω), which is mode matched to the TEM$_{oo}$ cavity mode, the field transmitted by the cavity E_{out} (Ω) is found from the response function F (Ω), where

$$F(\Omega) = E_{out} (\Omega)/E_{in} (\Omega)$$
$$\alpha \frac{(1/2\Gamma)(1+i\Delta) + i\Omega}{(\lambda_+ - i \Omega) (\lambda_- -i\Omega)}, \tag{7}$$

and where Ω gives the frequency of the probe field relative to ω_0. The eigenvalues $\lambda_{\pm}$ can be deduced directly from measurements of F(Ω). In the language of cavity Q.E.D., these eigenvalues represent the "vacuum radiative" linewidths and level shifts for the atom-cavity system. In the language of the quantum statistical theory of optical bistability (4) Eq. (7) is simply the state equation for weak-field excitation. From either perspective, the domain $g\sqrt{N} > (\kappa,\gamma)$ with $|\omega_A-\omega_c| < g\sqrt{N}$ is of particular interest as a nonperturbative regime, unlike the cases $|\omega_c-\omega_A| >> g\sqrt{N}$ or $(\gamma,\kappa) >> g\sqrt{N}$. We emphasize that our measurements are made with weak excitation for which F (Ω) is independent of the strength of E_{in} (Ω); we probe the linear response of the system.

Samples of our measurements are given in Figure 1, with Fig. 1a displaying the response of the empty cavity (no intracavity atoms) and Fig. 1b giving the response of atoms plus cavity. The normal-mode splitting described by Im ($\lambda_{\pm}$) is clearly evident. The full curves in the figure are the theoretical prediction for our system, essentially as given by Eq. (6) with the overall vertical scale as a free parameter (9). We as well make slight adjustments in Δ around zero to account for our experimental uncertainty in atomic detuning (Δ=0 ±0.2). The quantity $g\sqrt{N}$ is known in absolute terms from separate measurements of the steady-state hysteresis cycle in absorptive bistability (10); the cooperativity parameter C is related to $g\sqrt{N}$ by $g\sqrt{N} = (\kappa\gamma C/\Gamma)^{1/2}$.

From a series of transmission spectra such as Fig. 1b taken in the limit of weak excitation, we have examined the absolute dependence of the structural features (line positions and linewidths) on intracavity atomic number and on atom-cavity detuning. The results of this investigation are documented in Ref. 8-9. Briefly, we have verified the absolute parametric dependences specified in Eq. (6) to a reasonable level of accuracy for both Re($\lambda_{\pm}$) and Im ($\lambda_{\pm}$). Our measurements of the normal-mode splitting are made over a range of intracavity atomic number $20 \leq N \leq 600$, where N = ρV, with V as the effective cavity volume and ρ the atomic density (which is inferred from the measured atomic cooperativity parameter C). In addition to the coupling-induced splitting given by Im ($\lambda_{\pm}$) for $\omega_A=\omega_c$, we have observed spectral linewidths that are 25% below the free-space

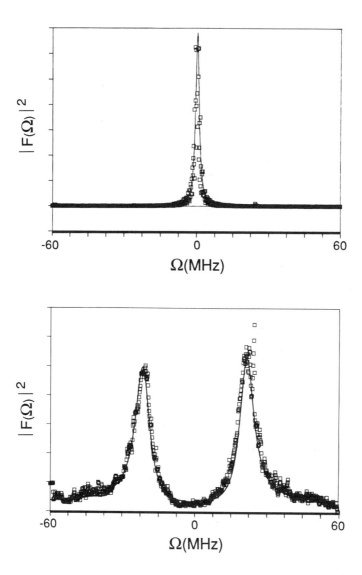

<u>Figure 1</u> - Transmission spectra $|F(\Omega)|^2$ (arb-units) vs. excitation frequency Ω with $\Omega=0$ corresponding to the position of the carrier frequency ω_0, which is approximately equal to the cavity resonance frequency ω_c and atomic transition frequency ω_A for these traces. (a) Empty cavity (N=0) response with Lorentzian fit of 1.8 MHz FWHM, (b) Cavity plus atoms for C=41, $\Delta=-0.07$, $\theta=0$, $\kappa/\gamma=0.09$ with the full line corresponding to the theoretical prediction as discussed in the text. Note that the normal-mode splitting in (b) is independent of the value of the intracavity field in the weak-field limit and that the linewidths of each of the two peaks are below the free-space radiative width of 10 MHz. Trace (b) is taken for N~400 atoms.

atomic width due to the process of linewidth averaging (that is, for $\omega_A = \omega_c$ and $g\sqrt{N} >$ $(\kappa, \gamma_\perp)$, $|\text{Re } (\lambda_\pm)| = (\kappa + \gamma_\perp)/2 < \gamma_\perp$ for $\kappa < \gamma_\perp$ as in our experiments). For $N \lesssim 20$ atoms, the normal mode splitting is not resolved and for small N we enter an overdamped regime with $\text{Im } \lambda_\pm = 0$ for $\omega_A = \omega_c$. In this domain we have reduced the effective intracavity atomic number to a level $N \sim 1$ and have acquired spectra that exhibit reduced peak transmission and broadened linewidth for the cavity-plus-atom system relative to the empty cavity. We thus infer that our method of observation has the potential sensitivity to observe the splitting for a single atom in the cavity if g can be made somewhat larger without an increase in κ.

Towards this general objective but with an emphasis on obtaining simultaneously a large value of C_1 and a small value of n_s, we have constructed a new apparatus employing a beam of atomic cesium. In the course of this work, we have investigated the characteristics of a cw Ti:Al_2O_3 laser operated in a traveling-wave configuration (11). By means of a piezoelectrically controlled mirror internal to the laser cavity, we have actively locked the laser frequency to an external reference cavity. We have then recorded noise spectra of frequency fluctuations with a second external cavity, which is optically isolated from both the reference cavity and the laser. Our initial observations indicate that the principal Fourier components contributing to the laser linewidth lie below a few kilohertz. Indeed with only our rather primitive piezoelectrically controlled mirror, we find an rms linewidth below 20 kHz for the Ti:Al_2O_3 laser (12).

This work was supported by the Office of Naval Research, by the Venture Research Unit of British Petroleum, and by the National Science Foundation. We gratefully acknowledge the support of the University of Texas at Austin, where the experiments involving the sodium atomic beam were carried out. T.L. Boyd is responsible for the frequency stabilization of the Ti:Al_2O_3 laser.

References

1. C.M. Savage and H.J. Carmichael, IEEE J. Quantum Electron. QE 24:1495 (1988).

2. P.R. Rice and H.J. Carmichael, IEEE J. Quantum Electron. QE 24:1351 (1988); R.J. Brecha and H.J. Carmichael, Sixth Rochester Conference on Coherence and Quantum Optics, Paper MCd3.

3. M. Tavis and F.W. Cummings, Phys. Rev. 170:379 (1968).

4. L.A. Lugiato, in Progress in Optics Vol. XXI, ed. E. Wolf (North Holland Publishing, Amsterdam), pp. 71-216.

5. J.J. Sanchez-Mondragen, et al. Phys. Rev. Lett. 51:550 (1983).

6. H.J. Carmichael, R.J. Brecha, M.G. Raizen, H.J. Kimble, and P.R. Rice, Phys. Rev. A (1989).

7. G.S. Agarwal, J. Opt. Soc. Am. B2:480 (1985); H.J. Carmichael, Phys. Rev. A 33:3262 (1986).

8. M.G. Raizen, R.J. Thompson, R.J. Brecha, H.J. Kimble, and H.J. Carmichael, Phys. Rev. Lett 63:240, 1989.

9. M.G. Raizen et al.,in preparation.

10.L.A. Orozco, A.T. Rosenberger, and H.J. Kimble, Opt. Commun. 62:54 (1987).

11. Schwartz Electro-Optics, Titan - CW Ti:Al_2O_3 Laser.

12.T.L. Boyd and H.J. Kimble, in preparation.

RADIATION BY ATOMS IN RESONANT CAVITIES

R.J.Glauber
and
Wenyue Hsu

Lyman Laboratory of Physics
Harvard University
Cambridge, MA 01238
U.S.A.

The field in a closed optical cavity with no absorption present can oscillate only at a sequence of discrete normal mode frequencies. When absorption is present, each of these normal mode frequencies is broadened into a band of finite width. An excited atom, placed within such a cavity, radiates only into those modes with which it posseses appreciable coupling. In an approximate sense then, if the frequency with which it radiates is v_0, and its own spectral decay width is Γ, it can only couple to modes possessing appreciable spectral density within the band from $v_0-\Gamma$ to $v_0+\Gamma$. (Those must furthermore be modes with non-vanishing amplitudes at the atomic position.)

The atomic decay width Γ, however, is not in general a fixed or predetermined quantity. Its value depends on the availability of modes into which a photon can be emitted. While the width Γ selects a narrow frequency band of cavity modes for the photon to enter, the number and amplitudes of those selected modes in turn determine the value of Γ. Radiative decay is, in this sense, a truly environmental effect. It depends sensitively on the boundary conditions obeyed by the electromagnetic field at vast distances from the excited atom.

The linewidth Γ depends, in general, on the position of the excited atom within the cavity and on the orientation of its dipole moment. It is only when these variables are averaged over that the decay width becomes proportional simply to the frequency density of the cavity modes (the "golden rule" result). Analogous results hold for the cavity-induced frequency shift that accompanies the atomic line-broadening.

We investigate the time-dependence of the excited atomic state by using Laplace transform methods. Its decay amplitude, we find, can be approximated by a sum of complex exponential functions $\exp(s_j t)$ in which the coefficients s_j can be determined by solving a complex generalization of the familiar secular equation for a system of coupled oscillators. This procedure applies to cavity couplings of all sorts, but two limiting cases are most easily dealt with. If the modes are not too closely spaced in frequency, the secular equation is algebraic and may be solved directly. If the modes are closely spaced, on the other hand, the atom will tend to interact with relatively few of the multiple reflections its radiated field generates within the cavity. In that case the terms of the secular equation can be approximated accurately by means of a multiple reflection expansion.

We shall discuss several sets of recent experiments[1-3] in which slow beams of Yb and Ba atoms undergo radiative decay while crossing the focal regions of concentric and confocal optical resonators. The atoms thus have the alternative of radiating into the discrete but damped resonator modes, or into the continuum of modes available outside the resonator. The line shifts and decay widths are found to oscillate as a function of resonator detuning, as do the photon intensities radiated inside and outside the resonator. The theory is shown to describe those oscillation phenomena quite satisfactorily.

In the most recent of these experiments[3], only a single Ba atom at a time is present within the focal region of a cavity of high reflectivity. The atomic oscillations are driven by a weak external laser beam that can be tuned through the atomic and nearby cavity

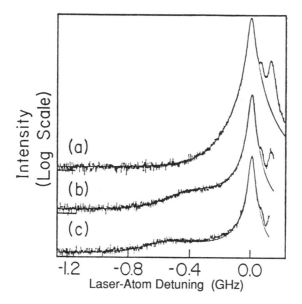

Fig. 1   Light intensity detected through an end reflector of the concentric resonator in Ref.3, compared with theory (continuous curve), for three fixed values of the cavity detuning Δ: (a) Δ=0, (b) Δ=-.475 GHz, (c) Δ=-.610 GHz. The peaks at positive values of the detuning are due to isotopic impurities.

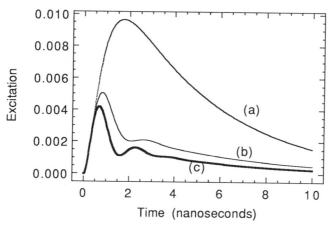

Fig.2   Time dependent photon counting rates predicted to follow pulsed excitation of the Ba atoms in an experiment with the same parameters as those of Ref.3. The three curves correspond to the same cavity detunings as those of Fig.1. The beating effect occurs because the photon may emerge through either of two atom-cavity modes with slightly different frequencies.

resonance frequencies. The driven atom is coupled strongly in this case to only a single resonator mode (apart from the continuum representing radiation outside the cavity). The atomic excitation and the cavity mode interact to form two linearly combined modes with altered frequencies and damping constants. These are the normal modes formed by linear combination of the states with atom excited and cavity empty, and with the atom unexcited but one photon present in the cavity.

The expression for the intensity emitted through the end reflectors of the resonator is found to depend on the driving frequency as the product of two Lorentzian functions centered at the two shifted mode frequencies. This expression, with the appropriate resonator parameters inserted, is found to fit the observed intensity function quite accurately (see Fig. 1). The photons observed through the reflectors are contributed by the two atom-cavity normal modes at slightly different frequencies. If the laser beam that excites the atom is pulsed rather than steady, the frequency difference should lead to a readily observed beating effect in the time-dependent photon counting rate (See Fig. 2).

Acknowledgement
    This work was supported in part by the Office of Naval Research and by the Department of Energy.

References
1. D. Heinzen, J. Childs, J. Thomas, and M. Feld,
       Phys. Rev. Lett. 58 , 187 (1987).
2. D. Heinzen and M. Feld,
       Phys. Rev. Lett, 58, 2623 (1987).
3. J. Childs, J. Hutton, M. Feld, and F. Dalby,
       (submitted to Phys. Rev. Lett.).

SINGLE ATOM-FIELD COUPLED QUANTUM OSCILLATORS

J. J. Childs, M. Donovan, D. J. Heinzen, J. T. Hutton, F. W. Dalby[*], W. Y. Hsu[+], R. Glauber[+]

and M. S. Feld

G. R. Harrison Spectroscopy Laboratory and Department of Physics

Massachusetts Institute of Technology

Cambridge, MA

[*]Department of Physics

University of British Columbia

Vancouver, Canada

[+]Department of Physics

Harvard University

Recently, much attention has focused on the study of a single atom interacting with a single resonator mode. Experimental observations of changes in the atomic spontaneous emission rates in the microwave as well as optical regimes have been reported.[1-4] Atomic line center frequency shifts induced by the presence of the cavity also have been observed.[5,6] All these effects have been attributed to cavity-induced changes in the density of modes of the surrounding vacuum. From this point of view, the cavity exerts an influence on the atomic oscillator only in so far as it modifies the density of modes. This approach considers only the influence of the cavity on the atomic oscillator, and not that of the atomic oscillator on the cavity.

From a more general point of view, however, a two-level atom coupled to an electromagnetic field mode of a resonator is fundamentally two damped, coupled oscillators, which can be viewed as a single system. By exciting either of these oscillators, we would expect to observe the transfer of energy to the other. The spectral response of such a system exhibits two peaks, one near the atom frequency and one near the cavity frequency. The frequency shifts and broadening of these peaks, as well as their relative heights, are determined by the atom-field coupling constant, g. The observation of this two-peaked spectrum reported here, demonstrates the symmetrical roles played by the atom and cavity oscillators.

If the electromagnetic field of the cavity is quantized, the atom-resonator system can be described in terms of dressed states. Under the assumption of weak excitation of this

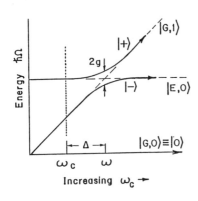

Fig. 1

system, we need only consider the lowest lying states of zero or one quanta of excitation with atom in the ground ($|G\rangle$) or excited ($|E\rangle$) state. This QED picture exhibits the two-peaked spectrum as a function of Ω (Fig. 1). Here ω is the atom frequency and ω_c the cavity frequency.

From a semiclassical point of view, these dressed-states are the normal modes of the atom-resonator coupled oscillator system. Resonator and atom damping are easily incorporated into this treatment. Using the coupled Maxwell-Schroedinger equations and neglecting saturation effects of the resonator field, we obtain an expression for the power spectrum emitted out the ends of the cavity as a function of exciting laser frequency, Ω:

$$P(\Omega) = N_E \hbar \, \omega g^2 \left(\frac{2\pi\gamma_c \, \gamma_p}{\Gamma_c \Gamma_p} \right) L_c L_p \quad , \qquad (1a)$$

where N_E is the excited state population and

$$L_j = \frac{(\Gamma_j / \pi)}{\left[(\Omega - \Omega_j)^2 + \Gamma_j^2 \right]} \quad . \qquad (1b)$$

In these equations, $2\gamma_p$ is the free space dipole decay rate (equal to the atom's spontaneous emission rate for pure radiative broadening), $2\gamma_c$ is the empty cavity decay rate, and g is the atom-cavity coupling constant. Ω_c and Ω_p are the frequencies of the two normal modes (dressed states) of the system, with Γ_c and Γ_p the corresponding decay rates, all of which can be expressed as functions of g, γ_c, γ_p and Δ ($= \omega_c - \omega_p$), the atom-cavity detuning (see Ref. 7 for details). In the limit of zero damping and $\Delta = 0$ the splitting, $\Omega_p - \Omega_c \to 2g$. Further predictions of this model for the atomic decay rate, $2\Gamma_p$, and radiative level shift, $\Omega_p - \omega$, agree with QED calculations.[5-8]

Our experiment studies the absorption lineshape of the system as a function of laser excitation frequency, Ω, by observing the fluorescence emitted out the ends of the cavity. An atomic beam of barium atoms is collimated into a beam $\approx 25\mu m$ in diameter and aligned to intersect at right angles with the optical axis (length = 5cm) of a concentric

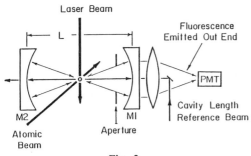

Laser Beam

Fluorescence
Emitted Out End

PMT

Cavity Length
Reference Beam

M2
Atomic
Beam

M1

Aperture

Fig. 2

optical resonator [free spectral range, FSR = 3.0 GHz and (empty cavity) FWHM = $2\gamma_c$ measured to be 440 MHz] (Fig. 2). The exciting laser light is focused to a beam waist of $\approx 20\mu m$ and aligned to intersect both the cavity's optical axis and the atomic beam axis at right angles, at the center of the cavity.

The beam density is such that no more than one atom interacts with a single resonator mode at a time. The laser is scanned through the $^1S_1 \rightarrow {}^1P_1$ transition of ^{138}Ba ($\lambda \approx 553nm$), with the polarization aligned normal to the optical axis of the cavity, so that only the $\Delta m = 0$ component of the transition (relative to the atomic beam axis) is excited. The scan range (1.5 GHz) and scan position of the laser are chosen to ensure excitation of both dressed states during each scan.

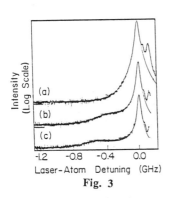

Fig. 3

Intensity (Log Scale)

(a)
(b)
(c)

-1.2   -0.8   -0.4   0.0
Laser-Atom Detuning (GHz)

Figure 3 shows the observed fluorescence as a function of laser frequency, for three values of Δ; (a) 0; (b) -475; (c) -610MHz. The peaks to the right of the $|+\rangle$ state are due to the presence of less abundant barium isotopes, and do not significantly affect the line shape.The smooth curves are fits using Eq. (1). For each curve an overall normalization factor was chosen, and the lineshape L_cL_p, properly modified to account for Doppler, transit time and misalignment broadening, was fitted to the data by adjusting g. The best fits which required small variations in γ_c and γ_p, gave g = 31 ± 8MHz. This is to be compared with the theoretical value for our resonator of g = 31.0MHz. Fits to the data for g = 0MHz were poor.

Observation of the two-peaked spectra, the widths, positions and relative peak heights all of which vary with Δ, verifies the equal roles played by atom and cavity oscillators and the extent to which the lineshape is affected by the coupling. Earlier experiments[5-6], which studied only the "atomic" oscillator peak, provide additional independant measurements of g. The average value obtained for all of the data together is g = 28.5 ± 3.0MHz. This is slightly below the predicted value, probably because of mirror surface aberrations not taken into account in the calculation of g.

ACKNOWLEDGEMENTS

We acknowledge valuable discussions with Michael Kash. This work was performed at the MIT Laser Research Center, a National Science Foundation Regional Instrumentation Facility. This work was supported by the National Science Foundation under Grant No. PHY-8706753.

REFERENCES

1. P. Goy, J. M. Raimond, M. Gross and S. Haroche, Phys. Rev. Lett. **50**, 1903 (1983).

2. R. G. Hulet, E. S. Hilfer and D. Kleppner, Phys. Rev. Lett. **55**, 2137 (1985).

3. W. Jhe, A. Anderson, E. A. Hinds, D. Meschede, L. Moi and S. Haroche, Phys. Rev. Lett. **58**, 666 (1987).

4. D. J. Heinzen, J. J. Childs, J. E. Thomas and M. S. Feld, Phys. Rev. Lett. **58**, 1320 (1987).

5. D. J. Heinzen and M. S. Feld, Phys. Rev. Lett. **59**, 2623 (1987).

6. D. J. Heinzen, J. J. Childs and M. S. Feld, Spectrochimica Acta **45A**, 75 (1989)

7. J. J. Childs et. al. to be submitted to Phys. Rev. A.

8. W. Hsu and R. Glauber, in these proceedings.

Part IV

Noise and Coherence

CHAOS AND COMPLEXITY IN QUANTUM OPTICS*

F.T. Arecchi

Dept. of Physics University of Florence and
Istituto Nazionale di Ottica, Firenze

ABSTRACT

This is a review of how deterministic chaos enters quantum optics,
introducing new features unexpected in the first twenty years of the
laser era. After a general introduction to chaos in lasers, attention
is focused to CO_2 lasers with feedback, which display the so called
Shil'nikov chaos.

1. Introduction[1]

Quantum optics from its beginning in 1960 with the first laser
was considered as the physics of coherent and intrinsically stable
radiation sources. Leaving out the peculiar statistical phenomena
which characterize the threshold points and which suggest a formal
analogy with thermodynamic phase transitions, the main point of
interest is that a single-mode laser provides a highly stable or
coherent radiation field.
Coherence is equivalent to having a stable fixed point attractor and
this does not depend on details of the nonlinear coupling, but on the
number of relevant degrees of freedom. Since such a number depends on
the time scales on which the output field is observed, coherence
becomes a question of time scales. This is the reason why for some
lasers coherence is a robust quality, persistent even in the presence
of strong perturbations, whereas in other cases coherence is easily
destroyed by the manipulations common in the laboratory use of lasers,
such as modulation, feedback or injection from another laser.
Sect. 2 is a general presentation of low-dimensional chaos in lasers;
Sect. 3 describes features of a new type of chaos, the Shil'nikov
chaos.

2. Deterministic chaos

Until recently the current point of view was that a few-body
dynamics was fully predictable, and that only addition of noise
sources, due to coupling with a thermal reservoir, could provide
statistical fluctuations. Lack of long-time predictability, or
turbulence, was considered as resulting from the interaction of a
large number of degrees of freedom, as in a fluid above the critical
Reynolds number (Landau-Hopf model of turbulence).
On the contrary, it is now known that even in systems with three

*Invited at Nicols '89

degrees of freedom nonlinearities may give rise to expanding directions in phase space and this, together with the lack of precision in assigning initial conditions, is sufficient to induce a loss of predictability over long times.

An example of chaotic motion is offered by the Lorenz model of hydrodynamic instabilities which corresponds to the following equations:

$$\dot{x} = -10x + 10y$$
$$\dot{y} = -y + 28x - xz \qquad (1)$$
$$\dot{z} = -\frac{8}{3}z + xy$$

If we consider N atoms confined in a cavity, and expand the field in cavity modes, keeping only the first mode E which goes unstable, this is coupled with the collective variables P and Δ describing the atomic polarization and population inversion as follows (Maxwell-Bloch equations),

$$\dot{E} = -kE + gP$$
$$\dot{P} = -\gamma_{\perp}P + gE\Delta \qquad (2)$$
$$\dot{\Delta} = -\gamma_{\parallel}(\Delta - \Delta_o) - 4gEP$$

For simplicity, we consider the cavity frequency at resonance with the atomic resonance, so that we can take E and P as real variables and we have three coupled equations. Here, k, $\gamma_{\perp}$, $\gamma_{\parallel}$ are the loss rates for field, polarization and population, respectively, g is a coupling constant and Δ_o is the population inversion which would be established by the pump mechanism in the atomic medium, in the absence of coupling.

The similarity of Maxwell-Bloch equations with Lorenz equations would suggest the easy appearence of chaotic instabilities in single-mode, homogeneous-line lasers. However, time-scale considerations rule out the full dynamics of (2) for most of the available lasers. Equations (1) have damping rates which lie within one order of magnitude of each other. On the contrary, in most lasers the three damping rates are wildly different from one another.

The following classification has been introduced

Class A (e.g., He-Ne, Ar, Kr, dye): $\gamma_{\perp} \simeq \gamma_{\parallel} \gg k$. The last equations of (2) can be solved at equilibrium (adiabatic elimination procedure), thus the laser dynamics is described by one single nonlinear field equation. N = 1 means fixed point attractor, hence coherent emission.
Class B (e.g., ruby, Nd, CO): $\gamma_{\perp} \gg k \gtrsim \gamma_{\parallel}$. Only polarization is

159

adiabatically eliminated (middle equation of (2)) and the dynamics is ruled by two rate equations for field and population. N = 2 allows also for period $\sim$ic oscillations.

Class C (e.g., FIR lasers) $\gamma_{\shortparallel} \simeq \gamma_{\perp} \simeq k$. The complete set of (2) has to be used, hence Lorenz like chaos is feasible.

We have carried a series of experiments on the birth of deterministic chaos in CO_2 lasers (class B). In order to increase, by at least 1, the number of degrees of freedom, we have tested the following configurations:

i) Introduction of a time dependent parameter to make the system non autonomous. Precisely, an electro-optical modulator modulates the cavity losses at a frequency near the proper oscillation frequency Ω provided by a linear stability analysis, which for a CO_2 laser happens to lie in the 50-100 KHz range, providing easy and accurate sets of measurements.

ii) Injection of a signal from an external laser detuned with respect to main one, choosing the frequency difference near the above mentioned Ω . With respect to the external reference the laser field has two quadrature components which represent two dynamical variables. Hence we reach N = 3 and observe chaos.

iii) Use a bidirectional ring rather than a Fabry-Perot cavity. In the latter case the boundary conditions constrain the forward and backward waves, by phase relations on the mirror, to act as a single standing wave. In the former case forward and backward waves have just to fill the total ring length with an integer number of wavelengths but there are no mutual phase constrains, hence they act as two separate variables. Furthermore, when the field frequency is detuned with respect to the center of the gain line, a complex population grating arises from interference of the two counter-going waves, and as a result the dynamics becomes rather complex, requiring N 3 dimensions.

iv) Add an overall feedback, besides that provided by the cavity mirrors, by modulating the losses with a signal provided by the output intensity. If the feedback has a time constant comparable with the population decay time, it provides a third equation sufficient ot yield chaos.

Notice that while methods (i), (ii) and (iv) require an external device, (iii) provides intrinsic chaos. In any case, since feedback, injection or modulation are currently used in laser applications, the evidence of chaotic regions puts a caution on the optimistic trust in the laser coherence.

3. Experimental characterization of Shil'nikov chaos by statistics of return times

The dynamic behavior of a single-mode CO_2 laser with feedback is characterized by global features in the phase space, related to the presence of three coexisting unstable fixed points. As a control parameter is monotonically increased, one can observe transitions from a Hopf bifurcation to a local chaos and eventually to regular spiking and Shil'nikov chaos. Furthermore, one can find evidence of competition among these different kinds of instability.[5] The phase-space trajectories are affected differently by each of the three unstable points, and by adjustment of the control parameters they can be characterized by the dominant role of only one, or a pair of them. A linear stability analysis shows the local features at each fixed point.

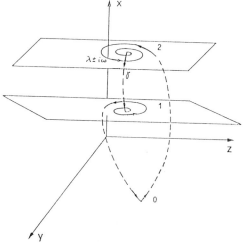

Fig. 1. Schematic view of a trajectory in the phase space when the dynamics are affected by all three unstable stationary points.

Precisely, point 0 (at zero intensity) is a saddle node with two stable directions and one unstable; point 1 has a plane unstable manifold with a focus and a stable third direction; point 2 has a stable manifold with a focus and an unstable third direction. Shil'nikov chaos is related to the saddle focus character of point 2. Around a saddle focus the motion consists of a contracting spiral $\exp(-\lambda t)\cos(\omega t)$ on the stable manifold and of an exponential expansion $\exp(\gamma t)$ along the unstable manifold. The presence of the other two unstable points ensures that the diverging flow is reinjected into the neighborhood of the saddle focus. Shil'nikov showed that for $|\lambda| < \gamma$ there exists a countable set of unstable trajectories close to the homoclinic one.[3]

The temporal behavior of laser output intensity in this regime is characterized by pulses almost equal in shape but with chaotic recurrence times[4]. The regularity in shape means that the points at any Poincaré section are so closely packed that impossibly precise measurements of their position would be required if the relevant features of the motion were to be found. Instead, there is a large spread in the return times to a Poincaré section close to the unstable point. For this reason, the statistics of the return times appears to be the most appropriate characterization of Shil'nikov chaos.

Our experimental setup consists of a single mode CO_2 laser with an intracavity electro-optic modulator. A signal proportional to the laser output intensity is sent back to the electro-optic modulator[5]. For a fixed pump, our system has two control parameters: the bias voltage B applied to the electro-optic modulator and the gain r in the feedback loop.

Adjusting the control parameters in order to have a dominance of the saddle focus 2, we obtain a motion consisting of a quasi-homoclinic orbit asymptotic to it. In this regime, the laser output is characterized by pulses with regular shapes but chaotic in their recurrence. Based on such a consideration the iteration map of return times (τ_{i+1} versus τ_i) displays an extremely regular structure that is in agreement with that arising from Shil'nikov theory.

In figures 2 and 3 we show numerical and experimental iteration maps respectively. Fig. 3 also shows the spread related to the enhanced sensitivity to fluctuations.

To summarize, a CO_2 laser with feedback shows different dynamic regimes depending on the dominant role of one or two of three coexisting unstable stationary points. In particular, in the regime of Shil'nikov chaos the iteration maps of return times display a statistical spread owing to a transient fluctuation enhancement phenomenon peculiar to macroscopic systems, which is absent in low-dimensional chaotic dynamics.

REFERENCES

1. Rather than giving specific references to well known fundamental papers on laser physics, I refer to a recent general review of mine plus References listed therein.
    F.T. Arecchi, in "Instabilities and Chaos in Quantum Optics", F.T. Arecchi and R.G. Harrison, eds., Vol. 34 of Springer Series in Synergetics (Springer-Verlag, Berlin, 1987), p. 9.

2. F.T. Arecchi, R. Meucci, and W. Gadomski, Phys Rev. Lett. **58**, 2205 (1987).

3. L.P. Shil'nikov, Dokl. Akad, Nauk SSSR **160**, 558 (1965); L.P. Shil'nikov, Mat. Sb. **77**, 119, 461 (1968); **81**, 92, 1213 (1970).

4. F.T. Arecchi, A. Lapucci, R. Meucci, J.A. Roversi and P. Coullet, Europhys. Lett. **6**, 677 (1988).

5. F.T. Arecchi, W. Gadomski and R. Meucci, Phys. Rev. A **34**, 1617 (1986).

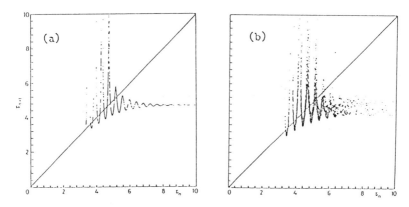

Fig. 2. Numerical iteration maps for Shil'nikov chaos. (a) and (b): maps without and with noise respectively.

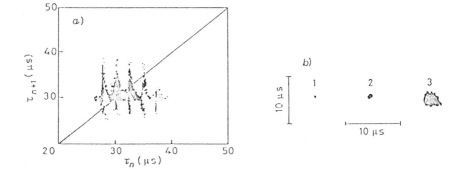

Fig. 3. Experimental iteration maps of the return times. (a): Shil'nikov case. (b): maps for periodic situations, namely, 1, an electronic oscillator; 2, the laser in a periodic regime; 3, the laser at the onset of the instability but still with a regular period.

BISTABILITY AND CHAOTIC INSTABILITIES OF LASER BEAMS COUNTERPROPAGATING THROUGH SODIUM VAPOR

R.W. Boyd, A.L. Gaeta, D.J. Gauthier, and M.S. Malcuit

The Institute of Optics
University of Rochester
Rochester, NY 14627

Recent studies of nonlinear optical systems have shown that the conceptually simple process of mutually interacting laser beams counterpropagating through a nonlinear medium can lead to extremely complicated dynamical behavior, including bistability and chaotic instabilities in the intensities and polarizations of the interacting waves. Silberberg and Bar-Joseph[1] analyzed the stability of laser beams counterpropagating through a Kerr medium and found that the intensities of the transmitted waves fluctuate in time above a critical threshold intensity. It was also determined that the intensities fluctuate chaotically far above the threshold for instability. They attributed the origin of the unstable behavior to gain arising from a process analogous to stimulated Rayleigh-wing scattering and to distributed feedback induced by the interfering input waves. Gaeta et al.[2] generalized the analysis of Silberberg and Bar-Joseph to include the polarization properties of the incident fields. It was found that the states of polarization of the transmitted waves are dynamically unstable at a threshold well below the threshold for intensity fluctuations. In addition to a Kerr medium, these types of instabilities have been predicted to occur for the case of a two-level atomic medium[3] and a Brillouin active medium.[4] These results suggest that instabilities of this type are not unique to the Kerr nonlinearity but can occur whenever counterpropagating beams interact in a nonlinear medium. Recent experimental observations have shown that the intensity[5] and polarization[6] of laser beams counterpropagating through sodium vapor are dynamically unstable.

In this paper we describe an experimental investigation of the stability characteristics of near-resonant laser beams counterpropagating through sodium vapor and find that the states of polarization of the transmitted waves are unstable to the growth of temporal fluctuations. Counterpropagating laser beams are derived from the output of a single-mode, continuous-wave dye laser tuned near to the sodium $3S_{1/2} \rightarrow 3P_{1/2}$ atomic transition. The laser beams are weakly focused at the center of a 5 cm long sodium interaction region to a spot size of ~750 μm and are carefully aligned to ensure that the beams are counterpropagating. Polarizing beam splitters are placed in each beam so that the input polarizations are linear and parallel. We find that there are broad spectral regions near to the atomic resonance for which the states of polarization are unstable for equal input powers of ~50 mW and an atomic number density of ~1×10^{13} atoms cm^{-3}.

We find that the temporal evolution of the radiation generated in the orthogonal polarization is sensitive to the detuning from the atomic resonance, to the intensities of the input beams, to the atomic number density, and to the buffer gas pressure. For a detuning of 730 MHz to the high-frequency side of the $3S_{1/2}(F=1) \rightarrow 3P_{1/2}(F=2)$ transition it is found that the states of polarization of the transmitted waves change abruptly to a value different than the input polarizations when the intensities of the waves are above a well defined threshold. The power of one of the input beams (denoted the forward beam) is held fixed

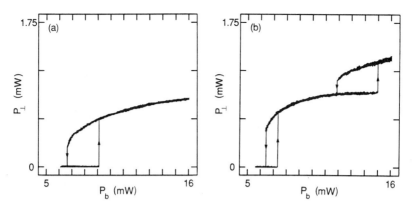

<u>Figure 1</u> Power emitted in the forward direction in the polarization component orthogonal to that of the input waves as a function of the backward wave power. The power of the forward wave is fixed at 74 mW and the number density is (a) 1.6×10^{13} atoms cm^{-3} and (b) 1.9×10^{13} atoms cm^{-3}.

at a value of 74 mW and the power P_b in the other beam (the backward beam) is ramped from a value of 5 mW to 16 mW and back again using an acousto-optic modulator. Figure 1a shows the power of the light generated in the polarization component orthogonal to that of the input beams as a function of the power in the backward beam for a number density of 1.6×10^{13} atoms cm^{-3}. It is seen that the power in the orthogonal polarization $P_\perp$ jumps from zero to ~0.6 mW for a backward wave power of ~9 mW and that the power in the orthogonal polarization displays hysteresis as the backward wave power is decreased. As the nonlinear coupling strength is increased by raising the number density to 1.9×10^{13} atoms cm^{-3}, it is seen from Fig. 1b that the threshold for instability is decreased and that there exists a new stable state. These results constitute the first experimental confirmation of the prediction[7] of polarization bistability due purely to the effects of the nonlinear coupling and the effects of propagation. For higher coupling strengths there are no stable states of the system and the fields fluctuate in time.

We have observed rich temporal behavior for different detunings from the atomic resonance. For a detuning of 490 MHz below the $3S_{1/2}(F=2) \rightarrow 3P_{1/2}(F=2)$ transition and a forward wave power of 39 mW, the threshold for instability corresponds to $P_b = 5.7$ mW. Above the threshold for instability, the polarizations of the transmitted waves fluctuate periodically. For higher backward wave powers, the temporal evolution is highly erratic, as shown in Fig. 2a. It is seen in Fig. 2b that the power spectrum of the fluctuating time series contains little structure and from Fig. 2c it is seen that the autocorrelation function decays rapidly. These results indicate that the temporal evolution is that of random noise or chaotic motion. An inspection of Fig. 2d which shows the attractor describing the dynamical evolution of the system constructed in a time-delay phase space reveals that attractor is a strange attractor because it does not uniformly fill phase space nor does it form a closed loop. We have determined the correlation dimension and order-2 entropy

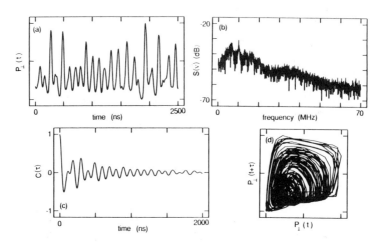

<u>Figure 2</u> Temporal evolution (a) of the light generated by the instability and the associated power spectrum (b), autocorrelation function (c), and attractor (d) constructed in a time-delay phase space for a backward wave power equal to 9.8 mW. The attractor is constructed by plotting the power emitted in the orthogonal polarization at time t+τ *vs.* that emitted at time t for τ equal to 20 nsec.

characterizing the attractor and find that they are equal to 4.2 and 60 Mbits sec^{-1}, respectively. For further increases in the backward wave power the fractal dimension of the attractors describing the evolution of the system increases initially and then decreases due to the effects of saturation.

REFERENCES

1. Y. Silberberg and I. Bar-Joseph, Phys. Rev. Lett., **48**, 1541 (1982); J. Opt. Soc. Am. B, **1**, 662 (1984)
2. A.L. Gaeta, R.W. Boyd, J.R. Ackerhalt and P.W. Milonni, Phys. Rev. Lett., **58**, 2432 (1987).
3. I. Bar-Joseph and Y. Silberberg, Phys. Rev. A, **36**, 1731 (1987).
4. P. Narum, A.L. Gaeta, M.D. Skeldon and R.W. Boyd, J. Opt. Soc. Am. B, **5**, 623 (1988).
5. G. Khitrova, J.F. Valley and H.M. Gibbs, Phys. Rev. Lett., **60**, 1126 (1988).
6. D.J. Gauthier, M.S. Malcuit and R.W. Boyd, Phys. Rev. Lett., **61**, 1827 (1988).
7. H.G. Winful and J.H. Marburger, Appl. Phys. Lett., **36**, 613 (1980); R. Lytel, J. Opt. Soc. Am. B, **1**, 91 (1984); A.E. Kaplan and C.T. Law, IEEE J. Quant. Electron., **QE-21**, 1529 (1985); and A.L. Gaeta, R.W. Boyd, P.W. Milonni and J.R. Ackerhalt, in *Optical Bistability III*, edited by H.M. Gibbs, P. Mandel, N. Peyghambarian, and S.D. Smith (Springer-Verlag, Berlin, 1986), pp. 302-305.

POLARISATION INSTABILITIES AND CHAOS IN A He-Ne MONOMODE LASER

Albert LE FLOCH
Laboratoire d' Electronique Quantique, Université de Rennes,
F-35042 Rennes Cédex (France)
Pierre GLORIEUX
Laboratoire de Spectroscopie Hertzienne, Université de Lille 1,
F-59655 Villeneuve d'Ascq Cédex (France)

Up to now, most studies of optical instabilities have concentrated on intensity or frequency measurements of the light coming out of an optical system (1). In the experiments reported here, the vectorial character of the electric field is taken into account and plays an important role in the dynamics of the particular system under investigation. We have observed experimentally instabilities and chaos in a monomode $3.39\mu m$ He-Ne laser with an isotropic cavity in which some slight loss or refractive index anisotropies have been introduced. They induce preferred directions for the state of polarization of the laser electric field and it was shown that this results in a single or double potential well respectively for this field (2). In addition, the active medium is subjected to static B_0 and sinusoidal B_1 magnetic fields colinear with the laser axis and these fields induce some evolution of the polarization because of the Faraday effect.

In the case of loss anisotropy and a static axial magnetic field, the polarization is linear and fixed as long as B_0 is less than some critical value B_c where the polarization direction begins to oscillate at a rate ω_0 fixed by the combined effect of B_0 and of the potential well created by the loss anisotropy, while the total intensity remains constant within a few per cent. When a small amplitude sinusoidal field B_1 is added in this configuration, frequency locking of the polarization motion occurs as soon as its frequency ω_1 is such that ω_1 / ω_0 is a rational number. As B_1 is increased, the widths and the position of the frequency-locked regions follow the Farey tree hierarchy. Between Arnold tongues corresponding to the frequency-locked regimes (3), quasiperiodicity is usually observed. However for large values of B_1, the laser output appears to be erratic and its Fourier spectrum suddenly becomes broadbanded. This occurence of a broad spectrum following a quasiperiodic regime is an indication of chaos through the destruction of a 2-torus.

When phase (refractive index) anisotropies are introduced, different dynamics occur since the potential has two wells with different depths and the system displays hysteresis between the two laser eigenstates(2). As an example, let us consider the evolution of the laser intensity along say the x-direction when the amplitude of the driving field B_1 is increased and its frequency ω_1 is kept fixed. When B_1 is zero, B_0 induces an hindered rotation of the polarization in a way similar to that observed with loss anisotropy as soon as B_0 exceeds a critical value. At low B_1 values (see Fig.1.b $B_1=0.13$ Gauss), the motion of the polarization is quasiperiodic and its spectrum displays peaks with a main component at ω_p and sidebands at $\omega_p \pm n \, \omega_1$. This motion is essentially a vibration of the polarization around an average position given by the rotation induced by B_0. This motion becomes erratic when B_1 is increased above 1.10 Gauss and is again periodic with a period equal to twice the modulation period when $B_1 = 1.33$ Gauss. Then it becomes 4T periodic, erratic and eventually T periodic.

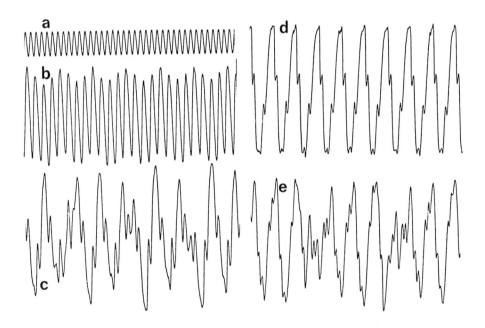

Fig 1 : Evolution of the dynamical behavior of the laser with phase anisotropy with the amplitude of the driving magnetic field. $\omega_1=2\pi\times220$kHz (a) driving field reference, (b) $B_1=0.13$ G, quasiperiodicity, (c) $B_1=1.10$G, chaos, (d) $B_1=1.33$G, 2 T periodicity, (e) B1=1.71G, chaos.

A simple model has been designed to describe the dynamics of the He-Ne laser with loss anisotropy and another one for the laser with phase anisotropy is under development. In both models, the evolution of the laser electric field is governed by three factors : the Faraday rotation inside the active medium, the rotation and attenuation due to the gain difference for the different states of polarization of the field and have different amplification factors due to alignment and orientation in the atomic vapor (4).

In fact the theoretical model of this slightly anisotropic laser has to be written with an electric field developed onto a set of eigenfunctions of the laser cavity containing the active medium (2). Depending on the laser line on which the laser is operating and on the nature of the anisotropy, these functions may have different forms. For a cavity containing a loss anisotropy, the eigenmodes of the laser cavity have linear polarization and the same frequency. On the other hand, in the cavity with index anisotropy, the eigenmodes have in general different polarizations and frequencies but there can exist situations in which they are so strongly coupled that there exist a kind of "supermode" in which both eigenmodes are locked to eachother (5). The model developed in a supermode picture in which the field is described in terms of amplitude, azimuth and ellipticity. This model provides a reasonable agreement with most of the experimental datas.

References
1. J. Opt. Soc. Am., special issues, B2, January 1985 and B5, May 1988
2. A. Le Floch, G. Ropars, J. M. Lenormand and R. Le Naour, Phys. Rev. Lett.,52, 918 (1984)
3. M. Jensen, P. Bak and T. Bohr, Phys. Rev. Lett. 50, 1637 (1983). J. Stavans, F. Heslot and A. Libchaber, Phys. Rev. Lett. 55, 596 (1985). J. Stavans, Phys. Rev. A35, 4314 (1987)
4. W. Culshaw, J. Kannelaud and F. Lopez, Phys. Rev., 128 1747 (1962). W. Culshaw and J. Kannelaud, Phys. Rev., 133, 691 (1964), 141, 228 and 237 (1966), 145, 257 (1966). W. J. Tomlinson and R. L. Fork, Phys. Rev., 164, 466 (1967). H. de Lang, Physica, 33, 163 (1967). H. de Lang, D. Polder and W. Van Heiringen, Phil. Techn. Rev. 32, 190 (1971)
5. L. A. Lugiato, C. Oldano and L.M. Narducci, J. Opt. Soc. Am.B5, 879 (1988). L. A. Lugiato, F. Prati, L. M. Narducci, P. Ru, J. R. Tredicce and D. K. Bandy, Phys. Rev. A37, 3847 (1988). L. A. Lugiato, G.-L. Oppo, M. A. Pernigo, J. R. Tredicce, L. M. Narducci and D. K . Bandy, Opt. Comm., 68, 63 (1988)

COMPOUNDED CHAOS

————————————————

R.Holzner, L.Flepp, R.Stoop, G.Broggi, E.Brun
Physik -Institut der Universität Zürich, Switzerland

ABSTRACT

The NMR-Laser can be driven into chaos by either forced parameter modulation or delayed feedback. Distinct chaotic response is observed when both mechanisms act seperately or simultaneously. The corresponding strange attractors have been investigated experimentally and numerically with regard to fractal dimensions and Lyapunov exponents.

NMR-LASER AND CHAOS

The free NMR-Laser [1] is modelled by a three dimensional system of nonlinear differential equations (Bloch-Kirchhoff equations)

$$\frac{dB_u}{dt} = - k \cdot (B_u - B_k \cdot Q^{1/2}) - C \cdot M_v$$

$$\frac{dM_v}{dt} = - \frac{B_u}{T_2} + 9 \cdot g \cdot M_z \cdot B_u$$

$$\frac{dM_z}{dt} = - \frac{(M_z - M_e)}{T_e} - 9 \cdot g \cdot M_v \cdot B_u$$

with B_u the transverse magnetic field in the rotating frame, k the cavity decay constant, B_k the mean thermal noise field, Q the cavity quality factor, C the coupling constant, M_v the transverse nuclear magnetization, M_z the longitudinal nuclear magnetization, M_e the equilibrium magnetization, T_2 the transverse relaxation time, T_e the longitudional relaxation time and g the gyromagnetic ratio.

To obtain chaotic behavior we either modulate the inhomogeniety of the static magnetic field which amounts at a modulation of T_2, effectively, or drive the quality factor Q of the resonator by a delayed feedback signal from the NMR-Laser output. The two mechanisms are described by

$$\frac{1}{T_2} = \frac{1}{T_{2c}} + A \cdot \cos(\Omega \cdot t)$$

$$Q = Q_c - B \cdot Q(t-T) \cdot |M_v(t-T)|$$

with Q_c the maximum cavity quality factor, B the feedback amplification constant, T_{2c} the effective transverse relaxation time, A the modulation amplitude, Ω the modulation frequency.

EXPERIMENT AND RESULTS

Applying both mechanisms simultaneously results in a chaotic response shown in fig.1a) which we call compounded chaos.

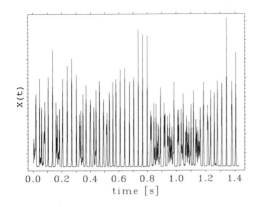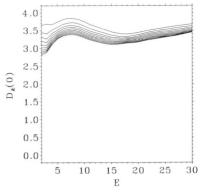

Fig.1   a) NMR-Laser output voltage x(t) proportional to the transverse nuclear
         magnetization M_V as a function of time for compounded chaos. Shown
         are 10^3 data points taken from a digitized time series of $5 \cdot 10^5$ data
         points sampled over a period of 700 s.

         b) fractal information dimension D(0) as a function of embedding space
         dimension E and mass k=5,10,...,40 from top to bottom line.

We determined the fractal dimension numerically [2] from several sets of experimental
time series of all three chaos mechanisms. For the set of which the compounded chaos is
shown in fig.1, the value for the information dimensions is D(0)=2.57 +/- 0.07 for chaos
by modulation, D(0)=3.15 +/- 0.1 for chaos by delayed feedback and D(0)=3.28 +/- 0.13
in the case of compounded chaos. The used numerical method produces excellent
convergence of the fractal dimension and hence a small error . For compounded chaos the
convergence is less obvious due to a missing broad plateau of constant D values as a
function of E (fig.1b). Therefore a larger error had to be assigned to D.
The numerical determination [3] of Lyapunov exponents from experimental time series
yields clearly one positive value in all cases which is expected for chaotic behavior. The
occurence of a second positive Lyapunov exponent has probably been found in the case
of compounded chaos indicating hyperchaos.
The application of delayed feedback does not increase the dimension of the system by
more than one. A larger increase would be possible in principle since the delay equation
corresponds to an infinite number of differential equations with an infinite number of
initial conditions. Chaos induced by delayed feedback seems to dominate the compounded
state as far as the fractal dimension is concerned.

REFERENCES

[1] E.Brun, B.Derighetti, D.Meier, R.Holzner, M.Ravani,J.Opt.Soc.Am.**B2**,156 (1985)
[2] M.Ravani, B.Derighetti, G.Broggi, E.Brun, J.Opt.Soc.Am.**B**,1026 (1988)
[3] R.Stoop, P.F.Meier, J.Opt.Soc.Am.**B5** ,1037 (1988)

Why Laser Beams Cannot Go Straight

M. D. Levenson,
 Research Division, IBM Corporation,
 San Jose, CA   95120, USA

Spontaneous emission by the laser medium into non-oscillating higher order transverse modes causes the centroid position of "single" mode laser beams to fluctuate.

When a TEM_{00} laser beam is centered on a split photodetector as in Fig. 1., the shot noise in the difference current between the two sides of the detector must be attributed to zero-point fluctuations of odd-symmetry spatial modes of the electro-magnetic field [1]. These fluctuating fields beat with the TEM_{00} local oscillator in a spatially nonuniform way that can also be interpreted as a stochastic motion of the centroid of the beam. These fluctuating fields also propagate along the propagation direction at the speed of light. Instantaneously, there are tiny quantum kinks in the beam direction! Quantum mechanics thus forbids a laser beam to go straight, except on the average.

Inside the laser cavity, these odd symmetry quantum fluctuations cause spontaneous emission into odd-symmetry cavity modes. Just as spontaneous emission into the fundamental mode broadens the laser spectrum, this odd-symmetry emission enhances the position fluctuations, but only at frequencies corresponding to higher transverse modes of the laser cavity [1,2]. Most laser cavities support such modes, although with a higher damping rate than the fundamental. According to Kogelnik and Li, the frequency of the TEM_{mnq} mode can be written as $\nu_{mnq} = q\,\nu_L + (m+n)\,\nu_T$ [3]. The heterodyne beat between the TEM_{01q} mode and the TEM_{00q} occurs at ν_T; that between the TEM_{01q-1} and TEM_{00q} falls at $\nu_L - \nu_T$. These two components appear in figure 2, which is a spectrum analyzer trace of the signal from a split photodetector. Also shown is the trace of the noise from the sum of the currents of the two sides of the detector which reveals a peak at ν_L due to the non-oscillating TEM_{00q+1} and TEM_{00q-1} modes.

The behavior of these spectral components illustrates the theory of regeneratively amplified spontaneous emission [2,3]. Near TEM_{01} threshold, the peaks can become quite large, but also very narrow. The total mean square fluctuation amplitude varies inversely as the net damping of the TEM_{01} mode (i.e loss minus gain), output power and fractional population inversion. Typically, the rms fluctuation is 0.01% of the laser beam radius [1,2].

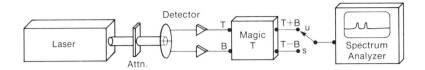

Fig. 1. An experiment to detect stochastic position fluctuations. The two current outputs from a split photodiode can be either summed or differenced by the hybrid junction magic T. The spectrum of current fluctuations was displayed on an R.F. spectrum analyzer.

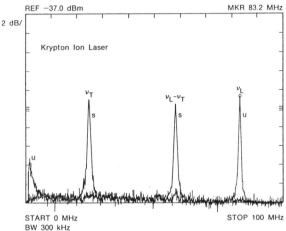

Fig. 2. Spectrum analyzer traces for uniform (u) and split (s) detectors produced by the apparatus of figure 1. The vacuum noise level is at bottom, and the Spectra 171-01 laser was operating at 647 nm.

Uniform   Split

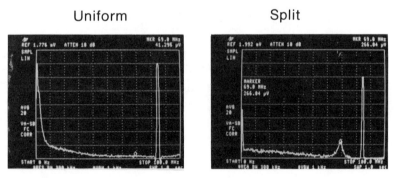

Fig. 3. Spectrum analyzer traces for uniform and split detectors illuminated by a Coherent 599-21 dye laser. The vertical scale is linear in noise power with the vacuum level at the bottom. The strong peak at 83 MHz results from mode beating in the pump.

Dye laser beams also show this kind of position fluctuation. Spectrum analyzer traces for the sum and difference of the currents from a split detector illuminated by a dye beam appear in figure 3. With care, it is possible to suppress this TEM_{01} fluctuation, but only at the price of enhancing the TEM_{02}!

The situation for semiconductor diode lasers is substantially more complicated. The nominal cavities of these devices are quite small, producing large longitudinal and transverse mode frequencies. However, these lasers are exquisitely sensitive to feed-back, and thus the effective cavity can include a reflector on the optical table meters from the diode itself. The flat facets of such a laser produce a degenerate cavity configuration where $v_T=0$. Gain-guiding and transverse confinement increase this frequency and also the damping rate for higher order modes. The net effect is to broaden the modes so much that position noise can be detected even at low frequencies. The small relative population inversion also enhances this effect.

Figure 4 compares the spectrum analyzer traces for split and unsplit photodiodes for a transverse junction semiconductor diode laser operated at 77K. Under the conditions of this experiment, the noise on the uniform photodiode should have been at or below the vacuum noise level, except for the action of traps. Fabry-Perot spectrum analyzer traces are also shown for comparison. We found that the split detector noise level could be either above or below that of an unsplit detector depending upon factors that were not under control. The noise spectra were essentially flat, reflecting only the finite bandwidth of the electronics. The Fabry Perot traces showed no change with noise level. Only in a narrow range near threshold did the position noise lie at the vacuum level. The amplitude noise found with a uniform detector could only be pushed below the vacuum level at much lower temperatures.

Thus it seems that laser beams - especially semiconductor diode laser beams - cannot and do not go straight. The deviations are small, but well within the range of detectability using modern apparatus. The effects of this phenomena on science and technology remain to be determined.

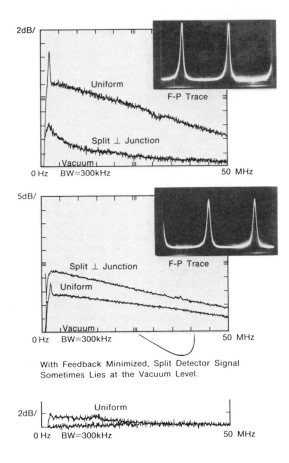

Fig. 4. Spectrum analyzer traces for a transverse junction semiconductor laser operated with a very quiet pump current. Fabry-Perot traces appear in the insets. The vacuum noise is again at bottom. Either the split or uniform detector noise can be larger, depending on fine details of the operating conditions. The roll-off of these traces reflects the finite electronic bandwidth.

References:

1. M.D. Levenson, W.H. Richardson and S.H. Perlmutter, *Optics Letters*, (to be published, 1989) and references therein.
2. M.D. Levenson, W.H. Richardson and S.H. Perlmutter, in *Quantum Optics V* (J.D. Harvey and D.F. Walls, editors) Springer-Verlag, Heidelberg 1989.
3. A.E. Siegman, *Lasers*, (Science Books, Mill Valley, CA, 1986) Chapters 11, 14-17.

Squeezing of Quantum Solitons in Optical Fibers in the Presence of Thermal Phase Noise

R.M. Shelby, M.D. Levenson, and S.H. Perlmutter
IBM Almaden Research Center
San Jose, CA 95120
P.D. Drummond and S.J. Carter
Physics Department, University of Queensland
St. Lucia, Queensland, Australia

1. Introduction

Squeezed light, in which fluctuations in one quadrature of the electromagnetic field are reduced to less than the vacuum or shot noise limit, has been produced and detected in several different laboratory experiments(1). This nonclassical light might be used to improve the signal to noise ratio in sensitive optical experiments, and perhaps in technological applications. Means of generation must still be developed which are simple, reliable and which produce squeezing over a useful bandwidth. In previous experiments, we have investigated the squeezing of light by self-phase modulation in single mode optical fiber(2). It was believed at the outset that single-mode fiber provided a low-loss broad-band nonlinear medium which could provide useful amounts of squeezing at easily attainable pump powers. However, excess phase noise partially obscured the squeezed quadrature, and less than 20% squeezing was observed.

This excess noise originates from thermally-induced fluctuations in the fiber refractive index. The thermal noise arises from two different origins: a structured spectrum from acoustic waves confined by the cylindrical fiber geometry (GAWBS light scattering(3)) and a power-law spectrum from thermally-activated relaxational modes of the amorphous silica matrix(4). The latter mechanism was the limiting one in cw squeezing experiments, and our recent investigations have led to a more thorough understanding of its origin and to new knowledge of the microscopic structure of fused silica.

The extent to which thermal phase noise limits squeezing can be characterized by a parameter proportional to the ratio of the light scattering cross section to the effective fiber nonlinearity. One approach to reducing this ratio is to consider using a train of short pulses, e.g. from a mode-locked laser. Distortion of the pump pulses due to self-phase modulation and group velocity dispersion led us to consider pump pulses in the form of optical solitons. We present below the results of computer simulations of quantum noise squeezing of optical solitons, including thermal phase noise. For sufficiently short pulses, significant squeezing can be expected.

2. Thermal Phase Noise in Optical Fibers

We have investigated in detail light scattering processes in fused silica optical fibers for frequencies between 100 kHz and 1 GHz. Guided Acoustic Wave Brillouin Scattering (GAWBS) produces a structured phase-noise spectrum beginning near 20 MHz and extending beyond our instrumental cutoff of 1 GHz. Phase noise peaks are observed, corresponding to the resonant modes of an elastic cylinder (i.e. the fiber)(3). These peaks are broadened principally by damping due to the plastic jacket which protects most fibers, and by variations in the fiber diameter. Above a few hundred MHz the peaks tend to merge, leading to a relatively structureless spectrum. One might expect a cutoff of the GAWBS scattering above a frequency where the sound wavelength becomes comparable to the optical mode diameter, i.e about 20-30 GHz. This cutoff has not yet been verified experimentally.

Glasses are nonequilibrium materials and the complex potential surface for the accessible configuration space of the glass is often modeled as a distribution of double-well potentials or so-called "two-level modes." Fluctuations in the refractive index result from the relaxational motions of groups of atoms modeled by thermal hopping over the barrier of the two-level mode. The mean time to hop to the other well is $\tau(E, T) = \tau_0 \exp[E/kT]$, where E is the barrier height which is distributed according to the function $D(E)$, and τ_0 is an attempt time $\sim 10^{-13}$ sec. Each scattering center produces a Lorentzian spectrum, yielding an overall power spectrum (4):

$$S(\omega, T) \propto T^{1.3} \int D(E) \frac{\tau(E,T)dE}{1 + \omega^2 \tau^2(E, T)} \,. \tag{1}$$

By fitting the observed temperature and frequency dependence of the phase noise spectrum, we have determined that $D(E) = \exp[- E/E_0]$ where $E_0/k \sim 350°K$, and varies somewhat depending on glass composition. The phase noise has a power-law spectrum and a temperature dependence with a maximum at about 80°K, given by

$$S(\omega, T) \propto \frac{T^{2.3} (\omega \tau_0)^{kT/E_0}}{\omega} \,. \tag{2}$$

Recent measurements have shown that this light scattering is quite highly polarized, with a polarization ratio on the order of 10. This means that the change in the polarizability induced by these relaxational motions is almost isotropic. At present the microscopic nature of the two-level modes in fused silica is unknown, and these observations should help to constrain possible models.

We now consider the effect of thermal phase noise on a periodic train of short pulses with repetition rate $\Delta\omega/2\pi$. The pump field can be written in terms of amplitudes ϕ_j:

$$E(\tau) = E_0 e^{-i\omega_0 \tau} \sum_j \phi_j e^{-ij\Delta\omega\tau}. \tag{3}$$

For soliton pulses, $E(\tau) = sech(\tau)$ and $\phi_j = [\pi\Delta\omega/4]^{1/2} sech[j\pi\Delta\omega/2]$. Each pump spectral component is modulated by the refractive index fluctuations, characterized by the amplitude $\rho(\delta)$. If the phase noise is homodyne detected with a pulsed local oscillator with the same normalized amplitudes, ϕ_j, but phase $\pi/2$, the detected noise power spectrum is given by:

$$S_G(\delta) = \zeta \left\{ \rho(\delta) + \sum_{n=1}^{\infty} [\rho(n\Delta\omega + \delta) - \rho(n\Delta\omega - \delta)][\frac{\pi n\Delta\omega}{2} csch(\frac{\pi n\Delta\omega}{2})]^2 \right\} \tag{4}$$

where δ and ζ are the noise frequency and propagation distance in normalized soliton units (see below and Ref. (6)).

From equation (4) it can be seen that the phase noise at a given frequency will have contributions from overlapping noise sidebands originating from the random phase modulation of each spectral component of the pulse train. The spectrum will be relatively featureless with no windows between GAWBS peaks, and GAWBS, with its order of magnitude larger cross section and larger bandwidth compared to the 1/f noise, will be the dominant noise source except very near zero frequency (well below 1 MHz).

3. Squeezing of Solitons

In general the spectral and temporal structure of a coherent pulse propagating in an optical fiber will be modified by self-phase modulation, which tends to generate new spectral components, and by the group velocity dispersion of the fiber which tends to broaden the pulse temporally. These processes can be accurately described by the nonlinear Schroedinger equation (NLSE), well-known in the study of nonlinear pulse propagation in fibers(6). For wavelengths in the anomalous dispersion region of a fiber, stationary soliton solutions to the NLSE exist in the form of hyperbolic secant pulses which for classical coherent light propagate in the absence of loss without distortion.

We have considered the role of quantum noise in soliton propagation. Our quantum treatment of this problem(5) consists of writing the field in terms of the positive-P representation and deriving stochastic differential equations for the associated c-number field amplitudes, $\phi(\tau, \zeta)$, where τ and ζ are the local time and propagation distance, scaled in the customary way.(6) These stochastic differential equations contain noise sources which take proper account of the quantum noise, and have a form similar to the NLSE:

$$\frac{\partial \phi}{\partial \zeta} = \left[\frac{i}{2} \left(\frac{\partial^2}{\partial \tau^2} - 1 \right) + i\phi^\dagger \phi + (i/\bar{n})^{1/2} \eta(\tau, \zeta) \right] \phi. \tag{5}$$

Here $< \phi^\dagger \phi >$ is the dimensionless photon flux and $\bar{n}$ the photon number per soliton. The noise source $\eta(\tau, \zeta)$ contains a delta-correlated quantum noise term and an exponentially decaying thermal noise term. Thus, phase noise due to GAWBS is modeled by a Lorentzian, i.e. $\rho(\delta) = (g\gamma^2)/(\delta^2 + \gamma^2)$, where the parameter g corresponds to the phase noise parameter used in the analysis of earlier cw fiber squeezing experiments (2). Based on our experience g = 5 is an optimistic average GAWBS level for an appropriate fiber cooled to 2°K. The value of the GAWBS bandwidth is unknown, but probably is near 30 GHz, yielding $\gamma < 1$ for a soliton pulse width of about 10 psec or less.

The stochastic nonlinear Schroedinger equation can be linearized about the classical soliton solution, yielding Fourier domain equations for the propagation of the stochastic part of the field and its correlations. Field correlations of the form $S_{i,j}(\omega_1, \omega_2, \zeta) \equiv < \delta\phi_i(\omega_1, \zeta)\delta\phi_j(\omega_2, \zeta) >$ where $\delta\phi_1 = \phi - <\phi>$ and $\delta\phi_2 = \phi^\dagger - <\phi^\dagger>$, can thus be numerically propagated.

Homodyne detection of the noise spectrum would be accomplished by phase shifting the coherent field of the pulse relative to the stochastic fields, e.g. by reflection from a phase-shifting interferometer (2) matched to the pulse repetition rate. This technique also eliminates local oscillator phase jitter due to phase noise within the interferometer bandwidth, e.g. from microphonic pickup by the fiber or the low frequency 1/f refractive index fluctuations.

If $\phi_j^0 \exp(i\theta)$ are the local oscillator spectral components, the detected photocurrent noise spectrum relative to the vacuum is given by:

$$V(j\Delta\omega) = \left(\frac{2\Delta\omega}{\pi} \right) \mathrm{Re} \left\{ \sum_{k, l} [(\phi_{j-l}^0 \phi_{k-j}^{0*} + \phi_{-j-l}^0 \phi_{k+j}^{0*}) S_{12}(l\Delta\omega, -k\Delta\omega) + e^{i2\theta} \phi_{j-l}^0 \phi_{k-j}^{0*} S_{11}(l\Delta\omega, -k\Delta\omega)] \right\}. \tag{6}$$

The squeezing spectrum resulting from such a calculation for g = 5 and $\gamma = 0.1$ are shown in figure 1. For short propagation distances the noise is squeezed over the entire soliton bandwidth, but as the pulse continues down the fiber, the squeezing is fully developed only near zero frequency, while noise at frequencies near unity or greater in normalized units is amplified for both quadratures. The noise reduction versus propagation distance is shown for zero frequency and various phase noise parameters in figure 2. When the GAWBS bandwidth is small compared to that of the soliton pulse, relatively little reduction of the squeezing is obtained, while for GAWBS bandwidth of 1 or greater, the squeezing is severely reduced. This results from a reduction in the characteristic propagation length which goes as τ_0^2 where τ_0 is the soliton pulse width in real units. As τ_0 is reduced (i.e. γ is reduced), the low frequency phase noise actually increases for a given fiber length due to the increase in the soliton photon number. However, the noise is squeezed more rapidly with distance so that the relative phase noise per normalized propagation length is diminished.

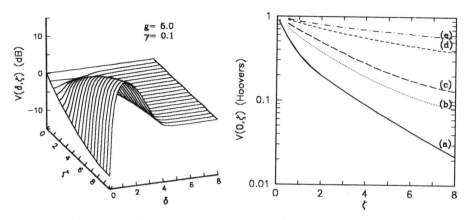

Figure 1. Soliton squeezing spectrum for optimum phase with g = 5 and γ = 0.1.

Figure 2. Noise variance relative to vacuum at $\delta = 0$ for (a) g = 0; (b), (d) and (e) g = 5 and γ = 0.1, 1, and 3, respectively. (c) g = 10 and γ = 0.1.

4. Conclusion

Probably the most important problem to be overcome to generate useful amounts of squeezing in an optical fiber is that of thermal fluctuations of the fiber refractive index and the resulting excess phase noise. The approach discussed here is to use soliton pulses which are short compared to the inverse of the bandwidth of the phase noise, allowing significant squeezing to be obtained in fibers short enough that the phase noise is not dominant. The phase noise bandwidth (and thus the necessary pulse width) is not known. It should be noted that for pulses significantly narrower than 1 psec, noise added by Raman scattering from localized acoustic phonons may begin to add noise. This has not been considered in detail.

References

1. See for example, R.E. Slusher, et.al., J.Opt.Soc.Am. B **4,** 1453 (1987); L.A. Wu, et.al., **ibid.,** 1465; M.W. Maeda, et.al., **ibid.** 1501.

2. G.J. Milburn, et.al., J.Opt.Soc.Am. B **4,** 1476 (1987).

3. R.M. Shelby, et.al., Phys.Rev. B **31,** 5244 (1985).

4. S.H. Perlmutter, et.al., Phys.Rev.Lett. **61,** 1388 (1988); S.H. Perlmutter, et.al., to be published.

5. S.J. Carter, P.D. Drummond, M.D. Reid, and R.M. Shelby, Phys.Rev.Lett. **58,** 1841 (1987); P.D. Drummond and S.J. Carter, J.Opt.Soc.Am. B **4,**1565 (1987); P.D. Drummond, S.J. Carter, and R.M. Shelby, Opt.Lett. **14,** 373 (1989).

6. L.F. Mollenauer, Phil.Trans.R.Soc. London Ser. A **315,** 435 (1985).

Squeezed Light: Progress and Perspectives

E. Giacobino, T.Debuisschert, C.Fabre, A.Heidmann, J.Mertz and S.Reynaud.

Laboratoire de Spectroscopie Hertzienne de l'ENS, Université Pierre et Marie Curie, 75252 Paris Cedex 05, France

The signal-to-noise ratio in many high resolution spectroscopy measurements is now limited by the standard quantum noise, or shot noise. This noise arises from a Heisenberg inequality relating the two phase quadrature components E_1 and E_2 of the electric field. E_1 and E_2 are operators which do not commute, and the product of their fluctuations verifies $\Delta E_1 . \Delta E_2 \geq E_0^2$, where E_0 is a constant. The ground state of the field, or vacuum state, has zero mean value and fluctuations $\Delta E_1 = \Delta E_2 = E_0$. In most light sources, there is no preferred phase for the quantum fluctuations and $\Delta E_1 = \Delta E_2$. For well stabilized sources and at high enough noise frequency, these fluctuations can be as small as the standard quantum limit E_0.

Following investigations in the field of quantum optics, initiated a decade ago, it has been proposed to generate "squeezed" states of the light in which the fluctuations in one quadrature are reduced with respect to the standard quantum limit E_0, at the expense of increased fluctuations in the other one[1]. The concept of squeezing can be extended to quantities involving several light beams. Among those, twin beams are of particular interest[2]. Twin beams ideally have perfectly correlated intensities: the noise in the difference between their intensities is zero. The conjugate quantity, which is the phase difference is then expected to have large quantum fluctuations.

1. Squeezed Light Generation

The schemes which have been shown to generate squeezed light make use of either nonlinear optical interactions, such as three- and four-wave mixing[3-4-5], or optoelectronic regulation of the photon flow of a laser by reduction of the pump fluctuations. Several of these processes rely on the creation of two signal photons while one or two pump photons are annihilated. When the process is "degenerate", the two photons are created in the same mode of the field, and squeezing of the emitted field occurs in some quadrature component. When the process is non degenerate, the signal photons are created in different modes, and the squeezed quantity is the difference of the intensities of the two fields: the system generates twin beams. In this section, we will give some more details about two experiments, based on parametric generation in a resonant cavity, which have yielded the largest quantum noise reductions to date.

In Ref[4], a *degenerate* parametric oscillator (OPO) made of a LiNbO$_3$ crystal in an optical cavity and pumped with a single mode doubled Nd:YAG laser is operated *below threshold*. To show evidence for the squeezing, the signal field exiting the cavity is combined with the fundamental component of the YAG laser (which acts as a local oscillator) in a balanced homodyne detector.

The fluctuations of the homodyned signal I are directly proportional to the fluctuations ΔE_S of one quadrature of the signal field: $\Delta I = \sqrt{I_{LO}} \, \Delta E_S$. If E_S is in a coherent state or in the vacuum state, $\Delta E_S = E_0$, and the balanced detector sees the usual shot noise. If the signal beam is in a squeezed state, the fluctuations in the difference signal can become smaller or larger than the shot noise, depending on which quadrature of the field is squeezed.

The occurence of squeezing in this case can be understood in terms of a semiclassical transformation by the parametric oscillator of the vacuum fluctuations[6] entering the cavity. The equation for the signal field α inside the optical cavity is:

$$\dot{\alpha} = -\gamma\alpha + g\alpha^* + t\alpha^{in} \tag{1}$$

where γ, the damping rate of the field in the cavity, is related to the amplitude transmision coefficient t of the output mirror by $\gamma = t^2/2$ (t is assumed to be small); g is the parametric gain, and α^{in} is the vacuum field entering the cavity through the output mirror.

Writing Eq. 1 and its complex conjugate in the frequency domain and using the phase quadratures $\alpha_1 = \alpha + \alpha^*$ and $\alpha_2 = \alpha - \alpha^*$ (with similar notations for α_1^{in} and α_2^{in}), one gets:

$$\alpha_1 = \frac{t\,\alpha_1^{in}}{\gamma - g + i\Omega} \qquad\qquad \alpha_2 = \frac{t\,\alpha_2^{in}}{\gamma + g + i\Omega} \tag{2}$$

The outgoing field α^{out} is related to the field inside the cavity and to the incoming vacuum field by:

$$\alpha^{out} = t\alpha - \alpha^{in} \tag{3}$$

Using Eqs. 2 and 3, one gets:

$$\alpha_1^{out} = \frac{\gamma + g - i\Omega}{\gamma - g + i\Omega}\,\alpha_1^{in} \qquad\qquad \alpha_2^{out} = \frac{\gamma - g - i\Omega}{\gamma + g + i\Omega}\,\alpha_2^{in} \tag{4}$$

Equations 4 clearly show that the vacuum field is amplified in quadrature α_1, while quadrature α_2 is deamplified and even goes to zero for $\Omega = 0$ and $\gamma \to g$: the vacuum field is squeezed at zero noise frequency when threshold is approached. The noise reduction obtained in Ref.[4] is 63% .

Squeezing also occurs in a parametric oscillator operated *above threshold* in the *non degenerate* regime[5]. The nonlinear medium is a type II KTP crystal pumped by an Ar+ laser. The emitted signal and idler beams , which have orthogonal polarizations, are separated at the output of the optical cavity with a polarizing beamsplitter. The intensities of the two beams are measured and subtracted; the noise in the intensity difference is found to be squeezed.

The semiclassical method used for the OPO below threshold can be extended to the above threshold case by linearizing the fluctuations around the steady state. The noise spectrum on the intensity difference I is defined as

$S(\Omega) \approx < \delta I(\Omega) \, \delta I(\Omega)^* >$ where $\delta I(\Omega)$ is a Fourier component of the fluctuations in I. One finds[2][5]:

$$S(\Omega) = \frac{\Omega^2}{\Omega^2 + 4\gamma^2} \tag{5}$$

where the shot noise is normalized to 1; γ is the cavity damping coefficient. At zero noise frequency, perfect squeezing is expected in the intensity difference. The experimental spectrum is found to be in good agreement with theory if losses are taken into account. Those are shown to degrade the maximum squeezing, explaining the measured 69% noise reduction[5].

2. Spectroscopy with Squeezed Light

Squeezed light having lower than shot noise fluctuations is expected to improve the signal-to-noise ratio in high resolution spectroscopy experiments. However, squeezed light is fragile and squeezing is easily destroyed by transmission losses. Losses can be modeled in a general way with a beamsplitter having amplitude transmission and reflexion coefficients t and r. The output field E results from the interference of the fields incident on the beamsplitter at its two input ports, the squeezed field E_S (assumed here to have a non zero mean value) and the vacuum field E_0:

$$E = t \, E_S + r \, E_0 \tag{6}$$

If the input field has an amplitude squeezing spectrum $S_A(\Omega) \approx < \delta E_A(\Omega) \, \delta E_A(\Omega)^* >$, the squeezing spectrum $S_I(\Omega)$ of the intensity I measured after the beamsplitter, e.g. the lossy element, is shown to be:

$$S_I(\Omega) = t^2 \, S_A(\Omega) + r^2 = \eta \, S_A(\Omega) + (1-\eta) \tag{7}$$

where η is the loss coefficient ($\eta = t^2$). It is clearly seen in Eq. 7 that the losses couple back some of the vacuum fluctuations into the squeezed light and tend to bring back the noise spectrum to shot noise (normalized to 1 in the above expression).

Quadrature squeezed light is particularly well suited to interferometric measurements of phase shifts[7]. A Mach-Zender interferometer is illuminated with coherent light of intensity I_1 in one port of the input beamsplitter and with squeezed vacuum in the other input port. The phase difference $\Phi(t)$ between the two arms is modulated at frequency Ω : $\Phi(t) = \Phi_0 + 2\delta \cos\Omega t$, with Φ_0 close to $\pi/2$. The interference signal is detected as the difference between the intensities in the two output ports. The Fourier component at frequency Ω is proportional to I_1 and δ:

$$J(\Omega) = \sqrt{2} \, \eta \, I_1 \, \delta \tag{8}$$

where η is the transmission loss coefficient. If non squeezed vacuum is incident at the second input port of the beamsplitter, the noise is the shot noise of a beam with intensity I_1. If a squeezed vacuum field with a noise spectrum $S(\Omega)$ ($S(\Omega) < 1$) is used instead at the second input port, Eq. 7 shows that the noise spectrum becomes:

$$R(\Omega) = \eta \, S(\Omega) + (1-\eta) \tag{9}$$

The signal-to-noise ratio is then improved by a factor $\sqrt{R}$. R factors on the order of 2 have been obtained experimentally[7].

Twin photons are very promising in difference measurements. The simplest one involves ultrasensitive absorption measurement[8]. The sample with an absorption coefficient $\gamma(t) = \gamma \cos\Omega t$ is placed on one of the twin beams, while the other one propagates freely to the detector. The signal at frequency Ω, measured on the intensity difference is proportional to γ:

$$J(\Omega) = 1/(2\sqrt{2}) \, \eta \, I \, \gamma \tag{10}$$

where I is the sum of the intensities of the two beams. If uncorrelated beams are used, the signal will be detected on a noise background equal to the shot noise of a beam with intensity I. In contrast, with twin beams, the noise spectrum is:

$$R(\Omega) = \eta \, S_I(\Omega) + (1-\eta) \tag{11}$$

where $S_I(\Omega)$ is the squeezing spectrum of the intensity difference between the twin beams. Here also the signal-to-noise ratio is improved by a factor $\sqrt{R}$.

When the twin beams have orthogonal polarizations and frequencies which differ by some value Ω, they can be used to measure a small *unmodulated* polarization rotation with an improved signal-to-noise ratio[9]. The polarizations of the beams are assumed to rotate by a small angle β due to some Faraday effect. The detection of the intensities is done after separating the beams with a polarizing beamsplitter whose axes are parallel to the initial polarizations. On each channel, a small part of each beam beats with the other one:

$$I_{\pm} = I/2 \, | \, 1 \pm \beta \exp(i\Omega t) \, |^2 \tag{12}$$

The Fourier component of $I_+ - I_-$ at frequency Ω is proportional to β. The background noise is reduced by the same factor $\sqrt{R}$ as before if twin beams are used.

References
1   D.F. Walls, Nature 306 141 (1983)
2   S. Reynaud, C.Fabre and E.Giacobino, J. Opt. Soc. Am. B 4 1520 (1987)
3   R.E. Slusher, L.W.Hollberg, B. Yurke, J.C.Mertz and J.F.Valley, Phys. Rev. Lett. 55 2409 (1985). A review of the recent experimental results on squeezing can be found in: H.J.Kimble in "Atomic Physics 11" Eds S.Haroche, J.C.Gay and G.Grynberg World scientific (1989) p467.
4   L.A. Wu, H.J. Kimble, J.L. Hall and Huifa Wu, Phys. Rev. Lett. 57 2520 (1986).
5   A.Heidmann, R.J.Horowicz, S.Reynaud, E.Giacobino and C.Fabre, Phys. Rev. Lett. 59 2555 (1987); T.Debuisschert, S.Reynaud, A.Heidmann, E.Giacobino and C.Fabre, Quantum Opt. 1 (1989); C.Fabre, E.Giacobino, A.Heidmann and S.Reynaud, J.Phys. France 50 1209 (1989).
6   S.Reynaud and A.Heidmann, Optics Comm. 71 209 (1989)
7   M.Xiao, L.A.Wu and H.J.Kimble, Phys. Rev. Lett. 59 278 (1987); P. Grangier, R.E. Slusher, B. Yurke and A. La Porta, Phys. Rev. Lett. 59 2153 (1987).
8   E.Giacobino, C.Fabre, S.Reynaud, A.Heidmann and R.Horowicz, ONR seminar on "Photons and Quantum Fluctuations" eds E.R.Pike and H.Walther, Adam Hilger (1988) p81.
9   J.J.Snyder, E.Giacobino, C.Fabre, A.Heidmann and M.Ducloy, to be published.

An experimental search for squeezing via four-wave mixing in Barium

H.-A. Bachor, P.T.H. Fisk[†], P.J. Manson, D.E. McClelland and D.M. Hope
Department of Physics and Theoretical Physics
†Laser Physics Centre, Research School of Physical Sciences
The Australian National University , P.O. Box 4, Canberra , ACT 2601, AUSTRALIA

It is well established both theoretically and experimentally [1-3], that four-wave mixing in an atomic medium is one of the nonlinearities which can be used to generate a nonclassical, squeezed state of light. We report experiments in Barium using the ^1S-^1P transition in ^{138}Ba. Our motivation was that this 2-level system might provide a stronger and more easily describable nonlinearity than the already successful case of sodium, which is complicated by optical pumping. However, at present we observe an apparent maximum in the magnitude of four-wave mixing in regard to atomic density, detuning from atomic resonance and laser pump intensity which would result in a suppression of the quantum noise of only 5% below the standard quantum limit. This is a degree of squeezing too small to be observed with our present heterodyne detector. Further theoretical work is required in order to explain these observations.

1. Experimental requirements for the generation of squeezing

The generation of squeezed light relies on the establishment of a suitable correlation between the fluctuations of two modes of a light field. This can be done by propagating both modes through a nonlinear medium [3]. The two modes and their correlation are best described in terms of their quadrature phases. By simultaneously detecting both modes with a heterodyne detector, the fluctuations in the two quadrature phases of the combined modes can be independently measured. The aim is to demonstrate that the fluctuations for one quadrature phase are suppressed below the standard quantum limit. In the case of four-wave mixing in an atomic system, the atoms are illuminated by two counterpropagating beams of light at frequency ν_p, the pump beams, which are tuned close to an atomic resonance in order to enhance the χ^3 nonlinearity. This optically prepared medium will amplify a probe beam at $\nu_{pr} = \nu_p - \delta$ and simultaneously generate a phase-conjugate beam , the signal, at $\nu_s = \nu_p + \delta$. In order to enhance the nonlinearity and to superimpose the probe and signal beam the atoms are enclosed by a cavity which is simultaneously in resonance with these two beams (fig. 1). The generation of squeezing requires maximizing the nonlinearity, whilst minimizing losses in the cavity, such as mirror losses and atomic absorption, and additional noise contributions, such as spontaneous emission and phase jitter.

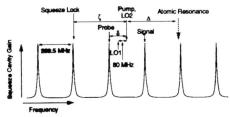

Fig.1 Location of the various beams in frequency space.

2. Experimental arrangement

The experiment, as shown in fig.2, used a high density barium beam with a residual Doppler width of 30MHz, an interaction length l of of 4mm and a line centre absorption of $1 < \alpha_0 l < 5$. The atoms were enclosed in an asymmetric, confocal cavity of length 0.5m with mirror intensity reflectivities R of 0.985 and 0.955. The cavity had a measured finesse of about 80 and the free spectral range (FSR) of the empty cavity was 298.5 MHz. It was locked, using the technique of sideband modulation, to a beam shifted by $\zeta = 1.5$ FSR from the pump frequency ν_p. The locking was sufficiently tight to obtain a frequency jitter of less than ± 1Mhz corresponding to a phase jitter of the empty cavity of less than $\pm \pi/12$. In this experiment two cavity modes at ± 0.5 FSR on either side of the pump have to be kept in resonance . The atomic absorption distorts and shifts the mode structure of the cavity. These shifts are different for the probe and the signal beam. However, by optimizing the frequency offset ζ and the probe detuning δ it is possible to keep both probe and signal in resonance with the cavity. The output beam is detected by a heterodyne detector using two different local oscillators LO1 and LO2 . LO1 is shifted 80 MHz from the pump frequency, and allows the separate detection of probe and signal. LO2 is at the pump frequency and thus superimposes the RF-signals produced by the probe and the signal beam. The output signal is measured and displayed by an electronic spectrum analyzer.

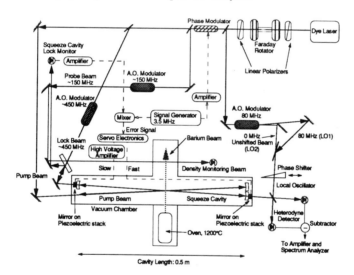

Fig.2 The experimental arrangement.

3. Results

Using the heterodyne detector with LO1, the ratio of the maximum amplitude of the four-wave mixing signal at $\nu_p + \delta$ to that of the probe beam transmitted by the cavity was measured to be 23%. This result is consistent with the measured amplitude of the local oscillator phase dependence of the signal obtained using LO2, where the transmitted probe and four-wave mixing signals are combined (fig. 3). This maximum occurred at pump field detunings of

300MHz $< \Delta <$ 450MHz. Neither by increasing the density nor by changing the pump intensity was it possible to increase this ratio. The transmission function of the cavity was measured by scanning the frequency of the probe beam acousto-optic modulator, and the single pass cavity loss (2% in intensity) and the shift of the modes due to the barium atoms (1.5MHz) at $\Delta =$ 300 MHz were determined by comparing results obtained with and without the barium beam. The model by Yurke et. al.[5] gives an analytic expression for the propagation of probe and signal through a cavity assuming identical gain and losses for both. Using this model we determined the single pass gain as approximately 0.5% in amplitude for our conditions and an expected degree of squeezing of approximately 5% below the standard quantum limit. We were able to observe spontaneous emission, centered around the pump frequency, which contributed an additional noise of less than 10% of the quantum noise at $v_p \pm \delta$. However, due to limitations in our detectors and amplifiers, the resolution of our noise measurements is currently no better than $\pm 5\%$ of the quantum noise level. Consequently no squeezing has yet been observed .

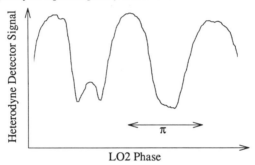

Fig.3 Example of the local oscillator (LO2) phase dependent four-wave mixing signal.

4. Analysis and future directions

We are surprised by our failure to achieve higher four-wave mixing efficiencies. Further theoretical and experimental work is required to explain this result. Improvements to our detectors and thus the resolution of the noise measurement should allow us to observe squeezing below the standard quantum limit. However, the degree of squeezing achievable in four-wave mixing does not compare favorably with the results produced using other nonlinearities[5].

References
1. H.P.Yuen and J.H.Shapiro: Opt.Lett. 4, 334 (1979)
2. R.E.Slusher, B.Yurke, P.Grangier, A.LaPorta, D.F.Walls and M.Reid:
   JOSA B 4, 1453 (1987)
3. R.Loudon and P.L.Knight: J.of Modern Optics 34, 709 (1987)
4. B.Yurke and R.E. Slusher: "Quantum Optics IV", Springer-Verlag (1986)
5. L-A. Wu, M. Xiao and H. J. Kimble: JOSA B 4, 1465 (1987)

Coherence in Random Multiple Scattering: Correlations in the Speckle of Multiply Scattered Light

Michael Rosenbluh, Isaac Freund and Moshe Kaveh
Department of Physics, Bar Ilan University, Ramat Gan, 52100, ISRAEL

When laser light traverses a multiply scattering disordered medium it emerges in all directions and appears to have irretrievably lost all information concerning the initial direction of the incident beam. Similarly, due to the total depolarization of strongly multiply scattered light, it would appear that information as to the initial polarization of the incident beam will also be totally scrambled as the light diffuses through the disordered medium. In spite of these intuitive expectations, we have been able to show that much information about the spatial variations of the incident waveform is retained in the transmitted or reflected speckle pattern. Even more complete information can be recovered concerning the polarization variations of the incident wave. Through the appropriate mathematical manipulation of only a small fraction of the intensity distribution of the scattered radiation pattern, information about the incident wavefront direction, phase, amplitude and polarization direction can be recovered. These observations thus suggest that, at least in principle, it is possible to image distant objects through a strongly scattering, highly random, static diffuser.

The reconstruction of the phase and amplitude variations of the incident wave from the scattered speckle pattern, known as the "memory effect", was predicted theoretically (1) based on diagrammatic calculations. Theory predicts (2) that transverse spatial variations of the incident wave that are on a scale smaller than L, the thickness of the scattering medium are indeed "forgotten". (L is the physical thickness of the scatterer for transmitted speckle patterns, and it is the transport mean free path, ℓ, for the reflected speckle pattern.) Variations on a larger scale, however, can be recovered by computing the intensity-intensity correlation function of a small portion of the speckle pattern. Thus for imaging applications, the incident waveform could in principle be reconstructed within a limiting resolution of the transverse spreading length scale of the light as it propagates through the diffuser.

The experimental confirmation of the memory effect (2-3) indeed demonstrates that the direction of the incident laser beam can be tracked via the correlation function of the speckle patterns. The experiments were performed by illuminating a solid multiple scatterer with a 5 mW, polarized, He-Ne laser, whose beam was expanded to 15 mm diameter and then apertured at the sample to 6 mm to provide a nearly uniform plane wave. The reflected or transmitted speckle pattern was recorded by a CCD video camera and computerized digitizer. The recorded image was stored as the reference pattern representing a "fingerprint" of the particular configuration of the scattering medium. Subsequently, a linear phase ramp was introduced on the incident wavefront by changing the direction of the incident laser or the sample

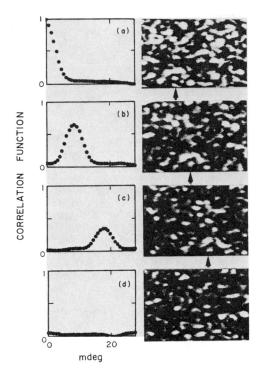

Fig. 1. The memory effect in tranmission for 370 μm opal glass. In images (b) through (d) the incident laser direction was rotated by 10 mdeg, (b), 20 mdeg, (c), and 1 degree, (d). On the right-hand side are images of the speckle patterns, with an arrow indicating a particular feature of the speckle pattern for visual reference. On the left are shown the correlation functions, for which the image in (a) was used as the "reference image". Figure (d) demonstrates that the correlation function for two unrelated images is zero.

orientation, and a new speckle pattern was recorded. The normalized intensity-intensity correlation function of the reference pattern with the new pattern was computed as;

$$C(\delta k) = | <E_1(K_1) \, E_2^*(K_2)> |^2 / <E_1(K_1)>^2 <E_2(K_2)>^2 \qquad [1]$$

where E_1 and E_2 indicate the electric field distributions in the reference and changed speckle patterns, the K's are outgoing wavevectors, and $\delta k = K_2 - K_1 = 2\pi\Delta/\lambda$, with Δ indicating the rotation of either the laser beam direction or the sample orientation.

In Fig. 1 we show the memory effect as obtained for transmission through a t=370 μm thick opal glass. On the right-hand side are shown a series of photographs of a small part of the scattered speckle pattern. As the laser direction rotates by 10 mdeg (in b) and by 20 mdeg (in c) the speckle pattern can be seen to track the laser direction. At a very large rotation angle (about 1 degree in d) where $\delta k t$ is much greater than unity, memory loss is clearly evident, and a totally unrelated speckle pattern is obtained. The correlation function of each of the patterns is shown on the left hand-side of the figure. In (a) this was obtained by multiplying the original speckle pattern (reference pattern) by its replica as the replica is shifted with respect the original. This results in a self correlation which has a width corresponding to the average angular extent of a single speckle spot. In (b) and (c) are shown the correlation functions of the reference pattern with the new pattern as a function of angular shift of the new pattern. This shows how the correlation function tracks the incident laser direction. In (d) the correlation function is close to zero since it represents the correlation with a pattern in which the memory has been lost.

188

The actual motion and change of the speckle patterns depends on whether the laser direction or sample orientation rotates, whether one is observing reflected or transmitted speckle, and the geometry of the incident and observation directions. In Fig. 2a we show the angular displacement of the peak of the correlation function as a function of angular rotation of either the laser $L(\theta_i,\theta_o)$ or the sample $S(\theta_i,\theta_o)$ for various laser incidence, θ_i, and speckle observation angles, θ_o, as measured from the sample normal. These measurements were obtained for a scatterer consisting of TiO_2 dispersed in polystyrene sheet, and show that for some geometries-[S(0,0)], the speckle pattern is stationary, while in others-[L(0,0)], it tracks the laser motion. In some cases-[S(0,60)], it even moves opposite to the rotation direction. The simple correlation function of Eq. (1), however, describes all of these cases. In Fig. 2b the correlation functions for the two extreme cases of Fig. 2a are shown to be identical. When sample surface reflections are accounted for, it can also be shown (4) that the theoretically predicted correlation function of Eq. (1) corresponds to that obtained in the experiment.

The complete memory of the incident wave polarization is demonstrated in Fig. 3. In these experiments (5) a linearly polarized laser was scattered from various samples having different values of the depolarization ratio, $\rho = I_\perp/I_\parallel$, where $I_\perp$ is the scattered intensity measured with a polarizer perpendicular to the input and $I_\parallel$ is the intensity with a polarizer parallel to the input. The correlation function in this case is computed by first recording a reference speckle pattern for a given incident polarization state, $I(\theta_o)$, and then rotating the input polarization to a new direction and recording a new pattern, $I(\theta)$. The correlation function is then given by

$$c(\theta_o,\theta) = \frac{[<I(\theta_o)I(\theta)> - <I(\theta_o)><I(\theta)>]}{\sqrt{[<I^2(\theta_o)> - <I(\theta_o)>^2][<I^2(\theta)> - <I(\theta)>^2]}} \qquad (2)$$

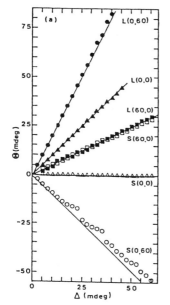

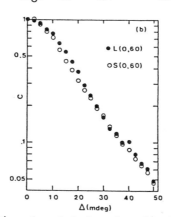

Fig. 2. (a) Angular displacement of the correlation function in transmission vs angular rotation, Δ, of either laser or sample, for various representative geometries. (b) Memory effect correlation function vs Δ for the two extreme cases shown in (a).

189

which for the case of a totally depolarized speckle pattern reduces to $C(\theta_o,\theta) = \cos^2(\theta-\theta_o)$. The general expression for arbitrary depolarization can also be shown to be $C(\theta_o,\theta) = (1-\beta)\cos^2(\theta-\theta_o)+\beta$, where $\beta=(1-\rho^2)/(1+\rho^2)$.

In Fig. 3 the polarization memory is demonstrated for three different samples with varying degrees of scattering depolarization, ρ. In all three cases the solid line is from Eq. 2, where we used the measured value of the average depolarization ratio. From this data it is obvious that given a reference speckle pattern, the incident polarization that was used in obtaining any other speckle pattern from the same scatterer configuration, can be immidiately determined. Correlation functions in the case when the detection is performed through a polarizer have also been derived (5), and shown to yield complete memory of the changes in the input polarization.

In conclusion, we have demonstrated that the apparently random and depolarized speckle patterns that are obtained upon strong multiple scattering of coherent light are not nearly as random as they seem. The intensity-intensity correlation function of different speckle patterns obtained from the same scatterers, can be used to reveal what changes occured in the incident illuminating beam. In principle, therefore, one can envision imaging and information transmission through strongly scattering media.

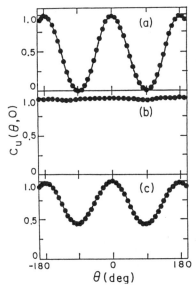

Fig. 3. Polarization correlation function in reflection without a polarizer at the detector. (a) $\rho = 0.94$ for $BaSO_4$ diffuse reflectance coating, (b) $\rho =0.002$ for bright stainless-steel surface scatterer, (c) $\rho =0.29$ for TiO_2 particles in polystyrene. The solid lines are from Eq. 2.

References

1. S. Feng, C. Kane, P. A. Lee, A. D. Stone, Phys. Rev. Lett., 61, 834 (1988).
2. I. Freund, M. Rosenbluh, S. Feng, Phys. Rev. Lett., 61, 2328 (1988).
3. I. Freund, M. Rosenbluh, R. Berkovits, Phys. Rev. B, 39, 12403 (1989).
4. I. Freund and R. Berkovits, submitted for publication.
5. I. Freund, M. Kaveh, R. Berkovits, M. Rosenbluh, submitted for publication.

RESONANCE FLUORESCENCE FROM ATOMS EXCITED BY STRONG STOCHASTIC LASER FIELDS

YEHIAM PRIOR, A.M.LEVINE[*], A.G. KOFMAN, and R. ZAIBEL
Department of Chemical Physics, Weizmann Institute of Science, Rehovot, Israel 76100.

Strong laser fields, when tuned on resonance to an atomic (molecular) transition, give rise to the well known triplet of resonance fluorescence (ResFl). The laser field is normally assumed to be monochromatic, even though the spectrum of real lasers, due to random processes in the laser itself, is never infinitely sharp . A very common way to treat laser field fluctuations has been to assume a stable amplitude, a well defined nominal frequency, and a fluctuating phase. Several models for phase fluctuations had been proposed, analyzed and applied to different problems in nonlinear optics. These include, among others, the Phase Diffusion Model, the uncorrelated phase jump model and the phase telegraph noise model[1]. Recently[2], we introduced the Generalized Jump Model (GJM) for phase fluctuations, which deals with **correlated** phase jumps. In this model, Markovian as well as non Markovian stochastic processes are discussed, and a general set of equations is derived for the treatment of nonlinear optical problems. In this paper we briefly review the model and discuss its application to the study of resonance fluorescence.

In the GJM, each phase jump β is defined to depend on the preceding jump β' according to the conditional probability density $h(\beta',\beta) = h(\beta-\gamma\beta')$. The parameter γ (obeying $|\gamma| \leq 1$) is a measure of the correlation between β and β'. The function that we used in this model is the Kielson Storer function[3], introduced originally for molecular collisions, but a much wider class of similar functions may be used instead. Thus, in the present work the conditional probability is given by

$$h(\beta',\beta) = \frac{1}{B\,[2(1-\gamma^2)]^{1/2}} \exp\,[\frac{-(\beta-\gamma\beta')^2}{2(1-\gamma^2)B^2}] \qquad (1)$$

In the limits, this model reduces to well known cases, and three obvious limits can be pointed out: For $\gamma=0$, the jumps are uncorrelated, and the model under consideration reduces to the Markovian case, studied in detail by Burshtein and coworkers[4], who derived the Burshtein equation for the time development of partially averaged density matrices. For $\gamma=1$ the jumps are fully correlated, i.e. the phase jumps at random times, but the jump size is fixed for any realization of the stochastic process. Under certain conditions, the case of $\gamma\approx1$ is almost analogous to frequency modulation. For $\gamma=-1$, the phase jumps are fully anticorrelated ($\beta=-\beta'$), and the system jumps (for any realization of the stochastic process) between two states. This case is equivalent to the random phase telegraph noise model of stochastic processes.

A laser field is defined by three stochastic parameters: a characteristic jump size B, the mean time between jumps τ_0 (taken

---

[*]   Permanent address: College of Staten Island, CUNY, Staten Island, NY, 10301.

from a Poissonian distribution), and the degree of correlation γ. The three together define the observed laser linewidth, but they should all be known in order to predict the results of a particular experiment. The analysis of the field is discussed in detail elsewhere[5], and is summarized here in fig. 1. The phase space spanned by $0<B<\infty$ and $-1\leq\gamma\leq1$ defines the range of the parameters, and in all the shaded regions we have obtained analytic solutions for the field correlation function and hence for the field spectrum. The figure includes cartoons depicting the negative logarithm of the correlation function, to show the typical properties of the field in the different regions.

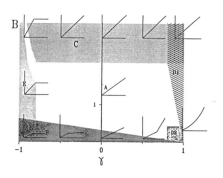

Figure 1. B-γ phase space. (with cartoons for the input field log correlation function)

Following the definition of the model, we have analytically derived a set of integro-differential equations, which replace the optical Bloch equations for stochastic input fields. These equations were solved for the case of resonance fluorescence of a Two Level System excited on resonance by the stochastic laser field defined above, under the assumption of a well resolved triplet. Analytic results have been obtained[5] in the shaded regions, and INDEPENDENT numerical solutions of the Bloch equations were obtained over the entire phase space and compared to the analytic results with excellent agreement. In this short paper only very few observations regarding the structure of the ResFl spectrum are given:

a. For uncorrelated jumps ($\gamma=0$, Region A), which is the Burshtein case of Markovian phase fluctuations, the lineshapes are Lorentzians, but the width ratio differs from the monochromatic case. Fig. 2 depicts the ratio of the width of the central (c) and side (a) components in a ResFl spectrum, and the ratio of the central component to the input field ratio as a function of the jump size.

b. In the small jump limit (B<<1, region B), the lineshapes are Lorentzians, but again, the ratios deviate form the predictions for the monochromatic case. In this limit a new effect is predicted: the ratio of the widths depend on the Rabi frequency (Fig.3). With increasing applied field, the c/a ratio decreases for anti-correlated jumps and increases for correlated jumps. An intuitive explanation may be given by observing that the two competing time scales are the Rabi oscillation period and the phase memory time of the system. If the field amplitude is such that the Rabi oscillations are faster (slower) than the memory time of the jump process, effective narrowing (broadening) occurs.

c. In the large jump limit (B>>1, region C), the lineshapes are Lorentzian and again the ratio of the widths is derived.

d. In the highly correlated region (region D_1, D_2) the observed lineshape changes from the "standard" Lorentzian into a Gaussian, with a transition region between them, both towards the small jump region and towards the large, highly correlated jump region.

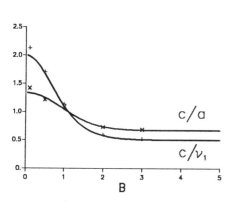

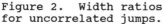

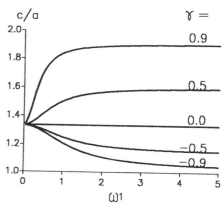

Figure 2. Width ratios for uncorrelated jumps.

Figure 3. Field dependence of the widths for small jumps

Three general comments should be made about the Generalized Jump Model:

1. A set of integro-differential equations has been derived, which enables the solution of any nonlinear optical problem within the assumptions of the GJM.

2. A procedure is defined for the extraction of the stochastic parameters, both from the input correlation function and from the ResFl spectrum. In some cases (non Lorentzian lineshapes) these parameters may be unambiguously derived from the input field alone, while in others, a nonlinear optical interaction (i.e. ResFl) is needed.

3. In all cases where the lineshape is Lorentzian, and in some more general cases[5], generalized relaxation rates Γ_1^* and Γ_2^* may be substituted for the usual relaxation rates in the optical Bloch equations.

In conclusion, a new model has been introduced for the treatment of laser field phase fluctuations, and applied to the case of Resonance Fluorescence. The model regains the known results in the proper limits, and makes several new predictions.

This work was partially supported by the US-Israel Binational Science Foundation.

1. A.I. Burshtein, Sov. Phys. JETP, **22**, 939 (1966); J.H. Eberly, Phys. Rev. Lett., **37**, 1387 (1976); G.S. Agarwal, ibid. 1383; P. Avan and C. Cohen Tannoudji, J. Phys. B, **10**, 155 (1977);J.H. Eberly, K. Wodkiewicz and B.W. Shore, Phys. Rev. **A30**, 2381 (1984).

2. A. G. Kofman, R. Zaibel, A.M. Levine and Y. Prior, Phys. Rev. Lett. **61**, 251 (1988)

3. J. Keilson and J.E.Storer, Quart. Appl. Math. **10**, 243 (1952).

4. A.I. Burshtein and Yu.S. Oseledchik, Sov. Phys. JETP, **24**, 716 (1967).

5. A. G. Kofman, R. Zaibel, A.M. Levine and Y. Prior, Phys. Rev. A (to be published).

Atomic Photon Statistics near a Phase Conjugator

B.H.W. Hendriks and G. Nienhuis[1]

Fysisch Laboratorium, Rijksuniversiteit Utrecht, Postbus 80 000,
3508 TA Utrecht, The Netherlands

1. Introduction

The atomic evolution near a phase-conjugate mirror (PCM) has interesting new properties in comparison to the evolution near an ordinary mirror in that it exhibits a phase-dependent decay of the coherences [1]. This is caused by the inherent phase of the PCM due to the driving pump fields. This essential phase-dependent evolution opens new ways to affect the spectral properties of the emitted radiation.

In this contribution we discuss the phase-dependent photon statistics of the fluorescence of an atom which is driven by a monochromatic field near a PCM. The nonclassical properties of the photon statistics can be enhanced due to the presence of the PCM [2].

2. Atomic evolution near a PCM

We consider the case of a two-level atom in front of a PCM that is based on four-wave mixing. We treat the fields coupling to the atom quantummechanically, while the four-wave mixer is driven by classical pump fields. Although classically the field $E_c(\mathbf{r})$ reflected by the PCM is related to the phase conjugate of the incoming field $E_i(\mathbf{r})$ with proportionality constant ν, hence $E_c(\mathbf{r}) = \nu E_i^*(\mathbf{r})$, this no longer holds quantummechanically. In the quantummechanical description of the PCM also the amplified vacuum field coming from the other side of the PCM contributes to the phase conjugate reflected field. In the Heisenberg representation the annihilation field operator a_c of the refected field can be expressed in the creation operator a_i^+ of the incoming field and the annihilation field operator a_v of the vacuum field of the other side of the PCM according to

$$a_c = -\nu a_i^* + \mu a_v .$$ (1)

To account for the effect of the PCM on the evolution, we have to make the substitution (1) in the interaction Hamiltonian $-\boldsymbol{\mu}\cdot\mathbf{E}$ for the field modes that couple to both the atom and the PCM and are directed from the PCM towards the atom. Using standard reservoir theory we obtain then the evolution equation for the density matrix ρ of the atom describing the modified spontaneous decay near a PCM. The solution in a standard rotating frame is found to be [1] (e=excited state, g=ground state)

$$\rho_{ee}(t) = \frac{\alpha|\nu|^2}{1 + 2\alpha|\nu|^2} + \left(\rho_{ee}(0) - \frac{\alpha|\nu|^2}{1 + 2\alpha|\nu|^2} \right) e^{-A(1+2\alpha|\nu|^2)t}$$ (2)

$$\rho_{eg}(t) = |\rho_{eg}(0)|e^{\frac{1}{2}i\psi}[\sin(\varphi - \tfrac{1}{2}\psi)\, ie^{-\gamma_+ t} + \cos(\varphi - \tfrac{1}{2}\psi)\, e^{-\gamma_- t}]$$ (3)

[1] Also at Huygens Laboratorium, Rijksuniversiteit Leiden, Postbus 9504, 2300 RA Leiden, The Netherlands.

with $\nu = |\nu|e^{i\psi}$, A the Einstein coefficient of spontaneous decay in the absence of the PCM, $\gamma_{\pm} = \frac{1}{2}A(1\pm2\alpha|\nu| + 2\alpha|\nu|^2)$, $4\pi\alpha$ the solid angle subtended by the PCM at the atom and φ the initial phase of the atomic coherences. These equations show that $\rho_{ee}(t)$ decays at an enhanced rate to a non-zero stationary value, reflecting that the atom near a PCM can also make transitions from the ground state to the excited state. The coherences decay with two different damping rates. The contribution of these damping rates is determined by the relative phase $\varphi - \frac{1}{2}\psi$. A similar phase-dependent decay is also found for the atomic evolution in a squeezed vacuum [3].

3. Atomic photon statistics

We now consider the case where the atom near a PCM is driven by a resonant monochromatic radiation field. Due to the presence of this radiation field the atomic density matrix will reach a steady state which is found to be [2]

$$\bar{\rho}_{ee} = \frac{\alpha|\nu|^2 A + \frac{1}{2}\Omega^2\Lambda_1(0)}{A(1 + 2\alpha|\nu|^2) + \Omega^2\Lambda_1(0)} \tag{4}$$

$$\bar{\rho}_{eg} = \frac{1}{2}i\Omega A^2 \frac{\frac{1}{2} + \alpha|\nu|^2 - \alpha|\nu|e^{i(\psi-2\varphi_L)}}{\gamma_+\gamma_-[A(1 + 2\alpha|\nu|^2) + \Omega^2\Lambda_1(0)]} \tag{5}$$

with Ω the Rabi frequency, φ_L the phase of the driving field and where we have defined

$$\Lambda_n(s) = \frac{\cos^2(\frac{1}{2}\psi - \varphi_L)}{(s + \gamma_+)^n} + \frac{\sin^2(\frac{1}{2}\psi - \varphi_L)}{(s + \gamma_-)^n}. \tag{6}$$

The driving field couples the evolution of the populations and coherences. Therefore, we find that the steady state populations and coherences depend on the relative phase $\frac{1}{2}\psi - \varphi_L$ due to the presence of the PCM. Furthermore, (4) and (5) demonstrate that the atomic evolution can be manipulated by changing the phase of the driving field.

The statistics of the detected photons can be derived entirely from the intensity correlation function $f(t)$, defined as the probability distribution for a photon detection at time t given that the atom is initially at $t = 0$ in the ground state [4]. The result for the Laplace transform $\tilde{f}(s) = \int_0^\infty e^{-st}f(t)dt$ is found to be [2]

$$\tilde{f}(s) = \kappa A \frac{\alpha|\nu|^2 A + \frac{1}{2}\Omega^2\Lambda_1(s)}{s[s + A(1 + 2\alpha|\nu|^2) + \Omega^2\Lambda_1(s)]} \tag{7}$$

with κ the detection efficiency and $\Lambda_1(s)$ defined in (6). We have assumed that the detector is positioned in such a way that it cannot see the PCM. A striking new feature in $f(t)$ compared with earlier work on photon statistics of free atoms is the explicit phase dependence contained in $\Lambda_1(s)$. This phase dependence originates from the phase-dependent decay of the coherences. Furthermore, even in the absence of the driving field ($\Omega = 0$) $f(t)$ is non-zero. This is not surprising since we have already shown that an atom in front of a PCM can get excited.

Finally we consider the Q-factor introduced by Mandel [4]. This quantity is a measure of the deviation from poissonian statistics and is defined by

$$Q(t) = (\langle n(n-1)\rangle - \langle n\rangle^2)/\langle n\rangle. \tag{8}$$

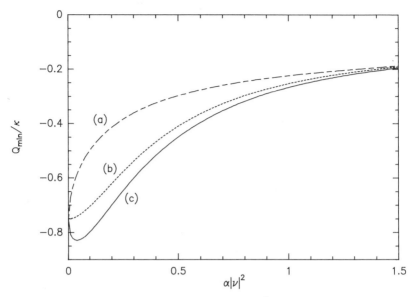

Fig. 1: Plots of the function Q_{min} as a function of $\alpha|\nu|^2$ for $\kappa \ll 1$, $1 - \alpha \ll 1$. The curve (a) corresponds to $\psi - 2\varphi = 0$, (b) to $\psi - 2\varphi = \frac{1}{2}\pi$ and (c) to $\psi - 2\varphi = \pi$.

Negative values correspond to sub-Poissonian light and positive values to super-Poissonian light. Since classical fields can only have non-negative Q-factors [4], sub-Poissonian statistics is an essential nonclassical property of light. For large detection times the Q-factor is found to be [2]

$$Q(\infty) = -\kappa A \frac{2\alpha|\nu|^2 A + \Omega^2[\Lambda_1(0) + A\Lambda_2(0)]}{[A(1 + 2\alpha|\nu|^2) + \Omega^2\Lambda_1(0)]^2} \qquad (9)$$

with Λ_1 and Λ_2 defined in (6). Since $\Lambda_1(0)$ and $\Lambda_2(0)$ are positive functions we find sub-Poissonian photon statistics for all sets of the parameters. In absence of the PCM ($|\nu| = 0$), $Q(\infty)$ has the well-known absolute minimum value $-\frac{3}{4}\kappa$ when $\Omega = \frac{1}{2}\sqrt{2}A$ [4]. To reveal the effect of the PCM, we minimize $Q(\infty)$ with respect to Ω^2 and plot these minimum values Q_{min} in figure 1 as a function of $\alpha|\nu|^2$ for different sets of the parameters. Figure 1 shows that the presence of the PCM makes the Q-factor phase-dependent. Furthermore, the Q-factor can have values below $-\frac{3}{4}\kappa$, demonstrating that the nonclassical properties of the photon statistics can be enhanced due to the presence of the PCM. This enhancement is due to the obstructed decay of the coherences.

References

[1] B.H.W. Hendriks and G. Nienhuis, Phys. Rev. A, to be published.
[2] B.H.W. Hendriks and G. Nienhuis, J. Mod. Optics, to be published.
[3] C.W. Gardiner, Phys. Rev. Lett. **56**, 1917 (1986).
[4] L. Mandel, Optics Lett. **4**, 205 (1979).

Part V

Quantum Size Effects

Spectral Hole Burning and Static Stark Effect in Quantum Confined Semiconductor Microcrystallites in Glasses

Ch. Flytzanis, F. Hache, D. Ricard and Ph. Roussignol
Laboratoire d'Optique Quantique du C.N.R.S., Ecole Polytechnique, 91128 Palaiseau Cedex, France

We observed spectral hole burning in very small $CdS_x Se_{1-x}$ crystallites in a glass matrix. The results clearly show the impact of the electron-phonon broadening and crystallite size distribution and are in agreement with the special conditions of the quantum confinement. In striking agreement with later also stand the measurements of the change $\delta\alpha$ of the absorption spectrum due to a static electric field ; the observed oscillations of $\delta\alpha$ as a function of the wavelength, are well explained as a Stark effect of the quantized electronic levels.

1. Introduction

When photoexcited electrons (e) and holes (h) in semiconductors are artificially confined into volumes whose dimensions are close to or below certain critical lengths their quantum behavior and dynamics undergo drastic changes with respect to those of the bulk ; the critical lengths we have in mind are the electron and hole radii, $a_e = h^2/m_e e^2$ and $a_h = h^2 \varepsilon/m_h e^2$ respectively. In particular the Bloch states with wavector close to or below $1/a$, a being the radius of the microcrystallite, are expelled and the band states coalesce [1,2] to a series of discrete states of electrons and holes on either side of the bulk energy gap E_g. As a consequence the continuum of valence to conduction band transitions observed in the bulk is now replaced (Fig.1) by a series of discrete transitions the lowest one being the 1s-1s transition located [1,2] at

$$h\Omega_o = E_g + h^2\pi/2\mu a^2 \qquad (1)$$

where $\mu = m_h m_e/(m_h + m_e)$.

On these grounds one expects [1,2,3] size dependent effects on the optical properties in general and on the nonlinear ones in particular. Such effects, also termed quantum confinement effects are of fundamental interest as they throw new light on some physical aspects which are suppressed in the infinitely extended bulk semiconductors, or other crystalline structures with well delocalized states.

In principle, these effects should be easiest to study in small microcrystallites of spherical shape. Unfortunately, this not the case in practice for two major reasons. First, because of their small size, a few tens of Ångströms, these particles must be embedded in a solid optically transparent matrix, usually glass, where they actually grow by a diffusion process subsequent to a heating of a whole material ; and second because of the underlying statistical growth process [4] their size distribution is not narrow but show a rather wide asymmetric spread which may substiantially broaden the quantum confinement features.

In the vicinity of the lowest energy structure, when it is prominent, each such quantum dot should behave as a two-level system whose saturation [3] is responsible for the optical nonlinearity. A large nonlinearity is anticipated since a large fraction of the oscillator strength is concentrated in a single structure. The spectral width of the structure therefore plays a

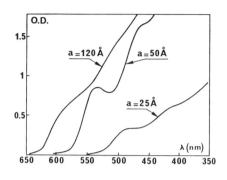

Fig.1

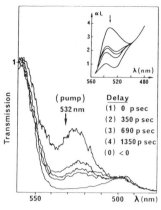

Fig.2

central role. In particular one must assess what part of the observed spectral width results from inhomogeneous broadening due to the overall size distribution as expected from (1) and what part is due to the homogeneous broadening intrinsicly related to the coupling of the localized electron states to other degrees of freedom in a single microcrystallite the electron-phonon coupling being the dominant one.

Here we summarize the experimental evidence [5] of the important role of this electron-phonon coupling in very small $CdS_x Se_{1-x}$ crystallites especially at room temperature. At lower temperatures ($\approx$ 12 K), however, spectral hole burning is observed [5]. In addition we report experimental evidence [6] of the occurence of Stark shift [7] in the quantized levels when a static electric field is applied. These observations are in agreement with the conditions of quantum confinement.

2. Hole burning

As stated above the electron-phonon coupling is the most important mechanism causing homogeneous broadening. In contrast to the electronic states, phonons are not affected by the confinement and the situation their is similar to the broadening of the localized electronic centers in crystals through their coupling to polar longitudinal optical phonons and other perturbations ; this problem has been extensively studied [8,9] in the literature and shown to be reducible to that of a displaced harmonic oscillator.

Assuming coupling [8,9] to single phonon of frequency ω_0, which would be valid for semiconductors, one finds that the line shape is :

$$G(\Omega) = \sum_{p=0} \frac{e^{-s} S^p}{p!} B_p(\omega)$$

(2)

which represents a series of sidebands of shape $B_p((\Omega)$ normalized to unity, spaced ω_0 apart and centred at $\Omega_0 + p\omega_0$ where $p = 0,1,2...$ indicates the number of phonons involved and Ω_0 is the zero phonon transition frequency; $S = \sum_q \left| M_{0q}/\hbar\omega_0 \right|^2$ is the Huang Rhys [8] parameter where M_{0q} is the electron-phonon matrix element for the phonon of quasi-momentum hq. Each p-phonon sideband is broadened via coupling to acoustic phonons and the broadening increases with p. For large values of S, because of the weighting factor $S^p\exp(-S)/p!$ in (2) the sidebands actually overlap to form a broad sideband centered at $\Omega_0 + S\omega_0$ with a linewidth $\cong S\Gamma_1$ for lorentzian shapes where Γ_1 is the linewidth for one phonon process ; the zero-phonon line appears isolated at Ω_0 with width Γ_0 $(\Gamma_0 < \Gamma_1)$ and its contribution decreases with increasing S since it appears with weight $\exp(-S)$.

For a distribution of Ω_0 usually assumed to be a gaussian shape [9] one must convolve (2) with this distribution function to obtain the complete line shape with inhomogeneous and homogeneous broadening. From this one can then obtain the shape and position of the hole eventually burnt in a hole burning experiment. For a symmetric distribution of Ω_0 and large S one finds [9] that the burnt hole is red-shifted when exciting in the high energy tail and is blue-shifted when exciting in the low energy tail.

From the above discussions it is clear that the Huang Rhys parameter plays the crucial role. It has been argued [3] that for III-V compounds S is small and the photon broadening should play a minor role. This, however, cannot be extended to all semiconductors. As can be seen from its definition S depends on the polarity of the semiconductor and for the II-VI's we may expect that S is large. Also S increases [3] with reduced microcrystallite size because of the increasing contribution of large quasi-momentum phonons. Finally the homogeneous phonon broadening increases with temperature.

The first experimental observation of hole burning performed [5] on specially prepared samples containing $CdS_{1-x}Se_x$ microcrystallites of appropriate size to exhibit distinctly quantum confinement features confirmed the above points ; it also revealed some new features which can be traced to the quantum confinement and the asymmetric size distribution and are absent in hole burning in the commonly [9] studied systems (ex. molecular systems with site distribution).

The samples provided by Schott Glasswerke were cut in different portions of a 15 cm long bar that was kept in a temperature gradient, the temperature ranging from 500°C to 700°C from end to end. Because the quantum confinement is more pronounced for a < a_e but at the same time S increases with decreasing a, as was previously pointed out, there is an optimum average size particles for hole burning.

A picosecond excite and probe technique was used to measure the nonlinear absorption and its temporal evolution. At room temperature no hole burning was observable but only an uniform saturation clearly indicating that phonon broadening is the dominant mechanism in II-VI compounds in contrast to what is expected in the III-V's. At low temperature ($\approx$ 12 K) no hole burning was observed when exciting on the high energy tail. However when the excitation occurred at the peak or low-energy tail of the structure a hole burning was clearly observed with the hole position shifted in the blue (Fig.2 ; inlet the same in absorption) ; the burnt hole disappears roughly within a nanosecond.

This asymmetric behavior is well explained [5] when the specific features of the microcrystallite size distribution and the variation of S with the size are properly taken into account. Before discussing the impact of these features, let us stress the evident fact that the high ernergy side of the structure is related to small size particles (a < $\bar{a}$) and the low energy side is related to large size particles (a > $\bar{a}$) ; this is obvious from (1). The two features that distinguish the microcrystallites with size distribution from molecular systems with "site" distribution [9] are :
- the size distribution is asymmetric,
- the Huang Rhys factor increases [3] with decreasing a which implies that homogeneous broadening is larger on the high energy side than

on the low energy side. Actually the phonon broadening on the high-energy side is so large that excitation there is inoperative [5] in hole burning; on the other hand it is sufficienlty reduced on the low-energy side so that excitation there burns a blue-shifted hole. A more quantitative analysis is complicated by superimposed higher energy features.

The above observation [5] is the first clear identification of the important role of phonon broadening in small microcrystallites at room temperature. At low temperature, it introduces an asymmetry in the hole burning process that is directly related to quantum confinement.

3. Static Stark Effect

A striking confirmation of the quantum confinement is also provided by the experimental study [6] of the electroabsorption in these samples i.e. the change in the absorption spectrum due to a static electric field ; the experimental results indeed can be very well accounted for in terms of a static Stark effect.

The samples were submitted to a 1 kHz sinusoidal voltage with V_{ms} = 450 V corresponding to an electric field inside the crystallite of 2.10^4 V/cm and the change $\delta\alpha$ in absorption of a white spectrum light beam was measured [6] after selecting the wavelength ; the measurements were done for nine different samples and since the particle density is constant for all the samples, we actually measure $\delta\alpha$ as a funciton of the size of the spheres.

In each case we observe an oscillating signal (Fig. 3 & 4) whose amplitude was measured to be proportional to the square of the applied voltage. The positions of the maxima and the zeroes are furthermore independant of the voltage as one should expect for the static Stark effect. When the size of the sphere is decreased the whole structure is broadened [6] and blue shifted as expected from the quantum size effect. The signal decreases with the size of the sphere but the oscillations weaken faster for the larger ones. For the larger spheres one also sees a replica of the structure separated by .33 eV which corresponds to the spin orbit splitt-off valence band (fig.3) ; this replica is blurred out [6] as we decrease the size.

The experimental observation can be well accounted for within the framework of the quantum confinement [1] where the electron and hole states are quantized within a spherical quantum well ; for infinite walls and in the absence of an electric field, the wavefunctions are given by the spherical Bessel functions and the total absorption is the sum of the elementary absorptions between the quantized electron and hole levels with same quantum numbers i.e. 1s-1s, 1p-1p... In the presence of the static electric field the changes of the absorption spectrum can be accounted for by perturbation expansion up to the second order with respect to the field strength ; this is justified because the static electric fields F is sufficiently weak. The main effect of the static electric field is the mixing of an nl state with the n'l $\pm$ 1 states and the displacement of the levels E_{nl} by an amount $\delta E \approx F^2$.

As a consequence [6,7] the elementary absorption peaks are shifted and the oscillator strengths of the transitions are decreased because the overlap of the electron and hole wavefunctions is less perfect. In addition, due to the mixing of unperturbed states, new transitions appear which give rise to additional peaks and compensated the loss of absorption on the allowed transitions. These modifications are proportional to F^2, leading to an absorption modification which varies as F^2, as experimentally observed. Calculating numerically the absorption coefficient without and with an electric field for the first nine elementary transitions we were able to reproduce very satisfactorily the experimental results (fig.4). In the calculation we introduced a finite width for these transitions ($\approx$ 30 meV for a =

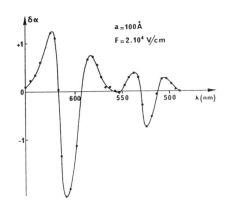

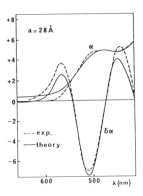

Fig.3 (ref.6)   Fig.4

28 Å) and used a log-normal distribution to account for the size dispersion in the samples (10-12 %). Thus the underlying mechanism of electroabsorption is the Stark effect and this is additional strong evidence of the quantization of the electron levels even for the larger spheres for which the quantum confinement effects in the unperturbated optical spectrum are rather incospicuous . We also note that the Stark effect is not in contradiction with the conventional Franz Keldysh effect [7] ; the latter in the low field limit in quantum wells [7] indeed gives a F^2 dependence as expected for the Stark effect.

As a concluding remark we wish to stress that the observation of the spectral hole burning and the Stark effect give striking confirmation of the quantum confinement that undergo the electron and hole states in very small semiconductor crystallites. The impact of this behavior may have important implications on the behavior of the nonlinear optical coefficients and particular in the optical Kerre effects and may open the way to a new class of nonlinear optical materials whose properties can be artificially modified.

References

1   Al.L.Efros and A.L. Efros, Sov.Phys.Semic.16, 772 (1952) (engl.ed.)
2   L.E. Brus, IEEE J.Quantum Electr. 22, 1909 (1986)
3   S. Schmitt-Rink, D.A.B. Miller and D.S. Chemla, Phys.Rev.B 35, 8113 (1987)
4   I.M. Lifshitz and V.V. Slezov, Sov.Phys.JETP 8, 331 (1959) (engl.ed.)
5   P. Roussignol, D. Ricard, C. Flytzanis and N. Neuroth, Phys.Rev.Lett. 62, 312 (1989)
6   F. Hache, D. Ricard and C. Flytzanis, Appl.Phys.Lett (to appear)
7   D.A.B. Miller, D.S. Chemla and S. Schmitt-Rink, Phys.Rev. B33, 6976 (1956)
8   K. Huang and A. Rhys, Proc.Roy.Soc.(London) A204, 406 (1959)
9   J.M. Hayer, J.K. Gillie, D. Tang and G.J. Small, Biochim.Biophys.Acta 932, 287 (1988)

High Resolution Nonlinear Laser Spectroscopy of Excitons in GaAs/AlGaAs Multiple Quantum Well Structures

D.G. Steel, H. Wang, J.T. Remillard, M.D. Webb
Departments of Physics and Electrical Engineering
University of Michigan, Ann Arbor, MI 48109 USA

P.K. Bhattacharya, J. Oh, J. Pamulapati
Department of Electrical Engineering and Computer Science
University of Michigan, Ann Arbor, MI 48109 USA

Frequency domain nonlinear laser spectroscopy is a powerful method for studying resonant optical excitation in solids. It provides a means for obtaining line shapes related purely to relaxation of optically excited states of the system. In addition, it often eliminates contributions from inhomogeneous broadening and enables studies of the homogeneous line shape and the effects of spectral diffusion.[1,2,3] In this paper, we discuss measurements of the exciton in a GaAs/AlGaAs multiple quantum well (MQW).

GaAs MQW's are semiconductor heterostructures of importance to electronic and optoelectronic devices[4]. They consist of a periodic structure of thin layers of GaAs and AlGaAs. The larger band gap of AlGaAs confines the optical excitation to the thin GaAs, resulting in quantum confinement of the exciton and an increase in the exciton binding energy. In this work, the MQW is grown by molecular beam epitaxy and consists of 65 periods of 96A wells of GaAs and 98A barriers of $Al_{.3}Ga_{.7}As$. The solid line in Fig. 1 shows the absorption spectrum at 5K. The two resonances are the heavy hole (HH1) and light hole (LH1) exciton which exhibit strong inhomogeneous broadening due to fluctuations in the well thickness.

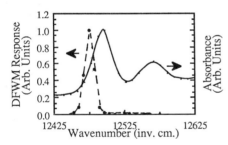

Figure 1. The absorption spectrum and the degenerate four-wave mixing response at 5K. The red shift in the nonlinear response shows the nonlinear response in this sample is due to localized excitons.

The primary backward four-wave mixing experimental configuration is discussed in reference 5. Two counter-propagating pump beams [identified by field amplitudes $\vec{\mathcal{E}}_f(\omega_f, \vec{k}_f)$ and $\vec{\mathcal{E}}_b(\omega_b, \vec{k}_b)$] interact with a third beam [$\vec{\mathcal{E}}_p(\omega_p, \vec{k}_p)$] through the third order susceptibility to produce a nonlinear polarization. The probe field k-vector $\vec{k}_p$ intersects the forward pump field k-vector, $\vec{k}_f$, at an angle θ. The signal field [designated, $\vec{\mathcal{E}}_s(\omega_s, \vec{k}_s)$], arises from the induced nonlinear response proportional to $\vec{\mathcal{E}}_f \vec{\mathcal{E}}_p \vec{\mathcal{E}}_b^*$. Since the frequencies are nearly degenerate, phase matching ensures that the signal is nearly counter-propagating with respect to the probe field. Physically, $\mathcal{E}_f \mathcal{E}_p^*$ form a traveling wave spatial grating associated with the excitation of the system. $\mathcal{E}_b$ scatters from this grating to produce the signal. The nonlinear response in these systems is due to band filling and exchange effects[6].

For $\vec{\mathcal{E}}_b \| \vec{\mathcal{E}}_f \| \vec{\mathcal{E}}_p$, measurement of $|\mathcal{E}_s|^2$ as a function of $\omega_p - \omega_f = \delta$ (for $\omega_b - \omega_f = \text{constant}$), designated the FWMp response, gives a measure of the relaxation rates associated with the optical field induced perturbation of the excited states and the ground state, provided there is no orientation signal (i.e., $E_s = 0$ if $\vec{\mathcal{E}}_b \| \vec{\mathcal{E}}_f \perp \vec{\mathcal{E}}_p$, which we have verified experimentally for low power cw excitation.) For fixed δ, measurement of $|\mathcal{E}_s|^2$ as a function of $\omega_b - \omega_f$ (designated the the FWMb response) gives the line shape associated with the nonlinear response in homogeneously broadened material, and a line shape related to the homogeneous line shape in inhomogeneously broadened material.[5]

At 300K, the exciton is ionized by LO phonons on a time scale of 300fsec. The FWMp response measures dynamics associated with the electron-hole (e-h) plasma. Figure 2a shows a typical FWMp response. The solid curve is a fit of a Lorentzian. Dynamics are determined by e-h recombination and spatial diffusion. The spatial diffusion rate depends on the grating spacing ($\Delta K = |\vec{k}_f - \vec{k}_p|$) and hence the angle, θ. Figure 2b show the angle dependence. The solid curve is a fit of the form $\gamma_{rec} + D(\Delta K)^2$. The data gives γ_{rec} (=5nsec) and D (=18cm^2/sec). The measurements of the

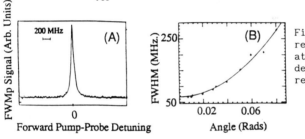

Figure 2. (a) The FWMp response at room temperature (b) The angle dependence of the FWMp response.

room temperature nonlinear response show the presence of a slow component. Figure 3a shows a high resolution display of the center of Fig. 2. The line shape shows a clear effect of interference in the scattering channel giving rise to the signal. Phenomenological modeling of the nonlinear response shows that the interference is likely due to a shift in the

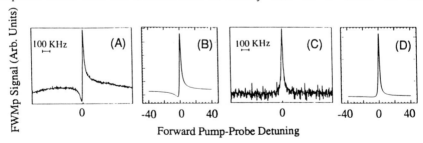

Figure 3. A high resolution scan of the FWMp response. (a) $\omega_b = \omega_f$. (c) $\omega_b \ll \omega_f$. (b) and (d) are the results of the model.

resonance. Figure 3b shows the predicted behavior, in good agreement with the experiments. Further confirmation of the explanation is seen by comparing Fig 3c and the corresponding model in 3d. In this case, the back pump beam is detuned to the low energy side of the resonance, eliminating the interference effect. The origin of the shift may be due to a renormalization of the exciton energy or a shift in the exciton energy due to thermal effects.

At low temperature, the exciton is stable against ionization, and the dashed line in Fig 1 shows the nonlinear response (measured by degenerate FWM) is red shifted with respect to the absorption spectrum. The recombination time of the exciton is 1.5 nsec, so the FWMp response would be expected to give a line shape with a width determined by the inverse life time. Figure 4a shows a much more complex structure. The line shape shows the presence of two clear decay channels, corresponding to 100 psec and 15 nsec. In addition, a high resolution scan of the center shows a narrow resonance similar to Fig. 3a. Figure 4b shows an expanded scan of

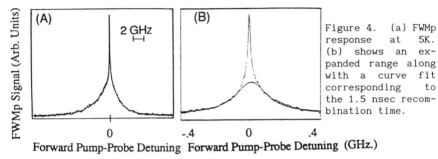

Figure 4. (a) FWMp response at 5K. (b) shows an expanded range along with a curve fit corresponding to the 1.5 nsec recombination time.

the intermediate region along with a curve fit corresponding to 1.5 nsec. We believe the 1.5 nsec decay is due to radiative recombination and the 100 psec decay corresponds to the decay due to spectral diffusion. Solutions of the optical Bloch equations for the FWMp response modified to describe spectral diffusion (MOBE)[3] show the FWMp response would have the form described by Fig 4b and the base of 4a. The origin of the 15nsec component is possibly due to the presence of residual electric fields which can result in an increase in the exciton lifetime.

Further evidence for spectral diffusion is seen in the FWMb response which eliminates inhomogeneous broadening and provides a line shape related to the homogeneous line shape along with contributions from spectral diffusion. Figure 5a shows the FWMb response obtained 28 cm^{-1} below the

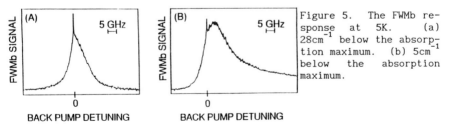

Figure 5. The FWMb response at 5K. (a) 28cm^{-1} below the absorption maximum. (b) 5cm^{-1} below the absorption maximum.

absorption maximum. This curve shows slight asymmetry on the high energy side. The asymmetry becomes more dramatic as the forward pump and probe frequency approach the absorption resonance, as seen 5b. It can be shown that in the absence of spectral diffusion, the FWMb response is symmetric about $\omega_b - \omega_f$ $(\approx \omega_p)$. The presence of asymmetry is consistent with excitation diffusion to other resonant frequencies, and such line shapes can easily be obtained from MOBE. A possible origin of the spectral diffusion is variable range hopping. Recent work has identified phonon assisted tunneling and thermal activation as two possible mechanisms. Figure 6 shows a comparison of the measured temperature dependence of the line width in Fig. 5a to a theory based on phonon assisted tunneling. The curve varies as $\Gamma \exp(B/kT^{-1.7})$ and is in general agreement with theoret-

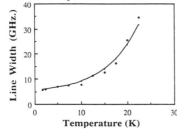

Figure 6. The FWMb line width in Fig. 5a as a function of temperature. The solid curve is a fit of the theory for phonon assisted tunneling.

ical expectations[7]. Finally the effects of spectral diffusion which we have observed are more clearly seen at higher temperatures. Figure 7a shows the FWMb response at 5K similar to Fig. 5b with an increase in scan.

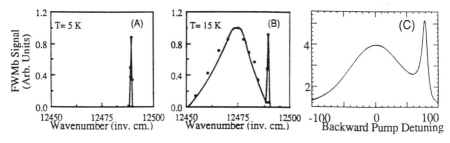

Figure 7. The FWMb response as a function of temperature. (a) 5K. (b) 15K. (c) Solution to MOBE.

At 15 K, we clearly observe the decay of the excitation to lower energy. In these measurements, we see that while the exciting radiation is centered at the peak of the absorption spectrum, the excitation clearly decays to the region of the localized exciton. The large broad peak is centered at the same wavelength as the peak of the nonlinear response shown in Fig. 1 and also corresponds to the peak in photoluminescence emission. Fig. 7c is a solution of the MOBE for the FWMb response in the presence of decay to a lower lying state.

This work is supported by the AFOSR and the ARO.

1. D.G. Steel and J.T. Remillard, Phys. Rev. A36, 4330 (1987).
2. P.R. Berman et. al., Phys. Rev. A38, 252 (1988).
3. P.R. Berman, J. Opt. Soc. Am. B3, 564 (1986).
4. D.S. Chemla and D.A.B. Miller, "Physics and Applications of Excitons Confined in Semiconductor Quantum Wells," in Heterojunctions: Band Discontinuities and Device Applications, F. Capasso and G. Margaritondo, eds., Elsevier Science Publisher B.V. 1987.
5. J.T. Remillard et al., Phys. Rev. Lett.
6. H. Haug and S. Schmitt-Rink, J. Opt. Soc. Am. B2, 1135 (1985).
7. T. Takagahara. Phys. Rev. B32, 7013 (1985).

Laser Spectroscopy of Semiconductor Quantum Dots

J.F. Lam
Hughes Research Laboratories
Malibu, CA 90265, USA

Recent advances in nanofabrication techniques have opened the door for the study of quantum confined semiconductor structures. The first observation of quantum dots was reported by Ekimov and Onushchenko[1] who measured the blue shift of the absorption lines of CuCl as the dimension of the dot decreases. The experimental data confirmed the theoretical model of Efros and Efros[2]. Theoretical studies in both the strongly[3] and weakly quantized[4,5] regimes indicated that semiconductor quantum dots might have enhanced optical nonlinearities. Hence applications of quantum dots in optical devices such as lasers[6] and modulators[7] appear to be promising, provided that the technology is sophisticated enough to produce well characterized samples of quantum dots.

In spite of its potential applications, little is understood concerning the spectroscopy of semiconductor quantum dots. Roussignol et al[8] have recently observed spectral hole burning in CdSSe microcrystallites using a pump-probe technique. Furthermore their measurements indicated that the phonon broadening dominated the spectral behavior of these materials. This article describes the optical Stark effect in these materials, and proposes that the technique of nearly degenerate two-wave mixing has the potential of providing a direct measurement of the energy relaxation and dipole dephasing times of semiconductor quantum dots.

In order to appreciate the regimes that are being considered, Table I gives typical semiconductor parameters, together with the estimates of the de Broglie wavelengths of the electron, λ_D, and the Bohr radius of the exciton, R_B. For example, a GaAs quantum dot with dimension of 200 A will enclose a quantum confined exciton. If the dimension decreases to 100 A, the quantum dot consists of electron and hole confined inside a box, independent of each other. Note that a 50 A GaAs quantum dot will possess optical transition in the visible regime. And if such a set of such dots can be inserted inside a pn junction, it will generate, in principle, coherent radiation at approximately 6000 A. This intriguing idea indicates that the optical properties of quantum dots require careful examination.

The study of optical processes in semiconductors relies upon the solution of the bi-local quantum transport equations, which takes into account both the coherent amplitude of the electron-hole pair and the number densities. We will restrict our studies to the case where excitonic processes are important. In this case the dot size is assumed to be large compared to the excitonic Bohr radius but its is limited by the de Broglie wavelength of the electron.

The ground state is determined by the Slater determinant of one-electron Bloch states, all located in the valence band

TABLE I

	ε	m^*/m_0	Eg (eV)	λ_D (Å)	R_B (Å)
Ga In As P x=0.27 y=0.60	12.4	0.053	0.95	276	137
In P	12.4	0.077	1.35	229	96
Ga As	13.1	0.067	1.42	245	118
Cd S	5.4	0.21	2.42	139	17
Cd Se	10.0	0.13	1.70	176	53

The first excited state is determined by the Slater derterminant of one-electron Bloch states, in which one electron and one hole have been created in the conduction and valence bands; respectively.

In the effective mass approximation, the interaction of light with excitons can be described in terms of

$$H = E_C (-i \nabla_e) - E_V (-i \nabla_h) + V (\vec{r}_e - \vec{r}_h)$$

$$- \vec{\mu} \cdot \vec{E} (\vec{R},t) \tag{1}$$

The state $|1\rangle$ is the eigenstate of the "unperturbed" effective mass Hamiltonian, H_0.

One can construct density matrix equations in the same manner as those found in atomic systems. That is,

$$it \frac{\partial \rho}{\partial t} = [H,\rho] + i\hbar \frac{\partial \rho}{\partial t} \Big|_{Relaxation} \tag{2}$$

where the elements of the density matrices have simple physical interpretation. $\langle 1| \rho |1\rangle$ is the probability of finding an exciton, and $\langle 0| \rho |1\rangle$ is the optical coherence amplitude. This is the starting point for the analysis of four-wave mixing processes in semiconductor quantum dots.

The diagonalization of the exciton Hamiltonian is achieved by finding the determinant of the following matrix

$$\begin{bmatrix} -U & -\vec{\mu}_R \cdot \vec{E} \left(\frac{L}{R_B}\right)^{3/2} \\ -\vec{\mu}_R \cdot \vec{E} \left(\frac{L}{R_B}\right)^{3/2} & U_{exc} - U \end{bmatrix} \tag{3}$$

TABLE II

Δ = 20 me V	L (Å)	I (W/cm²)	Stark Shift (meV)
GaAs-AlGaAs MQW	100	10^6	0.2
CdS dot	139	6×10^2	0.2

which gives the following result for the energy level of the exciton in the presence of a detuned light field

$$U = U_{exc} + \frac{|\vec{\mu}_B \cdot \vec{E}|^2}{U_{exc} - \hbar\omega_L} \left(\frac{L}{R_B}\right)^3 \qquad (4)$$

The first two terms are just the exciton binding energy and the last term is the A.C. Stark shift. Note that the Stark shift contains a geometric factor which gives the ratio of the dot size to the exciton Bohr radius. In order to appreciate the significance of this result, Table II provides a comparison of the intensities of the light field for the case of a GaAs Multiple Quantum Well and a CdS quantum dot. One sees that in order to achieve the same Stark shift, the power density for the multiple quantum well is four orders of magnitude higher than that of the quantum dot. This is intimately related to the geometric factor.

The application of the density matrix equations to the case of two-wave mixing can be carried out using perturbation theory. The result of the analysis provides an expression for the small signal gain coefficient

$$g = \alpha \cdot \left\{ 1 + \frac{\Gamma}{\gamma} \times \frac{\delta}{\Delta} \times \frac{\Omega^2}{\gamma^2 + \delta^2} \right\} \qquad (5)$$

where α_0 is the linear absorption coefficient, Γ is the exciton dephasing rate, γ is the exciton energy relaxation rate, δ is the frequency mismatch between the two input beam, Δ is the detuning of the strong pump from exciton resonance, and Ω is the exciton Rabi flopping frequency. This result is identical to the one obtained from resonant two-level systems.

Figure 1 gives the dependence of the gain-length factor as a function of the frequency mismatch δ, for the case of CdS quantum dot. In a manner identical to that encountered in resonant atomic system, amplification of the signal is achieved under certain conditions. The maximum of the gain coefficient gives a direct measurement of the exciton energy relaxation rate. While the amplitude gives a direct measurement of the exciton dephasing rate.

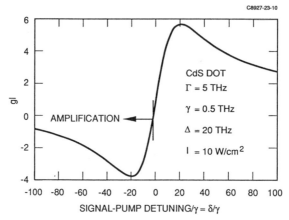

Figure 1.   Gain length product as a function of the normalized frequency mismatch.

* This work is supported by the ARO.

REFERENCES

1.   A.I. Ekimov and A.A. Onushchenko, Sov. Phys. Semicond. 16, 775 (1982)
2.   Al. L. Efros and A. L. Efros, Sov. Phys. Semicond. 16, 772 (1982)
3.   S. Schmitt-Rink, D.A.B. Miller and D.S. Chemla, Phys. Rev. B 35, 8113 (1987)
4.   E. Hanamura, Phys. Rev. B 37, 1273 (1988)
5.   F. Henneberger, U. Woggon, J. Puls and Ch. Spiegelberg, Appl. Phys. B 46, 19 (1988)
6.   Y. Miyamoto, M. Cao, Y. Shingai, K. Furuya, Y. Suematsu, K.G. Ravikumar and S. Arai, Jpn. J. Appl. Phys. 26, L225 (1987)
7.   See Reference 3
8.   P. Roussignol, D. Ricard, C. Flytzanis and N. Neuroth, Phys. Rev. Lett. 62, 312 (1989)

Part VI

Surface Spectroscopy

CONFORMATION, ORIENTATION AND INTERACTION IN MOLECULAR MONOLAYERS: A SURFACE SECOND HARMONIC AND SUM FREQUENCY GENERATION STUDY

R. Superfine, J. Y. Huang, and Y. R. Shen

Department of Physics, University of California, and
Center for Advanced Materials, Lawrence Berkeley Laboratory
Berkeley, California 94720 USA

Knowledge of the conformation and ordering of molecular monolayers is essential for a detailed understanding of a wide variety of surface and interfacial phenomena. Over the past several years, surface second harmonic generation (SHG) has proven to be a valuable and versatile probe of monolayer systems.[1] Our group has recently extended the technique to infrared-visible sum frequency generation (SFG) which has unique capablities for surface vibrational spectroscopy.[2-6] Like second harmonic generation, SFG is highly surface specific with submonolayer sensitivity at all interfaces accessible by light. The orientation of individual groups within an adsorbate molecule can be deduced by a polarization analysis of the SFG signal from the vibrational modes of the groups.

We have used SHG and SFG to study orientations and conformations of surfactant and liquid crystal (LC) monolayers and their interaction on a substrate. The interfacial properties of LC are of great interest to many researchers for both basic science understanding and practical application to LC devices. It is well known that the bulk alignment of a liquid crystal in a cell is strongly affected by the surface treatment of the cell walls. The reason behind it is not yet clear.[7]

The theoretical background and experimental arrangement of SHG and SFG have been described elsewhere.[4] In our setup, a 30 psec. Nd:YAG mode-locked laser system together with nonlinear accessories generates a visible beam at $.532\mu m$ and an infrared beam tunable about $3.4\mu m$. Both beams are focused to a common spot of $300\mu m$ dia. The typical signal off the surface from a compact ordered alkyl chain monolayer is ≈ 500 photons per pulse, easily detected with a photomultiplier tube.

In our experiment, we used the monolayer of pentadecanoic acid ($CH_3(CH_2)_{13}COOH$, PDA) on water as a standard system for the study of order in alkane chain monolayers. The SFG spectra for PDA at three different surface concentrations are presented in Fig.1. For the compact monolayer (Fig.1a), the peaks are assignable to the CH_3 symmetric stretch mode at 2875 cm^{-1} and the CH_3 fermi resonance at 2940 cm^{-1}. An analysis of the polarization dependence of the 2875 cm^{-1} peak gives an average tilt for the terminal methyl group of about 35° from the surface normal. This is consistent with the alkane chain being oriented perpendicular to the surface. The peaks due to the CH_2 stretch modes are hardly distinguishable. An infinite polymethylene chain in the all-trans configuration is centrosymmetric so that its vibrational modes should be strictly either Raman or infrared active.

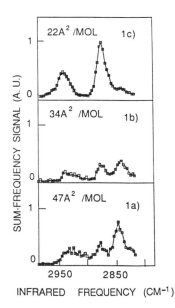

Fig.1 SFG vibrational spectra of PDA monolayers on the water surface.
1c) compact monolayer
 b) medium density
 a) low density

The steric interaction between alkyl chains that produces an all-trans configuration is weakened as the surface density of PDA is decreased. Indeed, as shown in Fig.1, at lower monolayer density the intensity of the CH_3 peaks becomes weaker, a sharp feature at 2850 cm^{-1} corresponding to the CH_2 symmetric stretch grows stronger and a broad background between 2880 and 2920 cm^{-1}, also attributable to CH_2 stretch modes, appears. These results are presumably due to the tilt and kink deformations of the alkyl chains.

We next studied a monolayer of n,n-dimethyl-n-octadecyl -3-aminopropyltrimethoxysilylylchloride (DMOAP) deposited on glass with a surface density of about 50 $Å^2$ per molecule. This system is expected to be similar to the low density case of PDA on water. Indeed, the SFG spectrum in Fig.2 (solid curve) is essentially identical to Fig.1c indicating a large degree of disorder in the alkyl chains.

The SFG spectrum for a full monolayer of liquid crystal molecules 4'-n-octyl-4-cyanobiphenyl (8CB) on glass, corresponding to a surface density of about 35 $Å^2$ per molecule, is presented in Fig.3 (solid curve). Two peaks due to biphenyl C-H stretch modes appear at 3070 and 3050 cm^{-1}. The relative peak heights at 2875 and 2850 cm^{-1} indicate that the 8CB alkyl chains have fewer kinks than the PDA chains at the same surface density. The CH_3 modes of 8CB appear at 2884 and 2946 cm^{-1}, showing a significant shift from those of the PDA monolayer. These higher peak positions are typical of a gas phase and indicate that the 8CB chains have little interaction with each other. This is understandable from the average distance between chains of about 6 Å and their short length. The SHG measurements determine that the biphenyl group orientation is about

213

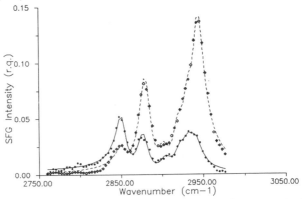

Fig.2 SFG spectrum of DMOAP monolayer (solid curve) is compared to SFG spectrum of same sample after deposition of 8CB (dashed curve).

70° from the surface normal with the alkyl chain end away from the surface.

For an 8CB monolayer on DMOAP coated glass, the DMOAP surface density is the same as described earlier, while the SHG signal calibrates the 8CB coverage to be about 70% of a full monolayer on clean glass. From the increased molecular surface density we expect greater steric hindrance of the chain conformation and stronger interaction between molecules. In Fig.2, we compare the SFG spectra of DMOAP before and after depositing 8CB and observe the 2850 cm^{-1} oscillator strength drop by 60%. This indicates a decrease in the number of kinks in the DMOAP chain. In Fig.3 we compare the SFG spectra of 8CB on clean glass and on DMOAP coated glass and observe shifts in the C-H stretch frequencies similar to those expected in changing from a gas to a more condensed phase. This is a clear sign of the existing interaction

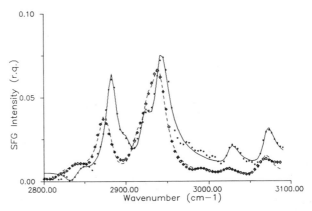

Fig.3 SFG spectrum of 8CB monolayer on clean glass (solid curve) is compared 8CB deposited on DMOAP coated surface ashed curve).

214

between the 8CB molecule and the DMOAP alkyl chains.

In conclusion, we have used SFG to study the order in a silane monolayer before and after the deposition of a coadsorbed liquid crystal monolayer. We observe an increase in the order of the chain of the silane molecule induced by the interpenetration of the liquid crystal molecules. By using SHG and SFG, we have studied the orientation and conformation of the liquid crystal molecule on clean and silane coated glass surfaces. On both surfaces, the biphenyl group is tilted by 70° with the alkyl chain end pointing away from the surface. The shift in the C-H stretch frequencies in the coadsorbed system indicates a significant interaction between molecules.

This work was supported by the Director, Office of Energy Research, Office of Basic Energy Sciences, Materials Sciences Division of the U.S. Department of Energy under Contract No. DE-ACO3-76SF00098. R.S. gratefully acknowledges support from the 3M Company and J.Y.H. acknowledges a postdoctoral fellowship from IBM.

References

1. Y. R. Shen, Annu. Rev. Mater. Sci. 16, 69 (1986).
2. X. D. Zhu, H. Suhr, and Y. R. Shen, Phys. Rev. B 35, 3047 (1987).
3. J. H. Hunt, P. Guyot-Sionnest, and Y. R. Shen, Chem. Phys. Lett. 133, 189 (1987).
4. P. Guyot-Sionnest, J. H. Hunt, and Y. R. Shen, Phys. Rev. Lett. 59, 14 (1987).
5. P. Guyot-Sionnest, R. Superfine, J. H. Hunt and Y. R. Shen, Chem. Phys. Lett. 144, 1 (1988).
6. R. Superfine, P. Guyot-Sionnest, J. H. Hunt, C. T. Kao and Y. R. Shen, Surface Sci. 200, L445 (1988).
7. J. Cognard, Alignment of Nematic Liquid Crystals and Their Mixtures (Gordon and Breach, London, 1982).
8. C. S. Mullin, P. Guyot-Sionnest, and Y. R. Shen, Phys. Rev. A 39, 3745 (1989).
9. M. Maroncelli, H. L. Strauss, and R. G. Snyder, J. Chem. Phys. 82, 2811 (1984).

Milli-Hertz Surface Spectroscopy

Eric Mazur, Doo Soo Chung and Ka Yee Lee

Department of Physics and Division of Applied Sciences, Harvard University
Cambridge, MA 02138, USA

A technique that has been repeatedly employed in high resolution light scattering experiments is that of light beating, or *heterodyne*, spectroscopy.[1] By detecting the beating signal between the scattered light and a 'local oscillator' derived from the same laser source, one can obtain ultrahigh spectral resolution, independent of the random fluctuations of the light source. We reported earlier of a novel Fourier transform heterodyne spectroscopy (FTHS) technique with high resolution[2] which is simpler and more direct than the conventional heterodyne technique; we have since improved our resolution ten-thousand fold to the 20-μHz range. We applied this technique first to study nonequilibrium phenomena at liquid-vapor interfaces. The ultrahigh resolution also enables one to observe the very small Doppler shift of a light beam reflected from a growing silicon crystal.

1. Fourier Transform Heterodyne Spectroscopy

In heterodyne spectroscopy, one detects the beating of the scattered light—with itself or with part of the incident light—and then analyzes the spectrum of the detector signal with a spectrum analyzer or autocorrelator. The main advantage of this scheme is that fluctuations in the phase of the incident laser field, which limit the resolution of spectroscopy, cancel out as long as the two fields reaching the detector are coherent.

One of the major drawbacks of conventional heterodyning, however, is that at near zero frequency the spectrum is usually distorted by high-pass filters used to block the DC component of the signal. The present detection scheme circumvents this problem by shifting the origin of the spectrum from zero to some optimal frequency. In this way, the spectrum at frequencies close to that of the incident light is still reliable, making the detection of very small Doppler shifts possible. As the beating of two signals is sensitive only to the difference frequency, the use of such frequency-shifted local oscillators has the additional advantage of spectrally separating the up-shifted and down-shifted beams—a feature which is indispensable in the study of nonequilibrium fluid interfaces. Figures 1a and 1b show two experimental setups that can be used for doing FTHS. In Fig. 1a the main beam and the local oscillator are mixed after the experiment, while in Fig. 1b they are mixed before incidence on the sample.

2. General Considerations

In each of these setups, the detector signal is sampled for a period T_s (the *sampling time*), and the digitized signal is then Fourier transformed to obtain the spectrum. According to sampling theory, the time interval between adjacent sampling points, δt_s (the inverse of the sampling frequency f_s), and the spectral range F are related by

$$F = \frac{1}{2\delta t_s} = \frac{1}{2}f_s. \tag{1}$$

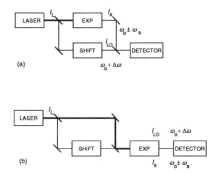

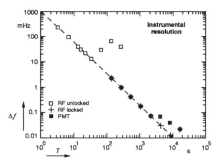

Fig. 2. Instrumental resolution measurements: electronic mixing with independent (□) and locked (+) RF drivers, and optical mixing with locked drivers (■).

Fig. 1. Experimental schemes for FTHS: (a) post-sample mixing; (b) pre-sample mixing.

We have used a variety of sampling frequencies, between 1 Hz and 20 kHz, resulting in a spectral range from 0.5 Hz to 10 kHz. In general, the upper limit of the range is determined by either the detector response time, or the speed of the signal digitizer; in practice with a 1-ns response time and a fast transient digitizer, the spectral range can be extended to the GHz regime. With a streak camera this could even be further extended by at least two orders of magnitude. Because of this broad range, the technique is applicable to a wide variety of fields of research.

Similarly the sampling time, T_s, and the frequency interval between adjacent spectral points, δf_s, are related by

$$\delta f_s = \frac{1}{T_s}. \tag{2}$$

Although this shows that the resolution can be made infinitely small for an infinitely long sampling time, in reality the resolution is limited by the electronic and mechanical stability of the setup.

To determine the instrumental resolution, the local oscillator was mixed directly with the main laser beam. The local oscillator was frequency-shifted by a few kHz using a combination of two acousto-optic modulators. The carrier frequency of the modulators was in the 20–40 MHz range, with a frequency difference equal to the desired frequency shift. When the two modulators are driven with two independent radio-frequency (RF) drivers, the resolution decreases with increasing sampling time as expected from Eq. (2). For sampling times in excess of 60 s, a loss of resolution starts to occur (see open symbols in Fig. 2). By comparing the optical beat spectrum with the one obtained by directly beating the two RF signals, it was found that the resolution limit was caused by instabilities in the MHz modulator frequencies. With two Hewlett Packard frequency synthesizers, locked to the same time base, the situation could be improved by a factor of one thousand. Figure 2 shows the resolution of both optical and RF beat spectra. As can be seen, the RF resolution follows Eq. (2) down to 10 μHz. The optical resolution starts to show deviations at about 70 μHz, most likely because of mechanical instabilities of the optical setup during the half-hour sampling time. The highest instrumental resolution of the present setup, 20 μHz, was obtained with a four-hour sampling time.

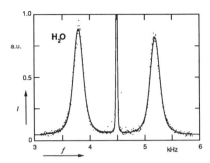

Fig. 3. Spectrum of light scattered from a nonequilibrium liquid surface fitted to three Lorentzian lines.

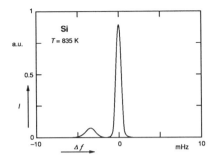

Fig. 4. Spectrum showing the Doppler shift of light reflected from a growing silicon crystal.

3. Application to Liquid Surface Studies

The above technique was first used to study nonequilibrium phenomena at a water-air interface. Thermal fluctuations on a fluid interface give rise to a Brillouin doublet in the spectrum of light scattered from the fluid interface.[3,4] In equilibrium, the Stokes and anti-Stokes Brillouin peaks have the same intensity since the populations, n_S and n_a, of capillary waves travelling in opposite directions are equal. Out of equilibrium, however, n_S and n_a are no longer equal and the spectrum becomes asymmetric. The FTHS, with its directional separation capability, renders the study of such asymmetric Brillouin doublet possible. A schematic diagram of the surface light scattering apparatus is shown in Fig 1a. A more detailed description of the apparatus can be found in Ref. 5. After scattering off a water-air interface subject to a temperature gradient, the scattered beam is combined with the shifted local oscillator. The detector signal is sampled for a certain period of time, stored in a microcomputer and then a fast Hartley transform[6] is applied to the sampled waveform. The resulting spectrum, an example of which is shown in Fig. 3, corresponds to the spectrum of the interfacial fluctuations.[7] These measurements confirm the existence of the predicted asymmetry in the Brillouin spectrum of light scattered from a fluid interface subject to a temperature gradient.[5] A more detailed discussion will appear in a forthcoming paper.

4. Application to Crystal-Growth Measurements

Because of the high instrumental resolution, one can use the FTHS technique to observe very small frequency shifts. A possible application is to measure extremely small Doppler shifts. To illustrate this we have applied this novel technique to study crystal growth at the crystalline-amorphous silicon interface in small silicon samples composed of 300-nm amorphous silicon on a substrate of crystalline silicon. When such a sample is placed on a heating unit, the amorphous material slowly crystallizes and the crystalline-amorphous interface moves up toward the air-amorphous silicon interface. The growth rate can be varied by adjusting the temperature; at 835 K it is approximately 0.2 nm/s. Light reflected from the moving interface will be Doppler shifted by a few mHz.

High temperature air turbulence around the sample and reduced mechanical stability of the heater holding the sample greatly lower the overall stability of the setup. It was therefore necessary to mix the main laser beam with the local oscillator *before* the sample, to ensure constant overlapping of the two (see Fig. 1b). To suppress reflection from the front surface (air-amorphous silicon), the incidence angle of the beam was kept near the Brewster angle. The main beam was p-polarized, the shifted local oscillator s-polarized. Most of the local oscillator beam is reflected at the air-amorphous silicon interface, while most of the main beam penetrates the amorphous layer and is reflected at the crystalline-amorphous interface. A beating signal between the two reflected beams was obtained with a 45° polarizer and recorded with a photomultiplier. Figure 4 shows a spectrum obtained this way. The large peak comes from the beating of the main beam and the local oscillator reflected at the air-amorphous silicon interface. The smaller peak is Doppler shifted by the motion of the crystalline-amorphous interface. The Doppler shift in this case is 3 mHz, corresponding to a speed of about 0.2 nm/s. This is in good agreement with interferometric measurements.[8]

In principle, one can obtain the *distribution* of speeds over the sampled area from the broadening of the Doppler shifted peak (the shift reflects only the *average* speed). The resolution of these measurements, however, is limited by the maximum sampling time, which in turn is limited by the ratio of the amorphous layer thickness to the growth rate. For a fixed growth rate this means one needs thicker samples to increase the resolution. Measurements on thicker samples are currently in progress.

5. Conclusion

This paper presents a simple heterodyne technique with ultrahigh resolution. Besides spectrally separating the up-shifted and down-shifted scattered light, the technique also allows one to obtain undistorted data at the low frequency end of the spectrum. This is essential both in low-frequency work as well as in cases where directional separation is called for, such as the study of nonequilibrium liquid interfaces. The ultrahigh resolution makes the technique also suitable to measure extremely small Doppler shifts. In air, a shift of 20 μHz corresponds to a speed of 7×10^{-12} m/s. Since a spectral range of up to 1 GHz can be covered, Fourier transform heterodyne spectroscopy can be applied to study a broad range of physical phenomena.

Acknowledgments

We are grateful to Professor M. Aziz for the silicon samples and many stimulating discussions and to Professor T. Chupp for lending us the HP synthesizers. We are also indebted to G.Q. Lu for his help in the silicon setup. This work was supported by NSF grant DMR-8858075.

References

1. H.Z. Cummins and H.L. Swinney, *Progress in Optics*, Vol. 8, Chapter 2 (North-Holland, Amsterdam, 1970).
2. E. Mazur and D.S. Chung, *Physica* **147A**, 387 (1987).
3. R.H. Katyl and U. Ingard, *Phys. Rev. Lett.* **20**, 248 (1968).
4. M.A. Bouchiat, J. Meunier and J. Brossel, *C.R. Acad. Sci. Paris* **266B**, 255 (1968).
5. D.S. Chung, K.Y. Lee and E. Mazur, *Int. J. Thermophysics* **9**, 729 (1988).
6. R.N. Bracewell, *The Hartley Transform* (Oxford University Press, New York, 1986).
7. M. Sano, M. Kawaguchi, Y.-L. Chen, R.J. Skarlupka, T. Chang, G. Zofrafi and H. Yu, *Rev. Sci. Instrum.* **57**, 158 (1986).
8. G.L.Olson and J.A. Roth, *Materials Sci. Rep.* **3**, 1 (1988).

ETHANOL ADSORPTION AND DESORPTION ON SILVER SURFACE PROBED BY OPTICAL SECOND-HARMONIC GENERATION

Le LI, Jia-Biao ZHENG, Wen-Cheng WANG and Zhi-Ming ZHANG
Lab of Laser Physics & Optics, Fudan University
Shanghai 200433, China

I. INTRODUCTION

The potentiality of nonlinear optical techniques used for surface and interface studies has been extensively explored and reviewed in recent years (1). Among which, the optical Second-Harmonic Generation (SHG) is the most developed and employed arts. Becuase of its extremely surface specific characteistics, its experimental investigations could be greatly simplified. One of the most interesting fields on the application of SHG as the surface probe is the study of adsorption and desorption of atoms and molecules with the studied surface to be situated under an ultrahigh vacuum environment for keeping the surface clean from contaminations. Results from such study then could be comparable to that of those obtained by the conventional techniques where electron is used as the probe. Information from both of these two techniques are complementary, which may result a more thorough understanding of the mechanism of such adsorption and desorption processes.

Our study is focused on the adsorption and desorption of ethanol on a polycrystalline silver surface probed by SHG. The reason for chosing ethanol as the adsorbates is based not only on its adsorption on silver is an important catalytical process in the chemical industry which demands detail understanding of its kinetics, but also, it is known from the previous studies that ethanol could be adsorbed either in a physisorbed state or in a chemisorbed state that depends on its environmental conditions of the studied surface met. Such a characteristic provides a benefit to study that whether the simple yet sophisticated SHG could be so sensitive to differentiate these two states of adsorption. As one might expect that for a chemisorption process, electron transferring occurs between the adsorbed molecules with the Ag surface, their effective surface susceptibility should be modified due to the variation of the surface electron density and hence will introduce some additional contribution to the SHG intensity. However, for the physisorption process, there does not exist such a charge transferring and the contribution for the SHG is mainly from the surface and the individual molecules (2). These two states of adsorption do reflect the difference in the intensities of the SHG signals. However, according to our study, we also found that some phase information contained in the SHG signals existed for the chemisorption of ethanol on Ag surface, which might be considered to be a characterization of the chemisorbed state as well.

An another interesting project that we have studied is to investigate that whether the surface electromagnetic field would be affected by the adsorbates. It is already known that the intensity of SHG at surface is extremely sensitive to the surface morphology and this will result an enhancement of the local fields. The local fields will be modulated by the transferred charges from the adsorbates under chemisorption and which is verified by the probing with SHG. This study gives us some very interesting and valuable information on the surface study with optical techniques such as SERS where the surface enhancement effect dominates.

Our experiments were carried out with three categories:

1. Adsorption and desorption studies of ethanol on a clean bare polycrystalline Ag surface -- a physisorption process;

2. Adsorption and desorption studies of ethanol on an Ag surface preadsorbed with atomic oxygen -- a chemisorption process;

3. Adsorption study of ethanol on a cold-evaporated Ag surface -- where surface enhancement effect will be affected.

By comparing the results of the first two experiments, we could evaluate the difference for the physisorbed and chemisorbed states, and by comparing the results of the first and the third experiments we could get some information about the influence on the local fields by the adsorbates.

II. EXPERIMENTS AND RESULTS:

All the exoeriments were carried out in a standard ultra-high vacuum system with the background pressure of 3×10^{-10} torr. The polycrystalline silver sample was made from a pure silver block and finely polished. The sample was then cleaned with the standard procedure of argon ion bombardment followed with a 400K annealing treatment, this procedure was repeated several times to ensure the surface to be free of any comtaminations. The cold-evaporated silver surface was prepared by cooling down the silver substrate to 120K and then a thin film

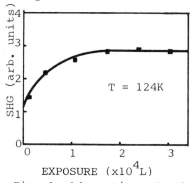

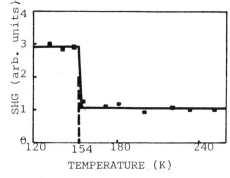

Fig. 1 Adsorption of etha- Fig. 2 Desorption of ethanol on a clean Ag surface   nol on a clean Ag surface

221

of pure silver of about 100A was deposited thermally onto
this substrate at a rate of about 2.5-5A/sec. During the deposition, the pressure rise in the UHV chamber never exceeded
over 1x10^{-9} torr. The ethanol was introduced into the UHV
chamber through a leakage valve at the pressure of 3x10^{-4} torr
that is the saturated vapour pressure of the ethanol.

Optical SHG probing of physisorbed or chemisorbed ethanol was performed by recording the SH signals from the cooled
silver surface as a function of the ethanol exposure. Desorption was inspected with a programmed temperature rising procedure began from 130K after the Ag surface was saturated with
ethanol molecules. SHG signal was measured from the reflection at the Ag surface with a p-polarized 1.06 um laser beam
from a Q-switched YAG laser. The beam energy was kept at a
low level to prevent thermal effects produced by the laser.
The incident angle was at 42° and the signals were averaged
by a boxcar and recorded.

Fig.1 shows the adsorption of physisorbed ethanol on a
clean Ag surface, the solid curve is fitted by the theoretical
formula of Langmuir but modified:

$$I_{SHG} \propto [1 + \frac{B}{A} \exp(-kpt/\theta_s]^2 , \qquad (1)$$

where A stands for the nonlinear susceptibility of the Ag substrate, B stands for the nonlinear susceptibility of the adsorbates, k is the adsorption coefficient and θ_s the saturated
coverage. From the fitting, we can get a real value of B/A =
0.61 and k/θ_s = 0.24x10^{-3}/layer. Fig. 2 shows the desorption
of this physisorbed ethanol that happened at 154K, which is
in consistency with the previous work of study on the ethanol
desorption on Ag(110) surface (3).

Fig. 3 and Fig. 4 show the SHG probed adsorption and desorption of ethanol on a Ag surface preadsorbed with atomic
oxygen, which will convert the adsorption to be a chemisorped
process. By fitting with the same formula, a complex value is
obtained with the value of B/A = 0.79x exp(i29°) and k/θ_s =
0.77x10^{-2}/layer. By comparing the result with the physisorbed
process, the adsorption coefficient for a chemisorption is 30
times greater. Besides, the complex quantity of B/A means

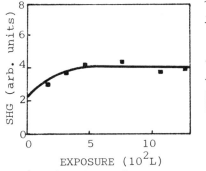 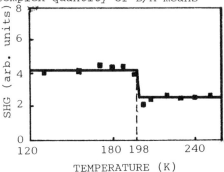

Fig. 3   Adsorption of etha-   Fig. 4   Desorption of ethanol on Ag surface pre-   nol on Ag surface preadsorbed with atomic O.   adsorbed with atomic O.

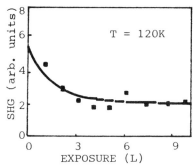

Fig. 5 Adsorption of ethanol on a cold-evaporated silver surface

that the contributions for the SHG from the Ag substrate and the chemisorbed molecules are no in phase where the physisorption does not show such a phase lag between them. This phase relationship might be an indication of the chemisorption process for the adsorption of ethanol on Ag surface where charge transferring occurs.

The result of ethanol adsorption on a cold-evaporated Ag surface is very interesting as shown in Fig. 5 where the SHG intensity decreases against to the coverage but rather increases as compared to a clean and smooth Ag surface. It is expected that the local field will be de-enhanced by those adsorbates that they filled the cavities of the rough surface. If the field for the cold-evaporated Ag surface is denoted as $E(\omega,t)$ and the adsorption will reduce the field strength to $(1-r)E(\omega,t)$, where r is the parameter specifies the de-enhancing effect. Let $r = r_o(\theta/\theta_s)$, where r_o is the de-enhancing factor defined for the value at saturated coverage, then the above theoretical equation of (1) can be modified as:

$$I_{SHG} \propto [1 + (\frac{B}{A_{cold}} - 2r_o) \exp(-kpt/\theta_s]^2 . \quad (2)$$

By fitting with the experimental data, we can get $(B/A_{cold}) - 2r_o = -0.40$, which is again a real value that originated as from physisorption. As a rough estimation, we take $(B/A_{cold}) \sim B/A = 0.61$ for the clean and smooth Ag surface, then we can deduce the $r_o \sim 0.5$. This value is actually large enough that could cause a decreasing of the SHG intensity by a factor of an order of magnitude, that means for those surface with enhancement effect, the effect of de-enhancement from the adsorbates should be taken into consideration.

III. CONCLUSION

The study of ethanol adsorption and desorption with the SHG technique gives that there exists phase relationship and intensity difference for the chemisorption process as compared to the physisorption process. As for the cold-evaporated Ag surface, the SHG technique also gives de-enhancement result that are interested.

REFERENCE:
(1) Y.R. Shen, Mat. Res. Soc. Symp. Proc. 51, 39, (1985)
(2) P. Guyot-Sionnest, et al, Phys. Rev. B33, 8254 (1986)
(3) I.E. Wachs, R.I. Madix, Appl. Surf. Sci. 1, 303 (1978)

Manipulation of Metal Particle Size Distributions on Surfaces with Laser Light

W. Hoheisel, U. Schulte, M. Vollmer, R. Weidenauer, and F. Träger
Physikalisches Institut der Universität Heidelberg
Philosophenweg 12
D-6900 Heidelberg 1
Federal Republic of Germany

The physics and chemistry of clusters and small particles is a rapidly developing interdisciplinary field [1,2]. Besides the goal of understanding the fundamental properties of such systems and their changes as a function of size, studies on clusters and in particular metal particles are stimulated by the prospect of possible applications, e.g. for catalysis or thin film production. For most of these purposes particles of a well defined size are required. Common methods, however, only generate mixtures of clusters with relatively broad distributions. This paper summarizes our experiments with the goal to manipulate such distributions with laser light and to tailor metal particles of a predetermined size and with a narrow distribution. As will be outlined below, this becomes possible by a novel laser desorption process [3,4] which can be controlled such that particles in a certain size range shrink selectively as a result of an ongoing removal of atoms from their surfaces.

In our experiments metal particles are formed on a LiF(100) single crystal surface in ultrahigh vacuum by deposition of sodium atoms and subsequent surface diffusion. The underlying Volmer-Weber growth mode of epitaxy is well understood and has been investigated repeatedly by electron microscopy [5]. It typically results in particle size distributions with a width of about one third of the mean cluster size [6,7]. The average particle size is determined by measuring the decreasing rate of inelastically scattered atoms during the deposition [8] and ranges between 10 and 80 nm. The particles are illuminated with light of an argon or krypton ion laser. As a result, individual atoms are ejected from the surface of the particles. They are detected with a quadrupole mass spectrometer as a function of irradiation time. Desorption starts immediately after the laser light is turned on and stops promptly when the beam is blocked. The desorption rate depends linearly on the light intensity in the range from 40 mW/cm^2 up to the highest available laser intensity of 160 W/cm^2. No threshold for the desorption signal is found. The photodesorption yield strongly depends on the laser frequency. For particles with a mean radius of 50 nm, for example, a resonance at a center wavelength of $\lambda = 490\pm5$ nm with a full width at half maximum of $\Delta\lambda = 90$ nm occurs. Similarly, a resonance as a function of particle size is observed for a fixed wavelength. The resonance of the desorption rate as a function of laser frequency can be explained by excitation of surface plasmons in the metal particles [3,4,9]. Desorption takes place as a *direct* result of such a collective excitation and is due to a non-thermal process. The quantum efficiency of 10^{-5}, however, indicates that photodesorption has to compete with other relaxation channels, particularly with energy transfer to the substrate. Further details of the desorption process can be found in several publications [3,4].

Illumination with red light preferably dissociates large clusters, whereas green or blue light interacts more strongly with small particles. Therefore, the laser photodesorption process can be controlled such that particles in a certain size range shrink selectively. As a result, the size distribution changes. The ablation naturally comes to an end if the size of the particles has decreased so much that they are shifted out of resonance. In general, a given distribution on a surface can be manipulated in different ways depending on the chosen laser frequency and on the average particle size. Most importantly, the distribution can be narrowed considerably.

A model has been developed to describe this size manipulation quantitatively. For a fixed laser wavelength λ the absorption resonance curve as a function of radius is approximated by a Gaussian centered at R_λ. The cluster size distribution f(R) is asymmetrical [6,7] and is approximated by a superposition of two half-Gaussians of different widths, which are centered at R_0. The number of desorption sites on a cluster of radius R is written as a·R, where a is a constant that depends on the shape of the cluster [10]. With these parameters the ablation rate dN/dt originating from particles in the size interval [R,R+ΔR] at time t can be expressed as

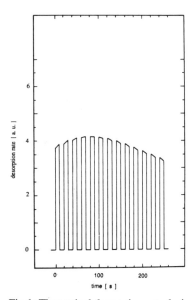

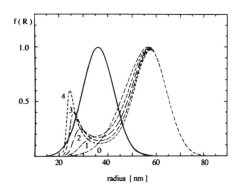

Fig. 2: Calculated change of the cluster size distribution during laser illumination corresponding to the desorption signals of Fig.1. The different traces refer to the distribution assumed initially on the surface (0) and obtained 60s (1), 120s (2), 180s (3) and 240s (4) after starting the illumination, respectively. Included is also the absorption profile (solid line) for light of $\lambda = 488$ nm.

Fig.1: Theoretical desorption rate during laser illumination ($\lambda = 488$ nm) versus time. The calculation was performed for an initial cluster size of 57 nm with a full width at half maximum of 20 nm.

$$\frac{dN}{dt} = I \, N_c \, a \, R \, Q \, \sigma_0 \, \exp\left[-\frac{(R - R_\lambda)^2}{2\,\beta^2}\right] f(R) \, \Delta R$$

σ_0 is the cross section for absorption of light at $R = R_\lambda$, I is the number of incident photons/cm^2·s, N_c the number of illuminated clusters on the surface, Q the quantum efficiency and β the width of the absorption profile. As a first step the desorption rate was calculated for each size interval ΔR. From this the resulting changes of the particle radii were computed. The time dependence of the distribution f(R) during continuous laser irradiation was obtained by integration over the actual size distribution and iterative application of the procedure outlined above.

The numerical computation has been carried out for different particle sizes, excitation frequencies and laser intensities. Here, an example with a laser wavelength of λ=488 nm will be considered. The mean particle size was 57 nm with a FWHM of 20 nm. Fig. 1 depicts the obtained theoretical desorption rate as a function of illumination time. Since the laser is turned on and off during the experiment at 10 s intervals in order to distinguish between signal and background, the same procedure was included in the calculation. An initial increase of the desorption rate by 10% to 15 % is found followed by a maximum before the signal finally drops off.

Fig. 2 shows the change of the cluster size distribution corresponding to the theoretical desorption rate of Fig. 1 at 0 s, 60 s, 120 s, 180 s, and 240 s after starting the illumination with laser light. Also included is the assumed absorption profile (solid line). Only clusters in the wing of the left side of the distribution strongly interact with the incident light and shrink in size. The effect is a narrowing of the initial distribution. In addition, however, those clusters that interact with the light shrink with illumination time *through* the size regime of the absorption profile. Therefore, a second peak of the size distribution grows. Actually, the process is quite similar to laser cooling of atoms in beams where the absorption curve is given

225

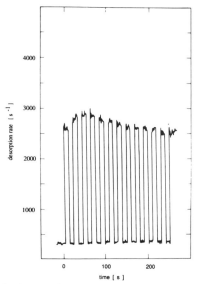

Fig. 3: Experimentally determined desorption rate as a function of illumination time with laser light.The measurement was made with $I \approx 100$ W/cm^2 and $\lambda = 488$ nm. The initial mean cluster size was $R_0 \approx 57$ nm.

by the Doppler profile [11].

The situation is quite different for clusters where the absorption profile and the particle distribution overlap more or less completely. In such a case the desorption rate decreases monotonically as a function of irradiation time. All of the clusters shrink in size until they are located in the left side wing of the distribution. This results in a considerable narrowing.

One means of monitoring these changes of the distribution is to study the desorption rate as a function of illumination time with laser light. Its variation with time contains all the information on the change of the initial size distribution. It is the interplay of two effects. First, the decrease in cluster size due to the desorption can result in an increase or decrease of the cross section for light absorption, depending on the position of the cluster radius with respect to the size dependent absorption profile. Second the number of atoms which can desorb decreases as a function of time due to the overall decrease of the surface coverage.The combination of these two effects determines the change of the desorption rate.

In order to test the validity of the theory, we have compared the calculated desorption rates with the experimental results. Fig. 3 shows such a measurement. As in the calculation the average initial cluster size was about $R_0 = 57$ nm. The clusters were illuminated by laser light of $\lambda = 488$ nm with a laser intensity of $I \approx 100$ W/cm^2. The time dependence of the experimental signals is in good agreement with the theoretical predictions. Furthermore, the theoretical total desorption yields agree with experiment.

The excellent agreement of the theoretical data with the measurements provides convincing evidence for the validity of the presented model and the manipulation of the size distribution with laser light. Especially a considerable narrowing of a distribution can be achieved. Further experiments with the goal to test the predicted manipulation of the size distributions with in situ electron microscopy are planned.

References
[1]   "Metal Clusters", eds. F. Träger, G. zu Putlitz (Springer, 1986)
[2]   Proc. 4th Int. Symp. Small Part. Inorg. Clusters, Z. Phys. D, in press
[3]   W. Hoheisel, K. Jungmann, M. Vollmer, R. Weidenauer and F. Träger, Phys. Rev. Lett. **60**, 1649 (1988)
[4]   W. Hoheisel, U. Schulte, M. Vollmer, R. Weidenauer, and F. Träger, Appl. Surf. Sci. **36**, 664 (1989)
[5]   J.A. Venables, G.D.T. Spiller, M. Hanbücken, Rep. Progr. Phys. **47**, 399 (1984)
[6]   H. Poppa, J. Vac. Sci. Techn. **2**, 42 (1965)
[7]   H. Schmeisser, Thin Solid Films **22**, 83 (1974)
[8]   M. Vollmer, F. Träger, Z. Phys. D3, 291 (1986)
[9]   K. Selby, M. Vollmer, J. Masui, V. Kresin, W.A. de Heer, W.D. Knight, Phys. Rev. B, in press
[10]   M. Vollmer, F. Träger, Surf. Sci. **187**, 445 (1987)
[11]   see e.g. Laser Spectroscopy VIII, eds. W. Persson, S. Svanberg, Springer Ser. Opt. Sci. **55**, Springer (1987)

Part VII

Laser Light Sources

Ultranarrow Linewidth Solid State Oscillators

Robert L. Byer
Department of Applied Physics, Stanford University, Stanford, CA 94305

Abstract

The combination of a diode pumped nonplanar ring oscillator in Nd:YAG with injection seeded operation of a 5 Watt cw Nd:YAG ring oscillator, which preserves the linewidth of the master oscillator, and SHG to the green followed by cw optical parametric oscillation in lithium niobate, has led to single axial mode tunable output from the OPO with a 20kHz linewidth.

Introduction

Diode laser pumped solid state lasers are efficient, all solid state sources of coherent optical radiation that have application to laser spectroscopy[1]. The linewidth of these sources has decreased from 10KHz for free running standing wave Nd:YAG oscillators[2] to less than 3kHz for nonplanar ring resonator laser oscillators[3]. Operation in a single axial mode has permitted the study of injection locking as a means of increasing the power of the laser source while preserving the narrow linewidth. Single frequency operation has also allowed resonant enhanced harmonic generation for efficient doubling of low power cw laser sources[4] and recently has led to cw operation of a lithium niobate optical parametric oscillator[5]. Diode laser pumped solid state lasers, frequency extended by nonlinear techniques, are compact highly coherent sources of optical radiation that are useful for spectroscopic applications.

30Hz Linewidth nonplanar ring oscillator

We have achieved a heterodyne linewidth of less than 30Hz for the beat note between the outputs of two 282THz(1.062μm) Nd:GGG nonplanar ring oscillators(NPRO's)[6]. The laser oscillators were independently locked to adjacent axial modes of a high finesse interferometer(Newport Research Corporation model SR-150 Cavity) using Pound Drever frequency modulated locking with two lithium niobate phase modulators operating at 40MHz and 61.5MHz. The transmission width of the 22,000 finesse, 6.327GHz free spectral range interferometer is 300kHz. The two NPRO laser oscillators were mode matched into the interferometer and isolated by a polarizing beam splitter and quarter wave plate. The resistance to optical feedback inherent in the NPRO design eliminated the need for further isolation.

Using a 123dB gain feedback loop with unity gain at 100KHz, the lasers remained locked to the interferometer for periods of longer than one hour. The error signals generated by the locking were detected and fed back to the PZT elements bonded directly to the laser cystals. The temperature stablized NPRO lasers had long term frequency drifts of less than 1 MHz per hour which was easily corrected by the voltage applied to the PZT crystals for strain tuning the NPRO frequency. The beat note was detected by a wideband photodiode and analyzed with an HP model 8566B 22GHz spectrum analyzer. The data showed a heterodyne linewidth of 30Hz for an integration time of 67 seconds.

In a subsequent experiment, the beat note was mixed down to the acoustic band at 20kHz and analyzed with a low frequency spectrum analyzer. The beat note showed modulation sidebands at 60Hz corresponding to interferometer length variations caused by residual voltage pickup through the interferometer piezoelectric leads. Upon clipping the leads, the linewidth was reduced to 2.9Hz. Figure 1 shows the beat note at a resolution bandwidth of 2.4Hz.

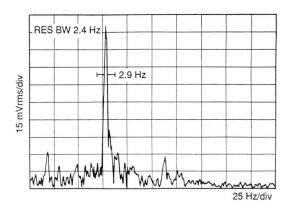

Figure 1. Heterodyne beat note between two NPRO's FM-locked in reflection to adjacent interferometer resonances separated by the free spectral range of 6.327GHz. The signal is mixed down to the acoustic band at 20kHz and analyzed by a low frequency spectrum analyzer.

Injection Locking

The extension of narrow linewidth, low power, master oscillator performance to higher power levels can be acheived by injection locking as first demonstrated more than forty years ago by Adler[7]. We have successfully injection locked a 5 Watt, cw, lamp pumped Nd:YAG ring oscillator using a 40mW NPRO master oscillator[8].

Injection locking is well known and has been demonstrated in a number of laser systems including ion lasers, dye lasers, and diode lasers. The problem has been extensively studied for laser oscillators[9]. The advantage of injection locking is the ability to impose on the slave oscillator, which may have useful characteristics such as high power, the desirable frequency properties of the master oscillator such a narrow stable linewidth. In this experiment, the master oscillator is a 20kHz linewidth, 40mW, diode laser pumped, nonplanar ring oscillator of Nd:YAG that is temperature stabilized. The injection locking forced single direction operation of the slave oscillator. The measured 0.3 radians of phase noise of the slave oscillator corresponds to a broadening of the linewidth of the master oscillator from 20kHz to 20.1kHz.

For this experiment, the slave oscillator was a lamp pumped cw Nd:YAG laser (Antares laser head from Coherent Inc.) modified to a ring resonator configuration with a cavity length of 133cm. The cavity length was controlled by a PZT actuator through an error signal derived from FM locking of the master oscillator to the slave resonator with a 50dB gain, unity gain bandwidth of 1kHz, feedback loop. The gain bandwidth was limited by the PZT actuator. The locking range of the injection locked system was measured to be 1.3MHz at 4 Watts of output power with a locking slope proportional to the square root of the ratio of master to slave oscillator powers of 13.8MHz as expected from theory.

The injection locked Nd:YAG laser system showed a slave to master power ratio of 125:1, an excellent frequency stability and very little added phase noise. In the future we plan to extend the injection locking method to an all diode laser pumped miniature slab Nd:YAG laser oscillator with an expected output power of 15 Watts when pumped by 60 one-Watt fiber coupled diode lasers.

20 kHz linewidth, cw lithium niobate OPO

Early efforts at operating doubly resonant OPO's were plagued by instabilities in power output and frequency due to the overconstrained simultaneous resonance condition of the signal and idler waves, and due to perturbations of the pump frequency. Since OPO's are of potential importance as tunable sources for spectroscopic applications, and for quantum optics applications we have worked to overcome the spectral and amplitude instabilities.

We have solved the instability problems of doubly resonant OPO's by using frequency stable, single-mode, diode laser pumped solid state lasers as the pump source. The OPO is a monolithic, ring-geometry, device with 10cm curvature mirrors polished onto the 12.5mm long crystal. Recent work[10] has demonstrated that the monolithic, ring resonator, lithium niobate OPO operates at a 12mW threshold with a pump depletion of 78% at at two times threshold. The cw output power is 8.2mW. The cw OPO is temperature tuned from 1007 to 1129nm limited by the bandwidth of the dielectric coatings that were applied directly to the crystal. The OPO operated in a single axial mode at the signal and at the idler. The OPO could be tuned over a 38nm(10THz) range near degeneracy for an applied electric field of 1050V. During frequency scanning, the OPO hops axial modes of spacing 5.4GHz, and also hops along a pair of modes known as clusters[11]. However, with the OPO servo locked for maximum output power, the frequency could be tuned over a 90MHz range without mode hoping. In the future, tuning along cluster mode pairs could be eliminated by operating the OPO in the higher threshold singly resonant configuration.

Single axial mode operation of the cw OPO was confirmed by observing the output near degeneracy with a 300MHz free spectral range scanning confocal interferometer. The OPO operated in a single frequency with a linewidth of less than 1MHz, the resolution of the scanning interferometer. The goal of the present work was to show that OPO linewidth would reproduce the linewidth of the pump laser source.

The successful operation of the cw, injection-locked, 5 W Nd:YAG laser oscillator offered the possibility studying the linewidth of the OPO by using a frequency beat note technique. The injection locked Nd:YAG oscillator was single pass frequency doubled in lithium niobate to generate 50mW of cw green radiation. The 532nm green source was used to pump the OPO which had a threshold power of 20mW.

The cw OPO operated in a stable single axial mode at the signal and idler wavelengths when pumped by the 20kHz linewidth 532nm source. The measure the linewidth, the OPO was tuned close to the 1062nm wavelength of an available Nd:GGG NPRO so that the beat note could be observed. Figure 2 shows the Double Resonant Oscillator(DRO) -Nd:GGG beat note spectrum which essentially reproduces the 20kHz linewidth of the pump laser source. One interesting observation, which may have implications for frequency chain measurements, is that the OPO signal and idler frequencies phase lock when tuned to degeneracy.

These observations confirm that the cw lithium niobate optical parametric oscillator is an efficient, narrow linewidth, source of tunable radiation that reproduces the frequency spectrum and stability of the pump laser. Work is in progress to extend the operation of the cw OPO to the three-to-one signal to idler spectral range and to investigate the microscopic tuning characteristics of the device.

Acknowledgements

This work was carried out by the graduate students and research associates of the Byer group with the support of the Office of Naval Research, NASA, and ARO. Support was also provided by Crystal Technology, Coherent Inc. and Lightwave Electronics Inc.

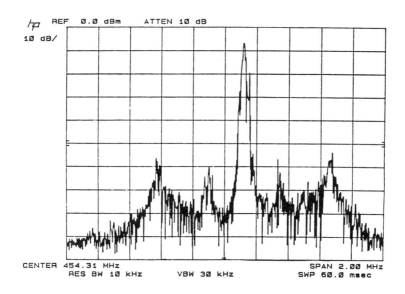

Figure 2. The cw lithium niobate doubly resonant oscillator - Nd:GGG beat note spectrum showing the 3dB linewidth of 20kHz. The spectrum essentially reproduces the pump laser spectrum of the injection locked Nd:YAG laser oscillator.

References

1. R. L. Byer, Science 239, 742 (1988)
2. B. Zhou, T. J. Kane, G. J. Dixon, and R. L. Byer, Opt. Lett. 10, 62 (1985)
3. A. C. Nilsson, E. K. Gustafson, and R. L. Byer, IEEE J. Quant. Electr. 25, 767 (1989)
4. W. J. Kozlovsky, C. D. Nabors, and R. L. Byer, IEEE J. Quant. Electr. 24, 913 (1988)
5. W. J. Kozlovsky, C. D. Nabors, R. C. Eckardt, and R. L. Byer, Opt. Lett. 14, 66 (1989)
6. T. Day, A. C. Nilsson, M. M. Fejer, A. D. Farinas, E. K. Gustafson, C. D. Nabors, and R. L. Byer, "30 - Hz linewidth, diode-laser-pumped, Nd:GGG nonplanar ring oscillators by active frequency stabilization," submitted to Electronic Letters, May 1989
7. R. Adler, Proc. IRE 324, 351 (1946)
8. C. D. Nabors, A. D. Farinas, T. Day, E. K. Gustafson, and R. L. Byer, "Injection locking of a 5 Watt Nd:YAG laser," submitted to Optics Letters, May 1989
9. Weng W. Chow, IEEE J. Quant. Electron. 19, 243 (1983)
10. C. D. Nabors, R. C. Eckardt, W. J. Kozlovsky, and R. L. Byer, "Efficient, single-axial mode operation of a monolithic MgO:LiNbO$_3$ optical parametric oscillator," to be published Optics Letters, 1989
11. J. A. Giordmaine and R. C. Miller, Appl. Phys. Lett. 9, 298 (1966)

Nonreciprocal Emissive and Absorptive Processes

S. E. Harris, A. Imamoglu, and J. J. Macklin
Edward L. Ginzton Laboratory, Stanford University
Stanford, CA 94305, USA

1. Introduction

Though it has long been believed that an inversion is critical to obtaining laser amplification and oscillation, this is not the case. Recently, we have shown that if two levels decay to an identical continuum, that this decay couples these levels and results in nonreciprocal emissive and absorptive profiles (1,2). Though previous work (1-3) has emphasized systems where the decay results from autoionization or photoionization, these ideas also hold for purely radiative decay (4). Figure 1 shows such a system. Two upper states of the same angular momentum radiatively decay to the same final level and are coupled through this decay. In a sense, one may look at this coupling as an "internal radiative trapping." When a photon is spontaneously emitted by an upper state, it has a probability of being instantaneously reabsorbed by the other upper state of the same atom. It is in this sense that the upper states are coupled.

When one examines the absorption profile of atoms which are in state |1>, one finds interferences in the absorption profile - which in this case can be viewed as interferences in the anti-Stokes scattering profile. These interferences are not present in the emissive profile of atoms which at t=0 are in state |2> or |3>.

2. Non-Cancelable Channels

In previous work, we have studied the ideal case, where both upper levels decay to the same final state. There are always other non-cancelable channels which we have neglected. In particular there is always spontaneous emission on the laser channel itself. The assumption is therefore that the decay rate of the cancelable channel is much larger than the decay rate of the non-cancelable channels. For this condition, it may be shown that the ratio of the stimulated emission cross section to the absorption cross section is equal to the ratio of the cancelable and non-cancelable decay rates.

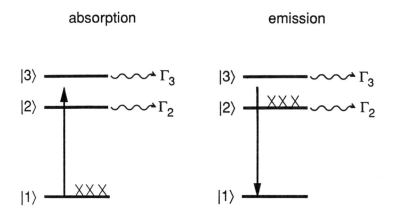

Fig. 1   Schematic of a system where the interference results from radiation broadening. The selection rule for this type of sytem is: states |2> and |3> must have the same parity, J, and m_J ; and must decay to a common final level (not shown in the figure).

3. Transient Response

It is important to recognize that the absorptive cancellation is a steady state effect that requires a time on order of the decay time of levels |2> and |3> for its establishment (5). In Figs. 2(a) and 2(b) we compare the absorption profile of a two level system and a three level system. The parameters of the three level system are chosen to cause a perfect steady state (Fano-type) interference. In the two level system we see an initial transient response followed by a Golden-Rule type steady state decay. In the ideal three level system we have only a transient absorption (zero slope); i.e. after the transient is over, the total number of photons which are absorbed by the atom does not increase with time.

If an atom at t=0 is put into an upper level it both decays and is stimulated to level |1>. The magnitude of the stimulated response is on order and usually less than the magnitude of the transient absorptive response.

When atoms are prepared in the lower level, the transient is invoked; and it is for this reason that, in most cases, that the rate of excitation of lower level atoms must be less than the rate of excitaion of upper level atoms if lasing without inversion is to occur.

The Fano type interference renders atoms transparent, but does so only after a time on order of the decay time of the upper levels.

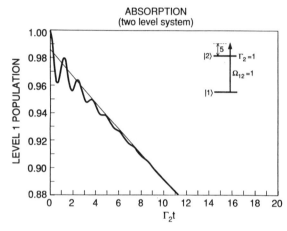

Fig. 2(a) Response of a two level system which at t = 0 is excited into level |1>. Boldface curve is numerical solution of the coupled equations (Ref.1) describing the evolution of the system. The straight line is the steady state solution from theory. The offset and slope of the straight line are the transient and steady state loss, respectively (Ref. 5).

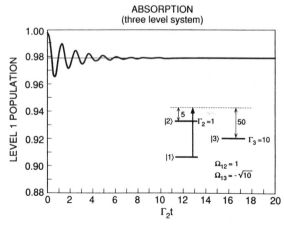

Fig. 2(b) Response of a three level system which at t = 0 is excited into level |1>. Curves have the same meaning as in Fig. 2(a). Zero slope means zero steady state loss, though a finite transient loss (non-zero intercept) occurs for the values of the parameters chosen.

4.   Density Matrix

We have developed a density matrix based computer model (6) to include the effects of dephasing collisions, excitation and de-excitation rates, non-cancelable decay channels, and to study the large signal behavior. We find that the effect of dephasing collisions is similar to that of non-cancelable channels: They become important as the dephasing rate approaches the cancelable lifetime decay rate. We also find (this is also shown by the single atom equations), that in the ideal case, that lasers of this type are immune to self termination. Figures 3(a) and 3(b) compare the response of a two level and three level system with the same parameters as that of Fig. 2. As expected the two level system self-terminates as the lower level population exceeds the upper level population. The three level system continues to lase with a lower level population that exceeds the upper level population.

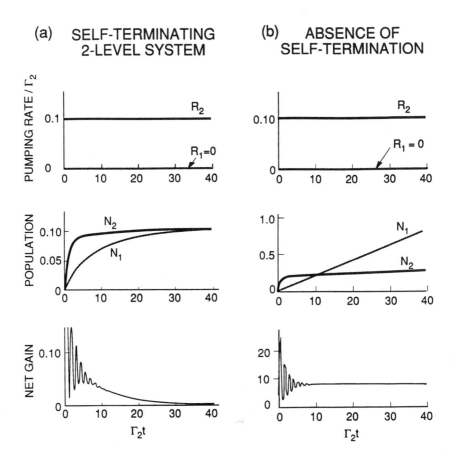

Fig. 3   Density matrix solution of a two level system (a) and a three level system (b) with the parameters of Fig. 2. It is seen that the three level system is immune to self termination. For the parameters used, level |3> population is much smaller than level |2> population.

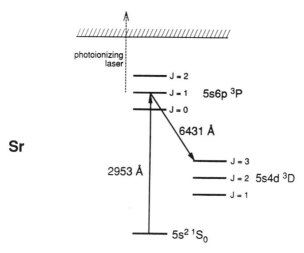

Fig. 4   Schematic of a continuum-coupled Raman process in neutral Sr.

5.   Continuum Coupled Processes

There are nonlinear processes that will operate by virtue of the coupling to the continuum. Figure (4) shows a Raman process in Sr vapor which is induced by photoionization of the upper levels. In the absence of a photoionizing laser there is no Raman scattering through the subset of levels which are shown; that is, as a result of the ΔJ selection rule, there is a zero oscillator strength in each path. In the presence of a photoionizing laser there is Raman scattering - with the new feature of partially cancellable loss to the continuum at the Stokes frequency. This type of scattering should be readily observable.

Acknowledgements

The authors thank K. H. Hahn and D. A. King for helpful discussions, and J. D. Kmetec for help with the numerical work. This work was supported by the U.S. Army Research Office, the U.S. Air Force Office of Scientific Research, and the U.S. Office of Naval Research.

References

1.   S. E. Harris, *Phys. Rev. Lett.* **62**, 1033 (1989).
2.   S. E. Harris, *OSA Proceedings on Short Wavelength Coherent Radiation: Generation and Applications*, R. W. Falcone and J. Kirz, eds. (Optical Society of America, Washington, DC, 1988), Vol. 2, pp. 414-417.
3.   V. G. Arkhipkin and Yu. I. Heller, *Phys. Lett.* **98A**, 12 (1983).
4.   A. Imamoglu, "Interference of Radiatively Broadened Resonances" (to be published as a Rapid Communication in Physical Review A).
5.   S. E. Harris and J. J. Macklin, "Lasers Without Inversion: Single Atom Transient Response" (submitted for publication).
6.   J. J. Macklin and S. E. Harris (unpublished).

External-Cavity Diode Lasers with High Brightness and High Spectral Purity[*]

W. F. Sharfin and A. Mooradian
Lincoln Laboratory, Massachusetts Institute of Technology
Lexington, Massachusetts 02173-9108, U.S.A.

We report the demonstration of a highly coherent, low-divergence, high-power, external-cavity diode laser suitable for applications in spectroscopy and nonlinear optics that require narrow linewidth in a diffraction-limited beam. Over 1.5 W of cw output power is measured at a wavelength of 856 nm, in a maximum spectral band of 0.02 nm. Greater than 0.35 W of single-frequency power is obtained in a diffraction-limited beam. A simple resonator is used, comprising a multielement spherical lens and a plane mirror in addition to the diode. The diode is antireflection (AR) coated on one facet and high-reflection coated on the other.

While present wide-stripe laser diodes and arrays can provide a few watts of continuous-wave power, they are multimode devices with poor temporal and spatial coherence. In this paper it is shown that the multimode, high output power from a wide device can be efficiently extracted in a diffraction-limited beam using a properly designed external resonator.

The diodes used in these experiments are GaAs/GaAlAs graded-index, separate confinement heterostructures (GRIN-SCH) grown by metalorganic chemical vapor deposition (1). Each has a single 5-nm-thick quantum well that is either 150 μm wide and 1200 μm long, or 60 μm wide and 600 μm long.

The geometry of the external-cavity laser is illustrated in Fig. 1. Figure 1(a) shows the dimension perpendicular to the p-n junction of the diode. The transverse mode in this (y) dimension is largely determined by the structure of the monolithic device. The lens is positioned so as to collimate the emission from the GRIN-SCH, thereby maximizing the collection efficiency of the lens and enabling the insertion of a tuning element (e.g., grating) if desired. Wavelength-selective elements are not required to achieve single-frequency operation in the lasers described here because there is neither spectral nor longitudinal spatial holeburning. The distance d_1 in Fig. 1 is fixed by the characteristics of the diode and the lens. The geometry of the lateral mode in the x-dimension, parallel to the diode junction, is shown in Fig. 1(b) and is set by adjustment of the distance d_2. The output coupler is placed so that the $(1/e^2)$ width $2w_{1x}$ of the fundamental external-cavity mode waist at the diode is larger than or equal to the width $2w$ of the active region. The latter condition occurs when d_2 satisfies the relation (1)

$$d_2 = d_2^{opt} = f\{1 + (nf/l)/[1 + (z_R/l)^2]\} \tag{1}$$

where we have defined $z_R = \pi w^2 n/\lambda$, f is the focal length of the lens, l is the length of the diode, n is the refractive index of the active region of the diode, and λ is the lasing wavelength in vacuum. Optimum relative overlap between the gain and desired mode profiles is then assured. The design principle is to saturate the gain with the fundamental mode.

---

[*] This work was sponsored by the Department of the Air Force and the Defense Advanced Research Projects Agency.

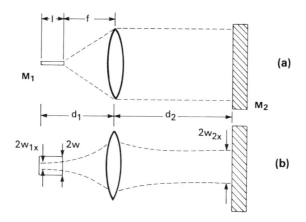

Fig. 1. Schematic diagram of external-cavity laser. (a) Dimension perpendicular to p-n junction of diode. Active region has length l. (b) Dimension parallel to junction plane. Active region has width 2w. (After Ref. 1.)

Figure 2 shows the cw output power of the external-cavity laser plotted as a function of the injected current. The slope efficiency is 0.68 W/A with the output coupler M_2 having 20% reflectivity. Overall power conversion efficiency of the laser at the maximum power shown is 32%. The spectrum of the external-cavity laser shown in Fig. 3 is considerably narrowed from that of the free-running diode. Figure 3(a) shows the spectral output of the external-cavity laser at over 1.5 W of total power (greater than five times threshold). The trace shown (power vs wavelength) is obtained from a scanned linear photodiode array in the plane of the exit slit of a 3/4 m grating spectrometer. The signal from a single array element (sharp central spike) predominates with a resolution of 0.02 nm (8 GHz). The output spectrum (power vs frequency) of the laser with an intracavity aperture is shown in Fig. 3(b), in which the transmission of a scanning Fabry-Perot interferometer with 1.5 GHz free spectral range is shown. Two peaks are displayed, separated by one free spectral range, showing the single-frequency spectrum of the laser at 110 mW of output power. The injection current (0.8 A) is just below the threshold current of the solitary, AR-coated diode. When the laser operates in a single frequency the beam divergence has been found to be diffraction-limited.

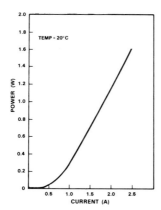

Fig. 2. Plot of output power vs injection current of cw external-cavity laser. (After Ref. 1.)

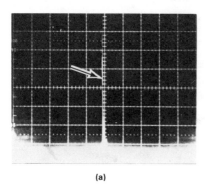 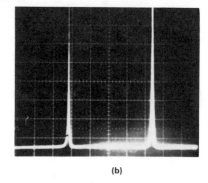

(a)   (b)

Fig. 3.   (a) Output spectrum of external-cavity laser at full power. Horizontal scale is 0.6 nm/div. Arrow indicates sharp spike with 0.02 nm width. (b) Transmission of scanned Fabry-Perot interferometer showing single-frequency spectrum of laser at 110 mW. (After Ref. 1.)

The unloaded external cavity provides little mode-selectivity by itself, as is apparent from a calculation of the Fresnel number N of the resonator using the following expression:

$$N = a_1 a_2 / \lambda d_0 \qquad\qquad (2)$$

The mirror diameters are $2a_1$ and $2a_2$, and $d_0 = d_1 + d_2 - d_1 d_2 / f$. Assuming $a_1 = 75$ μm (half the width of the active region of the diode), we find $N \sim 2 \times 10^2$ for the unapertured, 5-cm-diameter output coupler that is used. Additional selectivity is provided by an appropriate aperture at the output mirror (i.e., $a_2 = 250$ μm gives $N \sim 1$). This decreases the Fresnel number of the cavity and increases the losses of higher-order modes, effectively restricting the usable gain of the external-cavity laser to that of the fundamental mode. However, the unused gain remains available to the monolithic diode, resulting in the condition illustrated in Fig. 4. The power vs current is shown in Fig. 4(a) for an external-cavity laser with a mode-selective aperture. The laser comprises a 1200 μm-long diode with a 150-μm-wide-stripe, and a lens with a focal length of 5 cm. The same plot is shown in Fig. 4(b) for which the aperture has been removed.   A sharp decrease in the slope efficiency of the apertured laser is observed as the current passes through the threshold of the solitary diode (0.75 A). At currents less than this amount the laser operates in a single frequency. The change occurs because the intensity profiles of the diode and fundamental external-cavity mode overlap while both are lasing in a region of unsaturated gain. The diode modes and external-cavity mode compete for gain in the laser, thus depleting the available single-frequency power of the latter. This underscores the requirement for good antireflection coatings to achieve high-power, external-cavity-mode operation, thus maintaining desirable spatial coherence.

The difference $\Delta\gamma$ between the threshold gain coefficient γ_d of the solitary, coated diode and γ_{cc} of the external-cavity laser determines the range of operating current over which diffraction-limited operation is readily attained.   Consider a diode with an AR-coated facet with amplitude reflectivity r_2, in an external cavity with an output coupler having reflectivity r_3. Assuming $r_2 r_3 \ll 1$ gives the result

$$\Delta\gamma = \gamma_d - \gamma_{cc} = (1/l)[\ln (1 + r_3/r_2) - \alpha_{ext}L] \qquad\qquad (3)$$

238

where α_{ext} is the average loss coefficient of the passive section of the cavity of length L. Neglecting the loss in the passive section and using the actual values $r_3 = 0.45$ and $r_2 = 0.1$,

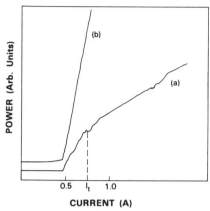

Fig. 4. Power vs current for external-cavity laser (a) with mode-selective aperture, and (b) without aperture. Aperture assures single-frequency operation below indicated threshold current I_t of solitary diode.

we have $\Delta\gamma \sim 14$ cm^{-1}. The gain coefficients in Eq. (3) are assumed to vary linearly with the injected current: $\gamma = \beta(I - I_0)$, where I_0 is the injection current required for transparency, and β is a function of temperature and device geometry. The operating range over which the performance of the compound-cavity laser is optimized is increased either by increasing the feedback from the external mirror, or by decreasing the reflectivity of the diode facet. Increasing r_3 decreases output coupling and is generally the less desirable alternative. Note that the operating range is independent of internal loss in the diode. The ratio of the solitary-diode and compound-cavity thresholds, however, decreases with increasing internal loss in the diode, while both thresholds become larger as this loss increases.

Stable, single-frequency operation of the external-cavity laser cannot be achieved at high power, without an aperture, when the external-mirror feedback is centered in the active region of the diode. If the feedback is displaced to one side of the active region, the laser operates without an aperture in a single-frequency, diffraction-limited mode when biased below the threshold of the solitary diode. We believe that this alignment configuration favors the single-lobed mode by increasing the relative losses of the higher-order modes due to absorption of the mode tails which penetrate the unpumped regions of the diode surrounding the active region. The output spectrum of the laser with 0.6 W of total power, at three times the threshold current of the solitary diode (1.8 A), continues to be dominated by the single-frequency spectrum of the low-power laser. The full spectrum shows an apparent superposition of a single-frequency signal and a broad hump characteristic of the monolithic device. The strong, single-frequency component, containing over 60% of the total laser power, and the broad hump originate from distinct regions of the diode. A single-frequency source producing about 0.35 W is made by using an aperture outside the cavity to mask part of the magnified image of the diode facet.

In summary, we have demonstrated a high-power, spectrally pure, external-cavity diode laser that is suitable for applications in spectroscopy and nonlinear optics. Improvement in antireflection coatings is expected to result in further increase in single-frequency power from these devices.

Reference

1.   W. F. Sharfin, J. Seppala, A. Mooradian, B. A. Soltz, R. G. Waters, B. J. Vollmer and K. J. Bystrom, Appl. Phys. Lett. **54**, 1731 (1989).

The Lamp-Pumped LNA Laser: a New Infrared Source for Optical Pumping of Helium

C.G. Aminoff*, C. Larat, and M. Leduc
Laboratoire de Spectroscopie Hertzienne de l'ENS,
24, rue Lhomond, F-75231 Paris Cedex 05, France

Continuous laser operation of a crystal of $La_{1-x}Nd_xMgAl_{11}O_{19}$ (LNA) transversely pumped by krypton arc lamps is described. This compact laser is tunable around the wavelength 1.083 μm and can be used to obtain high spin polarization of ^3He nuclei by optical pumping of the metastable 2^3S_1 level. It should therefore find applications in various experiments requiring an efficient production of spin polarized particles. For such applications a reliable source providing high cw power (several watts) at the helium wavelength is needed.

Previously, important progress in helium polarization by optical pumping was made using dye-laser-pumped colour-centre lasers and Nd:YAP lasers. A recently developed laser material presenting suitable spectral characteristics is the neodymium-doped LNA crystal [1]. This crystal has the advantage of shifting one of the laser transitions within the $^4F_{3/2}$ - $^4I_{11/2}$ multiplet in Nd^{3+} to the wavelength 1.082 μm while broadening the emission band, making the laser easily tunable to the helium wavelength. Although the transition cross section is smaller than in Nd:YAG, the LNA material can be doped to much higher concentrations, offering a potential for high gain in small-size active elements. LNA lasers have been pumped either by diode lasers for low power applications such as helium magnetometers [2], or by Ar^+ or Kr^+ lasers, which already has allowed interesting applications of polarized ^3He. With ion laser pumping, however, the limited available pumping power, the long-term instability and the high maintenance cost of these pump sources may be restricting factors for more demanding applications. There is therefore a present interest for developing alternative pumping techniques.

A problem in lamp-pumping of LNA is the strong heating of the crystal. The thermal conductivity of LNA is 2-3 times lower than in YAG, leading to strong thermal gradients that give rise to pronounced thermally induced focussing and birefringence. Moreover, the risk for cleavage of the anisotropic crystal under excessive thermal stress eventually puts an upper limit to the applicable pumping power.

The laser cavity configuration we have used is described in Fig. 1. The LNA rod, Nd-doped to 15 %, 10 cm long and 5 mm in diameter, is pumped by two krypton arc lamps in a commercial, water-cooled Nd:YAG laser head. The lamps are contained in a diffusing medium, providing homogeneous pumping of the rod. Our rod is cut along the crystal a-axis, the laser field then being spontaneously polarized and parallel to the a^*-axis. The end faces of the rod are curved with a 60 cm radius for partial compensation of the thermal focussing. We used a concave output mirror with the radius 50 cm in a cavity of total length 70 cm. The beam is schematically represented in Fig. 1 with an intracavity waist.

Equipped with a 0.5 % outcoupling mirror this laser gives 3 W in output power at 3.0 kW of electric power (without selective optics), the threshold being at 2.2 kW. The high threshold is partly due to the resonator configuration that is rather optimized for higher pump power. Losses from residual absorption, imperfect coatings and thermally induced birefringence also limit the laser performance.

*Permanent Address: Department of Technical Physics,
Helsinki University of Technology, SF-02150 Espoo, Finland

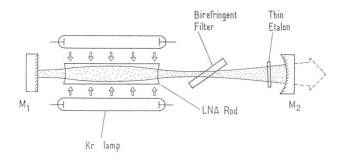

Fig. 1. Scheme of the LNA laser.

With a 6 mm single-plate birefringent filter at Brewster's angle in the cavity, the laser is broadly tunable within two wavelength bands around 1.055 μm and 1.082 μm, see Fig. 2. When a 2 mm birefringent plate and an uncoated thin etalon (0.7 mm) are used, the laser is easily tuned to one of the transitions within the helium line at 1.083 μm, yielding more than 800 mW at this wavelength in a transversely multimode output beam having a linewidth of 2-3 GHz. This linewidth is rather convenient for the purpose of optical pumping of the entire Doppler width of helium at room temperature.

In order to demonstrate the efficiency of the lamp-pumped LNA laser in optical pumping , we performed test experiments in ^{3}He. The metastable level was populated through a weak rf discharge in a cell containing 0.3 torr of ^{3}He. The 2^3S_1, F=1/2 - 2^3P_0 transition was optically pumped using a circularly polarized beam. In this process, oriented 1^1S_0 ground state atoms are subsequently produced by metastability exchange collisions [3]. The nuclear polarization was measured by analyzing the fluorescence at 668 nm. The results are shown in Fig. 3. With more than 600 mW of laser power in a beam with diameter 3 cm incident on the cell, we obtained a nuclear spin polarization of 66 % in the sample. This result is comparable to the highest polarization produced using more complicated colour-centre laser technology. More details of the present work will be published elsewhere [4].

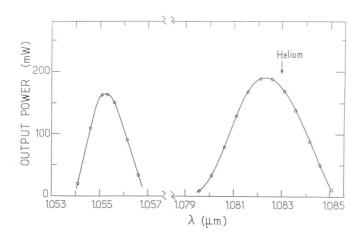

Fig. 2. Tuning curve of the LNA laser at 2.7 kW pumping power.

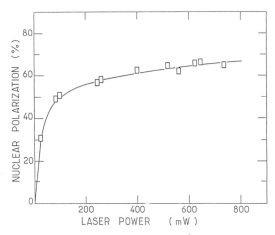

Fig. 3. Nuclear spin polarization in ^{3}He gas, measured as a function of LNA laser power incident on the cell.

Still higher nuclear polarization can be obtained by increasing the laser power. This calls for further improvements in LNA laser performance and optical pumping efficiency: growth of higher quality crystals, optimization of the doping concentration, growth of long rods along the crystal c-axis giving better laser efficiency and increased transverse thermal conductivity, etc. Another reason that limits the present laser efficiency is the rather poor fit of the krypton emission peaks to the LNA absorption spectrum. The use of high-power diode lasers emitting at 800 nm for pumping of LNA is expected to increase the power output in the future.

The lamp-pumped LNA laser is being used at ENS to polarize ^{3}He for studies of quantum effects at low temperature. The laser shows good free-running stability in power and frequency, allowing long experimental runs without readjustment.

In summary, lamp pumping of the LNA laser presents several features that are attractive for applications requiring efficient spin polarization of helium: simplified and robust technology at moderate cost, compact size, reliable long-term operation, and a potential for several watts of output power at the helium wavelength. The high spin polarization obtainable using LNA lasers renews the interest in the application of this technique to several fields of physics where an efficient production of spin polarized particles is desired, e.g. in the study of polarized quantum systems at low temperature, the preparation of polarized ^{3}He targets for scattering experiments in nuclear physics, the production of polarized electron beams, polarized dense ^{3}He used as spin filters for neutron beams, etc.. A future application might also be found in the enhancement of the cross section for fusion reactions by the injection of spin polarized nuclei. For a review see e.g.[5].

References

1. V.M. Garmash, A.A. Kaminskii et al., Phys. Status Solidi (a) **75**, K111 (1983); Kh.S. Bagdasarov, L.M. Dorozhkin et al., Sov. J. Quantum Electron. **13**, 1082 (1983); D. Vivien, A.M. Lejus et al., C.R. Acad. Sci. Paris **298**, 195 (1984).

2. J.J. Aubert, C. Wyon, A. Cassimi, V. Hardy, and J. Hamel, Optics Comm. **69**, 299 (1989).

3. F.D. Colegrove, L.D. Schearer, and G.K. Walters, Phys. Rev. **132**, 2561 (1963).

4. C.G. Aminoff, C. Larat, M. Leduc, and F. Laloë, Rev. Phys. Appl. **24**, August (1989), to appear.

5. D.S. Betts and M. Leduc, Ann. Phys. Fr. **11**, 267 (1986).

FREQUENCY TUNING METHODS FOR HIGH SPECTRAL PURITY DIODE LASERS SELF-LOCKED ON CONFOCAL FABRY-PEROT CAVITIES.

Ph. LAURENT*, D. BICOUT**, Ch. BREANT* and A. CLAIRON**.

* Laboratoire Lasers Ultra-Stables, E.T.C.A,16 bis av. Prieur de la Côte d'Or, 94114 ARCUEIL, FRANCE

** L.P.T.F., Observatoire de Paris, 61 avenue de l'Observatoire, 75014 PARIS, FRANCE

One major advantage of diode lasers is their frequency tunability as a function of injection current or temperature. Unfortunately these devices have very poor spectral characteristics in terms of frequency stability and spectral purity. One technique to reduce the frequency fluctuations is to use resonant optical feedback, illustrated on Fig.1, from a confocal Fabry-Perot (CFP) interferometer [1,2]. The laser spectral characteristics of the lasers frequency is then totally determined by 4 parameters :

1) the free-running laser frequency ω_N which is determined by the injection current
(stabilized to 10^{-6}) and the diode laser temperature (stabilized to the mK level)
2) the CFP length (L_p) or equivalently the round-trip time in the CFP (τ_p)
3) the distance between the diode laser (L_d) and the CFP or the round-trip time (τ_d)
4) and the optical feedback level (β), ratio of the feedback power to output power.

At this point the high spectral purity of the CFP coupled-cavity laser is of the order of a few kiloHertz but can only be tuned a few MegaHertz. The aim of this work is to present different ways to allow large continuous frequency tuning of the diode laser optically-coupled to the CFP without altering the laser linewidth reduction. From our theoretical model [2] the laser frequency ω is given by :

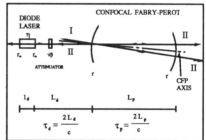

Fig. 1 . Schematic of the optical feedback locking system.The diode laser is optically locked to a tilted confocal Fabry-Perot resonator.

$$\omega_N = \omega + K \frac{\sin[\omega(\tau_d+\tau_p)+\theta] - r^4 \sin[\omega(\tau_d-\tau_p)+\theta]}{1+F^2 \sin^2 \omega \tau_p} \qquad (1)$$

with $\qquad F = \frac{2r^2}{(1-r^4)} \qquad$ and $\qquad K = (1+\alpha^2)^{1/2} \sqrt{\beta} \frac{c}{2\eta l_d} \frac{1}{2} \frac{\mathscr{F}_{cfp}}{\mathscr{F}_d}$

where $\alpha = \tan(\theta)$ is the phase-amplitude coupling factor and $\mathscr{F}_{cfp}$, $\mathscr{F}_p$ represent respectively the CFP and laser cavity finesse. It can be shown [2] that the minimum linewidth is obtained when the following two conditions are satisfied : **The laser frequency ω and the free-running laser frequency ω_N equal the CFP resonance frequency ω_0.**

To continuously tune the laser frequency while maintaining the maximum linewidth reduction the three parameters ω_N, τ_d, τ_p must be changed at the same time. This can be done by simultaneously changing these 3 parameters but it is very difficult to achieve in a reproductible manner. We have also investigated other methods using electronic servo-loops. The CFP transmitted intensity can be used as an error signal to lock the laser frequency to the maximum transmitted intensity and satisfy the first condition $\omega = \omega_0$. We investigated two different ways to obtain the error signal. First by modulating the laser frequency using the injection current, the CFP length or the distance from diode to CFP. In this case the signal coming from the CFP cavity is sent to a lock-in amplifier which provides an error signal proportionnal to the first derivative of the transmitted intensity T :

$$\varepsilon = \frac{dT}{dx} = -2T\Delta\frac{d\Delta}{dx} \quad \text{where } T = \frac{1}{4}\frac{\Gamma^2}{\Gamma^2 + \Delta^2}, \Gamma \text{ is the HWHM of the CFP resonance,}$$

$\Delta = \omega - \omega_0$ is the detuning and x is any of the three parameters ω_N, τ_d, τ_p . The second technique detects the imaginary part of the amplitude reflected by the CFP by using the Hänsch-Couillaud polarization method [3]. In this case the error signal is given by

$$\varepsilon = \frac{1}{2}\frac{\Gamma\Delta}{\Gamma^2 + \Delta^2}$$. The main advantage of this method is the absence of FM on the laser

frequency. In both cases the first servo-loop uses the error signal to ensure that $\omega = \omega_0$. As a result, when the first loop is locked, the maximum tunability of the system is of the order of hundreds of MHz, limited by frequency jumps out of the CFP resonance.

From expression (1) it can be shown that amplitude modulation of the feedback coefficient K induces FM modulation of the laser frequency ω, except if the 3 eigen-frequencies ω, ω_N and ω_0 are identical. When the first servo-loop is closed the induced FM is proportionnal to $(\omega_N - \omega_0) / K$. This FM modulation is easily detected by lock-in detection from the error signal of the first loop. This second error signal is used, after proper integration, to control the diode injection current. This second servo-loop imposes $\omega_N = \omega_0 = \omega$ and simultaneously it suppresses the FM modulation on the laser frequency created by the AM modulation of K.

The experimental set-up of the Hänsch-Couillaud method is shown on Fig. 2. Commercial, single GaAlAs type diode lasers emitting near 852 nm (Hitachi HLP 1400) were used in the present study. The polarization sensitive Fabry-Perot is made by inserting 2 Brewster plates or a dielectric polarizer inside the CFP resonator. The reflected beam from the CFP is sent through a polarization analyzer. This generates a dispersive shaped error signal which is used by a servo-loop to control the CFP to diode laser distance. The AM modulation of the feedback level is generated by an acousto-optic modulator used in the 0 order. After lock-in detection the error signal is fed to the diode laser current supply to control ω_N

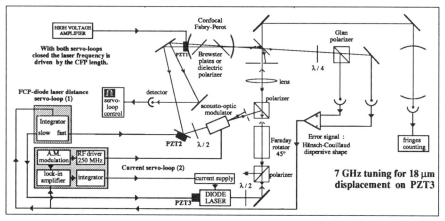

Fig. 2. Experimental set-up for continuous frequency tuning of optically CFP locked diode lasers.

and achieve $\omega_N = \omega_0$. With both servo-loops closed we were able to sweep continuously 21 modes of a 375 MHz FSR cavity, leading to a frequency tuning of 7 GHz. This scan range is only limited by the 18 μm displacement of the PZT translating the diode laser.

For very long scans the mode jumps of the diode lasers longidutinal modes would eventually become a problem. To illustrate the potential of this method we recorded the hyperfine structure of the D2 line of Cesium (Fig.3) to show at the same time both

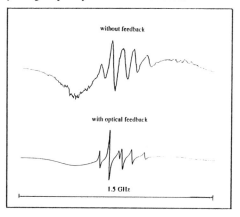

Fig. 3 . First derivative hyperfine structure of the $6\,^2P_{3/2}$ level of cesium (D2 line).

the frequency tuning and the high spectral purity of the coupled cavity laser. This technique will be useful for long term operation of optically pumped Cesium clocks. The combination of frequency tuning capabilities (> 7 GHz) and high spectral purity (linewidth of a few kHz) of these coupled cavity diode lasers, make this kind of laser system very attractive for ultra high resolution spectroscopy.

The authors are thankful to L. Hollberg for stimulating discussions. This work was supported by Bureau National de Métrologie and Direction des Recherches, Etudes et Techniques.

[1]   B. Dahmani, L. Hollberg and R. Drullinger, " Frequency stabilization of semiconductor lasers by resonant optical feedback", Optics Lett., vol. 12, n° 11, pp. 876-878, 1987.

[2]   Ph. Laurent, A. Clairon and Ch. Bréant, " Frequency noise analysis of optically self-locked diode lasers", IEEE, J. Quantum Electron., vol. 25, no.6, pp. 1131-1142, 1989.

[3]   T.W. Hänsch and B. Couillaud, Opt. Comm., vol. 35, pp. 441, 1980.

Optical Parametric Oscillators of Bariumborate and Lithiumborate: New Sources for Powerful Tunable Laser Radiation in the Ultraviolet, Visible and Near Infrared

A. Fix, T. Schröder, J. Nolting and R. Wallenstein
Institut für Quantenoptik
Universität Hannover, Hannover, FRG

Since the first demonstration of an optical parametric oscillator (OPO) by Giordmaine and Miller[1] in 1965 the OPO has been subject to detailed theoretical and experimental investigations[2,3]. The OPO is considered as a source of powerful, broadly tunable coherent radiation. The development and the scientific application has been hampered, however, by the scarcity of nonlinear optical materials with suitable optical and mechanical properties.

Recently new nonlinear materials - Bariumborate (BBO) and Lithiumborate (LBO) - became available. With their unique properties (i.e. high nonlinearity, wide transparency range and high damage threshold) BBO and LBO should be very useful materials for an OPO. These OPO's combine the advantages of an all solid-state tunable source with a wide tuning range in the ultraviolet, visible and near infrared, high peak and average power and high conversion efficiency.

Pumped by the harmonics of the Nd-YAG laser, the BBO-OPO as well as the LBO-OPO generate tunable laser radiation in the spectral range between 300 and 3000 nm. Fig. 1 shows the calculated tuning regions of the signal and idler waves as function of the phase-matching angle. The OPO's are pumped by 532 nm, 355 nm and 266 nm Nd-YAG laser radiation.

In a first experimental demonstration a BBO OPO was pumped by the 532 nm second harmonic of a Nd:YAG laser[4]. The wavelength tuning (940-1220 nm) was limited by the used mirrors. The tuning range could be extended substantially into the UV and the near infrared (330 nm - 2550 nm) by pumping with the 355 nm third or the 266 nm fourth harmonic of a Nd:YAG laser or a 308 nm XeCl excimer laser. This was demonstrated in several experiments first reported in 1988 at the CLEO conference[5][6][7]. Details of these investigations have meanwhile been published[5][6][7].

In the experiments performed at the Stanford University[5] the BBO-OPO was pumped by single axial mode 355 nm third harmonic Nd-YAG laser radiation. The Nd-YAG pump laser with unstable resonator (Spectra Physics, Model DCR3) was injection seeded to obtain single-mode operation (Spectra Physics Model 6300 Injection Seeder). The doughnut shaped output beam was spatially filtered in the far field by a suitable pinhole and frequency tripled with KD*P. The 355 nm radiation with almost Gaussian intensity distribution (diameter d=2.8 nm) provided 6 ns long light pulses with an energy of 30 mJ/pulse and a repetition rate of 30 Hz.

The 3.2 cm long resonator considered of two flat mirrors with a reflectivity of 98 percent (input mirror) and about 80 percent (output mirror). The transmission at 355 nm and at the idler wavelength exceeded 80 percent. The BBO

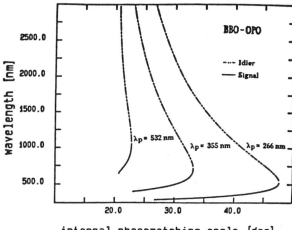

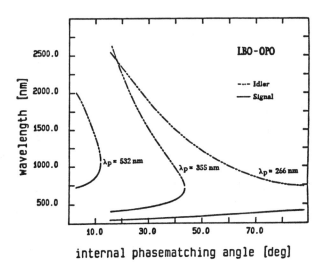

Fig.1: Wavelength of the signal and idler wave of the BBO-
OPO and LBO-OPO as function of the phase-matching
angle. The OPO's are pumped by the second, third
or fourth harmonic of a Nd-YAG laser.

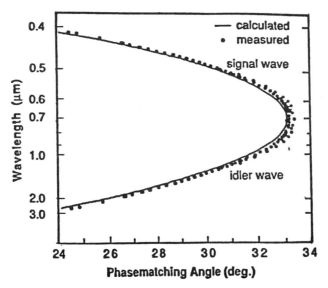

Fig.2: Measured and calculated wavelength of the signal and idler wave as function of phasematching angle of the BBO optical parametric oscillator (Ref.5).

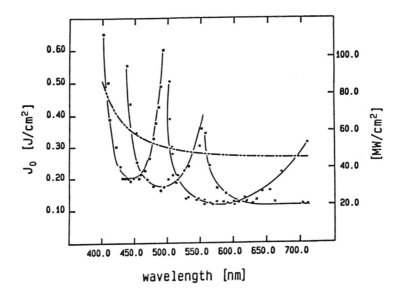

Fig.3: Energy density J_0 at threshold measured for four sets of mirrors. The beam diameter is 2.5 mm. The theoretical values of J_0 are calculated for a mirror reflectivity of 70 percent.

crystals (size 6 x 6 x 12 mm³) were cut at an angle of 25 and 35 degrees. Fig. 2 displays the measured wavelengths of the signal and idler radiation as function of the type I phasematching angle. The estimated experimental accuracy of each angle and wavelength measurement was about ±1 degree and ±1 nm, respectively. Within these uncertainties the measured values were in good agreement with those calculated from the Sellmeier formula reported by Kato [8]. The observed tuning range extended from 412 nm to 710 nm (signal wave) and 710 nm to 2.6 μm (idler wave). This operating range corresponds almost to the maximum possible tuning range which is limited by the increasing absorption of BBO at wavelengths larger than 2.5 μm [9]. As shown in Fig. 2 the measured phasematching angle varies between 24.5 and 33.2 degrees.

Besides the wavelength tuning, parameters like threshold power, conversion efficiency and the spectral width of the generated light are of special importance.

Fig.3 displays the energy density at threshold (J_o) measured as function of wavelength with an experimental set up similiar to the one described in ref. 5. The value of J_o depends - as expected - on the mirror reflectivity . The minimum values of J_o increase with decreasing wavelength from J_o=0.12 J/cm² to J_o=0.19 J/cm². The corresponding pulse energy is 5-7 mJ. The power density of 20 to 40 MW/cm² is well below the BBO damage threshold which is expected to be several GW/cm². Calculated values of J_o - also shown in Fig.3 - are larger by a factor of about 2. This difference between the theoretical and experimental results might indicate that the previously measured nonlinear coefficient of BBO is low [10].

Fig.4 displays the measured energy conversion efficiencies as function of the ration J_p/J_o, where J_p is the energy density of the pump radiation. With a 12 mm long crystal the conversion efficiency is close to 25 percent. For an 8 mm long crystal J_o is larger by a factor of 2.2. At a ratio of J_p/J_o=1.8 - which is limited by the available energy E_p=24mJ of the laser pulse - the conversion efficiency is less than 8 percent. The efficiency should increase with crystal length and larger values of J_p/J_o. However, walk-off between the pump and the generated radiation limits the useful BBO crystal length to about 25 mm.

For many applications narrowband operation is highly desirable. Pumping with a single-axial mode laser source [5] the output contained typically 6 axial modes of the 3.2 cm long OPO resonator indicating a linewidth of about 23 GHz. Single mode operation was achieved at 532 nm or 1,06 μm (and at the corresponding idler and signal wavelength) by injection seeding with light of the second harmonic or the fundamental of the single-mode Nd-YAG pump laser. The injection seeding not only provided single-mode operation but also reduced the oscillator build-up time from 4 to 2.2 nsec. Because of the reduced build-up time the pulse length increased from 2.5 nsec to 4 nsec. Simultaneously the OPO threshold power decreased by a factor of 3.

In addition to the systems mentioned so far, synchronously pumped BBO-OPO devices have been investigated. With the second harmonic of a mode-locked pulsed Nd-YAG laser [11] or

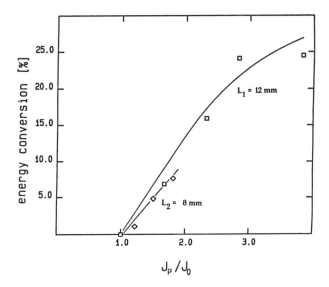

Fig.4: Total energy conversion measured as function of
the ratio J_p/J_o. J_p is the energy density of the
pump radiation. The length of the crystals are $L_1 =$
12 mm and $L_2 = 8$ mm. The corresponding values of J_o are
$J_{o1} = 0.12$ J/cm^2 and $J_{o2} = 0.26$ J/cm^2. The wavelength
of the signal wave is 620 nm.

the third-harmonic of a Nd-YAP system[12] the BBO OPO
produced 75 psec long light pulses tunable in the regions of
680 nm to 2.4 μm [11] or pulses of about 20 psec duration at
406 nm to 3.17 μm [12], respectively.

In addition to BBO LBO crystals are now available in
sufficiently large size and with high optical quality. As
seen in Fig. 1 the LBO-OPO should be tunable in the same
wavelength range as the BBO-OPO. In LBO the effective
nonlinearity is smaller, however. At 710 nm, for example,
d_{eff}(LBO)=0.39 d_{eff}(BBO). The values of J_o of the LBO-BBO
should thus be larger by a factor of 6.7. The values of the
ratio of J_o(BBO)/J_o(LBO) obtained in first measurements with
12 mm long crystals at the wavelengths of 630 nm, 560 nm and
440 nm provide ratios of 4.7, 5.2, 3.2, respectively [13].
These measurements confirm the higher energy density at
threshold of the LBO- OPO. The results are in agreement with
the theoretical predictions.

In LBO the tuning rate is considerably smaller compared
to BBO. Wavelength tuning in the range of 410-2600 nm
requires, for example, a change of the type I phase-matching
angle $\Delta\Phi = 28°$. In the UV tuning from 300 nm to 400 nm even
requires a change by $\Delta\Phi = 55°$.

The possible advantages of the LBO-OPO have still to be
investigated in particular in respect to narrow bandwidth
operation and short pulse generation.

In summary these first experiments clearly demonstrate the advantages of the BBO and LBO-OPO's as tunable coherent light sources. The most attractive features are the high power capability, high conversion efficiency, and in particular the large tuning range. The results obtained so far are certainly very promising for the further development of these new tunable all solid-state sources, which might be in the near future the radiation source of choice for applications like pulsed laser spectroscopy.

References:

(1) J.A. Giordmaine and R.C. Miller, Phys.Lett.14, 973 (1965)

(2) S.E. Harris, Proc. IEEE 57, 2096 (1969)

(3) R.L. Byer, in Treatise in Quantum Electronics, edited by H. Rabin and C.L. Tang (Academic, New York, 1973) pp.587-702 see also: Y.X. Fan and R.L. Byer in SPIE Proceedings Vol. 461, 27(1984)

(4) Y.X. Fan, R.C. Eckardt and R.L. Byer, Conference on laser and Electro-Optics (CLEO) 1986 (San Francisco), postdeadline paper ThT4

(5) Y.X. Fan, R.C. Eckardt, R.L. Byer, J. Nolting and R. Wallenstein, Conference on Lasers and Electro-Optics (CLEO) 1988 (Anaheim), postdeadline paper pd31; Appl.Phys.Lett.53, 2014 (1988)

(6) H. Komine, Conference on Lasers and Electro-Optics (CLEO) 1988 (Anaheim), postdeadline paper pd32; Opt.Lett.13, 643 (1988)

(7) L.K. Cheng, W.R. Bosenberg, D.C. Edelstein and C.L. Tang, Conference on lasers and Electro-Optics (CLEO) 1988 (Anaheim), postdeadline paper pd 33; Appl. Phys.Lett.54, 13 (1989)

(8) K. Kato, IEEE J.Quant.Electron., QE-22, 1013 (1986)

(9) D. Eimerl, L. Davis, S. Velsko, E.K. Graham and A. Zalkin, J.Appl. Phys. 62, 1968 (1987)

(10) C. Chen, B. Wu, A. Jiang, and C. You, Sci.Sin.Ser.B28, 235 (1985)

(11) L.J. Bromley, A. Guy and D.C. Hanna, Opt. Commun. 67,317 (1988)

(12) S. Burdulis, R. Grigonis, A. Piskarskas, G. Sinkevicius, V. Sirutkaitis, J. Nolting and R. Wallenstein (submitted for publication)

(13) A. Fix, T. Schröder, Ch. Huang and R. Wallenstein (to be published)

EFFICIENT THIRD HARMONIC GENERATION AND FOUR-PHOTON MIXING SPECTROSCOPY OF LASER-PRODUCED PLASMAS NEAR METAL SURFACES

A.B.Fedotov, S.M.Gladkov, O.S.Ilyasov, N.I.Koroteev
and A.M.Zheltikov

R.V.Khokhlov Nonlinear optics Laboratory,
International Laser Center
Moscow State University, Moscow 119899, USSR

1. EFFICIENT T H G IN LASER-PRODUCED PLASMAS.

In this thesis we report the effective (up to 3 percent) third harmonic generation (THG) of picosecond Nd:YAG laser radiation with wavelength $\lambda = 1.06$ μm in the under-dense laser plasma in a coherent (i.e. collinear) geometry. Thus conditions of our experiments significantly differ from that ones usual for optical harmonics generation experiments in a dense, hot thermonuclear laser plasmas. In our experiments we have not only achieved high THG efficiency, but have also produced almost diffraction-limited, coherent THG beam which could be used in other nonlinear-optical experiments.

In our experiments the spark has been produced by the optical breakdown of an atmospheric gas near a metal surface by the additional nanosecond Nd:YAG laser source . In contrast with usual experimental approaches we have used the new experimental scheme (first described in Ref. [1]) which possesses the following peculiarities: 1) Two independent laser sources have been used for the optical breakdown and THG. That one used for plasma creation has had the pulse duration about 15 ns and the second one used for harmonic generation has had the pulse duration 40 ps, with variable time delay between them; 2) The energy of the picosecond light source was no more than 50 mJ; 3) The plasma density has not exceeded $5*10^{19}$ cm^{-3}, thus radiation frequencies were much higher than the plasma oscillation frequency and the medium was almost totally transparent for all optical pulses used. As a result, we have achived THG efficiency as high as 3%, determined as the ratio of energies of relative pulses. All the results have been summarized in fig.1 where third-harmonic efficiency is plotted vs pump intensity. Open boxes on x-axis mean self-breakdown.

We propose that adequate way to describe harmonic generation in a low-temperature plasma gas with highly intense pico- and femtosecond laser pulses is to consider this process as a result of optical electron scattering on the ion in the presence of a strong laser field. We suppose that interaction of an optical electron with the light beam is stronger than that with an ion, thus the influence of the latter should be considered as a perturbation (Ref.[2]). In this way we obtain the following expression for the component of the dipole momentum, oscillating at the frequency $n\omega$, along the light field vector E, for the hydrogen atom [2]. We can note that the experimentally achieved THG efficiency is well matched with our theoretical result.

We have also theoretically studied saturation behavior of THG in both classical and quantum pictures. It was found that when the pump intensity is increased, the harmonic intensity first grows up, then saturates and goes down.

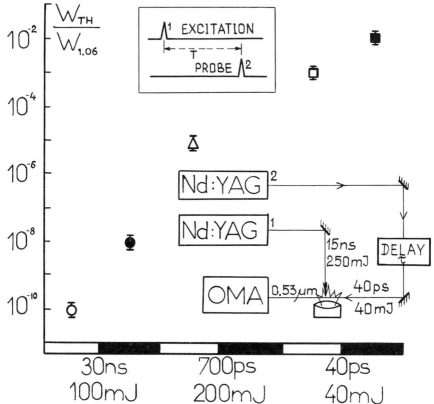

Fig. 1. Convertion efficiency and experimental set-up scheme for THG investigation [4].

2. RESONANCE FOUR-PHOTON MIXING SPECTROSCOPY OF EXCITED IONIC AND ATOMIC COMPONENTS OF LASER-PRODUCED PLASMAS.

Four-photon processes in "cold" atomic media are usually employed for frequency convertions and spectroscopy. For last ten years some papers appeared devoted to CARS in cold and excited atomic media [3]. Here we present our results on Coherent Anti-Stokes Light Scattering (CALS) experiments in the gas excited by the laser breakdown near the metal surface. In this scheme the coherent signal scattered via 4-photon process has the frequency ω which coincides with the frequency of dipole-allowed transition in the ion. In this case a strong resonance appears, usually 2...3 orders of magnitude higher than spontaneous emission lines. We have observed this kind of resonances in NII, InII and AlIII in the laser plasma produced on the metal surface in the air under pressures in the limits 0.1...1 atmosphere (Fig.2). In this figure the energy level diagrams and kinetics of CALS-signals are presented. Closed squares on the figure for NII belong to spontaneous emission on the same resonant transition. Coherent signal survives for much longer time

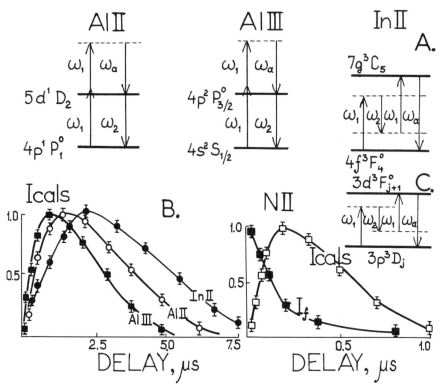

Fig. 2. CALS spectroscopy of ions   a) CALS in Al11,
Al111, In11   b) CALS intensity time dependence
c) CALS and spontaneous emission kinetics for N11

delays than a spontaneous one: the population of the lower
resonant level is growing up due to emission whereas that one
of upper level goes to zero. We have modelled the population
kinetics and also time dependence of both coherent and
spontaneous signals in the frame of a simple
collision-radiative model of plasma and achieved reasonable
agreement with the experiment. Resonances similar to those
described above have been observed in Al11, Al111, In11, Ar11,
Xe11 ions (see fig. 2, where normalized kinetics for some
resonances, and interpretation is presented). Thus we have
shown the possibility of metastable ionic states diagnostics
that hardly manifest themselves in spontaneous emission.
        In   conclusion   the   authors   acknowledge   professor
S. A. Akhmanov for valuable consultations and support.

R E F E R E N C E S

1. Gladkov S. M. , Koroteev N. I. , Rychev M. V. , Fedorov A. B.
    Pizma v Zhurn. Tekhn. Fiziki (USSR), 12, 1272 (1986).
2. Akhmanov S. A. , Gladkov S. M. , Koroteev N. I. , Zheltikov A. M.
    Moscow State University, dept. Physics, Preprint N5, 1988.
3. Gladkov S. M. ,   Koroteev N. I.   et al. Izvestija Acad. Nauk
    SSSR, ser. fizichesk. (USSR), 52, 443 (1988).
4. Gladkov S. M. , Zheltikov A. M. , Koroteev N. I. , Fedotov A. B.
    Pizma v Zhurn. Tekhn. Fiziki (USSR), 14, 1399 (1988).

ORIGIN OF OPTICAL NONLINEARITY IN D,A-DPH SYSTEMS

C.T. Lin, H.W. Guan, A.D.S. Marques and H.Y. Lee
Department of Chemistry, Northern Illinois University
DeKalb, Illinois 60115, U.S.A.

1. Introduction

It is known[1,2] that conjugated molecules generally have much larger hyperpolarizabilities than nonconjugated ones, which indicates a predominant contribution of the delocalized π-electrons. When a strong electromagnetic field of strength, E is applied, the molecular polarization response can be expressed as

$$P_i = \alpha_{ij}E_j + \beta_{ijk}E_jE_k + \gamma_{ijkl}E_jE_kE_l + \ldots$$

where α is the linear polarizability; β and γ are the second and third-order hyperpolarizabilities, respectively. While α and γ are nonvanishing for all types of molecules, a lack of a center of inversion in the molecule is required for nonvanishing of β.

The length dependences of the second- and third-order polarizabilities of conjugated molecules follow[1,2] as $\beta \sim L^3$ and $\gamma \sim L^5$, where L is the length of the π-electron delocalization chain. With the same length of the conjugated core, the optical nonlinearity depends on the choice of electron-donating(D) and accepting(A) substituents.[1,2] The largest values of β might be obtained when conjugated molecule contains substituents that lead to low-lying charge-transfer(CT) resonance states.

In this report, the optical spectroscopic techniques are employed to investigate the photophysical properties of D,A-DPH in solutions, where D and A are the groups $-OCH_3$, $-N(CH_3)_2$ and $-NO_2$. The intramolecular charge-transfer state in D,A-DPH are characterized as resulting from a twisted molecular conformation(i.e., a TICT state). The molecular nonplanarity in D,A-DPH are also computed theoretically. Finally, the optimum conditions for the optical nonlinearity in D,A-DPH systems are presented.

2. Results and Discussion

a. Characterization of TICT state in D,A-DPH

The left-hand portion of Figure 1 shows the 77K emission and 298K absorption spectra of (a) unsubstituted all-trans-1,6-diphenylhexatriend(DPH, referred as compound I) in n-heptane, (b) p,p'-disubstituted $(CH_3)_2N-C_6H_4-(CH=CH)_3-C_6H_4-NO_2$(compound II) in n-heptane, and (c) compound II in dioxane. The apparent origin of the absorption and emission in Figure 1(a) is at 388.2 nm(λ_{max} = 369.0 nm) and 398.5 nm, respectively. The electronic structure and molecular geometry of DPH was assigned[3] to have an all-trans planar configuration. This is in agreement with the X-ray crystallographic data[4], where the angle between the plane of the phenyl ring and the plane of the hexatriene chain in DPH is only 1.9°.

The absorption spectrum becomes broader and the spectral maxima are largely red-shifted in Figure 1(b)(λ_{max} = 439.0 nm) relative to those in Figure 1(a). Two types of emission are observed in Figure 1(b). One type of emission(referred as type I) located at 390-450 nm, which is similar in spectral position as that observed for the unsubstituted DPH in Figure 1(a). This emission has a lifetime, τ = 12 ns, and was assigned[3] to originate from a planar excited state configuration of $(DA)^*$ $\longrightarrow$ (DA). The other type of emission (type II) has an 0,0 band at 525 nm and τ = 6 ns, which was shown[3] to result from a "twisted

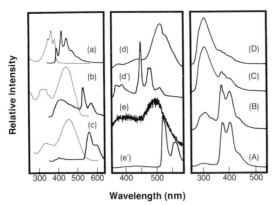

Figure 1. Absorption (---) and emission (——) spectra of DPH and D,A-DPH recorded for various conditions. See text for a detailed description. The left- and middle-portion spectra are taken with a concentration of $1 \times 10^{-4} - 1 \times 10^{-5}M$.

Wavelength (nm)

intramolecular charge-transfer"(TICT) excited state of $(D^{+}A^{-})^{*} \longrightarrow (D^{+}A^{-})$.

The spectroscopic properties of TICT state in D,A-DPH are very sensitive to the donor and acceptor substituents and also to the solvent polaity as shown in Figure 1(b) and 1(c). While the emission maximum of type I fluorescence at ~405 nm remains unchanged in Figure 1(c) as compared to that in Figure 1(b), the absorption maximum and the spectrum of type II fluorescence are largely red-shifted. Moreover, the intensity ratio of type II to type I emission increases as the solvent polarity increases. These results indicate that the charge-transfer character of the ground and TICT excited states in D,A-DPH are highly stabilized in polar solvents.

A thin film of D,A-DPH was also cast on porous Vycor glass substrates. The middle portion of Figure 1 displays the 77K emission of NO_2-C_6H_4-$(CH=CH)_3$-C_6H_4-NO_2(compound III), (d) embedded in Vycor glass, and (d') dissolved in n-heptane, and of compound II, (e) cast in Vycor glass, and (e') dissolved in n-heptane. The spectral features of D,A-DPH embedded in Vycor support are broader than those dissolved in solutions. The type II emission maximum in Figure 1(d) is ~40 nm red-shifted from that of spectrum 1(d'), while λ_{max} in spectrum 1(e) is ~20 nm blue-shifted from that of Figure 1(e'). This suggests that the surface Brønsted acid sites in a calcinated Vycor glass[5] tend to stabilize the TICT state of compound III and destabilize that of compound II. Presumably, the surface-adsorbate interaction is via the -NO_2 group of D,A-DPH, i.e., the electron-accepting power of -NO_2 group decreases in compound II and increases in compound III while D,A-DPH is embedded in Vycor substrates.

b.   Stability of TICT state and molecular geometry of D,A-DPH

The right-hand portion of Figure 1 displays the emission spectra of compound III in ethanol at various concentrations: (A) $1 \times 10^{-5}M$, (B) $1 \times 10^{-6}M$, (C) $1 \times 10^{-7}M$, and (D) $1 \times 10^{-8}M$. The relative emission intensity of type II to type I fluorescence is very sensitive to the solution concentration. The dilution of compound III in ethanol solution leads to a destabilization of the twisted geometry of D,A-DPH and decreases the type II fluorescence of the TICT state. This is presumably due to the fact that the intermolecular distance becomes longer and the intermolecular charge interaction(or hydrogen-bonding) becomes weaker when the concentration of D,A-DPH is reduced.

The lowest singlet excited state, S_1 of compound II is assigned as a strong TICT state at ~2.36 ev involving a transition from the n orbital of -$N(CH_3)_2$ to the π^{*} orbital of -NO_2 group, and the locally excited state is detected as a slightly weaker transition at ~3.12 ev resulting

256

Figure 2. The computed geometries and plotted by Alchemy II (Tripos Associates, St. Louis, MO). Top: DPH. Bottom: Compound II.

from the π and π^* orbitals of the hexatriene chain. The molecular orbital programs of HAM/3[6] and MOPAC[7] were used to optimize the twisted geometry of compound II by comparing the calculated transition energies with the observed values. The X-ray crystallographic data of unsubstituted DPH as shown in the top structure of Figure 2 was used as the conjugated backbone for D,A-DPH. The best set of calculated transition energies for the TICT state of compound II is ~2.03 ev with an oscillator strength, f = 2.04, and the planar locally excited state is at ~3.04 ev and f = 0.39. It is important to note that a twisted geometry as computed in the bottom structure of Figure 2 is required for a "direct" D to A transition and a stable TICT state in D,A-DPH systems. The computed angle between the plane of the phenyl ring and the plane of the hexatriene chain in compound II is ~25° which is slightly larger than that of 1,6-Di-o-methoxyphenyl-1,3,5-hexatriene(~ 15°) determined[4] by X-ray crystallography.

3. Remarks

The molecular second-order hyperpolarizability, β-value of $(CH_3)_2N$-C_6H_4-$(CH=CH)_3$-C_6H_4-NO_2 is determined as $770 \pm 60 \times 10^{-30}$ cm^5/esu by the electric-field-induced second harmonic generation technique, which is ~1700 times larger than that of urea(β(urea) = 0.45 X 10^{-30} cm^5/csu, note that the macroscopic second-order hyperpolarizability, $\chi^{(2)}$ of urea $\cong$ $\chi^{(2)}$ of KDP).[2] The required structural characteristics for the organic materials with large optical nonlinearities are the molecular length of delocalized π-electron chain, the molecular planarity and the asymmetric electronic environment leading to charge-transfer character. These conditions are clearly illustrated as the origins of optical nonlinearity in D,A-DPH systems.

References

1. Nonlinear Optical Properties of Organic Molecules and Crystals, D.S. Chemla and J.Zyss, Eds., Vols 1 and 2, Academic Press, Orlando, 1987.
2. D.J. Williams, Angew. Chem., 23, 690(1984).
3. C.T. Lin, H.W. Guan, R.K. McCoy and C.W. Spangler, J. Phys. Chem., 93, 39(1989).
4. T.J. Hall, S.M. Bachrach, C.W. Spangler, L.S. Sapochak, C.T. Lin, H.W. Guan and R.D. Rogers, Acta Crystallographica, in press.
5. C.T. Lin, W.L. Hsu, C.L. Yang and M.A. El-Sayed, J. Phys. Chem., 91, 4556(1987).
6. L. Åsbrink, C. Fridh and E. Lindholm, Chem. Phys. Letters, 52, 63(1977).
7. M.J.S. Dewar, E.G. Zoebisch, E.E. Healy and J.J.P. Stewart, J. Amer. Chem. Soc., 107, 3902(1985).

VUV AND XUV EXCIMER LASERS USING JET DISCHARGES WITH
SUPERSONIC COOLING

B.P. Stoicheff and T. Efthimiopoulos
Department of Physics, and Ontario Laser and Lightwave
Research Centre, University of Toronto, Toronto, Ontario,
M5S 1A7, Canada.

1. Abstract

Stimulated emission has been observed at 126 nm in
electrical discharges of Ar followed by supersonic jet
expansion of the plasma. The measured gain in an active
length of a few millimeters is approximately 3/cm. In
discharges with Ne, intense emission with line narrowing has
been found at 71 nm.

2. Introduction

It is well known that stimulated emission has been
achieved with the excimers Xe_2, Kr_2, and Ar_2 only by electron
beam excitation (1). The shortest wavelength emission is
generated by Ar_2, at 126 nm. This was first reported by
Hughes et al (1) in 1974, and more recently by Wrobel et al
(2) and Uehara et al (3). Output powers >2 MW have been
obtained with the Ar_2 lasers operated at gas pressures   20
atm and electron energies >500 keV. It is also known that at
these necessarily high pressures, it is difficult to maintain
a uniform glow discharge electrically in the pure rare gases;
usually only streamers are observed between the electrodes in
a discharge tube. Thus, in contrast to the rare-gas halide
lasers, electrical discharge pumping has not been successful
in generating stimulated emission from the Xe_2, Kr_2 and Ar_2
excimers up to the present time.

Here, we wish to review the results of recent
investigations using pulsed corona discharges in Ar and Ne
followed by supersonic expansion and cooling of the plasma.
Such schemes have been in use by spectroscopists in the past
decade as a means of obtaining metastable atomic species (4),
and of simplifying complex electronic spectra of radicals and
polyatomic molecules. In the latter application, the low
rotational temperatures achieved even for molecules in highly
excited electronic and vibrational states has facilitated the
analysis of intricate emission spectra (5). Various designs
of such sources have been described in the literature (4-6).

In the present application, the supersonic expansion
through a nozzle of an electrical discharge plasma was
considered as a means of producing rare gas dimers in excited
electronic states, and of concentrating population in low
rotational and possibly vibrational levels of the excimer
states. Such a mechanism could provide maximum population
inversion leading to lower thresholds for stimulated emission
and optimum gain in a variety of excimer and exciplex
systems. Argon was selected for the initial experiments
since the radiative lifetime of Ar_2 in the lowest vibrational
levels of the first excited state $(A1_u)$ is ~3µs (7),

considerably longer than that of Kr_2 (~250 ns) (8) or of Xe_2 ~100 ns) (9).

3. Experimental Method and Results

The experimental arrangement used in the present investigations is shown in Fig. 1 and described in Ref. 10. Briefly, a pulsed discharge takes place between a pointed tungsten electrode held in a quartz nozzle (after the design of Engelking (6)) and a linear nickel electrode, ~15 mm below the nozzle. A capacitor of 5.5 nf is connected across the electrodes and charged from 0-30 kV, producing peak discharge currents of ~1 kA at 12 kV in a rise time of ~50 ns. The expansion chamber is of aluminum (10x10x10 cm) and is provided with two viewing ports, an extension arm for a resonator mirror and an evacuated connection to a vacuum spectrometer. Mechanical pumps rated at 75 l/s maintain the chamber at a pressure of 10^{-4} Torr when Ar gas at ~2 atm pressure is introduced at a pulse rate of 1 Hz.

Preliminary experiments with Ar indicated a steep increase in emission intensity at 126 nm with discharge voltages >8kV, and a series of relaxation oscillations. When a cavity was aligned with Al + MgF_2 mirrors of ~80% reflectivity, the front mirror was ablated in a circular area of 6 mm diameter after only a few shots with voltage increasing to 16 kV. From earlier damage thresholds reported for Ar_2 lasers (1,3) we estimate a cavity energy of 20-30 mJ. This in turn leads to a population inversion of ~10^{16} for the emission of 10 eV photons and a gain >1/cm. A somewhat longer cavity resulted in a measured energy output of 0.2 mJ at 15 kV, through a 2 mm aperture in the front mirror.

A direct measurement of the gain of the Ar plasma was made by using a tunable VUV radiation source developed in this laboratory (11). With Hg vapor as the nonlinear medium, a probe beam of ~10 mW and 0.3/cm line width was tuned over

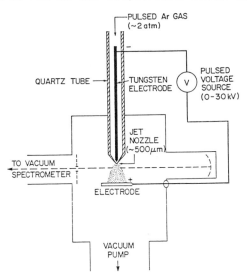

Fig. 1. Schematic diagram of the apparatus.

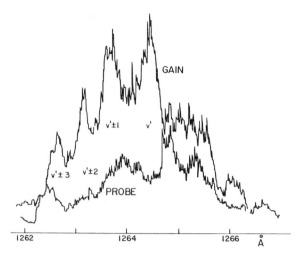

GAIN

v'±1   v'

v'±3   v'±2

PROBE

1262   1264   1266   Å

Fig. 2. Spectra of the VUV probe and of the gain recorded with the Ar discharge off and on, respectively.

the wavelength region of the Ar_2 emission, from 126-127 nm. A solar-blind photomultiplier and vacuum spectrometer were used for intensity measurements. Spectra were recorded with the argon discharge off and on. The results of the present measurements including the probe beam and gain spectra are shown Fig. 2. Substantial gain is indicated by the four intensity maxima which occur in the region 126.2-126.5 nm. The two main peaks (v' and v'±1) each lead to values of

$$I/I_0 = \exp[\gamma L] \sim 3.0$$

Here I and I_0 are amplified intensity and incident (probe) intensity, respectively, γ is the gain per cm, and L is the length of the amplifying (or gain) medium. With L ~3 mm (at 2 mm below the nozzle), the measured ratio I/I_0 yields for the gain, a value of $\gamma = 3$/cm.

With a discharge in Ne, intense emission was observed at ~71 nm, and this exhibited a sharp increase in intensity at voltages > 15 kV and line narrowing. The insertion of a rear mirror coated with osmium having a reflectivity of ~22% indicated a 50% increase in intensity, again leading to a high gain of ~3/cm.

4. Discussion

The coincidence of the emission at 126 nm with that due to Ar_2 lasers, and the high gain, suggests that this source produces excimers with high population inversion. The reaction may be represented by

$$Ar^* + Ar + Ar \rightarrow Ar_2^* \; (v', J' \; high) + Ar$$

Argon atoms excited into high states cascade down to metastable states where collisions with ground state atoms lead to the formation of Ar_2 excimers in high vibrational and rotational levels of the Al_u^2 and BO_u^+ states. In the supersonic expansion, collisions lead to the interaction

260

$$Ar_2^* (v',J' \text{ high}) + Ar = Ar_2^* (v',J' \text{ low}) + Ar$$

followed by emission at 126 nm and dissociation of the Ar_2

$$Ar_2^* (v',J' \text{ low}) = Ar + Ar + h\nu \ (\sim10eV)$$

While this simple and straightforward process may account for the present results, the observation of several other intense VUV and XUV emission lines due to Ar atoms and ions suggests the occurrence of a complexity of excitation processes.

The emission at 71 nm obtained with the Ne discharge does not occur at the expected wavelength (~84 nm) for Ne_2. According to theoretical potential curves (12), the transition may originate from a highly excited state of Ne_2, but confirmation is difficult since spectroscopic data are scarce. Obviously further experimental and theoretical studies are necessary to establish the physical processes taking place in this source.

References

1. W.M. Hughes, J. Shannon, A. Kolb, E. Ault, and M. Bhaumik, Appl. Phys. Lett. 23, 385 (1973): J. B. Gerardo and A. W. Johnson, J. Quantum Electron. QE-9, 748 (1973): P.W. Hoff, J.C. Swingle, and C.K. Rhodes, Opt. Commun. 8, 128 (1973); Appl. Phys. Lett. 23, 245 (1973).

2. W.-G. Wrobel, H. Rohr, and K.-H. Steuer, Appl. Phys. Lett. 36, 113 (1980).

3. Y. Uehara, W. Sasaki, S. Saito, E. Fujiwara, Y. Kato, M. Yamanaka, K. Tsuchida, J. Fujita, Opt. Lett. 9, 539 (1984).

4. J.Q. Searcy, Rev. Sci. Instrum. 45, 589 (1974).

5. A.T. Droege and P.C. Engelking, Chem. Phys. Lett. 96, 316 (1983); S. Sharpe and P. Johnson, ibid 107, 35 (1984); K.P. Huber and T.J. Sears, ibid 113, 129 (1985).

6. P.C. Engelking, Rev. Sci. Instrum. 57, 2274 (1986); K.G.H. Baldwin, R.Paul Swift, R.O. Watts, Rev. ibid 58, 812 (1987).

7. J.W. Keto, R.E. Gleason, Jr., T.D. Bonifield, and G.K. Walters, Chem. Phys. Lett. 42, 125 (1976).

8. T.D. Bonifield, F.H.K. Rambow, G.K. Walters, M.V. McCusker, D.C. Lorents, and R.A. Gutcheck, Chem. Phys. Lett. 69, 290 (1980).

9. T.D. Bonifield, F.H.K. Rambow, G.K. Walters, M.V. McCusker, D.C. Lorents, and R.A. Gutcheck, J. Chem. Phys. 72, 2914 (1980).

10. T. Efthimiopoulos, B.P. Stoicheff, and R.I. Thompson, Opt. Lett. 14, 624 (1989).

11. P.R. Herman, P.E. LaRocque, R.H. Lipson, W. Jamroz, and B.P. Stoicheff, Can. J. Phys. 63, 1581 (1985).

12. F. Grein, S.D. Peyerimhoff, and R.J. Buenker, J. Chem. Phys. 82, 353 (1985).

Picosecond X-Ray Sources

R.W. Falcone, M.M. Murnane and H.C. Kapteyn

Department of Physics
University of California at Berkeley
Berkeley, CA 94720

Short wavelength light pulses with a duration of less than a picosecond can be produced by focusing high-intensity, ultrafast laser pulses onto a solid. The intense, ultrafast laser pulses are absorbed by a plasma formed inside the surface of the solid before ablation occurs. The resulting short wavelength emission spectrum is characteristic of a high-temperature, short-lived, high-density plasma.

We recently proposed[1] and demonstrated[3-7] this technique using a 160 fsec laser heating pulse and a silicon target. Radiation from the resulting plasma is emitted at photon energies ranging from the visible spectrum up to x-ray energies exceeding one keV. Following the laser pulse, x-ray emission terminates when the plasma rapidly cools due to electron energy loss to ions, thermal conduction of energy into the solid and expansion of the plasma into the surrounding vacuum. We also demonstrated the importance of using low pre-pulse energy for the production of short pulse x-rays from laser-produced plasmas. Energy deposited before the arrival of the short pulse can cause ionized material to be ablated, leading to the formation of a lower density plasma in front of the solid surface. The ultrashort laser pulse will be absorbed in this lower density region and can lead to relatively long x-ray emission times, as discussed below.

Experiments were performed using a high-powered femtosecond dye laser system described in detail elsewhere.[8] Briefly, the system includes a colliding-pulse mode-locked laser which produces 100 fsec pulses at a wavelength of 616 nm. The pulses are amplified in dye laser amplifiers to an energy of 4 mJ and a pulse length of 160 fsec. These pulses are focused on target to power densities in excess of 10^{16} W cm^{-2} using a reflecting mirror. Peak plasma temperatures of several hundred eV are produced at the surface of the solid.

Reflectivity measurements [2,3] have confirmed that the short-pulse laser is coupled into solid or near-solid density plasma with an associated high, metal-like reflectivity. The experimental reflectivity values are in agreement with the predictions of a Drude model, assuming equilibrium ionization in the plasma. At a plasma temperature of 300 eV, the expansion velocity into the vacuum is about 1 Å per fsec. Since the laser-heated region in the solid is on the order of several hundred Angstroms (following rapid thermal conduction), no significant expansion will occur during the leading edge of the laser pulse.

Temporally resolved experiments [2,7] were performed using a modified commercial x-ray streak camera having a 100 μm input slit, 35 kV cm[-1] extraction field, fast voltage ramp and potassium bromide photocathode. A CCD camera and computer record the x-ray data for later analysis; the system is sensitive to single photoelectrons and has a dynamic range of about 10^2. The temporal resolution of the streak camera is determined by several factors including the spatial resolution of the electron imaging system and the energy distribution of the emitted photoelectrons. Instrumental calibration was determined using 300 fsec, 308 nm laser pulses. The x-ray measurements [2,7] indicate that emission is less than 1.5 psec in duration with the short laser pulse heating. Improvements in x-ray streak camera technology may eventually result in instrumental resolution better than 1 psec.

The experimentally observed x-ray emission pulsewidth is in reasonable agreement with theoretical predictions using both a simple heat-flow model [1,3] and the plasma simulation code LASNEX.[9] Both codes predict a rapid risetime for the electron temperature. The predicted total x-ray emission pulsewidth for these conditions is approximately 1 psec at full-width-half-maximum intensity, although shorter duration emission is expected at the shorter wavelengths. Shorter total x-ray emission pulses are predicted for higher atomic number targets such as tantalum.

Care was taken to keep pre-pulse energy, in the form of amplified spontaneous emission (ASE) from the dye laser amplifiers, less than 0.03% of the total short laser pulse energy. This ensured that no preformed plasma would be present when the 160 fsec pulse arrived on target. Very different results were obtained with increased ASE levels, with the measured x-ray pulse width from the plasma increasing to several tens of picoseconds. This long decay time in the presence of prepulse ASE can be interpreted as emission coming from a lower density plasma formed in front of the target with reduced cooling resulting from longer scale lengths and decreased temperature and density gradients.

Spectrally, the major component of the plasma x-ray emission under these conditions is expected to be broadband radiation.[3,7] At high electron densities, ionization depression reduces the number of bound levels to the first few excited levels and Stark broadening is severe; this can lead to rapidly varying line broadenings. Continuum emission (due to recombination, bremsstrahlung and broadened low-lying line radiation) is expected to dominate until the plasma cools and expands. Resonance line emission from long lived states has been observed to radiate for tens of picoseconds longer than the short pulse integrated emission observed using the streak camera. [4]

Future directions for this work include the extension of x-ray measurement techniques to sub-picosecond time scales, the development of more efficient and shorter wavelength laser plasma x-ray sources, the use of the short pulses for time-resolved x-ray scattering from rapidly evolving materials and the pumping of new types of x-ray lasers.[7]

This work was supported by the National Science Foundation, AT&T Bell Laboratories, E.I. Du Pont de Nemours and Company, Newport Corporation and through a collaboration with Lawrence Livermore National Laboratory under the auspices of the U.S. Department of Energy under contract #W-7405-ENG-48.

References

1. R.W. Falcone and M.M. Murnane, "Proposal for a Femtosecond X-Ray Light Source" in Short Wavelength Coherent Radiation: Generation and Applications, D.T. Attwood and J. Bokor, eds., 81 - 85 (AIP 147, New York, 1986).

2. M.M. Murnane, H.C. Kapteyn, R.W. Falcone, "High-Density Plasmas Produced by Ultrafast Laser Pulses," Phys. Rev. Lett. 62, 155 - 158 (1989).

3. M.M. Murnane, "Sub-Picosecond Laser Produced Plasmas," Ph.D. Thesis, University of California at Berkeley (1989).

4. M. M. Murnane and R. W. Falcone, "Short pulse laser interaction with solids," in High-Intensity Laser-Matter Interactions, SPIE Vol. 913, pp.5-8 (1988).

5. M.M. Murnane, H.C. Kapteyn, R.W. Falcone, "Sub-Picosecond Laser Produced Plasmas," to be published in Nuclear Instruments and Methods B: Beam Interactions with Materials and Atoms.

6. M.M. Murnane, H.C. Kapteyn, R.W. Falcone, "Picosecond streak camera measurements of short x-ray pulses," to be published in the proceeding of the SPIE Conference, August 1989.

7. M.M. Murnane, H.C. Kapteyn, R.W. Falcone, "Generation and Application of Ultrafast X-Ray Pulses," submitted to the IEEE Journal of Quantum Electronics.

8. M.M. Murnane and R.W. Falcone, "High-power femtosecond dye-laser system,"J. Opt. Soc. Am. B 5, 1573 (1988)

9. M.D. Rosen, Lawrence Livermore National Laboratories, private communication.

PROSPECTS FOR PARTICLE BEAM PUMPED SHORT WAVELENGTH LASERS

D. E. Murnick
Department of Physics, Rutgers University, Newark, NJ   07102
A. Ulrich, B. Busch, W. Krötz, G. Ribitizki and J. Wieser
TU - München, Fakultät für Physik E12, D-8046, Garching, FRG

INTRODUCTION

Heavy ion beam pumping provides a laser excitation technique which is potentially significant for producing short wavelength high power lasers.   Heavy ion accelerators providing ions from several MeV to several GeV have primarily been used for single collision nuclear and atomic physics experiments.   Much of our data base on few electron ion transitions, transition rates and inner shell excitation have come from heavy ion beam experiments.   Recent accelerator advances now allow the production of very intense, well-focused beams and have stimulated research in the field of heavy ion inertial confinement fusion.[1]   Beams developed for these programs may also be used for x-ray laser physics studies.

The energy of a heavy ion beam is directly transferred to efficient ionization of atoms and ions; the specific energy loss is high in a cylindrical volume around the particle beam axis with the energy loss $\frac{dE}{dx}$ scaling as Z^2, implying higher energy densities with the highest Z beams.   High power is achieved with pulsed bunched beams from present accelerators, typical pulse widths being around 1 ns, and most accelerators have very attractive duty cycles compared to pulsed lasers.

Instantaneous electric fields in ion-atom collisions can be several hundred e/a_0^2 with correspondingly high time derivatives.[2] This leads to large multiple ionization cross sections with relatively cool recoil ions moving transverse to the beam.[3]   Specific levels may be preferentially populated in fast recombination.   The study of the physics of excited state formation and decay is the most important area of current research.

EXPERIMENT TO DATE

The first heavy ion beam pumped ir laser was reported in early 1983 using a 100 MeV quasi cw ^{32}S beam in a 99% He 1% Ar gas cell at 200 to 700 mbar total pressure.[4]   Many other ir lasers were pumped with a 3.5 MeV He beam as well as a 1.9 GeV Xe beam.[5]

Extensions to the visible, uv, vuv and xuv parts of the spectrum have been concentrated on rare gas excimers and multiply ionized species in rare gas targets. Beams of 3.3 GeV U^{29+}, 640 MeV Ar^{9+} and 100 MeV S^{8+} efficiently produced Ar_2, Kr_2 and Xe_2 excimer radiation at 130, 150 and 170 nm respectively in pure rare gas targets around one barr pressure.[6] Time resolved spectroscopy allowed lifetime measurements of singlet and triplet levels, and determination of collisional rate constants in gaseous and solid targets.[7]

The absolute population densities n_i of potential upper laser levels pumped by a particle beam have been determined for 100 MeV S^{8+} and 610 MeV Ar^{10+} in Ne gas. The population density in the upper level is related to the observed intensity S by:

$$n_i = \frac{S}{A_{ik} V \Omega r \Delta t} \tag{1}$$

where A_{ik} is the Einstein coefficient for the observed transition, V is the target volume, Ω is the solid angle observed and r is the spectral response of the detection system. For the 100 MeV S beam pumping 200 mbar neon the measured population densities of selected levels are tabulated in Table I. The densities are time averaged over a 5 ns time window, Δt. Several of the transitions should show gain with higher intensity pumping.

PROSPECTS AT SIS - ESR

The new heavy ion accelerator facility under construction at GSI, Darmstadt, West Germany consists of a heavy ion synchrotron (SIS), a storage ring with the possibility of beam cooling (ESR) and a high current injector. Intense beams of any element at energies to 1.8 GeV/amu will be available. The properties of the high power

ion	wavelength (nm)	n_i (cm^{-3})	transition	J
NeI	352.04714	2×10^8	$4p'[1/2] - 3s'[1/2]^0$	0-1
NeI	345.41942	5×10^7	$4p[1/2] - 3s[3/2]^0$	0-1
NeII	337.828	3×10^7	$3p^2P^0 - 3s^2P$	1/2-1/2
NeII	332.375	2×10^7	$3p^2P^0 - 3s^2P$	3/2-3/2
NeII	191.6082	2×10^7	$3p'^2P^0 - 3sP$	3/2-3/2
NeIII	209.554	5×10^5	$3d'^3D^0 - 3p'^3D$	3-3
NeIV	235.796	5×10^5	$3p^4D^0 - 3s^4P$	7/2-5/2
NeIV	228.579	8×10^5	$3p'^2F^0 - 3s'^2D$	7/2-5/2
NeVI	205.5593	1×10^5	$3p^2P^0 - 3s^2S$	7/2-1/2

Table 1: The number densities n_i of Ne atoms in the upper level of selected transitions in the 100 MeV S^{8+} experiment are tabulated. The values are averaged over a 5 ns time window prompt with the beam. The densities are estimated to be accurate within a factor of 4.

heavy ion beams from SIS compare favorably with the properties of laser drivers for two x-ray lasers already reported.[8] There will be 200 J per pulse with projected delivered power 2.5×10^{13} W/cm^3 in a solid target having approximately a one centimeter range for the heavy ion beam, which can be focused to a 0.2 mm diameter. There is essentially zero reflected power from heavy ion beams in matter, and the resulting ion temperatures should be substantially lower than are obtained with long wavelength laser drivers. The first experiments scheduled for December 1989, at less than full power, will involve time resolved vuv spectroscopy similar to previous experiments and an attempt to pump visible and uv ion lasers. It is expected that many of the different laser schemes which have been proposed in the literature, such as the isoelectrionic Ne and C like sequences, will be studied in further experiments. Specific targets, including solid, high pressure gas and pulsed discharges, will be developed.

Initial operations of the accelerator will be with a beam pulse of width 80-100 ns. Additional bunching magnets may be added in the future to substantially shorten the pulse while keeping the total energy fixed. The rise time, peak and time after the pulse are all of interest for laser experiments. During the rise time, specific target and beam combinations could lead to population inversion situations. During the peak of the pulse, a dense target will radiate strongly in the soft x-ray range of the spectrum -- possibly of use for optical pumping. Recombination lasers are possible during the cooling of the target plasma. Finally, an additional enhancement of the power density can occur in a specially designed target by an implosion in the hydrodynamic phase. Detailed calculations using codes which have been developed for laser produced plasmas must be modified for the heavy ion beam case.

REFERENCES

1 R. C. Arnold and J. Meyer-ter-Vehn, *Z. Phys.* **D9** 65 (1988)
2 K. Boyer, et. al., *IEEE Trasactions on Plasma Science* **16** 541 (1988)
3 H. F. Beyer, F. Folkmann and R. Mann, *Nuclear Instr. and Meth.* **202** 177 (1982)
4 A. Ulrich, H. Bohn, P. Kienle and G. J. Perlow, *Applied Physics Letters* **42** 782 (1983)
5 D. E. Murnick and A. Ulrich, *Nuclear Instruments and Methods* **B9** 757 (1985)
  A. Ulrich, J. W. Hammer and W. Biermayer, *Journal of Applied Physics* **63** 2206 (1988)
6 A. Ulrich, et. al., *Journal of Applied Physics* **62** 357 (1987)
7 B. Busch, A. Ulrich, W. Krötz and G. Ribitzki, *Applied Physics Letters* **53** 1172 (1989)
8 M. D. Rosen, et. al., *Phys. Rev. Lett.* **54** 106 (1985)
  S. Suckewer, et. al., *Phys. Rev. Lett.* **55** 1753 (1985)

Part VIII

Trapped Ion Spectroscopy

The Order-Chaos Transition of Two Trapped Ions

J. Hoffnagle[1], R.G. DeVoe[1], L. Reyna[2], J. Rosenkranz[1], and R.G. Brewer[1]

[1] IBM Research Division, Almaden Research Center, 650 Harry Road, San Jose, California 95120-6099

[2] IBM Research Division, T.J. Watson Research Center, Yorktown Heights, New York 10598

In one of the first experiments demonstrating electrodynamic confinement of charged particles[1], Wuerker, Shelton, and Langmuir observed that trapped particles could describe two very different kinds of motion: small oscillations about equilibrium points in an ordered array, and disordered, apparently random motion. They also noticed that transitions between these two types of motion could be induced by varying the trapping potential. Recently, ensembles of a few trapped ions, cooled by radiation pressure, have been seen to exhibit an ordered state[2-4] and a transition to a disordered state[2,4]. We interpret these phenomena as order-chaos transitions in a deterministic dynamical system. To reduce the problem to its simplest form we have considered theoretically and experimentally the case of two ions of equal charge e and mass m confined in a Paul trap, and we find good agreement between experimental observations and computational results that clearly point to chaos.

Consider the motion of a pair of ions in a Paul trap, which has an oscillatory quadrupole potential $V = V_{ac}(\cos \Omega t)(z^2-(x^2 + y^2)/2)/\bar{r}_0^2$, where $\bar{r}_0$ is the trap radius and V_{ac} the amplitude of the applied potential. Introducing dimensionless parameters for time, $\tau = \Omega t/2$, and potential, $q_3 = 4eV_{ac}/(m\Omega^2 r_0^2)$, one finds that the equations of motion separate into center-of-mass and relative coordinates. The former obeys a Mathieu equation, with well-known analytic solutions, while the relative coordinate $\vec{r}$ obeys a modified Mathieu equation

$$\ddot{r}_i + \Gamma\dot{r}_i + \left[2q_i \cos (2\tau)-\alpha/r^3\right]r_i = 0, \qquad (1)$$

where $q_1 = q_2 = -q_3/2$, $\alpha = e^2/(m\Omega^2)$ is a Coulomb coupling parameter, and the dots denote differentiation with respect to τ. The subscript $i = 1,2,3$ designates the ith Cartesian component of $\vec{r}$, so (1) is a set of three equations, with the Coulomb term providing the nonlinear coupling which is essential for the appearance of chaotic solutions[5,6]. Because the trap is axially symmetric, one degree of freedom in (1) is simply a conserved angular momentum which does not contribute to the chaotic dynamics. We have performed calculations both with the full 3-dimensional version of (1) and with a reduced set of equations having only one radial and one axial coordinate (corresponding to the special case of vanishing angular momentum), and find essentially the same results in both cases.

The damping term in (1) is a first approximation to laser cooling in the limit of small ion velocities, and also describes the viscous cooling used in the experiments of Wuerker et al. A more accurate model of trapped, laser-cooled ions replaces this term with the light pressure force in the heavy-particle limit. The average force in the interval $t_{j-1} \leq t \leq t_j$ between the j-1st and jth photon scattering events is

$$\vec{F}_j = \hbar(\vec{k} - \vec{k}_j)/(t_j - t_{j-1}); \quad j = 1,2,3... \tag{2}$$

where $\vec{k}$ and $\vec{k}_j$ are the wave vectors of the cooling laser and the jth spontaneously emitted photon, respectively. The scattering events labelled by j occur at random times t_j, determined by a velocity-dependent probability distribution that displays photon anti-bunching, derived by Schenzle and Brewer[7]. The resulting fluctuating force leads to cooling, an uncertainty in the initial conditions of the ion motion, and also the inhibition in laser cooling due to micromotion first noted in Ref. 8.

Numerical solutions of (1) depend sensitively on the initial conditions and on the control parameter q. For small q and initial conditions close to equilibrium, the ions execute small oscillations about the equilibrium value $z_0 = 0$, $r_0 = 2(\alpha/q^2)^{1/3}$, with frequency $\omega_z = \omega_r = q(3/8)^{1/2}$. This is quasiperiodic motion with two frequencies, the micromotion (Ω) and the secular frequencies ($\omega_z = \omega_r$). As q becomes larger, a long-lived transient beat results in hard collisions, strongly coupling the z and r motions and leading to chaos. At the same time the excursions of the ions become much larger than for quasiperiodic motion, indicating that energy is transferred from the trapping field to the ions.

Analysis of the extended trajectories shows the typical characteristics of deterministic chaos: the Fourier spectrum is continuous even though the system has only six degrees of freedom, and the largest Lyapunov exponent goes from zero in the quasiperiodic state to a positive value in the chaotic state. We have calculated all four Lyapunov exponents for the 2-dimensional version of (1) using the method of Shimada and Nagashima[9]; this allows a determination of the Lyapunov dimension of the attractor[10], the result being $d_L = 3.95$. This is distinctly different from the phase space dimension of 4, and suggests that the chaotic motion is described by a strange attractor. Other phenomena typical of nonlinear dynamics are also encountered, e.g. frequency locking to rational multiples of Ω and hysteresis.

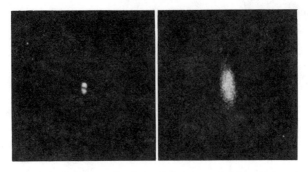

Fig. 1. Images of two Ba^+ ions in the ordered state (left) and chaotic state (right).

We note that (1), without any stochastic term, exhibits chaotic solutions. Any physical system will have some noise terms, e.g. the fluctuations in the light pressure force in (2), or Brownian motion in the experiments of Ref. 1. These displace the system from equilibrium and bring the nonlinear Coulomb interaction into play. The dynamics then lead rapidly to chaos in a deterministic way.

The transition between ordered and chaotic motion has been experimentally observed for two trapped, laser-cooled barium ions. The ions were detected by scattered laser light at 493 nm; together with a second laser at 650 nm this also cooled the ions. The residual pressure in the trap ($\bar{r}_0 = 0.25$ cm, $\Omega/2\pi = 3.55$ MHz) was less than 10^{-9} Torr. Fig. 1 shows images of the Ba^+ pair in the ordered and chaotic states, obtained with a position-sensitive photomultiplier. For low q the ions make small oscillations about their equilibrium position and therefore appear as two bright points separated by about $6\mu m$. When q is increased the ions describe chaotic motion, forming a featureless cloud of some $40\mu m$ in size; this dimension is in agreement with the calculated values. When q is decreased again, the ions recondense into the ordered state. In the chaotic state the fluorescence count rate drops dramatically, due to the large Doppler effect. Fig. 2 shows the count rate as the ion pair traverses one cycle from the ordered to the chaotic state and back, the transition to chaos occurring at about $q = 0.85$ and the recondensation at about $q = 0.82$. The former value is in good agreement with calculations based on (1) with the fluctuations in the light pressure force included, as shown by the dashed line in Fig. 2, and also with calculations based on (1) alone with the phenomenological damping term $\Gamma = 2 \times 10^{-4}$ corresponding to our experimental conditions. The calculation of the condensation is much more difficult, since the weak laser cooling means that the process is very slow and numerical errors in the calculation are more critical.

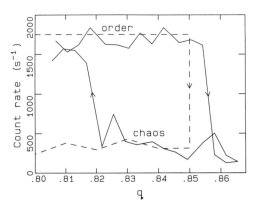

Fig 2. Hysteresis loop of the photon count rate v the control parameter q. Solid line: experimental; dashed line: simulation with $\Gamma = 2 \times 10^{-4}$ and fluctuations in the light pressure force included.

Finally, we note that Wuerker et al. reported the "melting" transition at q values ranging from $q = 0.645$ for 100 particles to $q = 0.866$ for 3 particles, the precise values depending also on the damping Γ. This approximate agreement with the q value at which the two-ion system undergoes transitions in both numerical and experimental work suggests that all of these observations have a common explanation as order-chaos transitions.

We are grateful to Ken Foster for technical assistance. This work was partially supported by the U.S. Office of Naval Research.

References

1. R.F. Wuerker, H. Shelton, and R.V. Langmuir, *J. Appl. Phys.* **30**, 342 (1959).
2. F. Diedrich, E. Peik, J.M. Chen, W. Quint, and H. Walther, *Phys. Rev. Lett.* **59**, 2931 (1987).
3. D.J. Wineland, J.C. Bergquist, W.M. Itano, J.J. Bollinger, and C.H. Manney, *Phys. Rev. Lett.* **59**, 2935 (1987).
4. J. Hoffnagle, R.G. DeVoe, L. Reyna, and R.G. Brewer, *Phys. Rev. Lett.* **61**, 255 (1988).
5. P. Berge, Y. Pomeau, and C. Vidal, *Order Within Chaos* (Wiley, New York, 1984).
6. J. Guckenheimer and P. Holmes, *Nonlinear Oscillations, Dynamical Systems, and Bifurcations of Vector Fields* (Springer-Verlag, New York, 1983).
7. A. Schenzle and R.G. Brewer, *Phys. Rev. A* **34**, 3127 (1986).
8. R.G. DeVoe, J. Hoffnagle, and R.G. Brewer, *Phys. Rev. A* **39**, 4362 (1989).
9. I. Shimada and T. Nagashima, *Prog. Theor. Phys.* **61**, 1605 (1979).
10. J.D. Farmer, E. Ott, and J.A. Yorke, *Physica* **7D**, 153 (1983).

Hg$^+$ Single Ion Spectroscopy*

J.C. Bergquist, F. Diedrich,[†] Wayne M. Itano, and D.J. Wineland
Time and Frequency Division
National Institute of Standards and Technology
Boulder, CO 80303, USA

Introduction

A single ion in an electromagnetic trap can be localized in a small volume and held for a long period of time. A trapped ion also can be isolated from collisions and other perturbations and laser cooled to low temperatures. By monitoring the presence or absence of fluorescence from a strongly allowed transition, one can detect each transition to a metastable state. These characteristics make it possible to study the spectrum of a single ion with great precision. The vibrational energy states of the bound ion are quantized. We have used laser-sideband-cooling to drive the ion to the lowest energy level of the harmonic well of the trap.

Experiment

A mercury atom that is ionized by a weak electron beam is captured in the harmonic well created by an rf potential applied between the electrodes of a miniature Paul trap (1,2). The ring electrode has an inner diameter of about 900 μm and is placed symmetrically between two endcap electrodes that are separated by about 650 μm. The frequency of the rf potential is about 21 MHz. Its amplitude can be varied up to 1.2 kV. The classical motion of the ion consists of a small-amplitude oscillation, at the frequency of the applied rf potential, superimposed on a larger-amplitude harmonic motion, called the secular motion. The frequencies of the secular motion depend on the strength of the rf field (and any static field applied to the electrodes) and were typically 1-4 MHz. The ion is laser cooled to a few millikelvins by a few microwatts of 194 nm cw laser radiation (3) that is frequency tuned below the $5d^{10}6s$ $^2S_{\frac{1}{2}}$ - $5d^{10}6p$ $^2P_{\frac{1}{2}}$ first resonance line (1,4). In order to cool all motional degrees of freedom to near the Doppler cooling limit (T = $\hbar\gamma/2k_B$ $\simeq$ 1.7 mK) the 194 nm radiation irradiates the ion from 2 orthogonal directions, both of which are at an angle of 55° with respect to the symmetry (z) axis of the trap. The 282 nm radiation that drives the narrow $5d^{10}6s$ $^2S_{\frac{1}{2}}$ - $5d^96s^2$ $^2D_{5/2}$ transition is obtained by frequency-doubling the radiation from a narrowband cw ring dye laser. In the long term, the laser is either stabilized by FM optical heterodyne spectroscopy to a saturated absorption hyperfine component in $^{129}I_2$ (5) or directly to a stable high finesse reference cavity (6). The frequency of the laser is scanned by an acousto-optic modulator that is driven by a computer controlled synthesizer. Up to a few microwatts of 282 nm radiation could be focussed onto the ion in a direction counterpropagating with one of the 194 nm light beams.

Optical-optical double resonance (electron shelving) (1,2,4,7-11) with a net quantum amplification in excess of 100 at 10 ms is used to detect transitions driven by the 282 nm laser to the metastable $^2D_{5/2}$ state. The fluorescence from the laser-cooled ion is constant when it is cycling between the $^2S_{\frac{1}{2}}$ and $^2P_{\frac{1}{2}}$ states and zero when it is in the metastable $^2D_{5/2}$ state (1,4,7,8). Thus the $^2S_{\frac{1}{2}}$ - $^2D_{5/2}$ resonance

---

*Work of U.S. Government; not subject to U.S. copyright.
[†]Present address Gsänger Optoelektronik GmbH, Planegg,
Federal Republic of Germany

spectrum is obtained by first probing the S-D transition at a particular frequency near 282 nm, then turning off the 282 nm radiation, turning on the 194 nm radiation, and looking for 194 nm fluorescence. The two radiation fields are alternately applied to avoid light shifts and broadening of the narrow S-D transition. The absence of 194 nm fluorescence indicates that a transition into the metastable D state has occurred; its presence indicates that the ion has remained in the ground state. The 282 nm frequency is then stepped and the measurement cycle repeated. Each new result at a particular 282 nm frequency is averaged with the previous measurements at that frequency. The normalization of the signal is 1 for each measurement of high fluorescence and 0 for each measurement of no fluorescence. The high fluorescence level makes it possible to determine the state of the atom with almost no ambiguity in a few milliseconds. Thus, it is easy to reach the quantum noise limit of a single atomic absorber (1). Figure 1 shows the signal from an 8 MHz scan of the 282 nm laser through a Zeeman component of the S-D transition in $^{198}Hg^+$ (1). The Doppler-free central feature (carrier) and the motional sidebands due to the secular motion of the cold ion are fully resolved (12,13). The number and strength of the sidebands are a direct measure of the amplitude of the ion's motion and of its temperature.

In Fig. 2, we show a high resolution scan through the Doppler-free resonance of the $^2S_{\frac{1}{2}}(F = 0, m_F = 0) - ^2D_{5/2}(F = 2, m_F = 0)$ transition in $^{199}Hg^+$ which is first-order field independent at a magnetic field $B \approx 0$. The full width at half maximum (FWHM) is approximately 86 Hz at 563 nm (172 Hz at 282 nm). This corresponds to a fractional resolution of better than 2×10^{-13} ($Q \approx 5\times10^{12}$). For this trace the laser was spectrally narrowed by locking to a mechanically, acoustically and thermally quiet reference cavity that had a finesse of about 60,000. The

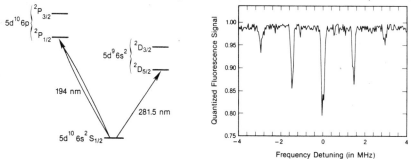

Fig. 1. On the left is a simplified energy-level diagram for $^{198}Hg^+$. The 282 nm transition can be observed by monitoring the 194 nm fluorescence. If the ion makes a transition from the $^2S_{\frac{1}{2}}$ to the $^2D_{5/2}$ level the 194 nm fluorescence disappears. For the figure on the right, the relative detuning from line center is plotted in frequency units at 282 nm. On the vertical axis is plotted the probability that the 194 nm fluorescence is on immediately after the 282 nm pulse. The Doppler-free recoilless-absorption-resonance or carrier (central feature) can provide a reference for an optical frequency standard. (From Ref. 1)

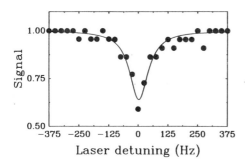

Signal

1.00

0.75

0.50

−375 −250 −125 0 125 250 375

Laser detuning (Hz)

Fig. 2. High resolution scan through the Doppler-free resonance of the $^2S_{1/2}(F = 0, m_F = 0)$ - $^2D_{5/2}(F = 2, m_F = 0)$ transition in a single laser cooled $^{199}Hg^+$ ion. A frequency doubled and stabilized 563 nm laser ($\nu \approx 5\times10^{14}$ Hz) is stepped through the resonance in 25 Hz increments. The full width at half maximum is about 86 Hz. The integration time per point is about 0.5 s.

frequency of the laser fluctuates less than 50 mHz relative to this reference cavity. This is determined from the measurement of the noise in the error signal. The actual frequency fluctuations of the laser are governed by the stability of the mechanically and thermally isolated reference cavity (6). The natural linewidth of the S - D transition is approximately 1.7 Hz (2). The measured linewidth in Fig. 2 is likely limited by low frequency fluctuations in the length of the stable reference cavity (14). Better mechanical isolation of the reference cavity might permit us to reach the 1.7 Hz resolution limit.

The possibility of cooling the Hg^+ ion further, to the zero point energy of motion, is intriguing for several reasons. Cooling so that the average occupational number $<n_v>$ for the motional energy of the bound atom is zero is a fundamental limit to laser-cooling for any bound particle (1,15) and forces a quantum mechanical treatment of the atom-trap system. Driving a single atom in a macroscopic trap to the zero point energy of motion exploits the benign environment near the center of an rf trap. Finally, if an ion is prepared in the lowest vibrational state, experiments such as squeezing the atom's position and momentum can be demonstrated (16).

In a rigorous treatment of the ion-trap system, stationary states do not exist since the trapping potential is time dependent. However, quasi-stationary states, obtained by solving for the eigenvalues of the Floquet operator, do exist (17,18). That these states correspond closely to the stationary states of the pseudo-potential of the rf trap is verified by the spectrum shown in Fig. 1. The carrier (at frequency ω_0) results from transitions in which the vibrational quasi-energy is unchanged. The upper and lower sidebands, spaced by multiples of the secular frequencies ($\omega_0 \pm n\omega_v$), correspond to transitions which increase or decrease the quasi-energy. Recently, we have cooled a single $^{198}Hg^+$ ion to the lowest vibrational state ($n_v = 0$) by a method called optical sideband cooling (15). First, the ion was laser-cooled to a few millikelvins by radiation scattered from the strong transition at 194 nm. This reduced the vibrational quantum number to a mean value of $<n_v> \approx 12$. The secular frequency was about 3 MHz. Laser radiation tuned to the frequency of the first lower vibrational sideband (at frequency $\omega_0 - \omega_v$) of the narrow S-D transition was then applied to the ion. For each photon re-emitted at the unshifted carrier frequency, the vibrational energy was reduced by $\hbar\omega_v$. After the sideband cooling, laser radiation of saturating intensity was applied at the lower sideband frequency.

Absence of absorption indicated that the ion was in the $n_v = 0$ state. The ion was found to be in the $n_v = 0$ state about 95% of the time.

Starting from the ground vibrational state ($n_v = 0$), the absorption of a **single** quantum of energy at a frequency corresponding to the secular vibrational frequency (ω_v) would raise n_v by one unit. This could be detected with an efficiency of nearly 100%. Also, it should be possible to produce squeezed states (16) of the atom's motion from the zero point energy state by a sudden, non-adiabatic weakening (and, in general, shifting) of the trap potential or by driving the atomic motion parametrically at $2\,\omega_v$. If after some time the atom could then be returned to the $n_v = 0$ state by reversing the above procedures, the zero point energy state could be detected by the absence of the lower sideband.

We acknowledge the support of the U.S. Air Force Office of Scientific Research and the U.S. Office of Naval Research. F. Diedrich thanks the Deutsche Forschungsgemeinschaft for financial support.

References

1. J.C. Bergquist, W.M. Itano, and D.J. Wineland, Phys. Rev. A **36**, 428 (1987).
2. J.C. Bergquist, D.J. Wineland, W.M. Itano, H. Hemmati, H.-U. Daniel, and G. Leuchs, Phys. Rev. Lett. **55**, 1567 (1985).
3. H. Hemmati, J.C. Bergquist, and W.M. Itano, Opt. Lett. **8**, 73 (1983).
4. J.C. Bergquist, R.G. Hulet, W.M. Itano, and D.J. Wineland, Phys. Rev. Lett. **57**, 1699 (1986).
5. J.L. Hall, L. Hollberg, T. Baer, and H.G. Robinson, Appl. Phys. Lett. **39**, 680 (1981).
6. J. Hough, D. Hils, M.D. Rayman, Ma L.-S., L. Hollberg, and J.L. Hall, Appl. Phys. B **33**, 179 (1984).
7. W. Nagourney, J. Sandberg, and H. Dehmelt, Phys. Rev. Lett. **56**, 2797 (1986).
8. T. Sauter, W. Neuhauser, R. Blatt, and P.E. Toschek, Phys. Rev. Lett. **57**, 1696 (1986).
9. D.J. Wineland, J.C. Bergquist, W.M. Itano, and R.E. Drullinger, Opt. Lett. **5**, 245 (1980); D.J. Wineland and W.M. Itano, Phys. Lett. **82A**, 75 (1981).
10. H. Dehmelt, J. Phys. (Paris) Colloq. **42**, C8-299 (1981).
11. H. Dehmelt, Bull. Am. Phys. Soc. **20**, 60 (1975).
12. R.H. Dicke, Phys. Rev. **89**, 472 (1953).
13. D.J. Wineland and W.M. Itano, Phys. Rev. A **20**, 1521 (1979).
14. D. Hils and J.L. Hall, in **Frequency Standards and Metrology**, edited by A. DeMarchi (Springer-Verlag, Berlin, Heidelberg, 1989) p. 162.
15. F. Diedrich, J.C. Bergquist, W.M. Itano, and D.J. Wineland, Phys. Rev. Lett. **62**, 403 (1989).
16. J.C. Bergquist, F. Diedrich, W.M. Itano and D.J. Wineland, in **Frequency Standards and Metrology**, edited by A. DeMarchi (Springer-Verlag, Berlin, Heidelberg, 1989) p. 287.
17. R.J. Cook, D.G. Shankland, and A.L. Wells, Phys. Rev. A **31**, 564 (1985).
18. M. Combescure, Ann. Inst. Henri Poincaré **44**, 293 (1986).

Quantum Jumps and the Single Trapped Barium Ion:
Lifetimes, Quenching Rates, and Anomalous Dark Periods

J.D. Sankey and A.A. Madej
Time and Length Standards
National Research Council of Canada
Ottawa, CANADA
K1A 0R6

1. Introduction and Experiment:

The realization of the confinement and observation of single, trapped ions has opened new opportunities for spectroscopy [1,2]. The long interrogation and confinement times for the particles allows the possibility to examine transitions with extremely high transition line Q's [4] without significant transit time broadening. Moreover, the use of laser cooling has enabled effective reductions in the ion kinetic energy into the millikelvin regime thus circumventing problems encountered with Doppler broadening and shifts. As a result, some transitions for trapped ions have been suggested as frequency standards [2]. One of the more fascinating results obtained with single trapped ions has been the observation of "quantum jumps" into and out of the ion's metastable level by monitoring the interruption of fluorescence from the ion cycling on a strongly allowed transition [2,3,5]. Not only has this effect been a source of considerable theoretical discussion [6-8], it also allows us the unique possibility of monitoring the single ion's internal state. It is thus possible to determine whether the ion has undergone an excitation into the metastable level via a weakly-allowed high Q transition (provided by excitation with a stabilized laser) with near unity detection efficiency [1,2].

In our present work, we have been examining the effect of the background gas environment on the ion by observing the quantum jump statistics of the barium ion subject to the presence of buffer gases at pressures (10^{-7}-10^{-5} Pa) where the mean collision time is on the order of the decay time of the 5d $^2D_{5/2}$ metastable level (35s). In this study, we have obtained a value for the natural radiative lifetime of the $^2D_{5/2}$ level, accurate quenching rates of the metastable level for the gases which typically make up the residual gas environment of our high vacuum system (H_2, N_2, CO_2, CO, CH_4, He, and H_2O), and we have observed evidence for the formation of molecular complexes of the Ba^+ with CO_2 and H_2O.

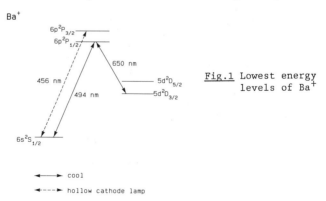

Fig.1 Lowest energy levels of Ba^+

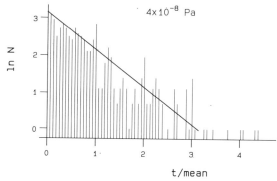

Fig.2 Distribution of
dark periods for a
single ion at low
pressure showing single
exponential character
(mean = 30s)

Figure 1 shows an energy level diagram of the lowest electronic
energy levels for Ba^+. As in other work [3,5], laser cooling and strong
fluorescence production is provided by exciting the ion on the $6p^2P_{1/2}$-
$6s^2S_{1/2}$ and $6p^2P_{1/2}$-$5d^2D_{3/2}$ transitions. Excitation into the metastable
$D_{5/2}$ level is achieved by indirect excitation of the ion to the $6p^2P_{3/2}$ level
which can relax into the $^2D_{5/2}$ level. Weak radiation from a commercial
barium hollow cathode lamp is used for this purpose. In this work, accurate
residual gas analysis and absolute pressures of the quenching gases were
provided by a calibrated mass spectrometer and ion gauge. This enabled us
to include the effects of other residual gases which can severely alter the
measurements of a particular quenching gas.

2. Quantum Jumps and Quenching Rates:

For our case of weak, incoherent excitation into the metastable level
[7], the distribution of the length of observed "dark periods" in the strong
fluorescence due to excitation into the metastable level is given by:

$$W_{off} = R_- \exp(-R_-T) \tag{1}$$

where W_{off} is the probability density of a period of duration T and R_- is
the rate of decay out of the metastable level. This decay rate is the sum of
the natural radiative rate $1/t_o$, (where t_o is the natural radiative lifetime)
and the effect of the individual quenching gases:

$$R_- = 1/t_o + \sum R_{Q,i} \, P_i \tag{2}$$

where $R_{Q,i}$ is the quenching coefficient and P_i is the partial pressure of an
individual gas. Figure 2 shows the observed dark period distribution for the
ion at the lowest pressures utilized in our study (P = 4 X 10^{-8} Pa). The fit
to an exponential distribution is extremely good and chi-squared tests of
the data for these and other quantum jump data sets gave probabilities of
following an exponential dependence in excess of 0.95. Due to the presence
of a number of residual quenching gases at high vacuum, gas pressures of
all mass species were measured during an experimental run and a full
multivariate fit was performed on all our quantum jump data (about 10^4
jumps with 7 different quenching gases). The result of the quenching
measurements are shown in Fig. 3. As can be seen, CH_4 and H_2O possess
very high quenching rates due presumably to the many degrees of freedom
in these molecules while He has an extremely small rate (2400 s^{-1} Pa^{-1} $\pm$
25%). From the multivariate fit, it was also possible to obtain an accurate
value of the natural radiative lifetime of the 5d $^2D_{5/2}$ level of t_o = 34.5 $\pm$
3.5 s, in good accord with previous measurements [3,9].

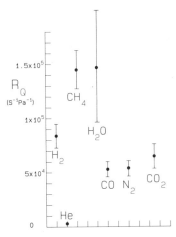

Fig.3 Summary of $^2D_{5/2}$ quenching coefficients

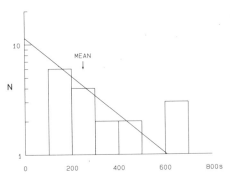

Fig.4 Distribution of anomalous dark periods observed in the presence of CO_2 (3×10^{-6} Pa) showing long period exponential character

3. Evidence for the Formation of Molecular Complexes:

In the presence of CO_2 the quantum jump durations were observed to have an additional distribution of anomalously long dark periods, several times the natural radiative lifetime of the $^2D_{5/2}$ level. In fact, durations of 11 minutes were observed! Although the rate of such anomalous dark periods was low (2000 s^{-1} Pa^{-1}), it was possible to collect a sufficient number for a study of the distribution of anomalous jumps. The distribution of all quantum jumps 10 times the duration of the quenched $^2D_{5/2}$ lifetime at the CO_2 pressure is shown in Fig.4. The distribution appears to be exponential. A chi-squared study for this data of 20 events indicates a probability of 0.91 that the distribution of durations is exponential. We are thus led to the conclusion that the Ba^+ is being "shelved" into a long lived metastable system from which it again decays. We have considered a number of possible mechanisms for this phenomena [10] and have concluded that this system is most likely a metastable, weakly-bound complex of the CO_2 and Ba^+. This molecule is then destroyed by subsequent collisions and fluorescence returns.

Experiments with H_2O also showed anomalous dark states for the ion. however due to presumably more permanent binding, the ion does not return from the dark state. As with the results for CO_2, the rate of formation of the dark state was an increasing function of partial pressure. An order of magnitude estimate of the formation rate coefficient has been determined using our observations [10] and is 6000 s^{-1} Pa^{-1}. Finally, Fig.5 shows the similarity of the observed ion in a permanent dark state with H_2O and shelving of the ion in a metastable state. In a) of Fig.5, we observe the fluorescence from three barium ions being held in the trap. Initially, all ions are pumped (via $^2P_{3/2}$-$^2S_{1/2}$ lamp radiation) into the $^2D_{5/2}$ level and the various levels of fluorescence at later times indicate the fluorescence from 1,2, and 3 fluorescing ions. At t=0 in a) the lamp is off and H_2O at 10^{-6} Pa is added. The step wise decrease in fluorescence is due to Ba^+ and H_2O binding. Note that these fluorescence levels agree with those observed

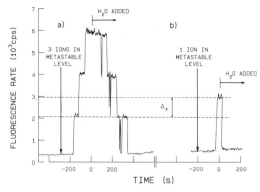

Fig.5 Observed equality of fluorescence levels for a set of ions pumped into the $^2D_{5/2}$ level (t<0) and later suffering molecular formation with H_2O (t>0) for 3 ions a), and 1 ion b).

when the ions were shelved into the metastable electronic level. Fig.5, b) shows the fluorescence response due to a single trapped barium ion. Note that in the case of the 3 ions, the individual fluorescence is lower (Δ_F). Due to the mutual repulsion of ions, micromotion increases the effective temperature and decreases the scattering intensity per ion. When we introduced H_2O, each ion in turn ceased to scatter radiation but the fluorescence from the remaining ions did not increase indicating that the trap still contained the ions perturbing the fluorescence efficiency of those remaining visible.

Although the observation of single atom molecular formation may be something that is wished to be avoided if the barium ion electronic transition is to be used as a frequency standard, there exists the exciting possibility of observing molecular formation dynamics and creation of molecular species which would otherwise be destroyed outside the near free space environment of the single trapped ion.

4. Acknowledgements:

We would like to thank W. Berger for suggestions and expert technical assistance in construction and operation of the experiment. The insightful suggestions and discussions provided by G.R. Hanes were greatly appreciated. Also, we would like to thank A.H. Bass for expert assistance in the calibration of our pressure measurements.

5. References:

1. H.G. Dehmelt, Phys. Scr. **T22**, 120 (1988).
2. W.M. Itano, J.C. Bergquist, and D.J. Wineland, Science **237**, 612 (1987).
3. W. Nagourney, J. Sandberg, and H. Dehmelt, Phys. Rev. Lett. **56**, 2797 (1986).
4. H. Lehmitz, J. Hattendorf-Ledwoch, R. Blatt, and H. Harde, Phys. Rev. Lett. **62**, 2108 (1989).
5. Th. Sauter, W. Neuhauser, R. Blatt, and P.E. Toschek, Phys. Rev. Lett., **57**, 1696 (1986).
6. R.J. Cook and H.J. Kimble, Phys. Rev. Lett. **54**, 1023 (1985).
7. G. Nienhuis, Phys. Rev. A **35**, 4639 (1987).
8. R. Blatt and P. Zoller, Eur. J. Phys. **9**, 250 (1988), and references therein.
9. F. Plumelle, M. Desaintfuscien, J.L. Duchene, and C. Audoin, Opt. Commun. **34**, 71 (1980).
10. J.D. Sankey and A.A. Madej: submitted to Appl. Phys. B.

Parametric Cooling of a Trapped Ion

Peter E. Toschek
Universität Hamburg, I. Institut für Experimentalphysik
D-2000 Hamburg 36, F.R. Germany

A single ion in a trap which spatially confines its motion is a system characterized by three vibrational degrees of freedom and various internal excitations. It has been demonstrated, however, that the ionic motion may reduce to one-dimensional harmonic oscillation |1|. If the laser-excited resonance line does not branch, only one internal degree of freedom is active. Nonetheless, even the dynamics of this system has intricate features. It is shown that the vibration of such a "two-level" ion is excited or deexcited by absorption and stimulated emission of *bichromatic* laser light. Two contributing mechanisms will be distinguished: (*i*) the energy flow between ion and field alternates. It is interrupted, when spontaneous decay stops the succession of stimulated light scattering, and a net transfer is left over. (*ii*) The ionic vibration phase-locks to the beat frequency of the light and gives rise to parametric energy transfer.

These mechanisms correspond to features in the dynamics of a *three*-level particle. In a recent experiment, vibrational excitation and deexcitation of such ion by stimulated interaction with laser light of two resonant colors have been demonstrated |2|. The resonance fluorescence of a Ba^+ion in a 1-mm electrodynamic ion trap was detected through two channels, one of which collected light from a small volume around the stationary position of the ion, whereas the collection volume of the other one was six times larger in diameter. The ratio of the two signals, when recorded upon tuning the wavelength of one of the light fields, is not sensitive to internal resonances but rather to the expansion and contraction of the ion orbit as a result of "heating" and "cooling" by the laser light.

The spectra of vibrational excitation show characteristic resonances. Their spectral positions agree with the locations of resonances in spectra calculated from a quantum-mechanical model for the light force exerted upon a three-level atom by two light fields of different frequencies |3|. These resonances have been explained in terms of a simple model which makes use of elementary interaction processes in the sense of perturbation theory. The characteristic Doppler shifts, averaged over repeated cycles of interaction, reveal the resonances as generated by either stimulated Raman-type two-photon transitions combined with spontaneous Raman scattering, or all-stimulated interaction of ion and light.

The light force on a two-level ion which vibrates in a trap at frequency ω is calculated by a semiclassical model |4| extended to include quantum fluctuations by the finite lifetime τ of the dielectric polarization. Bichromatic quasi-resonant laser light

$$Ee^{i\omega_L t} = 2E_o \cos(\delta t - \varkappa x) \, \text{expi}(\omega_L t - k_L x) \qquad (1)$$

with $\delta = \omega_1 - \omega_2 \ll \omega_{1,2}$, $\varkappa = k_1 - k_2 \ll k_{1,2}$, $2\omega_L = \omega_1 + \omega_2$, and $2k_L = k_1 + k_2$.

The trap absorbs momentum in the interaction of ion and field. The dispersive component of the light force is non-zero, and the energy transfer per time unit is at high saturation

$$\dot{w} \simeq <F_{disp}> v \simeq \hbar k v \delta \, tg(\delta t - \varkappa x) \qquad (2)$$

So far, the direction of energy transfer alternates, and the net transfer is zero. However, with a single particle, the dielectric polarization fluctuates on the order of its mean value. We replace the expectation values of the optical coherence and of the force by the weighted time average $|5|$

$$<F_{disp}> \rightarrow \int_0^\infty <F_{disp}>P(t)dt / \int_0^\infty P(t)dt, \qquad (3)$$

where the weight function $P(t) = e^{-t/\tau}$ describes the probability for a particular duration t of the coherence excited in the individual ion. In the lab system, and with $\varkappa x$ neglected in the exponential, the rate of energy transfer is

$$\dot{W} = \dot{W}_o \tau^{-1} \int_0^T f(t)e^{-t/\tau}dt \qquad (4)$$

with $f(t) = tg\delta t \cos(\omega t + \varphi(t))$ and $\dot{W}_o = \hbar k v_o \delta$, where $\varphi(t) = \omega \varkappa x(t)/\delta = \mu \sin \omega t$. The average extends over time $T = T_1 \gtrsim \delta^{-1} > \tau$. When the sidebands of the phase modulation are neglected, i.e. in the Lamb-Dicke regime,

$$\dot{W} = \dot{W}_o \rho \, J_o(\mu)(\zeta \cos \xi t + \eta \sin \xi t) \qquad (5)$$

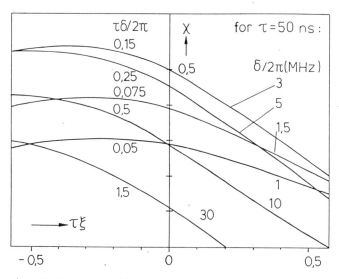

Fig. 1. Net energy flow ion - field (or $v.v.$) in units of "Doppler quanta" $\hbar k v_o$ per beat period/$2\pi = \delta^{-1}$, $vs.$ detuning of beat note from vibration frequency ω.

283

where $\xi = \omega - \delta$, $\zeta = \dfrac{\delta/\tau}{\delta^2 + \tau^{-2}}$, $\eta = -\dfrac{2\delta^2}{(2\delta)^2 + \tau^{-2}}$, and $\rho = \sqrt{\zeta^2 + \eta^2}$.

If $\dot{W}$ is also averaged over the time scale $T_2 \gtrsim \xi^{-1} > T_1$,

$$\overline{\dot{W}} = \dot{W}_0 \, J_0(\mu) \cdot \chi(\delta, \xi) \tag{6}$$

where J_0 is a Bessel function, and $\chi = \rho \left[\zeta (1 + \tau^2\xi^2)^{-1} + \eta\tau\xi \right.$
$\left. \times (1 + \tau^2\xi^2)^{-1} \right]$ which reduces to $\chi \simeq \zeta$ for $\xi^{-1} \gg \tau$. The
phase average χ is shown $vs.$ the detuning ξ of the beat note
in Fig. 1. It specifies the net energy transfer from the
alternating and damped instantaneous flow rate $\dot{W}$. The corresponding "cooling" or "heating" is related to stimulated
light scattering off the atomic coherence; from ω_1 to ω_2 or
$vice\ versa$. The coherence is predominantly generated at the
beat phase of maximum amplitude. Its destruction by spontaneous emission saves the non-zero net effect which arises from
the dominating small phases.

This process of energy transfer seems related to "stimula-
ted cooling" of resonant atoms in a standing light wave |6,
4, 7|.

The instantaneous flow rate $\dot{W}$ shows a particular evolution
on a longer time scale. Since $\mu = \varkappa v_0/\delta \ll 1$, $J_0(\mu)$ is approximated by $\cos\mu$. Moreover, $x_0 = x_0(t)$ and $v_0 = v_0(t)$ as required by the energy transfer. Then,

$$\dot{W} = \dot{W}_0(\rho/2)\,(\sin\Psi_- + \sin\Psi_+) \tag{7}$$

with $\Psi_\pm = \xi t \pm \mu(t) + \Phi$, and $\Phi = \operatorname{arctg}\eta/\zeta + \Phi_0$. The second
term oscillates on time scale T_2 and is averaged over on the
time scale $T_3 > (\xi - \varkappa\dot{x}_0)^{-1} > T_2$. Now with $\Psi_- \equiv \Psi$,

$$\dot{\Psi} = \xi - \varkappa\dot{x}_0 \ , \tag{8}$$

and

$$\dot{W} = \sqrt{\overline{W}m}\ \dot{v}_0 \ . \tag{9}$$

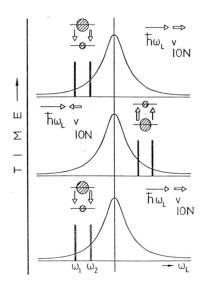

Fig. 2. Synchronization of
vibration and "coherence
pulsations" makes stimulated
light scattering one-way
(say, $\omega_1 \to \omega_2$) and keeps it
from alternating.

284

Inserting $\dot{v}_o$ from the derivative of eq. (8), and $\dot{W}$ from (7) shows that Ψ satisfies the pendulum equation

$$\ddot{\Psi} = - \Omega^2 \sin \Psi, \tag{10}$$

where $\Omega^2 = (\varkappa\rho/2\sqrt{Wm})\ \dot{W}_o = \dfrac{\varkappa\rho}{mx_o\omega}\ \hbar k v_o \delta$.

This equation of motion also governs the dynamics of electrons in a free-electron laser |8|, the prototype of a hybrid parametric oscillator.

The rate of energy flow from ion to field (or v.v.) is calculated from eq. 10 by, e.g., perturbative expansion in $(\Omega t)^2$ and subsequent weighted averaging over both t and Φ_o.

The derivation, analogous with Colson's procedure |9|, models the locking of the phase of the ionic vibration to the beat frequency. If locking sets in, continuous one-way energy transfer - from ion to light field, e.g. - is feasible, as outlined in Fig. 2. Its origin is motional phase modulation of the light which alternates favoring ω_1 or ω_2 for stimulated transitions, combined with synchronous population pulsations - or more precisely: coherence pulsations. This process does not require spontaneous emission for a non-zero net effect to result.

This work was supported by the Deutsche Forschungsgemeinschaft.

1. Th. Sauter, H. Gilhaus, W. Neuhauser, R. Blatt, and P.E. Toschek, Europhys. Letters 7, 317 (1988)

2. H. Gilhaus, Th. Sauter, W. Neuhauser, R. Blatt, and P.E. Toschek, Opt. Communic. 69, 25 (1988)

3. M. Lindberg and J. Javanainen, J. Opt. Soc. Am. B3, 1008 (1986)

4. J.P. Gordon and A. Ashkin, Phys. Rev. A 21, 1606 (1980)

5. R.J. Cook and H.J. Kimble, Phys. Rev. Lett. 54, 1023 (1985)

6. A.P. Kazantsev, JETP 39, 784 (1974)

7. J. Dalibard and C. Cohen-Tannoudji, J. Opt. Soc. Am. B 2, 1707 (1985)

8. See, e.g., G. Dattoli and A. Renieri, in: Laser Handbook. M.L. Stitch, M. Bass (eds.), Vol. 4, p. 1. Amsterdam, North-Holland, 1985

9. W.B. Colson, Phys. Lett. 64A, 190 (1977)

The Role of Laser Damping in Trapped Ion Crystals

R. G. DeVoe, J. Hoffnagle, and R. G. Brewer

IBM Almaden Research Center, San Jose, CA 95120, USA

Several groups[1-3] have recently observed ordered arrays of ions (ion crystals) in radio frequency quadrupole traps. Cooling by laser radiation pressure is essential to reach the milliKelvin temperatures at which crystallization can occur. We have recently studied[4] laser cooling in ion arrays and shown that cooling rates (1) depend on array size and on the position of the ion in the array, (2) can reverse sign so that heating occurs below resonance and cooling above, and (3) depend sensitively on trap parameters, for example, trap voltage and frequency. This is in contrast to the earlier view that the laser cooling of ion arrays is qualitatively the same as of a single ion[5].

Understanding the cooling of ion arrays is important for two reasons. First, it identifies a physical mechanism (a laser damping instability) underlying the recently observed[1] phase transitions between ordered and disordered states, which have now been analyzed as order-chaos transitions[3]. Second, it explains the failure of recent experiments to crystallize more than 100 ions as arising from inhibited laser cooling, and points the way to cooling techniques specifically designed for ion arrays.

This inhibition of laser cooling is due to Doppler shifts arising from radio frequency micromotion. Ions in an rf quadrupole trap undergo micromotion oscillations about their equilibrium position at the frequency of the rf trap voltage Ω. Micromotion has been ignored in previous laser cooling theories because it vanishes for a single ion at trap center. In ion arrays, however, substantial micromotion is always present to balance the mutual Coulomb repulsion of the ions. We have found that, for the conditions of recent experiments [1-3], micromotion amplitudes are comparable to the wavelength of the cooling radiation and strong Doppler shifts result. In the rest frame of the ion, they frequency modulate the cooling laser radiation and distribute it among dozens or even hundreds of FM sidebands which can, in a large array, span several GHz. These micromotion Doppler shifts can reduce or even reverse the sign of laser cooling in ion crystals relative to single ions.

In this paper we report further experiments which demonstrate spectral broadening and the inhibition of laser cooling by micromotion. As before[4] we use a single Ba$^+$ ion and bias it with an external DC potential to move it away from trap center and simulate its position in an ion array. The Ba$^+$ ion is excited so that its cooling resembles that of a two level system to facilitate comparison with theory. A laser at 493 nm drives the $6^2S_{1/2} \rightarrow 6^2P_{1/2}$ transition and is responsible for the cooling.

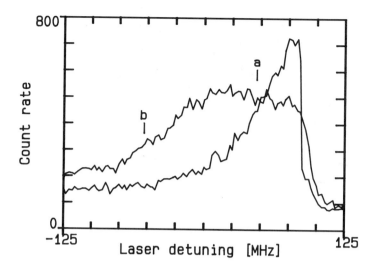

Fig. 1. Flourescence intensity of a single laser-cooled ion at trap center (peaked curve) and biased 6 μ off center (broad curve).

It is turned on for 45 out of every 50 ns by a microwave electrooptic modulator. Optical pumping into the $5^2D_{3/2}$ state is prevented by a 5 ns wide pulse of 649 nm radiation which repopulates the $6^2P_{1/2}$ level. The low duty cycle of the red light reduces any red cooling forces, while the "toggling" of the red and blue light suppresses three-level effects, such as the two-photon resonance. The validity of this approach was confirmed by analytic calculations and by numerically integrating the three-level optical Bloch equations for pulsed excitation.

Fig. 1 shows the fluorescence intensity as a function of blue laser tuning for the ion at trap center and with 6 microns of bias. The origin of the laser tuning scale is arbitrary. Note that the fluorescence of the biased ion peaks about 90 MHz lower than that of the unbiased ion and that the curve closely resembles the micromotion Doppler shift theory of Fig. 1, Ref. 4, for a micromotion index $\beta = 22$ and a trap frequency $\Omega = 3.55$ MHz. Thus the spectral broadening by micromotion predicted in Ref. 4 can be directly observed.

The effect of micromotion Doppler shifts on laser cooling can be determined from photomicrographs of the ion as the laser is swept through resonance in Fig. 1. The laser cooling rate Γ is, according to Ref. 4, proportional to the rate of change of the ion

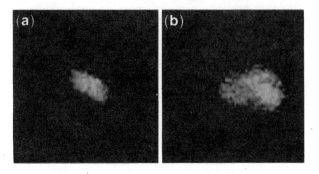

Fig. 2. Photomicrographs the biased ion for the detunings (a) and (b) in Fig. 1.

fluorescence rate versus frequency, or $\Gamma \propto \partial\rho_{22}/\partial\omega$, where ρ_{22} is the excited state density matrix element. Positive slope in Fig. 1 corresponds to cooling and negative slope to heating. Thus the biased ion should be cooled for detunings below 20 MHz, and heated above. In Figs. 2a and 2b we show pictures of the biased ion for the two detunings a and b in Fig. 1. Fig. 2a shows a well-cooled ion that exhibits micromotion along the trap z axis (horizontal axis in the figure). In Fig. 2b, all conditions are the same except the laser frequency has been increased by 100 MHz to point b of Fig. 1. This is a point of positive slope (cooling) for an unbiased ion, but negative slope (heating) for the biased one. Fig. 2b shows that the laser cooling has failed; the ion undergoes large secular oscillations and has left its equilibrium position at the center of the picture. We estimate its secular energy along the z axis as $> 1K$, in contrast to transverse energies $\sim 10mK$. The ion does not leave the trap because it is still strongly cooled in the x and y axes, which are perpendicular to the micromotion.

This work was partially supported by the Office of Naval Research.

1. F. Diedrich, E. Peik, J.M. Chen, W. Quint, and H. Walther, *Phys. Rev. Lett.* **59**, 2931 (1987).

2. D. J. Wineland et. al., J.C. Bergquist, W. M. Itano, J.J. Bollinger, and C. H. Manney, *Phys. Rev. Lett.* **59**, 2935 (1987).

3. J. Hoffnagle, R. DeVoe, L. Reyna, and R. Brewer, *Phys. Rev. Lett.* **61**, 255 (1988).

4. R. DeVoe, J. Hoffnagle, R. G. Brewer, *Phys. Rev.* A**39**,4362 (1989).

5. J. Javanainen, *Phys. Rev. Lett.* **56**, 1798 (1986); *J. Opt. Soc. Am. B* **5**,73, (1988).

TWO-PHOTON TRANSITION BETWEEN THE GROUND STATE AND A METASTABLE STATE IN MERCURY IONS CONFINED IN A RADIOFREQUENCY TRAP.   Theoretical study of the lineshape.

M. Houssin, M.Jardino, and M. Desaintfuscien
Laboratoire de l'horloge atomique. Unité propre de Recherche du CNRS.
Bat 221. Université de Paris-Sud. Orsay 91405. France.

Our experiment is the observation of a two-photon transition between the fundamental state $5d^{10}6s^2S_{1/2}$ and the metastable state $5d^96s^22D_{5/2}$ in mercury ions confined in a radiofrequency trap [1]. The lifetime of the metastable state is about 0,1s and so this transition is attractive as an optical frequency standard. Also the study of this line gives informations about the energetic properties of the confined particles. It can be used to measure the temperature of the cloud, and estimate the shift due to second order Doppler effect. That shift is an important limitation to frequency stability of mercury ions clocks.

First of all, the profile of this line has been calculated. A three level model has been used [2]. Without second order Doppler effect, the line is a Lorentzian curve with a linewidth Γ, where $1/\Gamma$ is the lifetime of the $^2D_{5/2}$ state.

In our experimental conditions, the temperature of the ions is between 4000 K and 8000 K. The characteristic constant of the second order Doppler effect is $\Delta\omega_q = \omega\ u^2/c^2 \simeq 10^4\ s^{-1}$ (where u is the quadratic speed of the ion and $\omega/2\pi$ the laser frequency).We have $\Delta\omega_q \gg \Gamma$ and then the second order Doppler effect cannot be neglected. The lineshape is also broadened and shifted by the electric and magnetic fields due to the confining voltages and by the laser used to excite the two-photon transition (light-shift). These effects are negligible compared with the second order Doppler effect. In these conditions, we calculate a broadened and red shifted line determined by (with $\omega_{ge}/2\pi$ the frequency of the transition) :

$$\Gamma_g = \Gamma_g(0)\ \frac{\Gamma_e\sqrt{\pi}}{\Delta\omega_q}\ \exp\left(\frac{2\omega - \omega_{ge}}{\Delta\omega_q}\right) \bigg/ \sqrt{\frac{2\omega - \omega_{ge}}{\Delta\omega_q}} \qquad \text{pour } 2\omega \leq \omega_{ge}$$

$$\Gamma_g = 0 \qquad \text{pour } 2\omega > \omega_{ge}$$

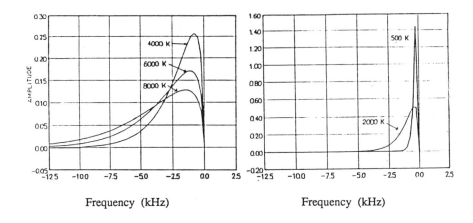

Frequency (kHz)   Frequency (kHz)

Fig 1. Calculated profile of the line, broadened and red shifted by second order Doppler effect, for different temperatures of the ions in the trap, with a monochromatic laser.

The profile of the line is asymmetrical. Its center is $\Delta\omega_q /4$ shifted, and its linewidth is $0,9\ \Delta\omega_q$ ($\simeq$ several kHz).

But, as described in [3], our laser has a Gaussian line profile with a linewidth of about 4 kHz (similar to the precedent linewidth). So, the expected signal is a convolution product of the laser line and the transition line. We calculate this product for different temperatures of the ions and different linewidths of the laser and plot the width at half maximum of these curves in fonction of the ions temperature.

Then, knowing the linewidth of the laser [3], we are able from the measurement of the width at half maximum of the observed signal to determine the temperature of the ions in the trap and so to estimate the shift due to second order Doppler effect.

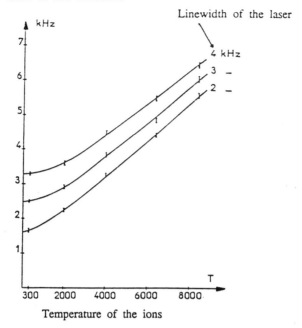

Fig 2. Width at half maximum of the observed signal in fonction of the temperature of the ions, for different linewidths of the laser.

References :

[1]   M. Houssin, M. Jardino, M. Desaintfuscien, proceedings of Laser Spectroscopy VIII, Are, Suede (ed. Springer verlag),p 77(1987)

[2]   B. Cagnac, G. Grynberg, F. Biraben : In J.de Physique, 34, 845(1973)

[3]   M. Houssin, M. Jardino, B. Gely, M. Desaintfuscien : In Optics Letters, Vol 13, n° 10, 823 (1988)

Stable $^{15}NH_3$ laser for cooled, single ion Ba^+
Frequency Standard

K.J. Siemsen, A.A. Madej and J.D. Sankey
 National Research Council Canada, Ottawa, Ontario, K1A OR6
J. Reid
 Lumonics Research Ltd., Kanata, Ontario, K2K 1Y3
G. Magerl
 Technische Universität, A1040 Wien, Austria

We propose to use an optically pumped mid infrared
ammonia laser for probing the fine structure transition of a
laser cooled single barium ion held in a radio frequency
trap. Elements of this project have now been completed and
are presented here.

The $^2D_{5/2} \leftarrow {}^2D_{3/2}$ magnetic dipole transition of $^{138}Ba^+$
lies at 800.974 cm^{-1} [1]. There exist several P-branch
transitions of the ν_2-band of $^{15}NH_3$ in close proximity to
the Ba^+ clock transition but none of them have been
previously observed in laser emission [2].

In this work, an experimental investigation has been
carried out to obtain laser action on the sP(8,7) line of
$^{15}NH_3$ at 800.95593 cm^{-1} and on the sP(8,6) line at
801.52895 cm^{-1}. The sP(8,7) line can be directly pumped via
the sQ(7,7) absorber line. As opposed to earlier findings
[3], it has now been shown that laser action does occur when
pumping via a Q-branch absorber line if the correct
polarization conditions are met. However, because of the
need of a radioactive $^{14}C^{18}O_2$ laser for pumping sQ(7,7) this
method was not pursued.

Laser action on the sP(8,6) line of $^{15}NH_3$ has now been
successfully obtained by direct pumping via the sR(6,6)
absorber line. In this case the pump laser operates on the
$R(40)_{II}$ line of $^{16}O^{12}C^{18}O$. The $R(40)_{II}$ laser frequency has
been calculated by Freed et. al. [4]; however, laser action
on this line was not previously observed. The isotope
$^{16}O^{12}C^{18}O$ does not exist in pure form due to dissociation
and subsequent scrambling in the gas discharge. Also, the
strong laser lines of $^{12}C^{18}O_2$ prevent laser action on
$R(40)_{II}$ of $^{16}O^{12}C^{18}O$ unless special precautions are taken.
We use the hot cell technique, known from the CO_2 sequence

laser work, in conjunction with a specially made high
resolution grating [5] to achieve lasing on $R(40)_{II}$. By
downshifting the $R(40)_{II}$ frequency with an acousto-optic
modulator and observing the saturated absorption dip in
$^{15}NH_3$, the line center of sR(6,6) was measured to be
72 ($\pm$ 1) MHz below $R(40)_{II}$. Lasing on sP(8,6) was observed
with a threshold pump power of 0.3 W in a nonselective high
Q cavity at room temperature. It is expected that the
sP(8,6) transition will lase with about the same passive
frequency stability that has already been observed for the
sP(7,0) line of $^{14}NH_3$ [6] which is ideal for testing the
intrinsically narrow clock transition in Ba$^+$.

The frequency offset between sP(8,6) and the Ba$^+$ clock
transition is 16.57 ($\pm$ .2) GHz. We propose to use the
generated sideband [7] from an electro-optic modulator to
probe the Ba$^+$ transition. The sideband power of such a
device is estimated to be ~ 130 μW per watt laser power and
per watt microwave power. The probe power saturation
requirements for Ba$^+$ are estimated to be a few nanowatts
provided the linewidth of the probe beam is below 1 kHz.
Thus with ~ 10 mW NH_3 laser power and 1 W microwave power,
the sideband power should be more than adquate to observe
the quantum jumps in the 494 nm fluorescence signal of Ba$^+$.
Trapping and laser cooling of single Ba$^+$ ions have been

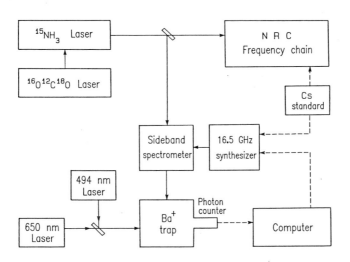

Fig. 1. Overview of proposed measuring scheme.

achieved at our laboratory, together with the observation of quantum jumps into the $^2D_{5/2}$ level using broad bandwidth indirect excitation [8]. In the initial spectroscopy on the $^2D_{5/2} \leftarrow {}^2D_{3/2}$ transition, we plan to rely on the passive short term stability of the NH_3 laser while tuning the microwave frequency of the modulator across the Ba^+ resonance.

References

1. C.E. Moore, "Atomic Energy Levels", Vol. III, NBS, Washington, (1958).

2. R. D'Cunha, S. Urban, K.N. Rao, L. Henry and A. Valentin, J. Mol. Spectr. **111**, 352 (1985).

3. K.J. Siemsen, J. Reid, D.J. Danagher, Appl. Opt. **25**, 86 (1986).

4. C. Freed, L.C. Bradley and R.G. O'Donnell, IEEE QE-16, 1195, (1980).

5. B. Bach, Hyperfine Inc., Boulder, Colorado 80301, USA

6. K.J. Siemsen, E. Williams and J. Reid, Opt. Lett **12**, 879, (1987).

7. G. Magerl, W. Schupita and E. Bonek, IEEE **QE-18**, 1214 (1982).

8. J.D. Sankey and A.A. Madej to appear in these proceedings.

Partial Laser Cooling and Saturation Spectroscopy
on 9 MeV ^{7}Li$^+$-Ions in a Storage Ring

G. Huber, S. Schröder, R. Klein, M. Gerhard, R. Grieser,
M. Krieg, A. Karafillidis
Institut für Physik, Universität Mainz, D-6500 Mainz, FRG
Th. Kühl
GSI Darmstadt, D-6100 Darmstadt 11, FRG
A. Faulstich, W. Petrich, D. Habs, R. Neumann, D. Schwalm,
A. Wolf and the TSR group
MPI Kernphysik and Universität Heidelberg, D-6900 Heidel-
berg, FRG

Laser cooling and spectroscopy in traps have reached im-
pressive perfections, as reported recently (1). We report
here laser cooling and partially Doppler-suppressed
spectroscopy on stored ions at 5.4 % speed of light in the
TSR heavy ion storage ring in Heidelberg. The requirements
for saturation spectroscpy with two counterpropagating
collinear laser beams are discussed.

In contrast to Penning and RF-traps the ions in a storage
ring move at a high longitudinal velocity with small transverse harmonic (betatron) oscillations around the central
orbit with just a few times the orbiting frequency in the
case of strong focussing. The injection of the high velocity
beam introduces however large betatron oscillations which
influence considerably the laser interaction. A proper tuning
of the storage ring, electron cooling and finally laser
cooling should condense the ion beam to a long lived beam
with $\Delta p/p$ well below 10^{-6}. The experiments at the TSR ring
yielded storage times for $^{12}C^{6+}$ at 140 MeV above 5 hours when
cooling with a well aligned electron beam (2). The stored
Li$^+$-ion

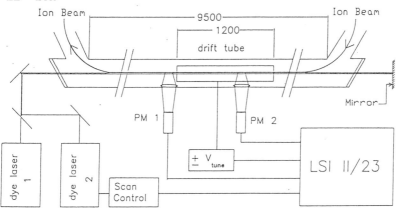

Fig. 1: Laser spectroscpy set-up at the TSR. In the collinear
section laser resonance can take place over 7.5 m. A drift
tube of 1.2 m allows for voltage tuning in addition to the
laser scans. Fluorescence is observed within and behind the
drift tube. The computer controlled recordings are
synchronized with the beam injection to the TSR.

beam (2 - 12 μA) is much more fragile due to stripping in the
rest gas collisions and has been stored at 10^{-11} hPa for
15 s. No substantial electron cooling could be obtained
within this short storage time.

The scheme of the experiment is shown in Fig. 1. The uncooled expanded ion beam is merged with two collinear laser
beams on a 0.5 cm² area. Experiments have been performed with
both counter-propagating and co- plus counter-propagating laser beams, delivered via a single mode fiber. The single mode
dye lasers were tuned to the red shifted σ = 17290 cm^{-1} and
the blue shifted resonance σ = 19217 cm^{-1} of the triplet line
$^3S_1$ - $^3P_2$, σ_0 = 18228.48 cm^{-1} of $^7Li^+$. The strong ring laser
beam was set to the F = 5/2 - F' = 7/2 two level system as
shown in Fig. 2.

The width of this line is a measure of the velocity spread
in z-direction for the inner zone of the ion beam. The corresponding momentum spread of $\delta p/p = 6 \cdot 10^{-6}$ from the electrostatic accelerator is exceeded by the transverse motions in
the 10^{-3} range. The longitudinal temperature of ~ 400 K is
very convenient whereas ~ 10^6 K for the transverse temperature must be reduced for precision experiments. These betatron oscillations can be reduced by electron cooling and
should be excited moderately at good settings of the ring.
Even without this reduction, Lamb dips have been observed
within the Doppler profile.

The second laser beam has been tuned to the F = 5/2 -
F' = 5/2 resonance, depleting a selected velocity class via
Optical Pumping to the F = 3/2 state. Very moderate intensities are sufficient for this scheme, where again the strong
laser is scanned as shown in Fig. 4.

The longitudinal velocity distribution shows the depletion
from the pumping laser, which exceeds by more than 150 the
natural width of 3.8 MHz. This broadening has a power component and a fixed pedestal. Moreover, the lifetime of this

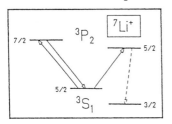

Fig. 2: Transitions in the triplet
line $^3S_1$ - $^3P_2$ relevant in this
work. The two level system on the
left for cooling and fluorescence
detection. The Optical Pumping
scheme is shown on the right.

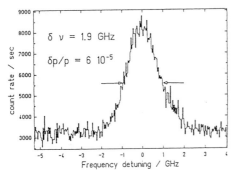

Fig. 3: Longitudinal Doppler
profile of the central ion
beam. The Gauß profile is
folded with the exponential
decay of the beam. The width
observed here is about half
of the Schottky signal width
for the total ion beam. The
optical signal refers to the
center part of the ion beam.

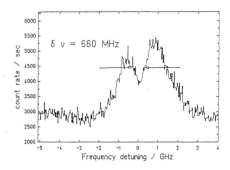

Fig. 4: Lamb dip in the v-system. A low power laser is set on the right transition in Fig. 2. Probing the modified velocity distribution with the strong line a broadened Bennett hole is observed.

pumping hole has been determined by mechanically chopping the pump beam at fixed laser settings. Relaxation times of about 140 ms have been observed. The variation of the dips' amplitude, width and lifetime with the pump laser power have been recorded. The width as a function of the light intensity is shown in a standard plot $(\delta v)^2/I$, assuming two independent contributions.

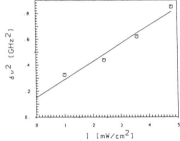

Fig. 5: Saturation width as a function of laser power. The square of the width is plotted to separate the power broadening from the Doppler pedestal. δv_D = 390 MHz is obtained and the power slope yields a_p = 0.277 GHz^2/mW.

Using a simple description of the ions motion in a storage ring the basic properties of the observed signals can reproduced, and stringent conditions for high resolution spectroscopy testing special relativity by measuring the one way speed of light can be constructed.

The ions position is given by

$$x, y = (\epsilon \cdot \beta_{h,v})^{1/2} \sin (Q_{h,v}\omega_0 t + \varphi_{h,v}) \qquad (1)$$

and its velocity by

$$v_{x,y} = (\epsilon \cdot \beta_{h,v})^{1/2} Q_{h,v} \omega_0 \cos(Q_{h,v}\omega_0 t + \varphi_{h,v}) \qquad (2)$$

with emittance ϵ, betafunctions β, betatune Q, orbit circular frequency and phase ω_0 and φ.

Via energy conservation v_z is modulated by the transverse motion from its unperturbed value V_0 to

$$v_z = V_0 - 1/2 \sum (v_{x,y})^2 /V_0 + \text{higher orders} \qquad (3)$$

The linear plot of $(v_{0_{x,y}})^2 \{1 - \sin^2(\ldots)\}/(x_0,y_0)^2 \sin^2(\ldots)$

gives the strong correlation between central position equal to laser interaction and maximum transverse motion. With β_x = 5.5 m, β_y = 2 m, $Q_{h,v} \sim 2.8$, ω_0 = 1.85 MHz, and a laser beam of 8 mm diameter the transverse motion - independent on its amplitude - induces a width δv_D = 25 MHz, whereas the minimum width from figure 5 is 390 MHz. This discrepancy is readily removed by the introduction of a small angle between laser and ion beam. The power slope in fig.5 can be expressed in terms of a relaxation time $\tau \gtrsim 160$ ms, that agrees with the direct observation of 140 ms, which may be related to intra beam and beam- ring perturbations. One can conclude that the residual width of the Lamb dip is a measure of transverse motion and/or beam misalignment.

Nevertheless has the present status allowed to record signals of laser cooling as shown in the last Fig. 6. In a laser scan represented on the left, ions are driven to the right end of the spectrum, mainly invisible since they stay out of resonance at the saturation S = 25. The fluorescence is due to ions with large transverse motion, thus escaping to the light force. The drift tube in fig.1 is kept during this time at a potential, shifting the resonance inside by 5.0 GHz. At the stop of the laser sweep, the drift tube resonance is scanned through the cooling part of the spectrum as shown on the right. At both PM tubes compressed and normalized to the beam life time enhanced signals are observed. Their width, 700 MHz at PM 1 and 440 MHz at PM 2 are different by the contribution of the stray field signals from the entrance and exit of the 200 mm diameter drift tube.

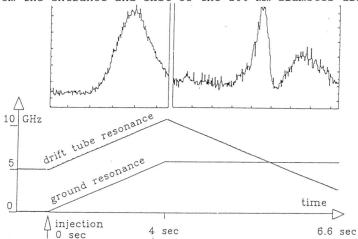

Fig. 6.: Laser cooling signal obtained in a combined laser and Doppler sweep. Only ions with moderate transverse motion are pushed, the others contribute to the unshifted profile on the right. This compressed signal from PM 1 has 700 MHz width, partly due to stray field broadening.

References
1. e.g. contributions to ELICAP '88 Paris and this conference
2. D.Habs et al. "First Experiments with the Heidelberg Test Storage Ring TSR" subm. to NIM B (1989)

Part IX

Ultrafast Spectroscopy

Excitation of Coherence in Molecular Systems With 6 Femtosecond Optical Pulses

Charles V. Shank, P. C. Becker, J. Y. Bigot and H. L. Fragnito

AT&T Bell Laboratories, Holmdel, NJ 07733, USA

Ultrashort optical pulses provide a unique tool for inducing coherent transients in large molecules. A 6 femtosecond optical pulse contains sufficient bandwidth to excite the entire manifold of S_0 to S_1 transitions in a molecule in solution. In this paper we are going to discuss two types of coherence that can be excited by the short pulse. In the first type, we will measure the dephasing of an induced optical polarization between vibronic levels using the two pulse photon echo technique. The second type of coherence arises as a consequence of nonequilibrium excitation of vibrational levels creating wavepackets that execute motion on both the excited and ground state potential surfaces. We can observe this wavepacket motion by measuring the evolution of the absorption spectrum as a function of time.

The photon echo technique has proven useful to study processes which influence dephasing of an induced polarization for a broad range of material systems (1, 2, 3). The use of coherent optical transients to study such processes in molecules in solution has been frustrated by the very fast dephasing rates in such systems. In this discussion we will describe the use of very short optical pulses to observe photon echoes and femtosecond quantum beats from organic dye molecules in solution.

In the experiments reported here we observe photon echoes using a two pulse sequence. Two pulses, one having a wave vector $\vec{k}_1$ and the other with wavevector $\vec{k}_2$, generate a photon echo in the momentum matched direction $2\vec{k}_2 - \vec{k}_1$. The echo is then separated spatially from the exciting pulses. The energy of the photon echo energy is then measured as a function of relative time delay between the two pulses.

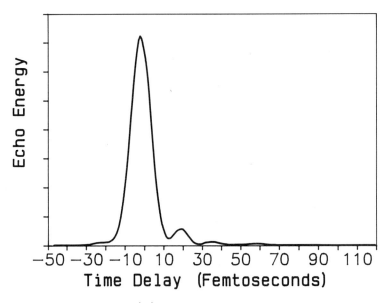

Fig. 1. Echo energy in the direction $2\vec{k}_2 - \vec{k}_1$ as a function of the relative time delay between the two excitation pulses, for the dye nile blue in ethylene glycol.

The primary utility of the photon echo is to measure the dephasing of the induced polarization which is characterized by a time, T_2. For a two level system which is purely inhomogeneously broadened the echo energy will decay with the relative time delay t_d between the two pulses as

$$E(t_d) \ \exp(-t_d/T_{echo}) \tag{1}$$

where $T_{echo} = T_2/4$.

The experimentally measured echo for the oxazine dye, nile blue, is plotted in Figure 1. Note that an initial rapid decay is observed followed by oscillations in time indicating the presence of quantum beats with a period of oscillation measured to be approximately 18 femtoseconds (4, 5). This indicates that a component of the echo is coming from transitions involving a pair of levels in the excited state with a spacing of 1850 cm^{-1}.

To understand this experimental result we need to consider the photon echo from a manifold of vibronic levels. Each level in the ground state is coupled to a manifold of vibronic levels having a Frank-Condon overlap with the excited state. For purposes of our discussion let us assume delta function in time excitation pulses, strong inhomogeneous broadening, with each level characterized by a single T_2. The echo energy is then given by

$$E(\tau) \propto \left[\sum_{i,j} \mu_{1i}^2 \ \mu_{1j}^2 \ \cos(\omega_{ji}\tau) \right]^2 \exp(-4\tau/T_2) \tag{2}$$

where the ground state is labeled 1 and the excited states i and j. Here ω_{ji} is the (angular) frequency difference between the excited states and μ_{1i} is the transition dipole moment between the ground state and level i.

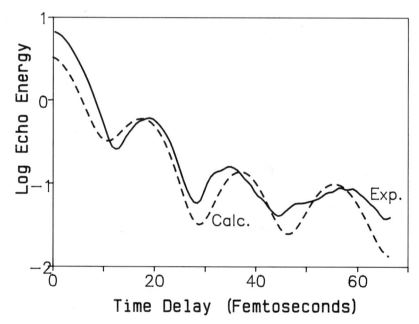

Fig. 2 Experimental (smooth curve) and calculated (dashed curve) logarithm of the echo energy for the dye nile blue in ethylene glycol. The parameters used for the calculated curve are two modes with frequencies 585 cm^{-1} and 1850 cm^{-1} with dipole moments in the ratio 1 : 0.15 : 0.2 and T_2 equal to 65 femtoseconds.

For a simple system with just a few levels this expression clearly predicts quantum beats. However for a system with a large number of modes the echo is rapidly dephased in a time corresponding to the reciprocal of the bandwidth of the absorption spectrum. Under this circumstance the echo proves no information about the dephasing of the coherent polarization. For the molecules discussed in this paper the absorption is dominated by a few modes having a large oscillator strength often referred to as system modes. We can approximate the sum in equation (2) as a sum over just the system modes.

In Figure 2 we have plotted the log of the photon echo energy as a function of relative time delay between the two pulses of the echo sequence. The dashed line was calculated using equation (2) assuming two system modes with frequencies $585 cm_{-1}$ and $1850 cm_{-1}$ with dipole moments in the ratio 1:0.15:0.2 and T_2 equal to 65 femtoseconds. This value of T_2 is very close that reported in recent hole burning experiments (6). Experiments of this type should be useful for gaining an understanding of the detailed nature of dephasing processes of molecules in liquids.

One of the consequences of exciting a molecule with such a short optical pulse is that virtually all of the Franck-Condon connected vibrational modes of the molecule are impulsively excited. When a molecule is excited with an optical pulse in a time short compared to a vibrational period the excitation occurs without nuclear motion. Nuclear motion is initiated to accommodate the new equilibrium position of the excited state potential surface. The nuclei will start to oscillate around a new equilibrium position. The nuclear wavefunction $\phi(Q,t)$ can be written as a sum of vibrational wavefunctions for the excited state potential surface $\phi_n(Q)$,

$$\phi(Q,t) = \Sigma_n a_n(Q) \exp(-i\omega_n t), \qquad (3)$$

where the ω_n's are the frequencies of the eigenstates and a_n's are coefficients. This expression represents a wavepacket moving on the S_1 surface. What is important for our discussion is that as the wave packet moves the absorption spectrum changes. Motion of the vibrational wavepacket on the potential surface can be observed by measuring the absorption spectrum of the molecule under study with a pulse short compared to a vibrational period.

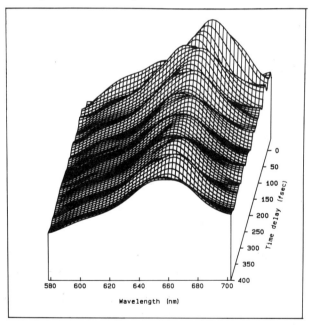

Fig. 3 Three dimensional plots of the differential absorption spectrum as a function of time delay and wavelength for the dye Nile Blue.

In Figure 3 we have plotted the induced transmittance change for the molecule nile blue using a 6 femtosecond pump and probe. The amplitude and the peak of the induced absorption spectrum are seen to oscillate with a period corresponding to 60 femtoseconds. The corresponds to the 590cm^{-1} ring distortion mode observed in dynamic hole burning measurements (6).

The implications of this experiment may be important to the study of molecular dynamics. We have shown that near delta function excitation pulses prepare a population of vibrational levels in a coherent state. The excited population may in some circumstances provide a means of observing the electronic potential surface as the electronic absorption spectrum is monitored as a function of time. The dephasing time for the vibrational polarization can be deduced from the decay of the amplitude of the observed oscillations. Note that this is not the dephasing time for the polarization of the vibronic transition. Either the hole burning method (6) or the photon echo technique described in this paper must be used.

References

1. N. A. Kurnit, I. D. Abella and S. R. Hartmann, Phys. Rev. Lett. 13, 567 (1964); W. H. Hesselink and D. A. Wiersma, Phys. Rev. Lett. 43, 1991 (1979); R. M. Macfarlane, R. M. Shelby and R. L. Shoemaker, Phys. Rev. Lett. 43, 1726 (1979).

2. C.K.N. Patel and R. E. Slusher, Phys. Rev. Lett. 20, 1087 (1968).

3. J. Hegarty, M. M. Broer, B. Golding, J. R. Simpson and J. B. MacChesney, Phys. Rev. Lett. 51, 2033 (1983).

4. R. Beach and S. R. Hartmann, Phys. Rev. Lett. 53, 663 (1984); R. L. Shoemaker and F. A. Hopf, Phys. Rev. Lett. 33, 1527 (1974).

5. L. Q. Lambert, A. Compaan and I. D. Abella, Phys. Rev. A4, 2022 (1971).

6. C. H. Brito Cruz, R. L. Fork, W. H. Knox and C. V. Shank, Chem. Phys. Lett. 132, 341 (1986).

AUTOIONIZATION AND FRAGMENTATION OF Na_2 STUDIED BY FEMTOSECOND LASER PULSES

T. Baumert, B. Bühler, R. Thalweiser and G. Gerber
Fakultät für Physik, Universität Freiburg, 7800 Freiburg, FRG

We report on first results obtained with femtosecond laser pulses applied to molecular beam studies of the dynamics and the pathways of ionization, autoionization and fragmentation of highly excited molecular states of Na_2. Electronic autoionization of doubly excited molecular states and fragmentation of highly excited neutral and ionic states of Na_2 are hardly investigated and generally not well understood. This is mainly because i) the final continuum states are usually not analyzed and ii) the dissociative ionization is not distinguished from neutral fragmentation with subsequent photoionization of excited fragments. Doubly excited states play a major role in the reaction dynamics of diatomic molecules, since these states directly couple the different continua of dissociation and ionization. Electronic autoionization and neutral dissociation are therefore competing processes, but information from both channels may be used to characterize the doubly excited molecular states.

To study the dynamics of multiphoton processes leading to excitation, autoionization and fragmentation we have applied ultrashort laser pulses to induce the transitions and Time-Of-Flight spectroscopy to determine the mass and initial kinetic energies of the fragments and the energy and angular distributions of ejected electrons. The experiments were carried out with "cold" molecular beams. Because of the strong cooling in supersonic expansions we predominantly produce Na_2 in the lowest vibrational state $v"=0$ and in very low J- states. The final continuum states can definitely be assigned from the measured electron- and ion-kinetic energy distributions.

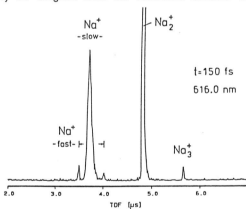

Fig.1: Time-Of-Flight spectrum of ions formed by the interaction of femtosecond laser pulses with a sodium molecular beam.

Application of femtosecond laser pulses considerably facilitates the interpretation of fragmentation processes, since the time duration of a laser pulse is much shorter than the fragmentation separation time. The terminal velocities of the separating fragments are typically 0.002 nm/fs for $W=3000$ cm^{-1} recoil energy. Therefore no further laser-induced excitation or ionization of the fragments can occur and all observed signals have to be related to processes which occur at small internuclear distances.

Femtosecond pulses are generated in a home-built colliding-pulse-mode-locked ring dye laser (CPM) with 4 intracavity prisms to adjust for the group velocity dispersion. The emission peak has been shifted to 616.0 nm by adjusting the DODCI absorber concentration. The output of the CPM dye laser was amplified at a rate of 100 Hz in a N_2-laser pumped dye amplifier to produce pulses of 10 nJ energy and t≈150 fs time duration. The pulse length was measured using the auto-correlation by second harmonic generation (SHG) in a nonlinear crystal.

The laser-molecular beam interaction region is placed between parallel plates and since the ions are extracted perpendicular to the beam with a low electric field, parent ions can be distinguished from fragment ions having initial kinetic energy. The released kinetic energy leads to a broadening or a double peak structure in the TOF spectrum due to fragments recoiling parallel and antiparallel to the extracting field. Therefore, from the observed difference in flight time the kinetic energy of the fragments can be inferred. The TOF-spectrum in fig.1 clearly shows the observation of Na_3^+, Na_2^+ and "slow" Na^+ as well as "fast" Na^+ fragment ions resulting from the fs-laser excitation at λ = 616.0 nm. "Fast" and "slow" Na^+ ions originate from fragmentation processes occuring at small internuclear distances of Na_2^*. Predissociation of Na_2^* and photoionization of Na^* as the origin of observed Na^+ ions can be ruled out considering the time duration of the fs-laser pulse. Based on this result, which is rather difficult to obtain from other experiments, and with the known molecular potential curves we completely determined for this model case the multiphoton excitation and fragmentation pathways.

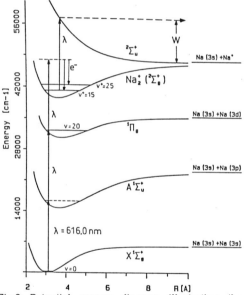

Fig.2: Potential energy diagram illustrating the origin of "fast" Na^+ ions.

With a pulsed tunable dye laser we observed in the wavelength dependent ionization spectra between 610 nm and 620 nm a very prominent peak at 616.08 nm in both the Na_2^+ and the Na^+ channel. Based on the known spectroscopy of molecular Rydberg states of Na_2 it is evident that this peak in the Na_2^+ spectrum is due to a resonance enhanced three-photon ionization of Na_2 This process is shown in the potential energy diagram in fig.2. The ionization enhancing intermediate level $^1\Pi_g$ is populated by the two-photon process Na_2 (X $^1\Sigma_g^+$,v"=0) + 2hv --> Na_2 (Ryd $^1\Pi_g$,v*=20). The Rydberg-molecule is then photoionized by absorption

of a third photon. The dimer ions are preferentially formed in the v*= 24,25 and v*= 14,15 vibrational states of the electronic ground state $X(^2\Sigma_g^+)$ due to favorable Franck-Condon factors. This direct ionization of the $^1\Pi_g$ Rydberg state leads to electrons having kinetic energies of E=810 +/- 10 meV and E=940 +/- 10 meV which are actually observed in the TOF-electron spectrum shown in fig.3. In addition to these strong peaks the electron spectrum shows less intense broad structures in the range between 300 meV and 500 meV and well below 200 meV extending to very low energies. The created dimer ions may now undergo a bound-free transition by absorption of one more photon from still the same fs-laser pulse:

$$Na_2^+ \ (X \ ^2\Sigma_g^+, v^+) + h\nu \longrightarrow Na_2^{+*} \ (^2\Sigma_u^+) \longrightarrow Na^+ + Na(3s) + W$$

Taking into account the populated v^+-levels and the known potential curves for the ionic ground and first excited states the corresponding bound-free transitions lead to recoil energies W between 10000 cm^{-1} and 11000 cm^{-1}. The energy W=10500 +/- 500 cm^{-1} obtained from the analysis of the TOF-ion spectrum perfectly agrees with that. We therefore conclude that "fast" Na$^+$ ions are produced by the ionization and fragmentation process shown in fig.1.

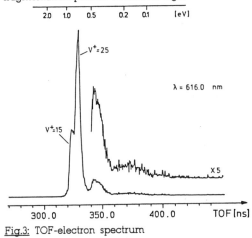

Fig.3: TOF-electron spectrum

The observation of "slow" Na$^+$ ions however cannot be explained within this framework since photoionization of the vibrational level v^*=20 of the $^1\Pi_g$ (3s + 3d) Rydberg state directly into the $^2\Sigma_u^+$ - continuum is energetically forbidden for the applied laser wavelength. Based on the measured electron energy distribution and the recoil energy W=900 +/- 500 cm^{-1}, obtained from the analysis of the "slow" Na$^+$ fragment ions, a consistent explanation is found by considering the excitation of doubly excited molecular states of Na$_2$ as it is shown in fig.4. The doubly excited states Na$_2^{**}$ (nl,n'l') form a Rydberg series converging versus the $^2\Pi_u$ state of Na$_2^+$, whose potential curve is known theoretically. Assuming that the shape of a $^1\Pi_u$ state potential curve, correlated to Na(3p)+Na(4s), is similar to the ionic $^2\Pi_u$ curve and that it has a potential barrier at large internuclear distances like the B $^1\Pi_u$ state from Na(3s)+Na(3p), we believe that in the resonance enhanced three-photon process vibronic levels close to the dissociation limit of the doubly excited $^1\Pi_u$ (3p+4s) state are excited. The wavefunctions of these vibronic levels extend from 3Å to approximately 10Å. These doubly excited levels may autoionize into the X($^2\Sigma_g^+$) ground state of Na$_2^+$ giving rise to electron energies between 260 meV and 500 meV which are actually observed. For internuclear distances greater than 6Å the vibronic levels cross into the continuum of the repulsive $^2\Sigma_u^+$ state of Na$_2^+$. Therefore for R ≥ 6Å there is a second open autoionization channel which is responsible for the observed electrons having energies in the range from 0 meV to 160 meV. This autoionization process

and the subsequent fragmentation

$$Na_2^{**} \ ^1\Pi_u \ (3p+4s) \longrightarrow Na_2^{+*}(^2\Sigma_u^+) + e^- (E_{kin})$$
$$Na_2^{+*} \ (^2\Sigma_u^+) \longrightarrow Na^+ + Na(3s) + W$$

produce "slow" Na$^+$-ions whose kinetic energies depend on the internuclear distance R where the autoionization takes place.

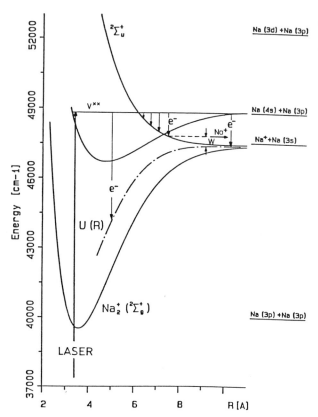

Fig.4: Excitation and autoionization processes of doubly excited Na_2^{**}.

In conclusion, this is the first reported experiment where in a molecular beam experiment a fs-laser has been used in combination with ion- and electron spectroscopy to clarify the ionization and fragmentation pathways of molecules excited by multiphoton processes.

On the Origin and Optical Properties of the Supercontinuum Generation

I. Golub

Centre d'Optique, Photonique et Laser (COPL), Université Laval, Québec, CANADA G1K 7P4

Intense picosecond pulses propagating in nonlinear media can produce frequency broadened output beams with a nearly white spectrum, called supercontinuum (SC)[1-3]. SC generation is correlated with self-focusing (SF) and self-trapped filament formation. The effect was demonstrated in solids, liquids and gases. Self-phase modulation (SPM), four-wave mixing (FWM) and SPM accompanied by FWM and Raman process have been proposed to explain the SC. All these models have difficulties and none of them is universal enough to explain the common characteristics observed in such a variety of media. For example, FWM process is inconsistent with the observed spatial order of the coloured rings; Raman effect can not explain SC in liquid Ar or Xe gas; and SPM predicts a spectral broadening smaller then the observed one and does not explain the spatial properties of SC/ring emission. The spatial properties are among the most striking features of SC: it is either accompanied by a ring emission or itself evolves in the form of a divergent cone/ring emission in the forward direction. We will refer to this emission as super-continuum cone emission -SCCE. We present new results on the spatial and polarisation properties of the SC generated in H_2O or D_2O and point out to the similarity between the optical properties of the SCCE, the conical emission observed in alkali metal vapors[4-5] and class II Raman radiation produced in liquids in a form of cones[6]. These properties are characteristic to an emission formed at the surface of self-trapped filaments. As a surface phenomenon, the SCCE obeys the condition of the Cherenkov-type emission process where the transverse component of the linear momentum is not conserved[7]. Surface radiation is expected at an angle satisfying the Cherenkov relation when the laser induced nonlinear polarisation travels faster than light[5,7].

A Quantel YG-471 mode locked laser which produced 22 psec duration pulses at 1.06μm of up to 35 mj energy or its second harmonic 15 psec duration pulses of 12 mj was focused into cells containing H_2O or D_2O. The intensity in the focal spot was up to 10^{12} W/cm². The SC was spread in a "circular rainbow", and for 1.06μm excitation the generated photon energy increases with the off-axis angle, while for 0.53μm excitation the pattern is more complicated.

Firstly, we found that the SCCE emission had the same circular or linear polarisation as that of the input laser beam. Any light of circular polarisation contains a small portion of counter rotating polarisation radiation. The weaker polarisation should be self-trapped before the stronger one, resulting in light changing its polarisation into a linear. This change occurs inside the filaments where the saturation degree is maximal and it was observed for Raman

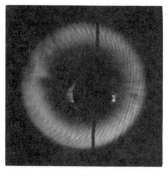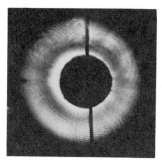

a   b

Fig. 1   The   SCCE generated in D_2O. The laser is focused
into the cell by a spherical lens (a) or by a combination of
cylindrical and spherical lenses (b). The focal line of the
cylindrical lens is vertical.

Stokes emission which originated in the self-trapped filaments[8].   Since SCCE preserves the circular polarisation of
the input laser, we conclude that it is generated in a non-
saturated region, i.e. at the surface of the self-trapped
filaments.

Next the laser light was focused into the D_2O by either a
spherical lens or a combination or cylindrical and spherical
lenses. The emission from the interior saturated parts of
different filaments should be spatially coherent. Emission
from the surfaces of different stochastically formed filaments, however, does not preserve the coherence because of
varying local conditions such as the saturation degree at the
surfaces of the filaments.   Thus the emission from interior
of the filaments should interfere to form an ellipse for the
cylindrical focusing mode corresponding to the spatial distribution of the filaments, as observed for usual (or class
I) Raman radiation[6].   The uncorrelated emissions from the
surfaces of the filaments should sum up incoherently resulting in a ring halo. The experiment (Fig. 1) shows that SCCE
spatial pattern is independent on the focusing mode leading
us to a conclusion that the origin of this emission is at the
surface of the filaments.

We also found that the SCCE angle is independent on the
lens focal length in the range 5-20 cm. The same effect was
observed for class II Raman radiation[6].   In contrast usual
Raman emission cone angles are inversely proportional to the
lens focal length[9].   All these observations support the
conclusion that the SCCE, similar to conical emission in
alkali metal vapors and class II radiation, is a radiation
formed on the surface of the self-trapped filaments - a
process involving longitudinal momentum conservation only.

Conical emission of 780 cm^{-1} bandwidth was observed in
sodium vapor irradiated with picosecond dye laser pulses and
it was explained in terms of Cherenkov emission[10]; this is

seen by us as a strong argument in favor of a common mechanism for SCCE and conical emission in alkali vapors. We propose a Cherenkov-type process to explain the generation of the SCCE. In our model, the laser induced medium excitation - nonlinear polarisation - propagates with the laser pulse group velocity, V_{gr}, resulting in an emission at frequencies fulfilling the Cherenkov condition at conical angle Θ_c given by $\cos\Theta_c = V_{ph}/V_{gr}$. V_{ph} is the phase velocity of the emitted light. In a region of normal dispersion $\cos\Theta_c = n_L/n_c(\omega)$, where n_L and $n_c(\omega)$ are the refractive indexes at the laser and the emitted radiation frequencies, respectively. The Cherenkov condition, $V_{ph} < V_{gr}$ or $n_L < n_c(\omega)$, is satisfied for excitation by a laser at 1.06µm and light emitted in the visible. It predicts light of higher frequencies to be emitted at larger angles as observed experimentally.

The pattern of SCCE for 0.53µm excitation is more complex. This can be a result of an alternation of the normal dispersion in the visible due to an induced Fano-type structure in the laser "dressed" continuum of the ionization levels.

In summary, we have shown that the supercontinuum conical emission is generated at the surface of self-trapped filaments and if shares its optical characteristics with other Cherenkov-type processes - conical emission in alkali metal vapors and class II Raman radiation. A model in which the surface radiation is emitted at angles satisfying the Cherenkov relation explains qualitatively the spectro-spatial properties of SCCE.

1. R.R. Alfano and S.L. Shapiro, Phys. Rev. Lett. <u>24</u>, 584, 592, 1217 (1970).

2. W.L. Smith, P. Liu and N. Bloembergen, Phys. Rev. A, <u>15</u>, 2396 (1977).

3. P.B. Corkum, C. Rolland and T. Srinavasan-Rao, Phys. Rev. Lett., <u>57</u>, 2268 (1986); JOSA B, <u>13</u>, 256 (1987).

4. C.H. Skinner and P.D. Kleiber, Phys. Rev. A, <u>21</u>, 151 (1979).

5. I. Golub, G. Erez and R. Shuker, J. Phys. B, <u>19</u>, L115 (1986); Opt. Commun., <u>57</u>, 143 (1986).

6. E. Garmire, in: Physics of Quantum Electronics, eds. P.L. Kelley, B. Lax and P.E. Tannenwald (McGraw-Hill, New-York, 1966), p. 167.

7. A. Szoke, Bull. Am. Phys. Soc., <u>9</u>, 490 (1964).

8. D.H. Close et al, IEEE J. Quantum Electron., <u>QE-2</u>, 553 (1966).

9. R. Chiao and B.P. Stoicheff, Phys. Rev. Lett., <u>12</u>, 290 (1964).

10. V.I. Vaichaitis et al, JETP Lett., <u>45</u>, 415 (1987).

Time-Resolved Transition-State Spectroscopy

J.H. Glownia, J.A. Misewich, and P.P. Sorokin

IBM Research Division, Thomas J. Watson Research Center, Yorktown Heights, N.Y. 10598-0218

Recent advances in the generation of ultrashort laser pulses have made possible the investigation of chemical dynamics on a femtosecond time scale, revealing details of the "transition states" region between reactants and products. For example, A.H. Zewail's group has reported femtosecond laser-induced-fluorescence (LIF) studies of both directly dissociating[1] and predissociating[2] systems.

A recent focus of our group has been the development of a pump-probe apparatus (Fig. 1) capable of performing femtosecond UV kinetic absorption spectroscopy experiments[3,4]. The excitation pulses are obtained by amplification of femtosecond pulses in XeCl and KrF excimer gain modules. The probe pulses rely upon continuum generation in the gas phase. Details of this apparatus are presented in Ref. 4.

We first describe the principal result of an experiment probing the formation of Tl atoms resulting from femtosecond UV photolysis of TlCl and TlI vapors[3,4]. Figure 2a shows the appearance of time-resolved absorption spectra recorded in the vicinity of the 377.6-nm Tl $7S_{1/2} \leftarrow 6P_{1/2}$ resonance line, following the application of 160-fsec, 308-nm pump pulses to TlI vapor. The absorbance was observed to be linear with pump intensity. It is seen that roughly one picosecond elapses from the moment the atomic transition appears, to the point at which no further changes in the appearance of the atomic resonance line occur. From previously reported time-of-flight data, one can deduce that the asymptotic line shape is reached before the Tl-I separation has increased by 7.1Å. This represents only an upper limit, since acceleration effects have been neglected. Note the existence of spectral regions of apparent negative absorption in Fig. 2(a). For the 351.9-nm, $6D_{5/2} \leftarrow 6P_{3/2}$ line, recorded under the same circumstances, this transient feature of apparent gain was absent, but a transient asymmetry with a strong red wing was observed[4]. For TlCl, with 248-nm excitation, the 351.9-nm absorption develops a pronounced transient *blue* wing.

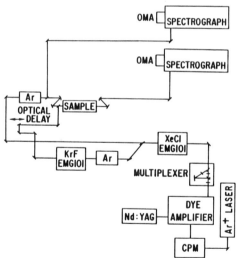

Fig. 1 Diagram of 160-fsec kinetic absorption spectrometer.

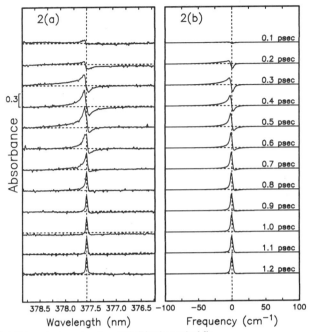

Fig. 2 (a) Tl transient spectra from TlI. (b) Theoretical fit.

We explain these spectra in terms of a model based upon the transient behavior of the radiating polarization induced by the probe continuum pulse as it interacts with the population of two-level atoms resulting from the photolysis pulse[3]. We assume that there are two contributions to the total output field emerging from the resonant vapor, the UV continuum probe pulse itself (assumed to occur at t = 0), and the field radiated by the polarization that the former induces in the vapor. The polarization is calculated from solutions to Schrödinger's equation for a two-level atom interacting with a continuum pulse. It is assumed that the photolytically produced atoms are formed in the ground state, with a time distribution that can be approximated by the pump pulse. Allowance is made for the existence of a continuous red (or blue) shift of the atomic transition frequency. The resultant expression for the time dependence of the induced polarization contains as a factor the phase integral

$$\Phi = \exp\left(\frac{-i}{\hbar} \int_0^t [\delta E_1(t') - \delta E_0(t')] dt' \right), \tag{1}$$

where the δE_i 's denote the deviations of the upper and lower atomic levels being probed from their asymptotic values. *This factor is the primary source of the asymmetry in the transient atomic absorption spectrum.*

An example of a sequence of calculated spectra based upon this model is shown in Fig. 2b. It is seen that a qualitative agreement exists between theory and experiment, with the particular feature of a transient spectral region of negative absorbance clearly displayed in the calculated spectra. To obtain a reasonable fit in the case of Fig. 2, it was necessary to assume that the Tl transition being probed is red-shifted from the corresponding isolated atom value by ~ 50 cm^{-1} at the instant the atoms are "born" (i.e. the instant the oscillator strength of the atomic transition being probed "turns on"), and that this shift then decays to zero in ~300 fsec. The simple theory outlined above also

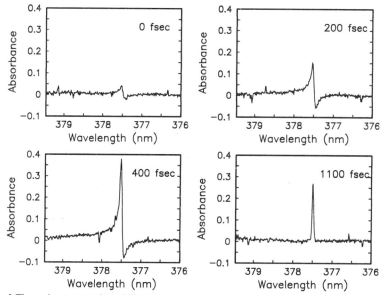

Fig. 3 Tl transient spectra from 308-nm photolysis of Tl-cyclopentadienyl.

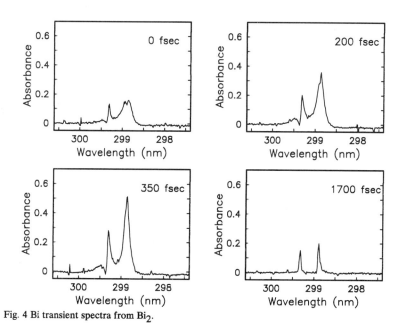

Fig. 4 Bi transient spectra from Bi_2.

shows that if a transition being probed is initially blue-shifted, the asymmetry will be reversed, that is, the apparent negative absorption will be on the low frequency side of the line, and a transient blue absorption wing will occur.

Additional examples of transient asymmetrical atomic absorption line profiles are found in other photodissociative systems, such as the organometallic compound Tl-cyclopentadienyl (Fig. 3). In the case of diatomic bismuth, photo-excited by application of UV femtosecond pulses into a dissociative continuum, both red-shifted and blue-shifted (Fig. 4) transient asymmetries appear. Diatomic bismuth is of special interest since very little energy is left over after absorption of a 308-nm photon to accelerate the separation of the Bi atoms from one another. A period of ~5 psec is observed to ensue after the pump pulse has been applied before the Bi atomic resonances suddenly appear (Fig. 5). A possible interpretation for this delay is that there is a critical distance the atoms must separate before the oscillator strength for the atomic transition "turns on". The transient enhancement of the atomic absorption coefficient that occurs at this point is seen to be a general feature in most of our spectra.

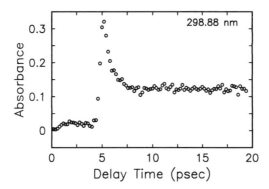

Fig. 5 Peak height vs time of transient Bi 298.9-nm absorption.

In summary, we have discussed experiments probing the "transition states" region resulting from photodissociation in which an atomic fragment is produced. Transient perturbations on atomic transitions during photodissociation have been shown to produce asymmetric line profiles whose interpretation requires use of a model for the transient polarization induced by the probe pulse.

We would like to acknowledge useful discussions with T.F. Heinz. This work was partially supported by the U.S. Army Research Office.

REFERENCES

1. M. Dantus, M.J. Rosker, and A.H. Zewail, J. Chem. Phys. <u>87</u> , 2395 (1987).

2. M.J. Rosker, T.S. Rose, and A.H. Zewail, Chem. Phys. Lett. <u>146</u> , 175 (1988).

3. J. Misewich, J.H. Glownia, J.E. Rothenberg, and P.P. Sorokin, Chem. Phys. Lett. <u>150</u> , 374 (1988).

4. J.H. Glownia, J. Misewich, and P.P. Sorokin, in: *The Supercontinuum Laser Source* , ed. by R.R. Alfano (Springer-Verlag, Berlin) 1989.

Optical dipoles in strong stochastic fields

A. Débarre, J.-C. Keller, J.-L. Le Gouët, and P. Tchénio
Laboratoire Aimé Cotton,
Centre National de la Recherche Scientifique II, Bâtiment 505
91405 Orsay Cédex, France

Resonant excitation of an optical transition results in coherent transients such as Photon Echoes, Optical Nutation, Optical Free Decay, Four Wave Mixing... These processes have been used for a long time to investigate relaxation mechanisms in vapors or in condensed materials. Much attention has been paid recently to some Time-Delayed Four Wave Mixing (TDFWM) schemes. They involve a step where the memory of excitation by a light pulse is stored in atomic dipoles which then interact with an other light pulse. It has been recognized that the time interval between those two interactions can be determined at the accuracy of the light source autocorrelation time τ_C . This ultimate time resolution is obtained when the two radiation fields are correlated. When excitation is achieved by broad band pulses, the time resolution τ_C, which coincides with the inverse spectral width of the pulse, may be much smaller than the pulse duration τ_L. Then the atomic dipole evolution may be monitored on a time scale much smaller than τ_L. One has taken advantage of this property to measure T_2-type lifetimes much shorter than τ_L [1]. In those experiments, the correlated pulses overlap. The decay process under investigation is thus monitored along a time interval which is not radiation free. The previously mentioned broadband experiments have been performed under the weak field condition which is fulfilled when the characteristic time of the field-induced atomic evolution T is much larger than the duration τ_L of the pulses. Then interaction with light does not interfere with the decay process under study. However, larger signal intensity would be attainable with stronger excitation pulses. It is therefore important to elucidate the strong field problem.

The study of coherent transients in the strong field regime also reveals new aspects of the fundamental problem of atomic interaction with a strong stochastic field. During the past ten years much effort has been devoted to determine time-dependent excitation probabilities and laser-induced resonance fluorescence spectra in the presence of a broadband field. The coherent transients differ from these processes through the nature of the relevant atomic quantities and through the specific character of excitation produced by time delayed correlated pulses.

The intensity of resonant multiphoton excitation and of resonance fluorescence signals is expressed in terms of the average population density which is transfered on an atomic level by radiative excitation. This quantity can be regarded as a single atom statistical moment. On the contrary, the coherent transient signal intensity is expressed in terms of the two atom statistical moments which represent the correlation between the atomic radiators.

The statistical properties of the driving field crucially determine the problem. In the frame of the works on resonance fluorescence, and on multiphoton excitation, the driving field is generally assumed to be a Markovian process. On the contrary, the formation of the TDFWM coherent transients involves a sequence of corelated time-delayed pulses which behave as a single composite Non Markov process. This property of the excitation process results in an unusual response of the atomic system, even in the situation when the characteristic time of evolution of the atom, T, remains much larger than the correlation time τ_c and the time delay between correlated pulses.

Both experimentally and theoretically, we have considered the model situation where the TDFWM transients result from the diffraction of a broadband probe pulse on a transient grating. This grating is engraved in the sample by two broadband mutually correlated pulses, time shifted with regard to each other by the delay t_{12}. In the system that we have chosen, the atomic relaxation mechanisms are negligible on the time scale of the TDFWM process. Besides the study is restricted to the case where only one pulse is strong. Two situations are considered. In the first one, the strong pulse is one of those which engrave the transient grating. In the second one, the grating which is engraved by weak pulses, is then probed by a strong pulse.

Experiments performed in the first situation raise the following question: how long is an atom which interacts with a given field at a given time able to keep memory of this interaction and to recognize the same instantaneous field at a later time? We have shown that

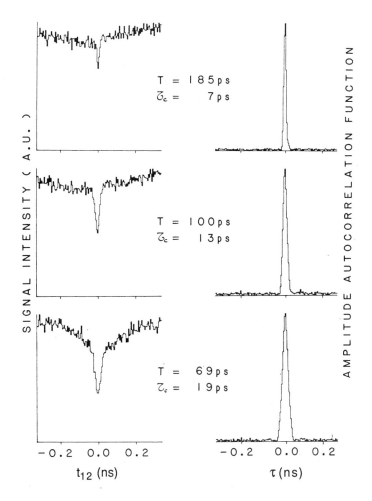

Figure 1- Experiment in Strontium vapor when one of the first two pulses is strong. Signal intensity and amplitude autocorrelation function of the strong pulse for three different values of τ_C.

under strong field conditions, the signal intensity is no longer dependent on the correlation between the first two excitation pulses as soon as $t_{12} > \tau_C$ [2,3] (Fig.1). In other words, the atoms do not keep memory of the correlation between excitation pulses over time intervals longer than τ_C, as soon as one of the driving fields is strong. The formation of the transient grating by two strong broadband correlated pulses has been recently explored theoretically [4]. Similar obliteration of the atomic memory over a time scale τ_C is then predicted.

Using a strong field to probe the grating engraved by two weak correlated pulses exhibits another property of a strong stochastic field, namely its ability to inhibit the inhomogeneous dephasing between optical dipoles. More specifically, let us consider two optical dipoles, detuned from each other by the frequency shift Δ. Then they do not build up any phase shift during the time interval t, provided Δ^2 Tt<<1, where T is the characteristic time of the strong field induced atomic evolution. A strong enough field is then able to lock the relative phase of all the dipoles to its initial value. Recording the signal intensity as a function of this initial value is a test of the phase locking property [5].

Connection with resonance fluorescence is achievable by detecting the signal amplitude instead of its intensity. Indeed the signal amplitude is expressed in terms of the first order statistical moments of atomic quantities like the resonance fluorescence signal. This quantity is detected through amplitude cross-correlation of the signal with the probe beam. Preliminary results have been obtained.

1- S. Asaka, H.Nakatsuka, M.Fujiwara and M. Matsuoka; Phys. Rev. A 29 2286 (1984). N. Morita and T. Yajima, Phys. Rev. A 30 2525 (1984).

2- P. Tchénio, A. Débarre, J.-C. Keller, J.-L. Le Gouët ;
Phys. Rev. Letters, 62 415 (1989).

3- P. Tchénio, A. Débarre, J.-C. Keller, J.-L. Le Gouët ;
Phys. Rev. A 39 1970 (1989).

4- V. Finkelstein and P. R. Berman; to be published.

5- P. Tchénio, A. Débarre, J.-C. Keller, J.-L. Le Gouët ;
Phys. Rev. A 38 5235 (1988).

Nonlinear Optical Diagnostics of Electronic Structure of Semiconductors and Metals.

S.A.Akhmanov, S.V.Govorkov, N.I.Koroteev, I.L.Shumay and V.V.Yakovlev

R.V.Khokhlov Nonlinear Optics Laboratory, Moscow State University, Moscow 119899, USSR.

Numerous experimental data have been accumulated in recent years on the use of nonlinear optical techniques for studing surfaces, interfaces and adsorbates [1]. These techniques are based on generation of optical harmonics (second and third), sum and difference frequencies in reflection. The main advantage of these methods besides nondestructive character of probing and high temporal resolution lies in the fact that nonlinear optical response governed by the corresponding nonlinear optical susceptibility tensor reflects the symmetry of the interface layers of solids or adsorbed molecules. The nonlinear optical response is formed essentially in an escape depth of the absorbing material unless surface contribution due to adsorbed molecules with high nonlinear optical susceptibility or reconstructed atomic layers of the atomically clean surface in high vacuum dominates [1]. Thus, by changing the wavelength of the probe radiation one can get information on the structure of the surface layer with a thickness of about α^{-1}, where α is the optical absorption coefficient. Naturally, the value of signal is small. However, using high-repetition rate picosecond laser sources the anisotropy of the second harmonic generation (SHG) in reflection from Si can be easily detected though this mate-rial belongs to m3m symmetry class and does not possess se-cond order nonlinearity in the bulk in electric dipole approximation due to the spatial symmetry selection rules for $\chi^{(2)}$. The source of SHG in this case is the quadrupole optical nonlinearity [1]. Under certain conditions electric dipole contribution to the second order nonlinearity can appear in the bulk of the near interface layer due to crystal symmetry distortion as a result of strong inhomogeneous mechanical deformation [2]. In what follows we discuss our recent results on mechanical stress detection in SiO_2/Si interfaces and real time control of its relaxation during thermal annealing.

Another interesting result under discussion is the observation of strong anisotropy of the optical SHG in reflection from Al monocrystal. This fact is theoretically interpreted in terms of anisotropy of 3s- and 3p-electrons nonlinear optical response in Al monocrystal.

1. Monitoring of stress in SiO_2/Si interface by optical SHG in reflection.

Inhomogeneous stress in the SiO_2/Si intrface manifests itself by the increase of optical SHG by a factor of about 20 and modification of its rotational dependence [2]. The origin of stress in this case is generally the mismatch of interatomic distances in Si and SiO_2 and also the mismatch of

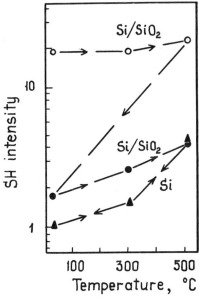

Fig.1

thermal expansion coefficients of the two substances. The latter is usually the dominant mechanism when thick (several tens of nm)oxide layer is grown by thermal oxidation and the sample is subsequently rapidly removed from the oven. Stress relaxation is known to occur when the sample is thermally annealed that is heated up to a temperature below but close to the temperature of thermal oxidation and gradually cooled down.

This process of stress relaxation can be easily controlled by optical SHG in reflection.The variation of the SH intensity during two thermal annealing cycles is shown in Fig.1. The arrows indicate the direction of temperature variation. The lower curve in Fig.1 corresponds to the Si wafer with native oxide layer. During the first annealing cycle SH intensity remains nearly constant. We attribute it to the interference of the two contributions, one being the decrease of stress as the result of annealing,and the second being the variation of the linear and nonlinear optical parameters of Si at the elevated temperatures as aresult of band structure modification. The latter mechanism tends to increase the SH intensity as can be seen from the lower curve in Fig.1 corresponding to the Si wafer with native oxide layer on its surface.

The decrease of the SH intensity as a result of thermal annealing is explained in terms of stress-induced electric dipole allowed contribution to the second order nonlinearity of Si: lattice distortions caused by stress relax during annealing procedure, so that bulk stress-induced electric dipole contribution to $\chi^{(2)}$ decrease leaving only bulk electric quadrupole contribution dominant in Si with native oxide. Thus, in the second annealing cycle, SH intensity dependence follows closely that of Si with native oxide.

2. The anisotropy of the Al monocrystal nonlinear optical response.

The first experiments on optical SHG in reflection from metal surfaces were made in the middle 60-s (see Ref.[3] and references therein). The results were theoretically interpreted in terms of hydrodynamical model of the free carrier plasma near the metal surface [4]. However the anisotropy of SHG in reflection from the monocrystals of Cu [5] and Ag [6] has been recently observed. The results on copper were interpreted in terms of contribution of strongly-coupled d-electrons to the nonlinear optical

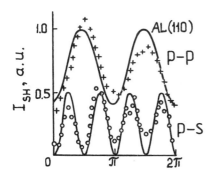

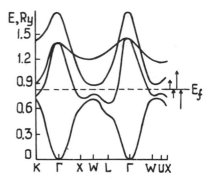

Fig.2 Rotational SH intensity
dependences for Al(110).

Fig.3 Calculated Al band
structure. Quanta energies
$\hbar\omega$ and $2\hbar\omega$ are shown by arrows.

response.

We have studied optical SHG in reflection from
mechanically polished chemically etched surface of Al
monocrystal. Note that optical response of Al is dominated by
s- and p-electrons.

We used passively mode-locked YAG:Nd^{3+} laser with a
pulsewidth $3 \cdot 10^{-11}$s as a pump sourse. Laser radiation
illuminated the sample at an angle of incidence $\theta = 45^\circ$. The
rotational dependence of the SH intensity was registered as
the sample was rotated against the surfacenormal.The observed
SH signal from Al had the intensity a factor of about 20
higher than that of Si monocrystal and had very strong
anisotropy. The SH rotational dependence for Al (110) surface
is shown in Fig.2. It is clearly seen that SH anisotropy
reflects the symmetry of the Al lattice (110) cut.

The linear optical properties of Al are isotropic as that
of any crystal with cubic symmetry. Crystalline lattice
influence on the nonlinear optical response of Al can be taken into account by summation of contributions of intra- and
interband electron transitions considering zone structure
anisotropy. We have numerically calculated nonlinear optical
quadrupole susceptibility of Al using quantum mechanical model [7]. The Al zone structure and electronic wave functions
have been calculated using pseudopotential approach. In a
single-particle approximation we have the following expression for the electron current density at 2ω in the case of interband transitions:

$$\vec{J}(2\omega) = -\frac{e^3}{2\,h^3\,c^2}\sum_{Pbb'b''}P(b,b'b'')\int(\vec{r}_{bb'})(\vec{\vec{rr}}_{b'b''})(\vec{r}_{b''b})x$$

$$(E_{b''}-E_{b'})(E_{b''}-E_b)(E_{b'}-E_b)\,R(\omega,\vec{k},\vec{q})\,d\vec{k}^3\,\vec{A}\,\vec{A}$$

Here E_b and $|b\rangle$ are the electron energy and the wave function
in a zone b with wavenumber $\vec{k}$. $\vec{A}$ is the vector potential of

the electromagnetic wave with the frequency ω and the wavevector $\vec{q}$ in the medium, $R(\omega, \vec{k}, \vec{q})$ is the resonant factor taking into account density of states, and $P(b, b', b'')$ is the permutation operator. Similar expression can be derived for intraband electron transitions.

The calculated Al zone structure is shown in Fig.3. Energy values at characteristic points of the Brillouin zone are in a good agrrement with other data. In calculating the Al optical quadrupole nonlinear susceptibility the integration over the Brillouin zone was made using tetrahedron procedure [8]. The calculated Al susceptibility tensor components in the case of pump wavelengh $\lambda = 1.06$ μm are

$$\chi^{(2)}_{1111}=7\cdot 10^{-13}\text{esu}; \quad \chi^{(2)}_{1122}=3\cdot 10^{-13}\text{esu}; \quad \chi^{(2)}_{1212}=\chi^{(2)}_{1221}=4\cdot 10^{-13}\text{esu}$$

Thus, the $\chi^{(2)}_Q$ tensor anisotropy is given by

$$\xi = \chi^{(2)}_{1111} - (2\,\chi^{(2)}_{1212} + \chi^{(2)}_{1122})$$

In isotropic material $\xi=0$, while in anisotropic material it governs the SH intensity rotational dependence. In our calculation $|\xi/\chi^{(2)}_{1212}| =1$ in excelent agreement with the experimental results (see Fig.2).

Thus, the anisotropy of the nonlinear optical response of Al monocrystal can be attributed to the interaction of the nearly free 3s- and 3p-electrons with the interatomic field of the crystalline lattice.Even in the absence of contribution of d-electrons nonlinear optical response of metal monocrystal contains information on crystal structure. This fact provides an opportunity of pico- and femtosecond diognostics of fast structural transformations in monocrystals of normal metals and superconductors using optical SHG in reflection.

Preliminary results unambiguously demonstrate the feasibility of using optical SHG for studying Y-Ba-Cu-O films structure. We believe nonlinear optical methods can be informative in analysing of the electronic structure dynamics of high temperature superconductors.

References:

1. H.W.K.Tom, T.F.Heinz, T.R.Shen: Phys.Rev.Lett.51, 1983 (1983); S.A.Akhmanov, V.I.Emel'yanov, N.I.Koroteev, V.N.Seminogov: Sov.Usp.Phys. 28, 1084-1154 (1985); G.T.Boyd, Y.R.Shen, T.W.Hansch: Opt.Lett.11, 97-99 (1986).
2. V.I.Emel'yanov, S.V.Govorkov, N.I.Koroteev, G.I.Petrov, I.L.Shumay, V.V.Yakovlev: JOSA B6 (1989) to be published.
3. N.Bloembergen, R.K.Chang, S.S.Jha, C.H.Lee:Phys.Rev.174, 813-827 (1968).
4. J.E.Sipe, V.C.Y.So, M.Fukui, G.I.Stegeman:Phys.Rev.B21, 3579-3586 (1986).
5. H.W.K.Tom, G.D.Aumiller:Phys.Rev.B33, 8818-8821 (1986).
6. V.L.Shannon, D.A.Koos, G.L.Richmond:Appl.Opt.26, 3579-3583 (1987).
7. S.S.Jha, C.S.Warke:Phys.Rev. 153, 751-759 (1967).
8. D.J.Moss, J.E.Sipe, H.M.van Driel:Phys.Rev.B36 1153-1161 (1987).

Desorption of Molecular Adsorbates from a Metal Surface by Subpicosecond Laser Pulses: NO/Pd(111)

J. A. Prybyla,* T. F. Heinz, J. A. Misewich, and M. M. T. Loy

IBM Research Division, T.J. Watson Research Center,
Yorktown Heights, NY 10598 USA

Introduction

The process of intact desorption of molecules from a surface is one of the most fundamental types of molecule/surface interactions. As a consequence, considerable effort has been devoted to understanding the nature of both thermal and nonthermal mechanisms for desorption.[1] Laser radiation has been widely applied as a means of initiating desorption. In essentially all of these investigations, the laser radiation has been of nanosecond duration or longer. From the point of view of examining the basic mechanisms of energy transfer, however, pulsed radiation of pico- or femtosecond duration is particularly attractive, since typical electronic and vibrational relaxation processes at surfaces occur on this time scale. Furthermore, excitation by ultrashort pulses may give rise to new mechanisms for desorption associated with highly non-equilibrium conditions.

In our studies, we have examined a representative chemisorbed system consisting of nitric oxide (NO) molecules adsorbed on a Pd(111) surface. We have previously investigated the nature of simple thermal desorption by characterizing the energy distribution of NO molecules desorbed upon heating the substrate with laser pulses of nanosecond duration.[2] Here we report on some recent studies of desorption induced by 200-fsec pulses of visible light for the NO/Pd(111) system. We describe briefly the salient features of the experiment and present results on the measured yield and the vibrational energy distribution of the desorbed NO molecules. These findings imply the existence of an efficient nonthermal desorption mechanism for subpicosecond optical excitation.

Experiment

The measurements of laser-induced desorption were performed in the ultrahigh vacuum system depicted in Fig. 1. Low-energy electron diffraction and Auger spectroscopy were employed to verify the cleanliness and order of the Pd(111) surface. A saturation coverage of adsorbed NO molecules was maintained by dosing the sample with a pulsed molecular beam. In the present investigation, the sample was held at a base temperature of 295 K. The 200-fsec (FWHM) laser pulses were supplied by a colliding-pulse modelocked dye laser operating at 616 nm. In order to induce desorption, these pulses were passed through a Nd:YAG-pumped dye amplifier. In this manner, 200-fsec pulses of mJ energy were produced at the 10 Hz repetition rate of the experiment. Detection of the desorbed NO molecules was accomplished by $(1+1)$ photon resonance-enhanced multiphoton ionization (REMPI) through the $A^2\Sigma^+$ intermediate state. The experimental apparatus for producing the required ultraviolet radiation, indicated schematically in Fig. 1, is described in more detail in Ref. 3. From the REMPI spectra, a complete description of the internal state distribution (rotational, vibrational, and spin-orbit level) of the desorbed NO molecules could be obtained. By varying the delay time between the laser pulse inducing desorption and the ionizing laser, we could also determine (state-specific) translational energy distributions in the form of time-of-flight spectra.

---

* Present address: AT&T Bell Laboratories, Holmdel, NJ 07733

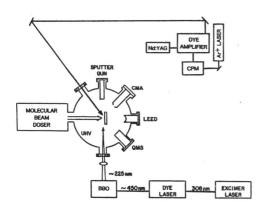

Fig. 1. Experimental apparatus. The ultrahigh vacuum chamber is equipped with low-energy electron diffraction (LEED), a cylindrical mirror analyzer (CMA) for Auger spectroscopy, and a quadrupole mass spectrometer (QMS). The 200-fsec pulses for desorption, produced by an amplified colliding-pulse modelocked dye laser, are directed onto the Pd(111) surface as shown. The NO molecules are detected above the sample with the frequency-doubled output of a nanosecond dye laser.

Since the NO molecules were detected at some distance from the surface, the issue of possible gas-phase effects had to be considered. For this purpose, measurements of the yield of both NO molecules and electrons were performed. The yield of desorbed NO molecules was established by calibrating the sensitivity of the REMPI detection scheme with the chamber back-filled with NO at a known static pressure. In all cases, the amount of NO desorbed during each laser pulse was restricted to $<<$ 0.01 ML. (1 ML corresponds to the surface density of Pd atoms, *viz.*, 1.5×10^{15} cm^2 .) The density of electrons emitted was determined from the charge transfer when the sample was irradiated in the presence of a strong static electric field. It was established that the yield of electrons was several orders of magnitude less than that of desorbed NO molecules. Under these conditions, neither NO-NO collisions nor NO-electron collisions in the gas phase are expected to perturb the nascent distributions.

Results

Desorption of NO molecules could be observed for absorbed laser fluences as low as 1 mJ/cm^2. For an absorbed fluence of 2.4 mJ/cm^2 , for example, the yield of desorbed NO molecules was roughly 10^{-3} ML. One may describe the process in terms of a quantum efficiency for NO desorption per absorbed photon. For an absorbed fluence of 2.4 mJ/cm^2 the quantum efficiency is found to exceed 10^{-4}. In fact, the NO yield exhibited a strong dependence on the fluence of the desorbing pulse, and still higher effective quantum efficiencies for desorption were observed under more intense laser excitation. We can compare the desorption yield with what is expected for a purely thermal desorption mechanism. If the 200-fsec laser pulse is considered simply to heat the Pd substrate under complete equilibrium conditions, the temperature rise can be computed by means of a (numerical) solution of the thermal diffusion equation. The maximum surface temperature is found to be 820 K and is seen to decay in a few picoseconds. On the other hand, previous investigations with nanosecond laser pulses indicated that an observable desorption yield could be obtained only when the surface temperature exceeded 1000 K for several nanoseconds.[2] This comparison shows immediately that a simple equilibrium thermal process cannot account for the desorption induced by the 200-fsec laser pulses. Indeed, if we apply the kinetic parameters for thermal desorption inferred from nanosecond measurements to the calculated equilibrium temperature profile for the 200-fsec laser pulses, the predicted desorption yield falls short of the experimental result by more than 3 orders of magnitude.

To gain more insight into the nature of the desorption process for ultrashort laser pulses, we have also examined the internal energy distributions of the desorbed NO molecules. One of the most striking features of the desorbed NO molecules is their high level of vibrational excitation. Integrating over other internal degrees of freedom, we found a ratio of molecules in the first vibrational state to the ground vibrational state of $N(v = 1)/N(v = 0) \sim 0.4$. For a thermal distribution of vibrational excitation, this would require a vibrational temperature in excess of 2500 K. In contrast to this observation, the vibrational temperature of NO molecules desorbed by nanosecond laser radiation was found to be similar to that of the surface at the time of desorption.[2]

Discussion

The results presented above offer strong evidence for the presence of a nonthermal desorption mechanism for NO/Pd(111) under excitation by visible subpicosecond laser pulses. These findings can be compared with recent reports of nonthermal desorption of NO from metal (and oxygen-covered metal) surfaces induced by nanosecond laser pulses. [4,5] While complete details of the mechanisms responsible for the desorption under nanosecond laser excitation have yet to emerge, it clear that non-equilibrium electronic excitation must be present. That the influence of non-equilibrium electronic excitation in the metal and metal/molecule complex should be at least as significant on a *subpicosecond* time scale is easy to understand: on this time scale, the energy initially deposited in electronic excitation has only begun to equilibrate with phonons.[6] In contrast to the case of irradiating a metal with nanosecond light pulses, the average energy of the electrons can be significantly greater than that of the lattice modes. This high density of hot electrons may account for the enhanced quantum efficiency of desorption observed in our measurement of NO/Pd(111) with respect to that for nonthermal desorption of NO/Pt(111) by nanosecond laser pulses.[5]

In order to develop a more quantitative interpretation of the experimental results, we have analyzed a model for excitation in the metal substrate in which the electrons are assumed to be equilibrated among themselves, but to have a finite coupling constant to lattice modes.[6] Under the relevant experimental conditions, we predict electronic temperatures in excess of 2000 K. Such a high level of electronic excitation can then lead to NO desorption through electronic states of the adsorbed molecules exhibiting repulsive molecule/surface potentials.[1] In this picture, the elevated vibrational temperature observed in the desorbed NO molecules may arise either from coupling of the molecular vibration to the hot electrons (and holes) in the substrate or from a sudden change in N-O bond length occurring upon excitation of an adsorbed molecule to a repulsive electronic state.

Partial support for this work by the Office of Naval Research is acknowledged.

References

1. See, for example, J. A. Barker and D. J. Auerbach, Surface Sci. Rep. 4, 1 (1985) and *Desorption Induced by Electronic Transitions- DIET III*, edited by M. L. Knotek and R. H. Stulen (Springer-Verlag, Berlin, 1987).

2. J. A. Prybyla, T. F. Heinz, J. A. Misewich, and M. M. T. Loy, to be published.

3. J. Misewich, P. A. Roland, and M. M. T. Loy, Surface Sci. 171, 483 (1986).

4. F. Budde, A. V. Hamza, P. M. Ferm, G. Ertl, D. Weide, P. Andresen, and H.-J. Freund, Phys. Rev. Lett. 60, 1518 (1988).

5. S. A. Buntin, L. J. Richter, R. R. Cavanagh, and D. S. King, Phys. Rev. Lett. 61, 1321 (1988).

6. H. E. Elsayed-Ali, T. B. Norris, M. A. Pessor, and G. A. Mourou, Phys. Rev. Lett. 58, 1212 (1987); R. W. Schoenlein, W. Z. Lin, J. G. Fujimoto, and G. L. Eesley, Phys. Rev. Lett. 58, 1680 (1987).

Part X

Fundamental Measurements

Precision Laser Spectroscopy of Positronium – Recent Progress

K. Danzmann, M.S. Fee, and Steven Chu
Department of Physics, Stanford University,
Stanford CA 94305

1. Introduction

Positronium, the bound state of electron and positron, is one of the most fundamental purely leptonic two-body systems available. It provides an ideal testing ground for bound-state quantum electrodynamics. Precision spectroscopy of this system seems particularly interesting in light of recently discovered discrepancies between theory and experiment for the 1s-2s energy difference,[1] the ground state lifetime,[2] the pair-creation cross section at threshold,[3] and the recent discovery of correlated electron-positron lines in heavy ion collisions.[4] It has been hypothesized that these results might be explained by the existence of a new light neutral particle, coupling to electrons and positrons.[4] The existence of such a particle would naturally shift the energy levels of positronium.[5] The most stringent limits on such a shift are derived from measurements of the anomalous magnetic moment of the electron (g-2).[6] They are on the order of 100 kHz to 1 MHz, depending on spin and parity of the presumed particle. Our experiment is aiming at a final precision of 10 kHz.

The 1s-2s two-photon transition is an especially promising candidate for precision spectroscopy because of its narrow linewidth (1 MHz) and large QED effects. This two-photon transition requires high laser intensities for its excitation. Because positronium has such a low mass, it moves with high velocities even at room temperature, and large beam diameters are required to suppress the transit time broadening to acceptable levels. This is a natural requirement for pulsed light, and the development of narrow linewidth pulsed lasers is necessary. As an alternative the use of cw laser light requires a build-up cavity to reach high intensities, creation of cryogenic positronium, and laser cooling of the positronium. We are making progress in both of the above areas, and hope to achieve a final precision of a few parts in 10^{11} for the 1s-2s transition.

2. Pulsed Spectroscopy

Precision measurements using pulsed light have always been plagued by rapid phase variations of the electric field (usually called phase chirp) caused by refractive index changes in the pulsed amplifier . This effect shows up as a broadening of the frequency spectrum beyond the Fourier-transform limit when the pulsed light is characterized in the traditional way with e.g. a Fabry-Perot spectrum analyzer. However, such measurements yield only the spectral intensity distribution and the phase information is lost. The non-linear response of an atom (like a two photon transition), though, is sensitive to the phase of the electric field.

Through computer simulation we have shown that the atomic response is not centered at the center of gravity of the spectral intensity distribution. Depending on the details of the phase chirp, the atomic response will be

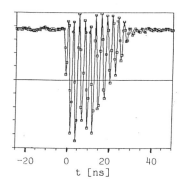

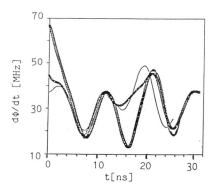

Fig. 1a: Beat-note between
pulsed and cw laser

Fig. 1b: Phase chirp of three
different laser pulses.

shifted by several MHz for typical two photon transitions and pulsed lasers. Filtering the pulsed light in a narrowband cavity reduces the linewidth, but the apparent shift remains.

To overcome this problem, we have developed a technique for complete characterization of the time-dependent phase and amplitude of the electric field. We heterodyne part of the pulsed light with light from a cw-laser that is frequency shifted by several hundred MHz and record a beat note. As an example we show in Fig.1a a beat note for the output of an injection seeded, excimer pumped, blue dye laser amplifier. This beat note contains information about the ASE background, the amplitude envelope, and the phase of the laser pulse. The low frequency Fourier components of the beat contain the ASE and amplitude information, whereas the high frequency part of the Fourier-transform is a convolution of amplitude and phase information. We separate the high frequency part, transform it back to the time domain, and square it (homodyning). After another Fourier-transformation, this procedure deconvolutes the amplitude and phase information and reconstructs the amplitude information in the low frequency part of the Fourier-transform. By application of a digital filtering algorithm, we can recover the time dependence of both amplitude and phase from a single beat note recording. In Fig.1b we show the phase chirp for three typical laser shots. It should be emphasized that the spectral intensity distributions of these three shots would appear nearly identical.

As a first application, we were able to correct the two most recent pulsed measurement of the 1s-2s energy difference in hydrogen and positronium. Using a computer simulation to approximate the known properties of their pulsed light, we calculate a correction that brings the pulsed[7] and cw[8] measurements on hydrogen into agreement. Our result for positronium[9] is now within 17 MHz (1.7 standard deviations) from the theoretical prediction.[10]

We have developed a pulsed tunable light source with 5 MHz linewidth and complete characterization of the time-dependent phase and amplitude of the electric field. The output of a Coherent 699-21 actively stabilized cw ring dye laser is amplified in a four-stage pulsed amplifier system transversely pumped by a Lambda-Physik EMG 202 MSC excimer laser. The 70-mJ, 24-nsec laser pulses have a bandwidth of 30 MHz and are sent through a confocal filter cavity to produce nearly Fourier-transform limited 5-mJ pulses of 5 MHz bandwidth. Part of the cw light is split off and passed through a stripline electro-optic modulator which creates sidebands with a variable spacing of 50 MHz to 1000 MHz. One of the sidebands is filtered out and frequency locked to a tellurium reference line. The error signal is generated via frequency-modulation spectroscopy and controls the fundamental frequency of the cw ring laser. A variation of the RF frequency driving the electro-optic modulator then effectively scans the cw fundamental frequency with an offset from the tellurium reference line that can immediately be read off from the RF signal generator.

3. CW – Spectroscopy

Cryogenic positronium at 50 K or below is now available from advanced positronium sources.[11] We are working on a 1s-2s measurement on such cold positronium using cw laser light. We have narrowed the linewidth of a commercial blue ring dye laser to the kHz level by insertion of an electro-optic phase modulator in the ring.[12] With 2 W of pump power from a krypton ion laser we achieve an output power of 200 mW. This light will be coupled into a long very high finesse build-up cavity to achieve internal circulating powers of kilowatts. The transmission frequency of the cavity is offset-locked to a frequency standard, presently a molecular tellurium line.

The intensity of 100 kW/cm^2 in a beam waist of a few mm diameter should be high enough to excite several percent of the positronium into the 2s state. The excited positronium atoms will be photoionized by a delayed laser pulse and the resulting positrons detected. The precision in this arrangement will be limited by transit time broadening. Because of its small mass, positronium at a temperature of 50 K moves with a most probable speed of 28 km/sec. In a laser beam of 3 mm diameter this leads to a transit time broadening of 6 MHz in the blue. We are planning to increase the beam diameter by placing the positronium source near one of the mirrors of a near concentric build-up cavity, but the signal to noise ratio at the smaller resulting intensity will set natural limits. The second order Doppler broadening at 50 K amounts to 2.5 MHz, but this can be reduced by delaying the photoionization pulse and selectively observing slow atoms

4. Laser Cooling of Positronium

To take advantage of the 1 MHz linewidth of the 1s-2s transition, we will laser cool the positronium to reduce all broadening mechanisms below the natural linewidth. A convenient cooling transition is the 1s-2p transition at 243 nm, which has a transition probability of 3x10^8 s^{-1} and a natural linewidth of 50 MHz. Laser cooling of positronium is quite different from normal laser cooling. Because positronium has such a small mass, it takes only 20 photons to stop atoms from an initial temperature of 50 K. This requires a cooling time of 100 nsec, which is fortunately shorter than the ground state annihilation lifetime of 140 nsec. But the one photon recoil momentum corresponds to a Doppler shift of 6 GHz, which shifts the atom completely out of resonance with a monochromatic cooling laser. The Doppler width at 50 K amounts to 200 GHz, too large to sweep a monochromatic laser beam in 100 nsec.

A solution is to use "white-light" cooling based on a modification of a scheme proposed by Hoffnagle.[13] We intend to create white-light molasses tuned to the entire red side of the Doppler profile with counterpropagating broadband beams, in order to cool the positronium to about 100 mK. The white light is made by filtering out a spectral piece of ASE from an excimer pumped blue dye, which then is amplified and frequency doubled. We estimate that 7 mJ

330

of broadband laser energy are required to cover the entire red Doppler wing at ten times the saturation intensity for a cooling time of 100 nsec.

At a final temperature of 100 mK the excitation probability in the 3 mm diameter internal laser beam of the build-up cavity approaches 100 %. The second order Doppler effect is 3.5 kHz, and the transit time broadening is reduced to 200 kHz. Under those conditions, the line is predominantly lifetime broadened to 1 MHz, and one could hope to measure the center to one percent of the linewidth. It will be an intriguing test of our current understanding of bound systems to perform a direct comparison of this line to, e.g., the nearby 2s-4s line in deuterium.

We would like to thank R.G. DeVoe and A.P. Mills for helpful discussions.

This work was supported in part by a grant from the National Science Foundation.

References

1.   S. Chu, A.P. Mills, and J.L. Hall, Phys.Rev.Lett. 52, 1689 (1984). D.H. McIntyre and T.W. Hänsch, Phys.Rev. A 34, 4504 (1986).

2.   C.I. Westbrook et al., Phys.Rev.Lett. 58, 1328 (1987).

3.   F.T. Avignone et al., Phys.Rev. A 32, 2622 (1985).

4.   T. Cowan et al., Phys.Rev.Lett. 56, 444 (1986).

5.   A. Schäfer et al., Mod.Phys.Lett. 1, 1 (1986).

6.   J. Reinhardt et al., Phys.Rev. C33, 194 (1986).

7.   E.A. Hildum et al., Phys.Rev.Lett. 56, 576 (1986).

8.   R.G. Beausoleil et al., Phys. Rev. A35, 4878 (1987).

9.   K. Danzmann, M.S.Fee, and S.Chu, Phys.Rev. A39, (1989) in press.

10.   T. Fulton, Phys.Rev. A 26, 1794 (1982).

11.   A.P. Mills, private communication.

12.   R.G. DeVoe and R.G. Brewer, Phys.Rev.Lett. 50, 1269 (1983).

13.   J. Hoffnagle, Opt.Lett. 13, 102 (1988).

331

Proposed Search for States of Forbidden Permutation Symmetry

D.E. Kelleher, J.D. Gillaspy, and K. Deilamian*
Center for Atomic, Molecular, and Optical Physics
National Institute of Standards and Technology
Gaithersburg, MD 20899, USA

All known systems of identical particles are well described by wavefunctions which are either symmetric or antisymmetric under particle interchange. The fundamental status of these symmetries is codified in the symmetrization postulate. More restrictive still is the spin statistics theorem, which requires that half-integer spin particles are antisymmetric; such states satisfy the Pauli exclusion principle (PEP). Early on, however, Pauli pointed out that quantum mechanics does not constrain physics to two permutation symmetries[1]. While the above symmetry principles are consistent with a broad range of physical experience, precision tests are practically nonexistent.

In 1953 Green[2] articulated the additional symmetries allowed in a generalized field theory in which the field operators obey trilinear commutation relations of the form $[a_i,[a_j^\dagger,a_k]]=\delta_{ij}a_k$. In such cases occupation numbers greater than one are possible for particles with half-integer spin. If these "parastatistics" exist at all, they would be on an equal footing with normal Fermi and Bose Statistics. The existence of unmitigated parastatistics thus appears incompatible with our physical world. Recently a number of theorists have proposed models[3,4,5] in which violations of PEP could occur on a small enough scale to be consistent with existing observations. However, general treatments[6,7] have shown that many such models suffer from the fact that they predict multiparticle states with negative probability. The general treatments do, however, reconfirm the consistency of Green's parastatistics with field theory. It has also been suggested[8] that if the dimensionality of space is greater than three, then permutation symmetries which are perfect in n-dimensions may be broken in three dimensions.

The axiomatic indistinguishability of identical particles requires that the Hamiltonian be symmetric to particle interchange. This precludes permutation symmetry-breaking terms in the Hamiltonian, and thus limits the type of observables one can utilize to search for a forbidden symmetry: There can be no coherent superposition of states of different permutation symmetry, and no interference or transition between states of different permutation symmetry. This is in contrast to parity violation, for example, which is attributable to the reflection-symmetry-breaking weak interaction term in the Hamiltonian. Other types of permutation symmetries which are neither symmetric nor antisymmetric also satisfy indistinguishability. Searches for violation involve looking for rare states of forbidden symmetry, rather than a small coherent admixture to states of normal symmetry. In two-particle systems indistinguishability allows only the two normal permutation symmetries.

*Permanent Address: Department of Physics, University of Wisconsin, Madison, WI 53706 (Bitnet: KAVOOS@WISCPHEN)

A number of experimental tests of PEP and symmetrization have been recently performed or are in progress[9,10,11]. We have chosen to begin by looking for permutation-symmetric states in helium. Drake[12] has calculated the energies that such states would have if they exist. Each symmetric state of spin zero (one) is nearly degenerate with the normal antisymmetric helium state having the identical spatial wave function but opposite spin, i.e. one (zero), (see Fig. 1). The only differences in energy are due to different fine structure and Breit terms. This means that spectral lines from forbidden symmetry states of low relative abundance would be swamped by the Doppler/instrument broadening of the allowed line emission. The method of photon-burst can be used in an atomic beam to enhance the discrimination from the allowed helium states by about ten orders of magnitude.

The photon-burst method involves repeatedly exciting a transition. The transition must be within an effectively two-level system, and the lower level must be long lived. As a resonant species crosses the laser beam, it can be excited and deexcited many times, giving off a burst of photons in fluorescence. The method is useful when single-atom sensitivity and high discrimination are required. If a rare-species line lies in the wing of an interfering-species line at a position where the absorption amplitude is down by a factor ϵ from the peak value, then a burst in which n photons are detected will allow discrimination of about one part in ϵ^{-n}. In our case $\epsilon=10^{-5}$, and our sensitivity is flux-limited for $n>3$. The metastable atomic beam, which has a collimated flux of 10^8 metastables per second, is generated via a differentially pumped discharge. The method requires a high photon-collection efficiency while maintaining very low scattered laser light, because the detected fluorescence has the same frequency as the laser light.

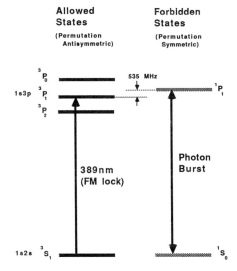

Partial Level Diagram of Helium

Fig 1. indicates the frequency-locking transition in normal helium, and the transition to be searched for using the photon-burst technique. We will FM lock the dye laser to the normal 2s $^3S_1$-3p $^3P_1$ transition in a helium discharge. We will then shift the laser frequency via an acoustic-optic modulator by 535 MHz to be resonant with the calculated transition energy of the symmetry-forbidden states (estimated uncertainty 20 MHz).

333

The laser beam will transversely excite several centimeters of the atomic beam. We plan to have two photon-burst detectors, one downstream from the other and separated by an aperture. The delay time between serial bursts will be measured, and coincidence will be registered if the delay time falls within the range expected from the distribution of longitudinal velocities. This allows discrimination against spurious scattering events and against normal atoms velocity-shifted into resonance with the laser. We will test the system using hyperfine components of ^{3}He.

A complication arises here due to the relatively large momentum of the excitation photons and low mass of the helium atoms. The atom recoil from each absorbed photon produces a Doppler shift of about 500 kHz, which is half the natural linewidth. Even in a standing wave configuration, the atom will eventually be knocked out of resonance with the laser. To overcome this we can detune the laser slightly to the red and/or broaden the laser bandwidth in a controlled manner using the acousto-optic modulator.

After the search for permutation symmetric states in helium, we hope to be able to look for the PEP-forbidden 1s^3 ground state of He$^-$. He$^-$ would be generated by double charge exchange of He$^+$. The stable ground state could be detected in a mass-sensitive trap.

References

1. W. Pauli, **General Principles of Quantum Mechanics**, (translated by P Achuthan and K. Venkatesan) Springer-Verlag, Berlin, 1980; Nobel Lecture, 1946 (Stockholm, 1948) p. 131.

2. H.S. Green, Phys. Rev. **90**, 279 (1953).

3. A.Yu. Ignat'ev and V.A. Kuz'min, Sov.J.Nucl.Phys. **46**, 444 (1987).

4. O.W. Greenberg and R.N. Mohapatra, Phys. Rev. Lett. **59**, 2507 (1987).

5. L.B. Okun, JETP Lett. **46**, 529 (1987).

6. A.B. Govorkov, Sov. J. Part. Nucl. **14**, 520 (1983).

7. S. Doplicher, R. Haag, and J. Roberts, Comm. Math Phys. **23**, 199 (1971); **35**, 49 (1974).

8. O.W. Greenberg and R.N. Mohapatra, Phys. Rev. Lett. **62**, 712 (1989).

9. E. Ramberg and G.A. Snow, Univ. of Maryland Rept. (1988).

10. V.M Novikov, A.A. Pomansky and E.H. Nolte, Moriond Workshop, Jan. 1989.

11. R.C. Hilborn, Research Proposal (1988).

12. G.W.F. Drake, Phys. Rev. **A39**, 897 (1989).

QED–Tests by Laser Spectroscopy

-- 2S Lamb shift measurement in hydrogen-like phosphorus and sulfur --

D. Müller, J. Gassen, D. Budelsky, L. Kremer, H.-J. Pross, F. Scheuer, P. von Brentano

Institut für Kernphysik, 5000 Köln 41, FRG

J. C. Sens, A. Pape

Centre de Recherches Nucleaires, 67037 Strasbourg, France

Regarding the Lamb shift experiments there are new attempts to measure the (1S - 2S) transition energy in hydrogen by two photon spectroscopy, which have bright perspectives /1/. But at present the best results are still the measurements on the $(2S_{1/2}\text{-}2P_{1/2})$ transition in hydrogen by Lundeen, Pipkin /2/ and Pal`chikov et. al. /3/. The test accuracy of these experiments is limited by the uncertainty in the proton structure, however. An important additional information is to measure the Lamb shift at high Z, where significant contributions of higher order terms, like $(Z\alpha)^n$, n = 6, 7, are tested.

We have measured the 2S-Lamb shift in $^{31}P^{14+}$ and $^{32}S^{15+}$ with a flashlamp pumped dye laser. The experiments were done at the MP tandem accelerator of the CRN-Institute in Strasbourg. From a beam of 100 MeV P^{8+} and 130 MeV S^{9+} hydrogen-like ions were prepared in the metastable $2S_{1/2}$ state with a beam-foil-technique. The ion beam intersects with the photon beam of the high power Laser. Thereby a transition from the $2S_{1/2}$ to the $2P_{3/2}$ state is induced, and the $2P_{3/2}$ state decays promptly to the $1S_{1/2}$ ground state. Counting the emitted x-rays as a function of the laser wavelength leads to the $(2S_{1/2}\text{-}2P_{3/2})$ splitting. By comparison with the fine structure $(2P_{1/2}\text{-}2P_{3/2})$ the 2S Lamb shift $(2S_{1/2}\text{-}2P_{1/2})$ is deduced. Experimental details can be found in ref. /4-7/. The latest results are:

Phosphorus $E_{LS}^{exp} (2S_{1/2}\text{-}2P_{1/2}) = 0.08349(12)\,eV$

Sulfur $E_{LS}^{exp} (2S_{1/2}\text{-}2P_{1/2}) = 0.10449(26)\,eV$ /7/

In fig. 1 these results are compared with other experiments and the theoretical calculations by Mohr /8/ and Johnson/Soff /9/. The open circles are indiviual runs corresponding to about 50000 laser pulses each. The black squares give the summarized results of our work. All experimental values are below the theory. The difference is 2.1 (1.7) experimental standard deviations for phosphorus (sulfur).

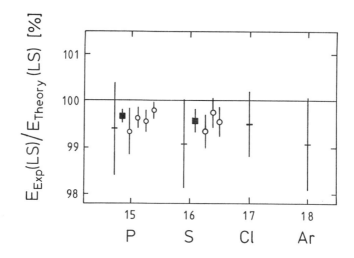

Fig. 1: Comparison of experiment and theory, references are given in ref. /4/.

The accuracy of the present laser resonance experiment is limited by statistical errors and by systematical errors arising from the large width Γ of the resonance ($\Gamma/E_{LS} \approx 1/4$) and the high velocity of the fast ion beam of hydrogen-like ions ($v_{Ion} \approx 0.1c$). Therefore a new laserspectroscopic method (Doppler shift Null experiment, DONUT) is under development at the Institut für Kernphysik, Köln /10/. The principle is illustrated in fig. 2.

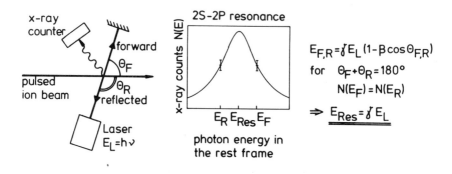

$$E_{F,R} = \gamma E_L (1 - \beta \cos\Theta_{F,R})$$
$$\text{for} \quad \Theta_F + \Theta_R = 180°$$
$$N(E_F) = N(E_R)$$
$$\Rightarrow \underline{E_{Res} = \gamma E_L}$$

Fig. 2: DONUT method, $\gamma = 1/\sqrt{1-\beta^2}$, $\beta = v_{Ion}/c$

We measure with a bidirectional fixed frequency laser beam using the two angles Θ_R and Θ_F relative to the ion beam, alternatively the count rate $N(E)$ at the doppler shifted energies $E_R(\Theta_R)$ and $E_F(\Theta_F)$. By chosing an appropriate laser frequency for which $N(E_R) = N(E_F)$ we find the centroid energy $E_{Res} = 1/2 \, (E_R + E_F)$. If the condition $\Theta_R + \Theta_F = \pi = 180^0$ is fulfilled to a high accuracy (10^{-5} rad), E_{Res} is independent of Θ_R and Θ_F and the resonance energy is given directly by $E_{Res} = \gamma \, E_L$.

The DONUT method can be used in principle for the following atoms and high power lasers: (P, S - dye laser, Cl - CO_2 laser, Ti - DF laser, As - Nd:YAG laser). Besides the reduction of the systematic errors, the statistics has to be increased. This can be done with pulsed high current ion beams which are matched to the laser pulse. Such beams are available from pulsed ion sources at electrostatic tandem accelerators and will be available from the heavy ion storage ring SIS/ESR at GSI, Darmstadt. A reduction of the present error by a factor 10-100 seems possible, in principle.

Gratefully acknowledged is the financial support by the Deutsche Forschungs-gemeinschaft (DFG) and the help from H.-D. Sträter and D. Platte.

References

1) Proceedings of the Symposium 'The Hydrogen Atom', ed. by G. F. Bassani, M. Inguscio, T. W. Hänsch, Springer-Verlag, Berlin Heidelberg, 1989.

2) S. R. Lundeen and F. M. Pipkin, Phys. Rev. Lett. 46 (1981) 232, Meterologia 22 (1986) 9.

3) V. G. Pal'chikov, Yu. L. Sokolov and V. P. Yakovlev, JETP Lett. 38 (1983) 418.

4) D. Müller, J. Gassen, L. Kremer, H.-J. Pross, F. Scheuer, H.-D. Sträter, P. von Brentano, A. Pape and J. C. Sens, Europhys. Lett. 5 (1988) 503.

5) P. von Brentano, D. Müller, J. Gassen, Proceedings of the Intl. Conf. 'Spectroscopy and Collisions of Few Electron Ions', Bucharest, Romania, 1988, to be published by World Scientific.

6) D. Müller, J. Gassen, F. Scheuer, L. Kremer, H.-J. Pross, D. Budelsky, P. von Brentano, J. C. Sens and A. Pape, in Atomic Physics 11 (edited by S. Haroche, J. C. Gay and G. Grynberg), World Scientific, Singapore, 1989.

7) A. P. Georgiadis, D. Müller, H.-D. Sträter, J. Gassen, P. von Brentano, J. C. Sens and A. Pape, Phys. Lett. A115 (1986) 108.

8) P. J. Mohr, At. Data and Nucl. Data Tables 29 (1983) 453.

9) W. R. Johnson and G. Soff, At. Data and Nucl. Data Tables 33 (1985) 405.

10) P. von Brentano, J. Gassen, D. Müller, H.-D. Sträter, J. C. Sens and A. Pape, Proceedings of the 'Workshop on Experiments and Experimental Facilities at SIS/ESR', GSI-Report 87-7, Darmstadt, FRG, (1987).

TOWARDS A FIELD-FREE MEASUREMENT OF
PARITY VIOLATION IN ATOMIC CESIUM

A.Weis,N.Schlumpf,V.L.Telegdi,D.Zevgolis and L.Zhao
Institute for High Energy Physics, ETH-Zurich, Switzerland

J.Hoffnagle
IBM Research Center Almaden, San Jose, U.S.A.

For several years we have been pursuing (in contrast to others) a <u>field-free</u> measurement of parity violation in the 6S → 7S M1 transition in Cs, exploiting the circular dichroism induced by the $E1_{pv}$-M1 interference (as originally suggested by the Bouchiats (1)). Note that, as compared to Stark-induced methods, field-free experiments have different systematics and absolute scales (calibrations) in terms of parity conserving amplitudes. A description of our experimental approach, which utilizes a high-gain thermionic diode, has already been published (2).

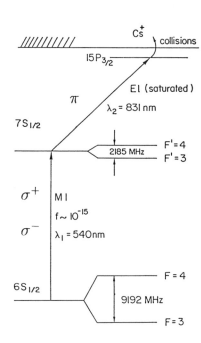

Fig. 1.
Relevant Cs energy levels

One important potential source of error are polarization imperfections due to unavoidable stress-induced birefringence of the diode windows.

In the same 2-photon sequence ($6S_{1/2}$→ $7S_{1/2}$→ $15P_{3/2}$; Fig. 1) as in the parity experiment, we have measured the parity-conserving asymmetry A_{CP} :

$$A_{CP} \equiv (I_s - I_o)/(I_s + I_o),$$

with the two beams having same (s) or opposite (o) circular polarizations.

A_{CP} shows a strong dependence on the power of the saturated (E1) transition (Fig. 2, (3)), due to differences in the saturation behaviour of the various (unresolved) h.f. components. The comparison of our measurements with rate equation model calculations sheds light on the dynamics of this unusual M1-E1 2-photon process.

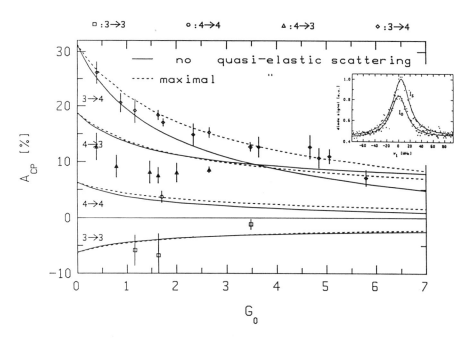

Fig. 2. Dependence of A_{CP} on the saturation parameter G_o of the $7S_{1/2} \to 15P_{3/2}$ transition. Insert: I_s and I_o vs. ν_1 with ν_2 fixed.

References

1. M.A.Bouchiat and C.Bouchiat, Phys.Lett. **48B**, 111 (1974).
2. P.P.Herrmann, J.Hoffnagle, N.Schlumpf, V.L.Telegdi, and A.Weis, J.Phys.B:At.Mol.Phys. **19**, 1607 (1986).
3. A.Weis, J.Hoffnagle, N.Schlumpf, V.L.Telegdi, D.Zevgolis, and L.Zhao, Eleventh International Conference on Atomic Physics (ELICAP), Paris, July 1988.

New Measurement of the Rydberg Constant
by Two-Photon Spectroscopy of Hydrogen Rydberg States

F.Biraben, J.C.Garreau, L.Julien and M.Allegrini[*]

Laboratoire de Spectroscopie Hertzienne de l'Ecole Normale
Supérieure, Université Pierre et Marie Curie, Tour 12 E01,
4 Place Jussieu, F-75252 Paris Cedex 05

1. Motivation

During the last fifteen years, the development of laser
spectroscopy applied to atomic hydrogen has led to an
improvement of three orders of magnitude on the precision of
the Rydberg constant R. Although this constant is the most
precisely known fundamental constant at the present time,
there are several reasons for the continued effort to further
improve its precision to one part in 10^{11} or better:

(i) R ties together the three fundamental constants m
(electron mass), e (electron charge) and h (Planck constant),
and plays a key role in the least-squares adjustment of
fundamental constants [1] for which it is used as an auxiliary
constant (i.e. with a fixed value). The agreement between
several results involving or not involving R, as in the case
of the fine structure constant determination, is a test of the
consistency between various domains of physics.

(ii) A very precise value of R is needed to test
predictions of QED on simple systems. Two examples are given
with the measurement of the 1S-2S transition frequency giving
the 1S Lamb-shift, and the study of the $1^3S_1 - 2^3S_1$ transition
in positronium which is a purely leptonic system. In the
latter, one observes a slight disagreement with theory which
indicates the need for further calculations.

(iii) The Rydberg constant can be deduced from any
transition frequency (from UV to microwaves) between two
hydrogen levels of different n; it gives then a test of the
1/r dependence of the coulombian potential, or, if this
dependence is assumed, it allows the eventual use of the
hydrogen atom itself for the realization of the meter in the
visible range.

2. Optical spectroscopy of hydrogen

During the last few years, three types of optical
transitions have been successfully investigated in atomic
hydrogen leading to determinations of the Rydberg constant:
the 2S-nP (n=3,4) Balmer transitions, the 1S-2S and the 2S-nD
(n=8,10,12).

Due to the progress in tunable lasers, the Balmer-α
absorption line at 656 nm has been extensively studied

[*] Permanent address: Istituto di Fisica Atomica e Molecolare
del CNR, Pisa, Italy

since the early 1970's. In 1978, the Stanford group used polarization spectroscopy with cw excitation to measure R with a precision of 3 x 10^{-9} [2]. The Yale group in 1981 used a 2S atomic beam crossed with a cw dye laser beam to determine the Rydberg constant with a precision of 10^{-9} [3]. More recently a sophisticated version of this experiment led to measurements of the Balmer-α [4] and Balmer-β [5] transitions with a precision of few parts in 10^{10}. Because of the natural lifetimes of the 3P and 4P levels, which are responsible for a relative linewidth of a few parts in 10^8 in these transitions, it seems that the ultimate limitation of this method has been reached.

A very promising transition is the 1S-2S two-photon transition (natural width 1.3 Hz). Two groups (Stanford-München [6] and Oxford [7]) have succeeded in measuring its frequency with cw radiation. Because of present limitations in the 1S Lamb-shift calculations and the experimental uncertainty of the proton radius, these quite exciting experiments give more a determination of the 1S Lamb-shift than a measurement of the Rydberg constant.

Our own way of measuring the Rydberg constant is in the study of Doppler-free two-photon 2S-nD transitions ($n \geqslant 8$) in hydrogen and deuterium. The advantage of these transitions, which avoid the broader nP levels, is their narrow natural linewidth (300 kHz for the 2S-10D) allowing one to observe linewidths of a few parts in 10^{10}.

3. Experimental method and results

In order to actually observe very narrow signals, we use a metastable atomic beam which is collinear with two counterpropagating laser beams [8]. Collisional and transit time broadenings are then negligible. Beyond the interaction region with the lasers, the flux of metastable atoms is measured: a quenching electric field is applied and two photomultipliers measure the induced Lyman-α fluorescence. When optically excited towards the nD states, atoms decay preferentially to the ground state so that the absorption signal can be detected in the corresponding decrease of the 2S beam intensity.

The transition wavelengths, in the range 730-780 nm, are compared with that of an I_2-stabilized standard He-Ne laser at 633 nm through a high finesse Fabry-Perot etalon.

We performed in 1986 a preliminary measurement of the Rydberg constant [9] derived from the study of three 2S-nD transitions (n=8 in hydrogen and deuterium and n=10 in hydrogen). Our result differed slightly from the preceding one and improved its precision by a factor 2.

Since 1986 several improvements to our set-up have been made. We have built a new metastable beam apparatus which is evacuated by cryogenic pumps and can be heated. Stray electric fields seen by the atoms are then reduced to less than 2 mV/cm so that we can excite n > 10 levels without appreciable broadening. A new system for control and measurement of the dye laser frequency has been developped: an acousto-optic device allows a sweep of the dye laser frequency over a

250 MHz range around any chosen frequency with a reproducibility better than 10^{-11}.

Numerical calculations of the line profiles have been made summing the contributions of all possible atomic trajectories in the metastable beam (where the atomic longitudinal velocity distribution is deduced from the 2S-3P absorption profile). They show that the main broadening effect is the inhomogeneous light-shift experienced by the atoms. The observed signals have been fitted with theoretical profiles to determine very precisely the line position (see fig.1). Each fit gives the experimental line center (half-maximum center HMC) and the light-shift corrected line position (CLP). For each studied transition, these two data have been studied using various light powers and extrapolated to zero power as shown in figure 2. It has been verified that the two series of data give consistent results (in figure 2 the slight difference between the two extrapolations is well explained by the saturation of the light-shift).

The interferometric wavelength comparison has been carried out very carefully, taking into account both the slight ageing of the silver mirrors during the experiment and the mirror curvature imperfections.

Using this method we have measured the frequencies of six transitions $2S_{1/2}-nD_{5/2}$ in H and D [10]. The resulting six independent determinations of the Rydberg constant are in very good agreement. Our final result is:

$$R = 109\ 737,315\ 709\ (18)\ cm^{-1}.$$

This determination of R is the most precise one at the present time. Most of the error (1.6×10^{-10}) comes from the standard laser whereas the precision with respect to this standard is 4.3×10^{-11}.

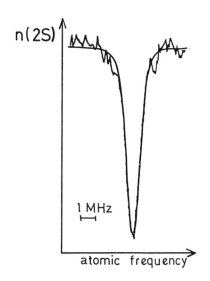

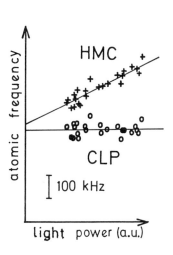

Figure 1: Typical signal recording ($2S_{1/2} - 12D_{5/2}$ transition in D).

Figure 2: Extrapolation of the line position versus light power.

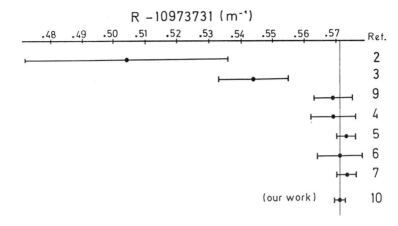

Figure 3: Comparison of the Rydberg constant measurements performed during the last few years, using cw laser excitation of 2S-nP 1S-2S or 2S-nD transitions.

Figure 3 shows that since 1986 there is a very good agreement between results obtained by various methods.

4. Perspectives

From our result, it is clear that the frequency of the He-Ne standard laser need to be remeasured, as is planned in Paris at the Laboratoire Primaire des Temps et Fréquences (LPTF) for the near future. However a second limitation of our measurement is the interferometric method itself whose accuracy is at best a few parts in 10^{11}. We plan to replace this by a direct frequency comparison between the 633 nm standard laser and the 2S-8S or 2S-8D transitions in hydrogen. As suggested by A.Clairon (LPTF), such a comparison would be possible using a CH_4-stabilized He-Ne laser at 3.39μm. We are making preparations for this comparison and are now developping an auxiliary frequency standard near our atomic transitions using a laser diode stabilized on a molecular line.

References

1. E.R.Cohen and B.N.Taylor, Rev. Mod. Phys. 59, 1121 (1987)
2. J.E.M.Goldsmith et al., Phys. Rev. Lett. 41, 1525 (1978)
3. S.R.Amin et al., Phys. Rev. Lett. 47, 1234 (1981)
4. Ping Zhao et al., Phys. Rev. A34, 5138 (1986)
5. Ping Zhao et al., Phys. Rev. Lett. 58, 1293 (1987)
6. R.G.Beausoleil et al., Phys. Rev. A35, 4878 (1987)
7. M.G.Boshier et al., Nature 330, 463 (1987)
8. F.Biraben et L.Julien, Opt. Comm. 53, 319 (1985)
9. F.Biraben et al., Europhys. Lett. 2, 925 (1986)
10. F.Biraben et al., Phys. Rev. Lett. 62, 621 (1989)

UV Laser Spectroscopy of ^{3,4}He Rydberg Series

W. Hogervorst, W. Vassen[*], and T. van der Veldt
Faculteit Natuurkunde en Sterrenkunde, Vrije Universiteit,
De Boelelaan 1081, 1081 HV Amsterdam, The Netherlands

The ^{4}He fine structure and ^{3}He hyperfine structure as well as the isotope shift between ^{3}He and ^{4}He in the $1s2s^3S_1 \rightarrow 1snp$ (n=5-79) have been measured in a laser-atomic-beam experiment [1]. Also the isotope shift in some $1s2s^1S_0 \rightarrow 1snp$ transitions was determined. In addition the linear Stark effect in $1snp^{1,3}P$ Rydberg states (n≈40) of ^{4}He was studied [2]. The evolution of angular momentum manifolds was followed up to the regime where Stark states originating from different n-values interact and narrow avoided level crossings were detected.

A beam of metastable (1s2s) helium atoms was produced with a DC electric discharge in an expanding, isotopically enriched gas mixture of ^{3}He (50%) and ^{4}He. To reduce Doppler effects a collimated beam was intersected perpendicularly with the focused light of an intracavity frequency-doubled CW ring dye laser. Tunable, narrowband UV laser radiation (1 MHz) in the wavelength region 260-330 nm was produced using various nonlinear crystals and dyes. UV power varied in between 3 and 14 mW. For n<10 the fluorescence was observed with a photomultiplier, whereas for n≥10 He-ions (from field ionisation) were detected with a quadrupole mass filter. Atomic and laser beam intersected in a metal shielding box to ensure field-free excitation for the studies of fine and hyperfine structure and isotope shifts. Two capacitor plates were mounted inside this box for the Stark effect investigations. The residual Doppler width of a field-free transition to a $1snp$ Rydberg level was about 15 MHz.

An example of the excitation $1s2s^1S_0 \rightarrow 1s40p^1P_1$ in a gas mixture of ^{3}He and ^{4}He at a wavelength of 313 nm is shown in Fig.1. Two hyperfine peaks of ^{3}He (nuclear spin I=1/2) are observed. The ^{3}He peak at the low frequency side is the hyperfine-induced excitation of the ^{3}P state, which in ^{4}He cannot be observed due to the near-absence of singlet-triplet mixing. From the ^{3,4}He spectra fine structure (for n<20) and hyperfine structure splittings were deduced. The splittings for ^{3}He as a function of n are shown in Fig.2 relative to the single F=5/2 hyperfine component.

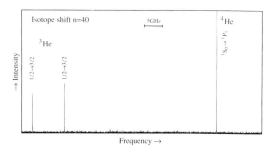

Fig. 1. The $1s2s^1S_0 \rightarrow 1s40p^1P_1$ transition at 313 nm showing the hyperfine structure of ^{3}He and the isotope shift between ^{3}He and ^{4}He.

[*] Present Address: Max Planck Institut für Quantenoptik, D-8046 Garching, W.-Germany

The observed fine structure in ^{4}He may be accurately reproduced with fine structure constants and a Slater exchange integral which follow a n^{-3} scaling law. To calculate the hyperfine structure of ^{3}He the Fermi contact interaction of the 1s-electron is added to the fine structure Hamiltonian. The calculated hyperfine splittings agree with experiment for n<20, whereas for n>20 deviations increasing with n were observed as a consequence of hyperfine-induced n-mixing effects. In a three-channel two-limit quantum defect model without any adjustable parameter the experimental results may be excellently reproduced.

The isotope shift between ^{3}He and ^{4}He (~50 GHz) mainly stems from the normal mass shift, which is easily calculated from the transition frequency. As volume shifts may be neglected the remaining part is the specific mass shift (SMS). SMS measures the correlation in the electron momenta $<\vec{p}_1.\vec{p}_2>$. From the experimental data as a function of n SMS for both the 1s2s^{3}S$_1$ and 1s2s^{1}S$_0$ metastable level could be deduced. The results are: SMS$_e$(1s2s^{3}S$_1$)=2189.6(2.5) MHz and SMS$_e$(1s2s^{1}S$_0$)= 2792.0(3.0) MHz. These data are in excellent agreement with recent mass polarisation calculations and confirm an isotope dependence of $<\vec{p}_1.\vec{p}_2>$ [3]. The calculated values are: SMS$_c$(1s2s^{3}S$_1$)=2190.29 MHz and SMS$_c$(1s2s^{1}S$_0$)=2791.24 MHz. These results show that relativistic effects cannot contribute more than a few MHz to the mass shift [4].

In the Stark effect studies main interest was on the evolution of angular momentum manifolds as a function of electric field strength. This linear Stark effect, most easily studied in 1snp^{1}P$_1$ levels, occurs when the field energy is not small compared to the zero-field energy separations between levels with different ℓ-values. In the ^{1}P$_1$-case using π-excitation only M=0 sublevels were populated from 1s2s^{1}S$_0$. At sufficiently high values of the field strength the merging of adjacent manifolds could be investigated in great detail. Unlike the hydrogen case the levels do not cross with increasing field strength and many narrow avoided level crossings were recorded. An example for the singlet case is shown in Fig.3 for the two outermost levels of the n=41 (the level |41,-40,0> in parabolic quantum numbers) and n=40 (|40,39,0>) manifolds for electric fields between 16 and 16.5 V/cm. This type of avoided level

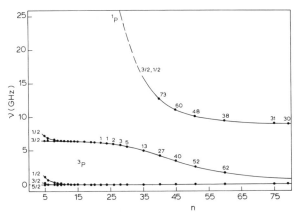

Fig. 2. Experimental hyperfine level energies of ^{3}He 1snp relative to the F=5/2 position as a function of n. For each hyperfine component the ^{1}P fraction in the wave function is as indicated (or smaller than 1%).

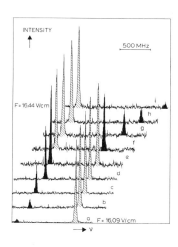

Fig. 3. The avoided crossing of the ^{4}He manifold levels ($^1P_1$-case) $|40,39,0>$ and $|41,-40,0>$.

crossing could only be reproduced in a calculation where the complete energy matrix in the presence of the field was diagonalised and at least four n-values (for the example shown n=39-42) were included.

References

1. W. Vassen and W. Hogervorst, Phys. Rev. A39 (1989) 4615
2. C.T.W. Lahaije and W. Hogervorst, Phys. Rev. A39, to be published
3. G.W.F. Drake, Nucl. Instr. and Meth. in Phys. Res. B31 (1988) 7
4. E. de Clercq, F. Biraben, E. Giacobino, G. Grynberg and J. Bauche, J. Phys. B14 (1981) L183.

347

Observation of an electronic correlation for double Rydberg states of barium

P.Camus, J.-M.Lecomte, P.Pillet and L.Pruvost
Laboratoire Aimé Cotton*, C.N.R.S. II, Bâtiment 505, Campus d'Orsay
91405 Orsay Cedex, France.
J.Boulmer
Institut d'Electronique Fondamentale, Bâtiment 200, Université Paris Sud
91405 Orsay Cedex, France.

We report a recent study of highly excited Nln'l' double-Rydberg states (D.R.S.) of barium atom, for N=22 to 34 and n'=39 to 60 as the principal quantum numbers of the two electrons are brought closer together in two ways : n' decreasing toward N fixed and N increasing toward n' fixed. In these conditions, the observed spectra trace the evolution from D.R.S. in which the electrons are uncorrelated (N<<n') to those in which they are supposed to be (N~n'). These autoionizing states lie only ~ 0.1 eV below the double ionization limit and their ionic products which are highly excited Rydberg states of Ba^+, are selectively microwave field ionized in Ba^{2+} before to be analyzed by time-of-flight (1).

In the first approach, typical spectra 6snp→26dn'p and 6snp→27sn'p for n=60to 39 (2) trace the evolution in which the electrons are uncorrelated showing double peak structures to the limit N^*~0.65n'* fixed by the emergence of an unstructure weak broad continuum instead of the double peak resonances. Similar studies for intermediate D.R.S. with N=9 to 11 (2)(3)(4) have shown the same phenomena for roughtly the same N^* and n'* relative values. First we observed a shift of the resonance structures to the red wavelength compared to the ionic parent lines (see the arrows on Fig.1(a)). These shifts are unexpected in the isolated core excitation model (I.C.E.) (5) which predicts that the observed structures should be centered around the ionic parent line (3).

In the second approach for the energy range above the 28s inner electron, we observed also the appearance of weak normally forbidden two-photon resonances in the neighborhood of the 6s→28p and 6s→25f ionic line position.

These two new features can be understood in terms of a fairly simple model based on the electrostatic interaction of the outer electron with the Ba^+ core which is neglected in the I.C.E. approximation. In this picture the outer electron polarizes the Ba^+ core in a manner qualitatively similar to the Stark effect inducing l mixing which allows the observation of the forbidden two-photon transitions.

A generalization of the isolated core excitation formalism taking into account dipole and quadrupole terms leads to quantitative understanding of the long-range electronic correlation observed (6). Since the outer electron moves slowly, we first assume it to be frozen in place at r_2 and calculate the polarized Ba^+ wavefunctions Ψ_k and the energies $E_k(r_2)$ taking into account the electrostatic interaction with the outer electron. Choosing r_2^{-2} as the abscissa leads to the energy-level diagram of the Fig.2 very similar to the energy levels in a static electric field. Second, we calculate the wavefunctions of the outer electron by considering its adiabatic evolution in a supplementary potential corresponding to the polarized core Ba^+ ion. They can be considered as Coulombian wavefunctions dephased by $\delta_k(r_2)$. This dephasing leads to the shift δ for the center of gravity of the structure enveloppe which corresponds to the overlapping integral between the initial and final Rydberg states of the outer electron due to the second step two-photon excitation. The calculated shifts are in good agreement with the experimental values reported on Fig.3.

On Fig.1, calculated spectrum from the theoretical data (trace (b)) fits the corresponding observed spectrum (trace(a)).

In conclusion, we have reported the observation of an electronic correlation between two excited Rydberg electrons which is can be understood as the polarization effect of the excited core electron by the outermost electron.

* Laboratoire associé à l'Université Paris Sud

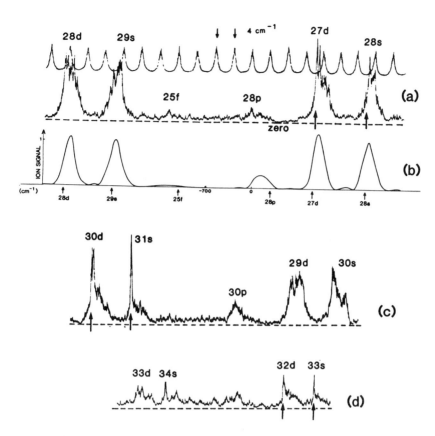

Fig.1 Continuous two-photon absorption spectra 6s45p→Nln'l' for different NI core electrons :
(a) 28s→28d.
(c) 30s→30d.
(d) 33s→33d.
(b) Calculated spectrum from the theoretical data for 28s→28d.
Narrow resonance lines correspond to ionic resonances of Ba+ due to coincidences between atomic (Ba) and ionic (Ba+) resonances.

References

1. J.Boulmer, P.Camus, J.-M. Gagné and P.Pillet, J. Phys. B 20, L143 (1987).
2. J.Boulmer, P.Camus, J.-M.Lecomte and P.Pillet, J. Opt. Soc. Am. B 5, 2199 (1988).
3. P.Camus, P.Pillet and J.Boulmer J.Phys. B 18, L481 (1985).
4. R.R.Freeman, L.A.Bloomfield, J.Bokor and W.E.Cooke in Laser Spectroscopy VII, T.W.Hänsch and Y.R.Shen, eds. (Springer-Verlag, Berlin, 1985), p.77.
5. W.E.Cooke, T.F.Gallagher, S.A.Edelstein and R.M.Hill, Phys. Rev. Lett. 40, 178 (1978).
6. P.Camus, T.F.Gallagher, J.-M.Lecomte, P.Pillet and L.Pruvost, Phys. Rev. Lett. 62, 2365 (1989).

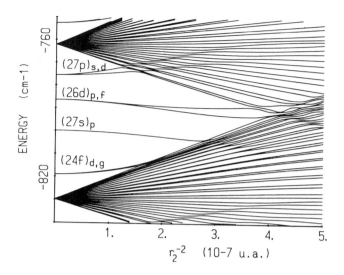

Fig.2 Energy diagram of the Ba+ target vs r_2^{-2} in the vicinity of N=23 and 24 manifolds. The total angular momentum is equal to l. The spins of both electrons are neglected.

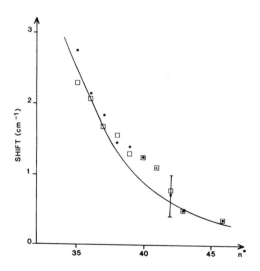

Fig.3 Shifts of the centers of gravity compared to the ionic parent line positions for the double-Rydberg resonances. Observed data for 6snp→26dn'p (□) and 6snp→27sn'p (●) vs n* ; n=39-47 and 50. Theoretical curve is the full line.

OBSERVATION OF SUB–DOPPLER LASER INDUCED NUCLEAR ORIENTATION OF ^{85m}Rb

W.W. Quivers, Jr., J. Mackin, J.T. Hutton, M. Lercell,
M. Otteson, G. Shimkaveg, R.R. Dasari, C.H. Holbrow,
M.S. Feld, and D.E. Murnick.

George R. Harrison Spectroscopy Laboratory
Massachusetts Institute of Technology
Cambridge, MA 02139

I. Introduction

We report the first observation of sub–Doppler resolved gamma anisotropy produced by Laser Induced Nuclear Orientation (LINO) [1]. As discussed below, the anisotropy was produced by laser optical pumping the D1 transition of the $1\mu s$ ^{85m}Rb isomer.

Sub–Doppler LINO is uniquely suited for resolving closely spaced hyperfine structure (hfs) of very shortlived isomers, whose lifetimes are between 50 nanoseconds and 1 milli–second. In addition, sub–Doppler LINO allows more precise measurements of hf splittings to be made. For example, this leads to more accurate measurements of nuclear magnetic dipole and electric quadrupole moments, as well as isomer shifts.

LINO uses laser optical pumping to produce electronic alignment. This is transferred to the excited nucleus via the hyperfine interaction. When the oriented nuclei decay, anisotropy is observed in the spatial distribution of the emitted gamma rays. The hfs of the isomer is determined by monitoring the anisotropy as a function of laser detuning.

II. Laser Optical Pumping & Velocity Changing Collisions

Many applications of LINO require the use of a buffer gas. For example, on–line cell studies use buffer gas to slow down and confine the species of interest. Also, as is in our case, the buffer gas can be the parent of the species of interest.

The main effect of the buffer gas is to induce velocity changing collisions (vcc's) between the isomers and the buffer gas atoms.

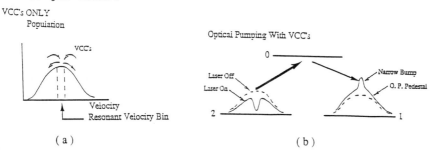

Fig. 1   Laser Optical pumping with VCC.

As Fig. 1a shows, vcc's knock atoms out of the resonant velocity bin and distributes them over the remainder of the Doppler distribution. Simultaneously vcc's are also knocking atoms into the resonant bin. The net effect produces increased Doppler coverage [2]. The effect of the vcc's on the optical pumping process is shown in Fig. 1b. Anisotropy is caused by the excess population transferred to level 1. This population consists of a broad (~Doppler width) optical pumping pedestal and a narrow (sub-Doppler width) feature. The pedestal is produced by those isomers that undergo vcc's before they decay, whereas the narrow bump represents isomers that decay before any vcc's. For a given intensity and increasing buffer gas pressure the pedestal will grow at the expense of the bump. At high enough pressures the bump will disappear and the experiment enters the Doppler limited regime [2]. So to achieve sub-Doppler resolution it is crucial that the buffer gas pressure be low enough for the narrow feature to be present.

III. Sub-Doppler LINO

As the above discussion indicates, sub-Doppler resolution requires low buffer gas pressures. This is a disadvantage because there is a decrease in signal size due to incomplete Doppler coverage. A further reduction in signal size also occurs if the buffer is the parent.

However, since the saturation intensity is proportional to the vcc rate [3], it becomes easier to saturate a par-ticular velocity group. Hence less intensity is required to saturate and larger cell volumes can be pumped, with a corresponding increase in signal size.

These effects can be combined in such a way as to result in resonably sized signals. The key is to determine the optimum cell design (i.e. volume, pressure, etc.) for a given laser power.

IV. LINO and Saturation Spectroscopy

In our experiments the saturation effect of two counter-propagating pump beams (Lamb-dip configuration) is used to produce the sub-Doppler anisotropy signal.

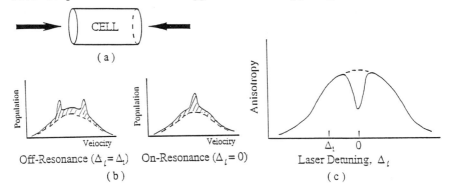

Fig. 2 Saturation and Anisotropy

Figure 2 shows this effect. The anisotropy is proportional to the area under the excess population transferred to the unpumped level, as depicted by the shaded portion of Fig. 2b. Due to the nonlinearity of saturation, the area off resonance is larger (more efficient pumping) than the area on resonance (less efficient pumping). So, as Fig. 2c shows, as the laser is tuned through the hf transition, the anisotropy is greater off resonance than near or on resonance. This decrease manifests itself as a sub-Doppler anisotropy signal, much in the same way as the usual Lamb-dip does in ordinary spectroscopy.

V. Experiments and Results

Figure 3a shows the production scheme for the $1\mu s$ ^{85m}Rb isomer, which decays with the emission of a 514 kev gamma ray. The anisotropy is produced by optical pumping on the D1 transition ($\lambda = 7947$ Å). The pumping scheme is depicted in Fig. 3b. A linearly polarized beam pumps the $F = 5$ to $F' = 4$ hf transition. The anisotropy is created by the excess population transferred to the end $M_F = +5$ sub-levels. Since the ground state hf splitting of 12.5 GHz is much larger than the Doppler width, population also accumulates in the $F = 4$ hf level. The overall effect is added to the isotropic background.

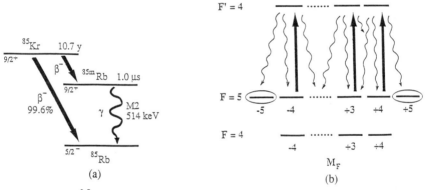

(a)   (b)

Fig. 3 ^{85}Kr decay abd optical pumping schemes.

The experimental setup is shown in Fig. 4. The cell contains 300 mTorr of krypton, which is the parent and also acts as a buffer gas, plus natural rubidium for resonant charge exchange. The number of isomers in the cell is much less than one. Two counterpropagating beams of equal power (150 mW) are incident on opposite faces of the cell. The beams are formed by splitting the output beam from a dye

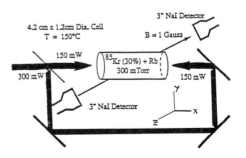

Fig. 4 Experimental set up.

353

ring dye laser utilizing LD700 dye and pumped by a 5 W krypton ion laser. The sub-Doppler anisotropy change signal is shown in Fig. 5a. The solid curve is a computer fit. The signal shows a narrow feature with a 90 MHz width (FWHM) resting on the beginnings of a broad optical pumping pedestal. Each data point represents five hours of counting time, done in ten minute intervals to correct for laser drift, etc. Figure 5b shows that the width of the sub-Doppler signal is approximately 10% of the Doppler width.

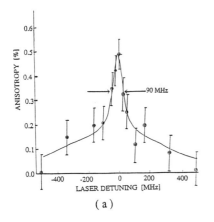

(a)

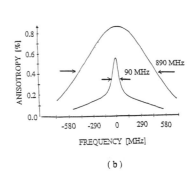

(b)

Fig. 5   Sub-Doppler anisotropy change signal

VI. Future Prospects

Our next step is to use sub-Doppler LINO to make the first measurements of the ^{85m}Rb electric quadrupole moment. This requires that we apply LINO to the D2 transition, where the estimated excited state hf splittings are less than the Doppler width. We will also utilize sub-Doppler LINO in precise on-line studies of chains of very shortlived isotopes and isomers.

In addition to the above projects, LINO will be applied to kinematic studies of the weak interactions. Generally speaking, in these kinds of experiments the recoil of the daughter is studied. One such experiment uses LINO to measure the electron-antineutrino angular correlation in ^{85}Kr beta-decay [4]. Although not necessarily an application of LINO, it may also be feasible to use lasers to search for neutrino mass in bound-state beta-decay studies [5] and electron capture.

[1] Physics Department Wellesley college, Wellesley, MA 02181.
[2] L-43, Lawrence Livermore Natl. Laboratory, Livermore, CA 94550.
[3] Dept. of Physics & Astronomy, Colgate University, Hamilton, NY, 13346.

Hamilton, NY, 13346.
[4] Department of Physics, Rutgers University, Newark, NJ.

References

1. M. Burns, P. Pappas, M.S. Feld, and D.E. Murnick, Nucl. Instum. Methods 141, 429 (1977).   •
2. G. Shimkaveg et al., Phys. Rev. Lett. 53, 2230 (1984).
3. W.W. Quivers, Jr., Phys. Rev. A34, 3822 (1986).
4. J.T. Hutton and W.W. Quivers, Jr., Phys. Rev. C40, 314(1989)
5. S.G. Cohen, D.E. Murnick, and R.S. Raghavan, Hyper. Int. 33, 1 (1987).

Laser spectroscopy of neutron deficient lead and thallium isotopes:

Systematics of nuclear radii in the Z = 82 region

S. Dutta, R. Kirchner, O. Klepper, T. Kühl, D. Marx, GSI Darmstadt,

U. Dinger, G. Huber, D. Knussmann, R. Menges, S. Schröder, Universität Mainz.

G. Sprouse, SUNY Stony Brook, USA,

Introduction

Inspected with the precision of high resolution optical spectroscopy, atomic transition energies reveal the influence of nuclear properties. The high sensitivity achievable in laser spectroscopic methods allows the extraction of nuclear quantities even of artificially produced shortlived nuclides far off from nuclear stability. By analyzing the hyperfine structure and the isotope shift magnetic dipole moments, electric quadrupole moments and the change of the nuclear charge radius can be studied [1]. The evaluation, however, depends to a large fraction on a complete description of the atomic physics involved. With the amount of available data constantly rising, atomic physics calculations are critically tested by the consistency or inconsistency with complementary data. We report here on experimental results concerning the isotope shift and hyperfine splitting of $^{190-197}$Pb isotopes [2] and $^{188-196}$Tl [3]. These measurements add a number of new isotopes to our earlier lead data [4] and the data of the Karlsruhe group on $^{197-214}$Pb [5] and the Oak Ridge group on $^{190-194}$Tl [6].

Experiment

The experiment was carried out at the GSI on-line mass separator using collinear fast atomic-beam laser spectroscopy. Details of the experimental set-up are described elsewhere [7]. Radioactive lead and thallium isotopes were produced by bombarding natural tungsten targets with oxygen beams of app. 500 particle nA at an energy of 9-10 MeV/u. For the lighter thallium isotopes a tantalum target was used. A special "bunched-beam" mechanism [8] allowed to enhance the signal-to-background ratio and, in the case of thallium, to study low-spin daughter nuclei from the decay of lead. In lead the $6p^2\ ^1D_2$ - $6p7s\ ^3P_1$ transition, λ = 723 nm, was excited. The metastable starting level was populated in the charge-exchange process which converts the ion beam of the mass separator into an atomic beam in flight. In thallium excitation with λ = 535 nm starts from the $6p\ ^3P_{3/2}$ state and leads to the $7s\ ^2S_{1/2}$ state. The decay to the $6p\ ^3P_{1/2}$ groundstate is detected.

Atomic physics and nuclear data evaluation

Nuclear moments

In case that the nuclear moments of at least one nuclide of a chain of isotopes is known, this information can immediately be transferred to all the other isotopes by taking the ratio of the hyperfine structure A - factors to calculate the unknown magnetic dipole moment and the ratio of the B - factors for the electric quadrupole moment. In many elements, however, the nuclear spins of stable isotopes are smaller than 1/2, so that no information on the quadrupole moment is given. Lindgren and Rosen [9] have calculated the B - factor for a large number of elements, including thallium, and compared the results, where possible. The accuracy is usually found to be within 1 - 5 %. In lead a calculation for the 6p6s $^3P_1$ using the effective operator approach is given by Dembcynski et al. [10]. We took the same approach for the 6p^2 $^1D_2$ state [11]. In fig. 1 the spectroscopic quadrupole moments evaluated in this way for the I = 13/2

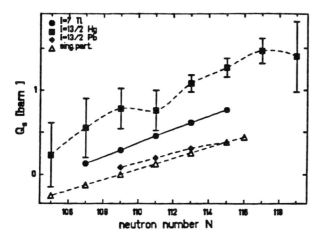

Fig. 1: Spectroscopic nuclear quadrupole moments of νi13/2 isomers evaluated from the 6p^2 $^1D_2$ hyperfine structure compared to calculated single particle moments

isomers in lead are compared with the values predicted for a pure single particle quadrupole moment, i.e. the case that collective nuclear deformation does not contribute. In lead the ex- tracted data agree well with this very reasonable prediction. For comparison data on isomers in thallium and mercury are included, where the same single particle moments contribute.

Change of the nuclear charge radius

The evaluation of the change in nuclear radius from the isotope shift depends both on the elec- tronic volume shift and the specific mass shift, which reflects the change in correlated motion of the electrons caused by the different specific mass of the multielectron-plus-nucleus system. Calculations by King and Wilson [12] on the specific mass shift in lead and thallium yield values smaller than the normal mass shift for lead and thallium. As discussed in [5], several inde- pendent sources of information are available on the radius change in the stable lead isotopes: Muonic data and electron scattering data have been combined to give a very accurate value, and in addition atomic x-ray data exist. Combining these data with the specific mass shift cal- culation the Karlsruhe group already extracted a value for the volume shift factor $F = \delta\nu/\lambda$. In the meantime a multi-configuration Dirac-Fock calculation of the electronic factor became available [13]. We used a three-dimensional King-plot to compare the combined result of the available experimental data with combinations of specific mass shift and values of F around the theoretical values. The optimum combination gives F = 24.6(5) GHz/fm and a specific mass shift 25(55) MHz/δA, compared to 24.9 GHz/fm calculated by Fricke [13] and 19 MHz/δA by King and Wilson [12] for the respective value. Encouraged by this result, we used theoretical values from the same authors to evaluate the neighbouring thallium isotopes. The results are shown to-gether with other available data in the lead region in fig. 2.

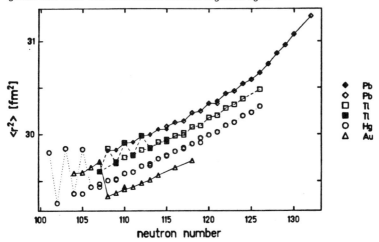

Fig. 2: Charge radii in the lead region. The absolute values are calculated, by adding the droplet value at a given N to the $\delta < r^2 >$ values obtained from isotope shift measurements. Full symbols: this work, other values see ref. [1,5,6]

Conclusion

In this region data from optical spectroscopy span a wide range with considerable variation in the change of the charge radius per isotope, going far off from stability. Using the calculations now available for the electronic factor and the specific mass shift, an evaluation of lead and thallium isotope shift data became possible, which is obviously consistent with a wealth of al-ready existing data in this mass region. This seems to indicate, that the different Hartree-Fock and Dirac-Fock calculations for the specific mass shift and the electronic factor F can reach a reliability similar to the calculations of hyperfine structure constants.

References

1. E.W. Otten, "Nuclear Radii and Moments of Unstable Nuclii", Treatise on Heavy Ion Science, D.A. Bromley(Ed.), Vol. 8(1988), 515; Plenum Press, New York
2. S. Dutta et al., to be published
3. R. Menges et al., to be published
4. U. Dinger et al., Z.Phys.A328 (1987) 253
5. M. Anselment et al., Nucl.Phys.A451 (1986) 471
6. J.A. Bounds et al., Phys.Rev.Lett.55 (1985) 2269
7. T. Kühl et al., Nucl.Instr.Meth.B26 (1987) 419
8. R. Kirchner et al., Nucl.Instr.Meth.A247 (1986) 265
9. I. Lindgren, A. Rosen, Case studies in atomic physics 4 (1974)
10. J. Dembczynski, H. Rebel, Z.Phys.A315 (1984) 137
11. U. Dinger, Thesis, Mainz 1988, GSI report 88-11
12. W.H. King, M. Wilson, J.Phys.G11 (1985) L43 and priv. comm.
13. B. Fricke, priv. comm.

How Lasers Can Help Probe the Distribution of Nuclear Magnetism

H.T. Duong[1], C. Ekström[2], S. Liberman[1][†], I. Lindgren[3], R. Neugart[4],
R. Pellerin[5], S. Penselin[6], J. Pinard[1], I. Ragnarsson[7], O. Redi[8],
H.H. Stroke[8], J.L. Vialle[5]
1) Laboratoire Aimé Cotton, Orsay, 2) Uppsala University, 3) Chalmers
University of Technology, 4) Johannes Gutenberg Universität, Mainz, 5)
Université de Lyon, 6) Universität Bonn, 7) Lund University, 8) New York
University, †) deceased

High-resolution atomic spectroscopy has played an important part in
the study of nuclear electric and magnetic structure.[1] Laser spectroscopy has been crucial, particularly for the measurement of isotope shifts,
which reflect the variations of nuclear charge radii and shapes: the
high sensitivity and frequency resolution have allowed experiments to be
done systematically over extensive ranges of stable and radioactive isotopes, with lifetimes as short as a few milliseconds. An extensive review has been made recently by Otten.[2] While the laser experiments also
yield results for nuclear multipole moments, no measurements are obtained
of the distribution of nuclear magnetization.

Nuclear structure properties can be probed by penetrating electrons
$(s_{1/2}, p_{1/2})$. Thus, for an atomic angular momentum $J = \frac{1}{2}$ and nuclear spin I,
the two total angular momentum states $F = I \pm \frac{1}{2}$ are separated by the
magnetic hfs interaction energy $h\Delta\nu \equiv (2I + 1)W/I$. Bohr-Weisskopf
theory[3] gives the expressions for $W = W_S + W_L$ (S,L spin and orbital contributions). We can write

$$W = W_{extended} \equiv W_{point} (1 + \varepsilon), \qquad (1)$$

where "point" denotes the point nucleus, and ε is the "hfs anomaly" or
"Bohr-Weisskopf effect". Non-relativistically

$$\Delta\nu = (8/3)\pi g_I \mu_o (2I + 1)|\psi(0)|^2 \equiv a(I + \frac{1}{2}), \qquad (2)$$

the Fermi-Segrè formula. Here μ_o is the Bohr magneton, g_I the nuclear g
factor, and $|\psi(0)|^2$ the electron probability at the nucleus. The magnetic
dipole interaction constant a appears in the Hamiltonian $\vec{H}_{hfs} = a\vec{I}\cdot\vec{J}$.
Corresponding to (1), we have

$$a = a_{extended} = a_{point}(1 + \varepsilon). \qquad (3)$$

Generally, for electrons, our knowledge of the theoretical value
a(point) is not sufficiently precise to determine the small quantity ε
by direct comparison with the experimental value a(extended). We can,
however, to good approximation, determine the isotopic difference $\Delta_{12} \equiv \varepsilon_1 - \varepsilon_2$

$$\frac{a_1}{a_2} = \left(\frac{a_1}{a_2}\right)_{point} \left(\frac{1 + \varepsilon_1}{1 + \varepsilon_2}\right) \approx \frac{g_1}{g_2} (1 + \Delta_{12}). \qquad (4)$$

In the last step, with reference to (2), we have set $(a_1/a_2)_{point} \approx g_1/g_2$.
For the two isotopes 1,2, we obtain the differential hfs anomaly

$$\Delta_{12} = \frac{a_1}{a_2} \frac{g_2}{g_1} - 1. \qquad (5)$$

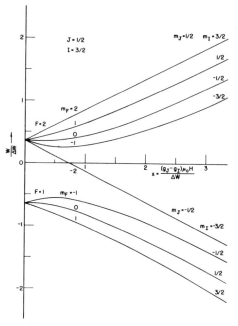

Fig. 1 Zeeman effect of the hfs energy for $J = \frac{1}{2}$ and $I = 3/2$.

To determine Δ_{12} we need four experimental quantities. The g-value measures the interaction of the magnetic moment with an external homogeneous field, while a does it with the slightly inhomogeneous field produced by the electron, and is hence sensitive to the nuclear magnetization distribution. A number of previous laser spectroscopy or rf experiments can often give a-values with adequate precision.[4] Except for stable isotopes, for which g-values can be measured by nmr, we must rely on atomic beam magnetic resonance (ABMR) techniques to obtain these for the very small quantities of radioisotopes generally available.[5]

In order to do systematic Δ measurements over large ranges of isotopes analogous to the isotope shift studies, work is now underway for on-line laser-rf ABMR experiments at project ISOLDE at CERN to obtain the required g-values. The determination makes use of the Breit-Rabi equation that evaluates the hfs Hamiltonian in an external magnetic field H

$$\mathcal{H} = a\vec{I} \cdot \vec{J} + g_J \mu_o \vec{J} \cdot \vec{H} + g_I \mu_o \vec{I} \cdot \vec{H}. \tag{6}$$

The result is

$$\frac{1}{h} W_{F=I+\frac{1}{2}, m_F} = - \frac{\Delta\nu}{2(2I+1)} + \frac{g_I \Delta\nu m_F x}{g_J - g_I} + \frac{\Delta\nu}{2}\left(1 + \frac{4m_F}{2I+1}x + x^2\right)^{1/2}$$

$$\frac{1}{h} W_{F=I-\frac{1}{2}, m_F} = - \frac{\Delta\nu}{2(2I+1)} + \frac{g_I \Delta\nu m_F x}{g_J - g_I} - \frac{\Delta\nu}{2}\left(1 + \frac{4m_F}{2I+1}x + x^2\right)^{1/2} \tag{7}$$

where

$$x = \frac{(g_J - g_I)\mu_o H}{h\Delta\nu}. \tag{8}$$

In Fig. 1 we show W(H) for $I = 3/2$. An analysis of (7) shows that the

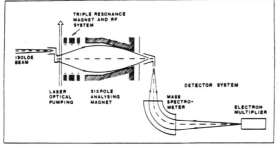

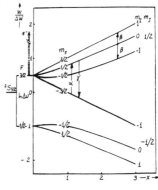

Fig. 2 Experimental setup for triple
resonance. The analyzing magnet
focusses atoms in $m_J = +\frac{1}{2}$ states.

Fig. 3 The rf transitions
induced are in or-
der α, β, γ. $\alpha = \gamma$.

energy difference for the same labelled pair of adjacent m_F levels in the
$F = I \pm \frac{1}{2}$ states equals $2g_I \mu_o H$, from which g_I can be determined. We have
used this "doublet method" to obtain Δ for ^{133,134,135,137}Cs by a Rabi-
type ABMR technique.[6] This does not work for $I = \frac{1}{2}$ or, with sufficient
precision, for very large $\Delta\nu$. We are thus developing a triple resonance
method that will allow measurements more generally for the cesium isotopes.
The experimental setup is shown in Fig. 2 and we describe the technique
for I=1 with reference to Fig. 3. The laser is used as the polarizer and
the multipole magnet as the analyzer. Use of σ^- laser light between the
ground F=3/2 and excited $^2P_{3/2}$ F=5/2 states pumps atoms into the ground F
=3/2, m_F=-3/2 levels, depleting the focussed m_J=+$\frac{1}{2}$ states. A decrease in
signal is produced. We now induce at low field an rf transition α followed by the reverse transition γ, leaving the same decreased signal. A
third rf region β is interposed. At resonance, the two Δm_I transitions,
with frequency difference $2g_I \mu_o H$, are induced. These depopulate the F=
3/2, m_F=$\frac{1}{2}$ state, so that the reverse γ transition is reduced. This results in a "flop-in" type experiment. At homogeneous transition region
fields of less than 1 T, adequate precision for measurements of the Bohr-
Weisskopf effect is expected in the cesium isotopes produced at ISOLDE.
Our earlier cesium magnetization distribution experiments already display
correlations with other known nuclear structure.

We acknowledge partial support from the US-France NSF-CNRS Cooperative
Research Program, NSF Grant INT-888815310 and NSF ECS-8819352.

References

1. See, for example, reviews P. Jacquinot and R. Klapisch, Rep. Progr.
   Phys. 42, 773 (1979); O. Redi, Physics Today 34, 25, 26 (Feb. 1981);
   H.H. Stroke, in Atomic Physics 8, edited by I. Lindgren, A. Rosén,
   and S. Svanberg (Plenum, New York, 1983).
2. E.W. Otten, in Treatise on Heavy Ion Physics, Vol. 8 (Plenum, NY,'87).
3. A. Bohr and V.F. Weisskopf, Phys. Rev. 77, 94 (1950).
4. G. Huber et al., Phys. Rev. Lett. 41, 459 (1978); C. Thibault et al.,
   Nucl. Phys. A367, 1 (1981); S. Liberman et al., Phys. Rev. A22,2732
   (1980); H.T. Duong et al., J. Physique 47, 1903 (1986).
5. C. Ekström et al., Nucl. Phys. A292, 144 (1977); Phys. Scr. 19, 516
   (1979), 34, 624 (1986).
6. H.H. Stroke, et al., Phys. Rev 105, 590 (1957), 123, 1326 (1961).

Nuclear Spin Dephasing in the Frozen Core

T. Muramoto[+], R. Kaarli[*] and A. Szabo
Division of Physics, National Research Council of Canada,
Ottawa K1A 0R6

1. Introduction

Optical Raman heterodyne detection (RHD) of nuclear
magnetic resonance (NMR) in low temperature solids has been
demonstrated [1] to be a sensitive and powerful spectroscopic
technique. Recently [2] we have extended RHD from its
original application to hyperfine (hf) spectra of rare earth
ions to superhyperfine (shf) spectra of Al in ruby. For shf
spectra, the radio frequency and optical photons are absorbed
by different atoms (Al and Cr respectively). The Raman
transitions are activated by the exchange and dipole coupling
between Al and Cr ions. Excellent agreement of theoretical
and experimental Al spectra was obtained using earlier
measured values of the various coupling and quadrupole
constants. We have now extended this work to observation of
Al NMR spin echoes. This is of interest, since for the first
time, we are able to measure dephasing times in the frozen
core. The concept of the frozen core or diffusion barrier
was introduced by Bloembergen [3] in 1949 to describe spin
diffusion and later dephasing phenomena. According to this
picture, nuclei close to a paramagnetic ion will not undergo
spin flips with those in the bulk because of detuning. Thus
we expect the dephasing time of Al ions in the frozen core to
be much longer than those in the bulk. Our echo measurements
confirm this picture.

2. Experimental

A very dilute sample of ruby (0.0034 Wt% Cr_2O_3) was used
to minimize effects of Cr-Cr spin flipping on the ^{27}Al
dephasing time. The sample was at 2K and had a magnetic
field of 3.6kG applied along the c axis. A single frequency
laser beam (2mm diameter, 50mw) from a Coherent ring dye
laser was propagated along the c axis. Circular polarization
was chosen to excite the $^4A(-3/2) \rightarrow \bar{E}$ $(-\frac{1}{2})$ transition at
14418.32 cm^{-1} for the Cr^{52} isotope. The inhomogeneous
linewidth was sufficiently narrow to also permit study of the
Cr^{53} isotope shifted blue [4] by 3.58 GHz. The excitation
and heterodyne detection scheme for the echo experiments was
essentially the same as described earlier [1]. 40 μsec rf
pulses were used for excitation with the laser and rf pulse
sequencing as well as data display and processing all under
computer control.

+Permanent address, Shiga University, Otsu 520, Japan.
*Permanent address, Academy of Sciences, Tartu, Estonian SSR.

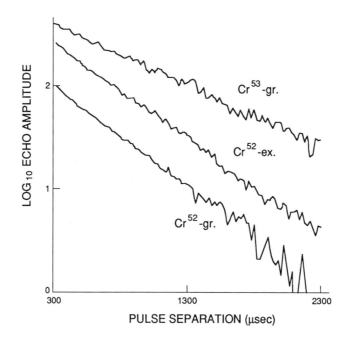

Fig. 1. NMR two pulse echo decay of the L'th nearest Al set of Cr in Al_2O_3. Transition frequency = 6.88 MHz at 3.60 Kg for ground state (Cr^{52}-gr, CR^{53}-gr) and 4.79 MHz for excited state (Cr^{52}-ex).

3. Results and Discussion

Plots of the two pulse Al NMR echo decay are shown in Fig. 1. Similar decays were observed for the other nearest neighbour sets I, J and K [5]. These four sets are well within the frozen core since their NMR frequencies are many linewidths (the NMR linewidth is ~ 10 kHz) different from those in the bulk. As expected, the dephasing time T_{2c} in the frozen core of ~ 1 msec is much longer than the bulk value of ~ 50 μsec [6]. We now address the mechanism of Al dephasing in the core and show how it is related to optical dephasing of the Cr R_1 line. Dephasing occurs by direct and indirect spin flips - between Al ions (nuclear spins) and also between Cr ions (electron spins). Direct spin flips between Al ions are ruled out as discussed earlier. We conclude as well that direct flips between Cr ions are not predominant in determining T_{2c}. The striking evidence for this is the identical T_{2c} seen for ground and excited levels, even though because of inhomogeneous broadening, the excited state population is some 10^5 times smaller than that in the

363

ground state (the ratio of optical inhomogeneous to homogeneous widths is 2×10^6 kHz/20 kHz). It should be recalled that, because of different g factors, the ground and excited Cr ions cannot undergo mutual spin flips. Also all ground state ions, regardless of their optical frequency, can mutually spin flip [7]. This leaves only indirect Al-Al and Cr-Cr spin flips to consider. Since T_{2c} is found to be still concentration dependent [8], it appears that both types of flips contribute. The indirect dephasing mechanism is confirmed by the measured Cr optical dephasing time $T_{2o} = 12 \pm 3$ μsec. Since $T_{2c} \sim 1$ msec is a lower limit on the Cr-Cr spin flip time, it is evident then the T_{2o} is determined almost completely by indirect Cr-Cr and Al-Al spin flips. If we assume that the fluctuating magnetic field at a Cr site is the same as that for a core Al, then we expect that $T_{2c}/T_{2o} = So/Sc$ where S is the magnetic splitting factor. We have $T_{2c}/T_{2o} = 83$ and $So/Sc = 312$ kHz/G/1.1 kHz/G $= 280$. We surmise that because the core Al are closer to the bulk Al than the Cr, a larger magnetic fluctuation appears at the Al sites. This is in qualitative agreement with the observed low T_{2c}/T_{2o} ratio. In conclusion we have directly verified the nuclear frozen core concept for solids and shown that for 0.0034% concentration, nuclear and optical dephasing in ruby is determined mainly by indirect Cr-Cr and Al-Al spin flipping. This conclusion is contrary to Compaan's earlier studies [9] on photon echo dephasing in ruby.

The technical assistance of J. Froemel is gratefully acknowledged.

4. <u>References</u>

1. J. Mlynek, N.C. Wong, R.G. DeVoe, E.S. Kintzer and R.G. Brewer, Phys. Rev. Lett <u>50</u>, 993 (1983).

2. A. Szabo, T. Muramoto and R. Kaarli, Optics Lett. <u>13</u>, 1075 (1988).

3. N. Bloembergen, Physica <u>15</u>, 386 (1949).

4. P.E. Jessop and A. Szabo, Optics Commun. <u>33</u>, 301 (1980).

5. P.F. Liao and S.R. Hartmann, Phys. Rev. <u>B8</u>, 69 (1973).
   P.F. Liao, P. Hu, R. Leigh and S.R. Hartmann, Phys. Rev. <u>A9</u>, 332 (1974).

6. R. Boscaino, F.M. Gelardi and R.N. Mantega, Phys. Letters <u>A103</u>, 391 (1984).

7. P.E. Jessop, T. Muramoto and A. Szabo, Phys. Rev. <u>B21</u>, 926 (1980).

8. T. Muramoto, R. Kaarli and A. Szabo (to be published).

9. A. Compaan, Phys. Rev. <u>B5</u>, 4450 (1972).

A LASER-DRIVEN SOURCE OF POLARIZED HYDROGEN AND DEUTERIUM

L. Young, R. J. Holt, R. A. Gilman, R. Kowalczyk, K. Coulter
Physics Division, Argonne National Laboratory, Argonne, IL 60439

A novel laser-driven polarized source of hydrogen and deuterium which operates on the principle of spin-exchange optical pumping is being developed[1]. This source is designed to operate as an internal target in an electron storage ring for fundamental studies of spin-dependent structure of nuclei[2,3]. It has the potential to exceed the flux from existing conventional sources[4] (3 x 10^{16}/s) by an order of magnitude. Currently, the source delivers hydrogen at a flux of 8 x 10^{16} atoms/s with an atomic polarization of 24% and deuterium at 6 x 10^{16} atoms/s with a polarization of 29%. Technical obstacles which have been overcome, with varying degrees of success, are: 1) complete Doppler-coverage in the optical-pumping stage without the use of a buffer gas, 2) wall-induced depolarization and 3) radiation-trapping. Future improvements should allow achievement of the design goals of 4 x 10^{17} atoms/s with a polarization of 50%.

Fig. 1 shows the prototype laser-driven polarized hydrogen/deuterium source. Two Ar+-pumped standing-wave dye lasers operating single-mode with Pyridine 2 dye provide the 770 nm radiation required to optically pump the D1 line of potassium. The spectral density of the lasers is tailored to match the Doppler-broadened absorption profile of the K vapor ($\approx$ 1.5 GHz) using a $LiTaO_3$ electro-optic modulator configured in a travelling-wave mode as a $\approx$50 ohm transmission line. The EOM is driven by a series of components which are used to generate high-power white noise. The modulated laser lineshape is characterized by an unbroadened carrier peak superposed on a noise-broadened background. The noise-broadened background has a spectral density which is reasonably well approximated by a Gaussian profile. Typically, the EOMs are run at a modulation index of 1.5 where roughly 10% of the power remains in the unbroadened carrier and the bandwidth (HWHM) is determined by the 3 dB roll-off of the amplifier (500MHz). The "tailored" laser output is sent through a linear polarizer, circular polarizer, and expanded to fill the cross-sectional area (4.5 cm^2) of the spin-exchange cell. At an incident intensity of $\approx$ 15mW/cm^2 a factor of ten enhancement is observed in the density of polarized potassium atoms when the laser is noise-modulated. This can be seen in Fig. 2 where the integrated polarization signal (as described below) is increased ten-fold by noise-modulation.

The spin-exchange cell is constructed as an integral unit with the rf dissociator (H/D source) and the K reservoir. It is placed in a static field of $\approx$ 10 G and heated to 230° C to prevent alkali condensation. The interior is coated with a polarization preserving material, drifilm[5]. The flux of the H/D atoms is controlled by a mass flowmeter in conjunction with a servo-driven needle-valve. The density of the K atoms is independently controlled through the reservoir temperature. Typically, K densities are 1-3 x 10^{11}/cm^3 and H/D densities are $\approx$ 10^{13}/cm^3. The usable flux of H/D atoms is determined by measuring, at the output, the fraction of H_2/D_2 which remains dissociated using a mechanical chopper, quadrupole mass spectrometer, and lock-in amplifier. A typical fraction for a 1000 bounce cell with a well-prepared drifilm surface is 70-80%.

The polarization is measured by optical detection of magnetic resonance transitions between the various Zeeman sublevels[6]. The transparency of the sample to the resonance radiation is decreased by inducing Zeeman transitions with a coil placed at right angles to the holding field. The scheme is shown in Fig.2 and measures polarization for both the alkali and deuterium, assuming a spin-temperature distribution for the populations of the magnetic sublevels: $N(m_F)=N_o exp(\beta m_F)$. Atomic polarizations of $\geq$80% are routinely obtained for the alkali, but hydrogen and deuterium polarizations range from 20% to 40%.

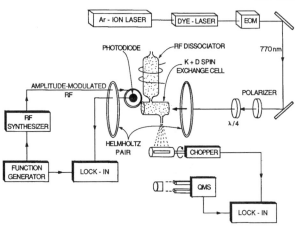

Fig. 1. Schematic of polarized source.

The foremost challenge in the continued development of this source is to enhance the efficiency of polarization transfer from the alkali to the H/D atoms. A rate equation model of the polarization transfer process between two spin 1/2 atoms can be used to understand the the relevant parameters[7]. At steady state, the polarization of the H/D, P_D, can be related to that of the alkali, P_A, by $P_D = P_A k_{se}/(k_{se}+k_d)$, where $k_{se} = n_A v \, \sigma_{se}$ is the spin-exchange rate, and $k_d = \alpha v(1+l/a)/2a$ is the geometry-dependent wall depolarization rate (α=depolarization probability/wall bounce, l=cell length, a=cell radius). For maximum efficiency of polarization transfer, one should maximize n_A and P_A while minimizing α.

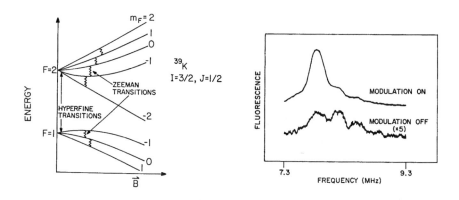

Fig.2. Polarization determination by optical detection of magnetic resonance. Ten-fold enhancement of polarized K density using noise-modulated laser.

For a given alkali and geometry, the maximum alkali density is limited by radiation trapping[8]. A critical density, n_o, can be defined as $(\sigma_o a)^{-1}$, where the alkali polarization is degraded to $\approx 50\%$. Radiation trapping tests on closed cells filled with an Ar buffer gas, have shown that the critical density for Na is a factor of two greater than that for K, and thus the attainable k_{se} using Na should be two times larger. With this consideration, we have modified the apparatus to use Na, rather than K as the spin-exchange intermediate. Preliminary results show that even with non-optimal optical pumping conditions for Na (an estimated 75% coverage of the Doppler profile) we were able to observe an increase from $\approx 25\%$ to $\approx 38\%$ in the deuterium polarization in the spin-exchange cell. Tests with improved laser power and Doppler coverage are currently underway. Future plans include optimization of the cell geometry, using both K and Na as spin-exchange intermediates and (if necessary) studies of wall coatings.

References

1.) R. J. Holt, Proceedings of the Workshop on Polarized Targets in Storage Rings, ANL Report No. 84-50 (1984); L. Young et al. Nucl. Inst. and Meth. B24/25 963 (1987); L. Young et al., Proceedings of the Topical Conference on Electronuclear Physics with Internal Targets, SLAC, Stanford CA Jan. 9-12, 1989.
2.) R. J. Holt et al., Nucl. Phys. A446, 389c (1985).
3.) D. H. Beck et al., Proceedings of the 3rd Conference on Intersections Between Particle and Nuclear Physics, Rockport ME May 14-19, 1988.
4.) H. G. Mathews et al., Nucl. Instr. and Meth. 213, 155 (1983).
5.) D. R. Swenson and L. W. Anderson, Nucl. Instr. and Meth. B28, 627 (1988).
6.) L. W. Anderson and A. T. Ramsey, Phys. Rev. 132, 712 (1963); R. Knize and J. Cecchi, Phys. Lett. 113A, 255 (1985).
7.) T. E. Chupp et al., Phys. Rev. C36, 2244 (1987).
8.) D. Tupa and L. W. Anderson, Phys Rev A36, 2142 (1987).

R.J.C. Spreeuw and J.P. Woerdman
Huygens Laboratory, University of Leiden, P.O. Box 9504,
2300 RA Leiden, The Netherlands

Concepts of electronic band structure theory can be used to describe the behavior of light waves in periodic dielectric media. An example of a generalized band structure is shown in Fig. 1. The periodicity involved can be microscopic as well as macroscopic. A well-known example of photonic band structure for the case of a microscopic dielectric periodicity is propagation of light through a multilayer dielectric mirror. An extension thereof to the three-dimensional case has been speculated upon (1). Following a recent theoretical prediction (2) we have now realized a photonic band structure with a *macroscopic* periodicity, namely an effectively rotating fiber ring (3) (Fig. 2). Dielectric structure and therefore backscatter is introduced by means of an air gap in the ring and rotation is simulated by means of the Faraday effect. We have mapped the "valence" and "conduction" bands of this photon band structure; band gaps are manifest as doublet and quartet splittings (Fig. 3). Due to the macroscopic nature of our system typical orders of magnitude of the quantities involved are very different from those in semiconductor physics, but the principles are the same. For our ring the periodicity length (i.e. the circumference) is typically ≈ 1 m and the band gaps (i.e. the splittings in Fig. 3) are typically $\approx 10^7$ Hz; for a semiconductor these quantities are $\approx 10^{-10}$ m and 10^{14} Hz, respectively. The light wave in the ring is a pure standing wave at the top of the "valence" band and also at the bottom of the "conduction" band. The difference between these two standing waves is that a node or an anti-node, respectively, is present at the air gap: the resulting difference in dielectric polarization energy determines the magnitude of the optical band gap. The macroscopic nature of our system has important advantages for basic study since it allows easy manipulation of the parameters of the band structure; this is not (yet) possible for the microscopic case.

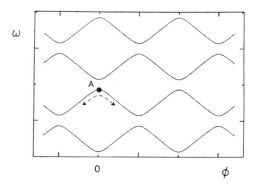

Fig. 1: *Generalized band structure for the case of wave propagation through a periodic medium. In the electronic case the control parameter ϕ is the electron wave number; in the case of an optical ring it is a nonreciprocal phase (Sagnac or Faraday).*

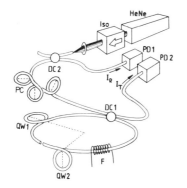

Fig. 2: Sagnac fiber-ring with
backscattering element R.
PC is polarization controller,
QW quarterwave fiber subloop,
DC directional coupler,
PD photodiode and
F Faraday element.

The vector character (polarization) of the photon doubles the number of bands as compared to the electronic case. Generally, the bands have degeneracy 1, 2 or 4. If the round-trip birefringence vanishes we have effectively the scalar case; experimentally this corresponds with singlet and doublet spectra. A net round-trip birefringence (linear or circular) lifts the remaining degeneracy and generally leads to quartet spectra. Level crossings in the photon band structure form a demonstration of Dyson's threefold symmetry classification (4); the three classes correspond to 2 x 2 matrices with real, complex and quaternion elements, respectively.

The photon band structure shows interesting dynamical aspects; as an example, recently we have observed Bloch oscillations (5). For a brief description of this phenomenon we refer again to Fig. 1, using the terminology relevant to photons circulating in an effectively rotating optical ring cavity. For slow changes of the nonreciprocal phase Φ, the optical field will follow adiabatically. In points like A this leads to a reflection at the bandgap and a reversal of the group velocity ($\propto d\omega/d\Phi$). Thus a slow variation of Φ may change the direction of the Poynting vector from clockwise (cw) to counterclockwise (ccw). However, if the change in Φ is sufficiently fast, the bandgap will be crossed, a phenomenon known in solid-state electronics as Zener tunneling. The criterion for adiabaticity is that the variation of Φ must be small on a timescale of the reciprocal bandgap. It is important to appreciate that Bloch oscillations (or bandgap reflections) are due to the coherent addition of backscattered fields. Thus the cavity loss rate should be sufficiently small

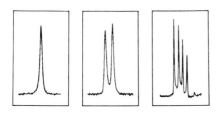

Fig. 3: Examples of singlet,
doublet and quartet spectra
detected by PD1 when the
laser frequency is tuned.

369

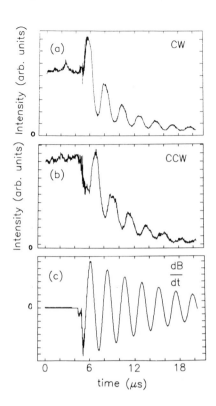

Fig. 4: Bloch oscillations in the photonic band structure which occur upon magnetically simulated rocking of the ring (see dashed arrow in Fig. 1). The cw and ccw intensities oscillate in antiphase (see (a) and b)) when an AC current passes through the Faraday coil. The voltage across a pick-up coil is a measure of the magnetic field in the Faraday coil (see (c)).

in order not to destroy this coherent addition: Bloch oscillations can only take place within the photon lifetime of the cavity. Experimentally, we have used an optical ring with an intracavity phase-preserving amplifier (HeNe) in order to realize a large value of the photon lifetime (5). By magnetically simulated "rocking" of this ring the optical wave follows adiabatically and its Poynting vector changes indeed periodically from cw to ccw (Fig. 4).

In comparison, the search for (electronic) Bloch oscillations in solid-state physics has been extensive but so far unsuccessful (6). The difficulties encountered in this case are possibly due to the role of inelastic losses (e.g. electron-phonon scattering) and due to the fact that electrons are fermions so that the motion of many electrons in a solid is incoherent. As a consequence of the latter, electronic Bloch oscillations are expected to be smeared out, if many electrons are involved. Photons are bosons and can form a coherent state ('classical light wave'), so that smearing does not occur; this also allows us to use an amplifier to cancel the losses.

One of the underlying motivations of our work is that we are intrigued by the question: "How deep is the analogy between a photon and an electron?". Often heard is the argument that the scalar wave equation of electromagnetism can be reduced, in the slowly varying envelope approximation, to the Schrödinger equation so that light-wave physics can be mapped on matter-wave physics. This mapping is extremely useful but has, nevertheless, severe limitations as already pointed out by Pauli and Ehrenfest (7,8,9): photons *are* different from electrons! The discussion above, on the vector character and on the boson/fermion issue, illustrates this point.

Also interesting is the connection between the photonic band structure of our ring cavity and frequency locking phenomena occurring in laser gyros (10). In both cases the physics is that of two coupled modes. In the case of a laser gyro one considers an active (i.e. oscillating) ring with minimized but finite backscatter; this may result in locking of the two counterpropagating modes, leading to a dead zone. In our case we deal with a passive ring with large backscatter; this may lead to an optical band gap. The two phenomena are intimately connected, yet distinct; the nature of the backscattering is crucial in this respect. Conservative backscattering leads to two normal modes with different frequencies, i.e. to an optical band gap. Dissipative backscattering leads two normal modes with different dampings but with the same frequency, i.e. to a dead zone.

References:

1. E. Yablonovitch, Phys.Rev.Lett. 58, 2059 (1987).
2. D. Lenstra, L.P.J. Kamp and W. van Haeringen, Opt. Commun. 60, 339 (1986).
3. R.J.C. Spreeuw, J.P. Woerdman and D. Lenstra, Phys.Rev.Letters 61, 318 (1988).
4. F.J. Dyson, J.Math.Phys. 3, 1199 (1962).
5. R.J.C. Spreeuw, E.R. Eliel and J.P. Woerdman, submitted for publication.
6. Physics Today, May 1988, p. 19.
7. P. Ehrenfest, Z. Physik 78, 555 (1932).
8. W. Pauli, Z. Physik 80, 573 (1933).
9. R. Peierls, *"Surprises in theoretical physics"* Princeton University Press, Princeton, 1979, p. 10.
10. H.A. Haus, H. Statz and I.W. Smith, IEEE J. Quantum Electron. 21, 78 (1985) and references therein.

Experimental Studies of Atom Diffraction and The Mechanical Forces of Light on Atoms

David E. Pritchard

Research Laboratory of Electronics and Department of Physics
Massachusetts Institute of Technology, Cambridge, MA 02139

In 1985 the Journal of the Optical Society of America published an issue on the "Mechanical Forces of Light" (JOSA-B85). Although the theory of light forces was well developed at that time, there were then only a few experimental studies of light forces on atoms owing to the smallness of these forces and the associated experimental difficulties. The review issue had only *one* paper which quantitatively compared measured and theoretical light forces, and that paper qualitatively contradicted earlier experiment.

In the four years since then, experimental study of light forces has progressed substantially, especially in our lab at M.I.T. and labs in Moscow and Ecole Normale Superieure in Paris. A great deal of the "standard" theory of light forces has been compared with our high resolution experiments, and several new effects have been studied at E.N.S. This paper is a brief summary of the M.I.T. work. Each different regime of the parameter space of light forces which we studied is reported separately, and a short relevant bibliography is presented for that regime.

Our experiment on the quantitative comparisons with light forces theory has produced significant results in five major areas. Each area represents a different combination of the basic parameters of the problem: Rabi Frequency ω_R, detuning δ, number of spontaneous decay photons N_S, interaction time τ, and Doppler shift due to the initial velocity of the atoms across the nodes of the standing light wave, $v_\perp$.

The basic interacting parameters of the laser field of strength ε_L and frequency ω_L with the two state atom Na (3S, F = 2, m_F = 2) are:

$\gamma_s \equiv 2\pi \cdot 10$	MHz spontaneous decay rate of atom
$\omega_R \equiv <3p \,\lvert\, e\vec{r}\cdot\vec{\varepsilon}_L \,\rvert\, 3s>/\hbar$	Rabi Frequency
$\delta \equiv \omega_L - \omega_{atom}$	Detuning
$\delta_D \equiv k\,v_\perp$	Doppler shift due to atom's motion
$t \equiv$	Interaction time of atom with field
$N_S = \dfrac{\omega_R^2}{4\delta^2 + \gamma_s^2}\,\gamma_s\, t$	Number of spontaneous decays in time t

All of our experiments were carried out in a standing wave (ε_L is the field of one travelling wave component) except for the photon absorption statistics (which used a travelling wave).

Our experiments on light forces were done in a high resolution apparatus where the momentum transferred by light to a monochromatic beam of atoms prepared in a single hyperfine state (GRM87) could be measured to ~ 5%. The momentum resolution was less than $\hbar k$, the momentum of a single photon, thus revealing the finest details which can be present. The optical interaction parameters were known well enough to permit quantitative comparisons with various theories of light forces at the $\leq 10\%$ level.

a. Kapitza-Dirac diffraction ($\delta_D = 0$, $\gamma_s t \approx 10$, $N_s < 0.2$).

We observed (GRP86) diffraction of atoms by a standing light wave in the regime where the laser and atomic beams were orthogornal the interaction time was short ($\tau \doteq 100$ nsec) and the detuning of the laser's frequency from the atomic transition frequency was much larger than the laser intensity (i.e. $\delta \gg \omega_R$) so that spontaneous emission was negligible ($N_s < 0.2$). This experiment resolved a previous discrepancy between experiment (ALO79, GKN81) and theory (BES81), in favor of the latter. It also provides a textbook example of wave-particle duality: matter waves diffracting from a light grating is the waves $\leftrightarrow$ particle dual of light waves diffracting from a normal material grating. This experiment is the first realization of an old suggestion that electrons could scatter from a standing light wave (KAD33).

b. Diffraction of moving atoms ($N_s < 0.2$, $\gamma_s t \approx 10$, $\delta_D \approx t^{-1}$).

We observed and measured diffraction of atoms moving through a standing light wave with the same parameters used for Kapitza-Dirac diffraction (except for δ_D). The effects of such motion had not previously been studied experimentally or theoretically. Fortunately we were able to generalize the theory, predicting for our experimental conditions that the rms deflection should behave as a Guassian in velocity space (with a width ≈ 1 m/s). Our data agreed with this prediction within error, and quantitatively verified our theory for this regime (MGO87, ABS81).

c. Bragg Scattering ($N_s < 0.3$, $\gamma_s t \approx 10^4$, $\delta_D \approx t^{-1}$).

We studied (MOM88) the diffraction of atoms by a standing light wave in the regime where the interaction time was $\sim 10^{-5}$ sec ($\sim \times 100$ larger than for Kapitza-Dirac diffraction), long enough for the atoms to slosh back and forth in the valleys of the standing wave potential. We observed restricted, unidirectional momentum transfer, whose scattering pattern was dramatically altered by tilting the atomic beam with respect to the laser beam at very small angles ($\theta = 30 \mu$rad; $v_1 = 0.03$ m/s). This regime is a new physical realization of Bragg scattering (BRA12) familiar in x-ray and neutron scattering from crystals. We were able to study it in a new area of parameter space which led to pioneering observations of 2nd and 3rd order Bragg scattering, as well as permitting experimental variation of the interaction strength of the grating (we changed the laser power to observe the Pendellosung effect (SHU68)).

d. Transition from diffraction to diffusion ($0.3 < N_s \leq 10$, $\gamma_s t \approx 10$, $\delta_D = 0$).

We explored the transition to the regime where spontaneous emission becomes important (GMR89). The momentum transfer function changed from diffractive (sharp peaks) to diffusive (structureless) as the number of spontaneous emissions per atom became $\gg 1$. These measurements represent the first stringent test of diffusive light forces theories (BFA78, COO80a, TRC84). These are generally based on a Fokker-Planck approach). The smoothed average of our experimental curves agreed satisfactorily with the predictions, but our measurements revealed an unexpected feature that theory has not yet adequately addressed - the persistence of diffractive structure into the supposedly diffusive regime.

e. Travelling wave photon absorption statistics ($0.3 < N_S < 10$, $\gamma_S t \approx 10$, $\delta_D = 0$)

This recent work (OMG90) is the first careful study of the statistical properties of photons absorbed using the momentum transfer in a travelling wave/atom interaction (COO80b). The statistics of the momentum transfer distribution displayed a less than stochastic variation of the number of absorptions by each individual atom. This is due to a regularization of the times of absorption for photons, an effect closely related to anti-bunching of subsequent fluorescence photons from a single atom which has been observed in several experiments [SHM83]. Our experiment is the first to measure the statistics of absorption, the first to measure the dependence on laser detuning, and has displayed the largest suppression of fluctuations relative to Poisson statistics (over 50%).

f. Diffraction of atoms by a matter transmission grating.

In the previous sections (esp. a) the cause of the diffraction of atoms was the light-induced change of the energy of the atom (ac Stark effect); this gave the phase of the atom wave a periodic dependence on position after passing through the light. In optical terminology the standing light wave would be termed a "transmission phase grating" since it alters the phase of the transmitted atom de Broglie wave which passes through, but not its amplitude. A more familiar type of optical transmission grating is an amplitude grating, which typically consists of a periodic array of opaque bars. We have recently demonstrated that if these bars can be made self-supporting (ie. with no glass plate used to support the bars) then they can diffract atom waves (KSS88). This demonstration suggests that it should be possible to construct a three grating atom interferometer (ALT73, CHE87), a possibility which we are actively pursuing.

The work described here was started about seven years ago and was done primarily by people other than the author. I would particularly like to thank Phil Moskowitz, Phil Gould, Peter Martin, Bruce Oldaker and David Keith who all spent several years working on these experiments. The work was funded by the National Science Foundation (PHY86-05893) with help from the Joint Services Electronics Program (DAAL03-86-K-0002) which supports the M.I.T. Submicron Structures Laboratory where the atom diffraction gratings were made.

Finally, I apologize to the many workers, especially theoretical ones, whose work has not been mentioned here in spite of its clear relevance — I wanted to keep the references shorter than the paper. We are preparing a longer review of this material which will be more inclusive — please write the author (Room 26-237/MIT) if you want a copy.

References

ABS81   E. Arimondo, A. Bambini, and S. Stenholm, Phys. Rev. A 24, 898 (1981).

ALO79   E. Arimondo, H. Lew, and T. Oka, Phys. Rev. Lett. 43, 753 (1979).

ALT73   S. Altschuler and Lee Frantz, U.S. Patent # 3761721, Sept. 25, 1973.

BES81   A.F. Bernhardt and B.W. Shore, Phys. Rev. A 23, 1290 (1981).

BFA78   J.E. Bjorkholm, R.R. Freeman, A. Ashkin, and D.B. Pearson, Phys. Rev. Lett. 41, 1361 (1978).

BRA12   W.L. Bragg, Proc. Cambridge Philos. Soc. 17, 43 (1912).

CHE87   V.P. Chekotayev et al., J. Opt. Soc. Am. B 2, 1791 (1987).

COO80a   R.J. Cook, Phys. Rev. A 22, 1078 (1980).

COO80b   R.J. Cook, Opt. Commun. <u>35</u>, 347 (1980).

GKN81   V.A. Grinchuk, E.F. Kuzin, M.L. Nagaeva, G.A. Ryabenko, A.P. Kazantsev, G.I. Surdutovich, and V.P. Yakovlev, Phys. Lett. <u>86A</u>, 136 (1981).

GRM87   P.L. Gould, G.A. Ruff, P.J. Martin, and D.E. Pritchard, Phys. Rev. A. <u>36</u> 1478 (1987).

GRP86   P.L. Gould, G.A. Ruff, and D.E. Pritchard, Phys. Rev. Lett. <u>56</u>, 827 (1986).

GMR89   P.L. Gould, P.J. Martin, G.A. Ruff, R.E. Stoner, J. Picque, and D.E. Pritchard, submitted to Phys. Rev. Lett.

JOSA-B85   Special Issue of J. Opt. Soc. Amer. <u>B2</u>, 1706-1860, 1985.

KAD33   P.L. Kapitza and P.A.M. Dirac, Proc. Cambridge Philos. Soc. <u>29</u>, 297 (1933).

KSS88   D.W. Keith, M.L. Schattenburg, Henry I. Smith, and D.E. Pritchard, Phys. Rev. Lett. <u>61</u>, 1580 (1988).

MGO87   P.J. Martin, P.L. Gould, B.G. Oldaker, A.H. Miklich and D.E. Pritchard, Phys. Rev. A. <u>36</u>, 2495 (1987).

MOM88   P.J. Martin, B.G. Oldaker, A.H. Miklich, and D.E. Pritchard, Phys. Rev. Lett. <u>60</u>, 515 (1988).

OMG90   B.G. Oldaker, P.J. Martin, P.L. Gould, M. Xiao, and D.E. Pritchard, (in preparation).

SHU68   C.G. Shull, Phys. Rev. Lett. <u>21</u>, 1585 (1968).

SHM83   R. Short and L. Mandel, Phys. Rev. Lett. <u>51</u>, 384 (1983).

TRC84   C. Tanguy, S. Reynaud, C.Cohen-Tannoudji, J. Phys. B. <u>17</u>, 4673 (1984).

Improved Kennedy-Thorndike Experiment - a preliminary report

Dieter Hils and J. L. Hall*
Joint Institute for Laboratory Astrophysics
National Institute of Standards and Technology and University of Colorado
Boulder, CO 80309-0440

The possibility of using lasers to improve the accuracy of the classical experiments (1-3) of Special Relativity (SR) was originally suggested by Javan and Townes and in fact they were the first to make a more precise Michelson-Morley experiment (4,1). The full potential of modern laser frequency metrology for length measurements was however not exploited until the more recent precision Michelson-Morley experiment (MM) of Brillet and Hall (5) which achieved a fractional frequency uncertainty of $\pm 2.5 \times 10^{-15}$ in showing the isotropy of space. They noted also the technical difficulties which would have to be overcome to achieve similar large improvements in a laser version of the Kennedy-Thorndike (KT) experiment (2), which compares the transformations of time and length in a moving frame. The present measurements yield a sensitivity of $\sim 2 \times 10^{-13}$ for a term with the expected 24 sidereal hr term, corresponding to a ≈ 300-fold higher accuracy than the original KT experiment.

Following Roberson (6), Mansouri and Sexl (7-9) have developed a useful framework for explaining what an experiment measures and how it relates to other experiments. They consider two coordinate systems in relative motion, and write the most general transformation between Σ (the preferred frame) and S (the moving frame). They introduce multiplicative kinematical parameters which might be determined by theory but, more importantly, can be determined by experiment. Because of isotropy in Σ, these transformation parameters are even functions, dependent only on $(v/c)^2$. Apart from synchronization, the transformation between the preferred frame Σ and a moving frame S is specified by these parameters. The velocity of a light ray in S in general depends on the direction of propagation. Special Relativity makes the unambiguous statement $\alpha = -1/2$, $\beta = 1/2$, $\delta = 0$, corresponding to $c(\theta) = c$ for all frames.

Modern astrophysical measurements (10) of anisotropy in the microwave background (μW) seem to define a universal standard of rest which reasonably could be taken to be the preferred frame Σ. In the following we will assume that the relevant velocity v is given by the observed motion of earth with respect to the μW frame. The signature of the effect we search for is its dependence on the sidereal modulation of v due the earth's rotation.

Present knowledge of the parameters α, β, δ (which also quantifies the degree of agreement of Einstein's SR and observation) comes principally from the second order MM and KT optical experiments and from optical and Mossbauer rotor experiments which determine the time dilation parameter a. (For a more detailed discussion of the many excellent experiments we refer to Refs. 7-9, 11.) The most accurate time dilation experiments (12-14) imply $\alpha = -(1/2) \pm 1 \times 10^{-7}$ and the most accurate MM experiment (5) determines $\beta - \delta = (1/2) \pm 5 \times 10^{-9}$. The original KT experiment (2) leads to $\alpha - \beta = -1 \pm 2 \times 10^{-2}$ which thus introduces the single greatest uncertainty in the transformation equations. For these reasons, the importance of improving the KT experiment has been stressed repeatedly (9, 15, 16).

In the KT experiment a differential comparison was made between a standard of time defined by a mercury lamp and a standard of length in the form of an unequal arm Michelson interferometer. Our laser experiment utilizes instead two He-Ne lasers, one locked to a molecular absorption line in I_2 (R(127) 11-5), while the other is locked to a very stable Fabry-Perot reference cavity. Their frequencies are compared by optical heterodyne detection. Let ν^S_2 denote the I_2-stabilized laser frequency. From Eqs. (6.15-18) of Ref. 8, we determine (17) the cavity locked laser frequency ν^S_1 in the moving frame S. The heterodyne beat of the two optical frequencies is

$$\nu^S_{beat}/\nu_c = 1 + (\beta-\alpha-1)(v/c)^2 + (\delta-\beta+1/2)(v/c)^2 \sin^2\theta - \nu^S_2/\nu_c \quad , \tag{1}$$

where $\nu_c = pc/(2L)$, p is an integer, and L is the length of the Fabry-Perot interferometer as measured in S. The precise MM result of Brillet and Hall (5) implies $\delta-\beta+1/2 = 0\pm5\times10^{-9}$ and so to this accuracy

$$\nu^S_{beat}/\nu_c = 1 + (\beta-\alpha-1)(v/c)^2 - \nu^S_2/\nu_c \quad . \tag{2}$$

Working in the preferred frame Σ one can show that $\nu^\Sigma_{beat}/\nu_c = [1 + \alpha(v/c)^2] \cdot (\nu^S_{beat}/\nu_c)$, a result to be expected due to the time dilation effect.

We next determine the velocity v of our laboratory with respect to the μW frame. Working in the earth equatorial frame we find for the main sidereal component:

$$(v/c)^2 = (u/c)^2 + 2(u/c)(\Omega R_\oplus/c)\cos\phi_L \cos\delta_\mu \sin[\Omega(t-t_S) + \Phi] \quad . \tag{3}$$

In this equation $u = 377\pm14$ km/s is the velocity of earth with respect to the μW frame (10,18), $\Omega = (2\pi/P_\oplus)$ with $P_\oplus$ being the sidereal period and $R_\oplus$ the earth radius, $\phi_L = 40°$ is the latitude of Boulder , $\delta_\mu = -6.4\pm1°$ is the observed declination of the μW velocity vector (10,18) and Φ is the phase at the start of the analysis epoch. From Eqs. (2) & (3) we see that determination of the factor $(\beta-\alpha-1)$ critically depends on our ability to measure the 24 hr sidereal variation of the fractional beat frequency.

The principle of our experiment is discussed with reference to Fig. 1. A He-Ne laser (λ=6328 Å) is locked (19) to a highly stable, isolated Fabry-Perot interferometer, thereby satisfying optical standing-wave boundary conditions. The servo system then transforms length variations of the cavity (of accidental or cosmic origin) into laser frequency variations. These can be sensitively detected by optically hetero-dyning some of the laser power with an optical frequency reference provided by an I_2-stabilized laser. The beat frequency (~160 MHz if the I_2-stabilized laser is locked to the "d" component of the R(127) 11-5 transition) is counted for 40 s and stored with negligible dead-time. We usually acquire 4320 beat frequency readings (=2 days) in memory before we store the data on disk and restart the measurement.

Our fundamental standard of length is the Fabry-Perot interferometer. It uses Zerodur "gyro" quality mirrors optically-contacted to the ends of a Zerodur spacer (length 30 cm, diameter 15 cm). The radii of curvature are R=575 cm and R=∞; transmission is T=30 ppm. The cavity fringe width=72 kHz fwhm giving a finesse F=6600, and a resonant transmission of ~2%. The interferometer is suspended by two stainless steel ribbons (1x0.01 cm^2) , one at each end, inside a thick-walled vacuum envelope. A low-drift servo loop with ac thermistor sensing stabilizes this aluminum wall temperature to better than 5 μK over a 1 day period. The experiment is located inside the "Quiet House" (20) which provides a >20 dB thermal and acoustic barrier. The inside air temperature is stabilized by Peltier-cooled panels which pump the laser discharge heat out via a slow water flow. This yields a stability better than 1 mK over times of 1 hour and better than 10 mK over a 1 day period, in the face of room temperature variations of ±1 degree.

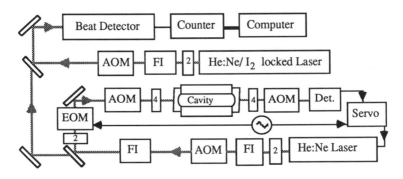

FIG. 1. Laser-based Kennedy-Thorndike experiment. A He-Ne laser is locked to a transmission fringe of a highly-stable Fabry-Perot cavity. The heterodyne beat between it and a second He-Ne laser, stabilized to I_2, is measured in 40 s intervals and stored in the computer memory. After 2 days (4320 points), the beat frequency and temperatures data are stored to disk and the experiment reinitialized. FI means Faraday Isolator. The thin waveplates are marked in inverse fractional waves.

The "hard-seal" He-Ne laser provides about 1mW of red light at 6328 Å. The beam passes through an isolation stage formed by a pair of Faraday isolators and one acousto-optic modulator (AOM). An ADP phase modulator crystal (EOM) is driven at ~1 kV pk-pk at a modulation frequency of 25 kHz via a resonant step-up transformer. Additional isolation (via frequency shift) of the laser beam reflected by the Fabry-Perot interferometer is provided by a second AOM. The transmitted light is frequency shifted by a third AOM to avoid fringes of the photo-detector's scattered light and the output mirror of the cavity. The ac output of the photodetector, after lockin detection, provides the error signal for the servo-system. Based on the 10 μW of fringe signal and a unity gain frequency of 5 kHz, the shot noise limit of the cavity-locked laser is expected to be ~1 milliHz, while the observed (19) frequency noise at short times (~1 s) is less than 50 mHz using first harmonic detection.

The I_2-stabilized He-Ne laser which serves as our optical frequency standard achieves a stability of ~500 Hz in 40 s due to the finite S/N of the ~0.1% I_2 saturation peak in ~100 μW of laser power. The long term stability is ~100 Hz. In contrast, the cavity length has excellent short-term stability, but long-term changes arise from internal processes (creep) and environmental thermal effects. Due to ageing effects (shortening) of the Zerodur spacer, the beat between the two lasers exhibits a long-term uniform drift of 1.65 Hz/s (3×10^{-10}/day or 1.1×10^{-7}/yr) beginning in July 1986, down to 1.06 Hz/s in March 1989. Our observed drift rates agree well with the ageing curve measured by Bayer-Helms et al. (21) for Zerodur gauge blocks. The cavity frequency is predictable within 1 Hz for 1000 s and within <300 Hz for 1 day.

Figure 2 shows a two day segment of the recorded heterodyne beat frequency. The uniform Zerodur creep, which is about 185 kHz, has been subtracted. We suppose residual temperature changes working via the I_2 pressure shift or the cavity expansion may cause part of the observed (small) variations.

From a three week record of the beat frequency data, we made predictions/correlations based on measured Quiet-House residual temperature variations. The model fitted (for the I_2) a small prompt term (-1.1 kHz/K) and (for the cavity) a "thermal integrator" of 24 hr time-constant (-320 Hz/K). The correspondence is good, but not perfect. It is clear that thermal control is still a problem. See Fig. 2.

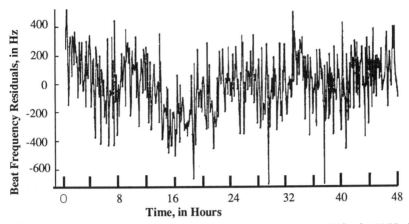

FIG. 2. Heterodyne beat frequency residuals. Uniform cavity drift of 1.08 Hz/s removed. Fast noise is from the I2 reference. Slow drifts may be partly thermal.

Our first KT data begin on July 27, 1986, and end September 22, 1986. This includes ~15 days of uninterrupted data. Another block of data begins October 31, 1988, ends April 31, 1989, and includes more than 90 days of uninterrupted data. Analysis proceeds as follows: We first average the 4320 beat-frequency samples (=2 days) in blocks of 9 which reduces the sampling rate to 10/hr. We next remove from the 2 day data set the uninteresting linear trend from cavity creep. We note this procedure may remove some power (<10%) from a hypothetical sinusoidal signal at 1 cycle per day. We then decimate the data set (like Fig. 2) to a final sample rate of 1/hr. This leaves us with the remaining frequency residuals due to uncontrolled environmental perturbations, as well as to a possible "aether effect."

We have examined the frequency residuals for a sidereal signal by several methods, with similar results. One method is back-folding the data to form a 1 year record of more or less continuous data. Instead, here we Fourier analyze the full nearly-3-year-long record by putting zero values into the data gaps. (The amplitude scale has been corrected x 11.3 for the attenuation due to the zero-fill procedure.) The amplitude spectrum for Fourier frequencies up to 3 cycles per day is interesting. For frequencies <1/2 cycle per day the observed amplitudes steeply decline due to the removal of the linear trend from the data. For Fourier frequencies above a few cycles per day, the spectrum falls off rapidly as expected for a model which has thermal disturbances acting on a reference cavity with a time constant of ~1 day.

Examination of Fig. 3A indicates an enhanced noise level near 1.0 cycles per day. It is easy to imagine a strong driving term at the solar frequency, perhaps phase-shifted and broadened by variables such as cloud cover, weekend work schedules, etc. With breaks in the data, the usual windowing procedure is ineffective, so even a bright line input will corrupt adjacent frequency bins. To find the transfer function, we tried adding a strong solar or sidereal signal in the time domain. The half-amplitude FT width was ±3 bins. Removing the best solar sinewave drops the sidereal amplitude from 31.7 to 14.9 Hz. See Fig. 3B.

To test if the increased noise level was due to leakage from a "real" signal at the sidereal frequency, we removed this best-fitting sinewave in the time-domain as before. The spectrum was essentially unchanged from Fig. 3A. We conclude that the broad noise buildup around 1 cycle/d is due to solar -- not sidereal -- input.

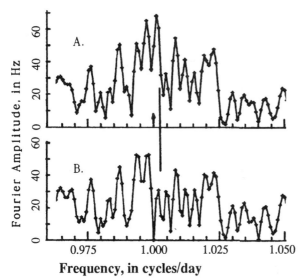

FIG 3. Beat spectrum near 1 cycle per day. Arrow shows solar frequency, 1 cycle/d, corresponding to bin 1097. Vertical line marks sidereal frequency, 1.00274 cycle/d corresponding to bin 1100.

A. FT of 3 years of data.
B. Solar term of 49.4 Hz amplitude removed from time domain data before FT. Sidereal amplitude drops from 31.7 Hz to 14.9 Hz. See text.

For this report we work conservatively with the unmodified data of Fig. 3A. Further, we ignore the phase information implicit in Eqs.(2) & (3). Working near (±4 bins) the sidereal bin, the quadrature amplitudes are found to be normally distributed, with standard deviation $\sigma_n = \sqrt{<A_c^2>} = \sqrt{<A_s^2>} = 29$ Hz. The sidereal amplitude, 31.7 Hz, gives a normalized measured value $x_m = 31.7/29 = 1.09$.

We now wish to analyze our results with the hypothesis that we have a sinusoidal signal of amplitude P in the presence of random noise. From tabulated values of the integral of the error function, we find there is 10% probability for finding a realized value beyond $x_{10} = 1.28$. Thus from a simple analysis we would conclude that there is less than 10% probability that there was a real signal as large as $X = x_{10} + x_m = 1.28 + 1.09 = 2.37$ units of σ_n, i.e. $P < 68.7$ Hz.

However inspection of Fig. 3A shows that one local peak (A=65 Hz) falls in bin 1094, just conjugate to the sidereal frequency of interest, bin 1100, ie. symmetric around the solar bin (1097). We offer the following, less optimistic scenario: Suppose that the modulation processes (weather, cloud-cover, etc) that spread out the solar forcing-function produce symmetrical "sidebands" around the solar bin. Then we can view the the upper "weather" sideband, also of A=65 Hz, as partially cancelling a putative sidereal amplitude to give the realized value of 31.7 Hz. A pessimistic estimate would assume ideal out-of-phase cancellation, producing the value 65 + 31.7 = 96.7 Hz for this sidereal amplitude. We regard this scenario as less than 10% probable, and can therefore believe that *our experiment shows, with a probability > 90%, that there is a no sidereal signal as large as 96.7 Hz (rounded to 100 Hz)*.

The present result, setting an upper limit to a possible "aether effect" amplitude of P<100 Hz, corresponds to a fractional frequency amplitude $\Delta v/v < 2 \times 10^{-13}$. From Eqs.(3),(4) our experimental result can be expressed in the form $2(\beta-\alpha-1) u < 50$ m/s. Using the value u = 377 km/s one obtains $(\beta-\alpha-1) < 6.6 \times 10^{-5}$. This limit enables us to deduce separately $\beta = (1/2) \pm 7 \times 10^{-5}$ and $\delta = 0 \pm 7 \times 10^{-5}$. Taken with the already known value (12-14) $\alpha = -(1/2) \pm 1 \times 10^{-7}$, we see that the three parameters defining the transformation equations are now known to agree with Einstein's SR values to better than 70 ppm. This value represents a 280-fold improvement over the original

KT experiment. Further analysis, taking phase into account, may sharpen our limit.

The present experiment is not easily improved by a large margin, at least not for earth-based experiments. Some cavity problems would be minimized with a space-based experiment, e.g. cavity distortion due to body forces would be essentially eliminated. For an orbital period of ~90 min the effective thermal isolation of the reference cavity would be ~20 times better. With thermal control to 1/10 K, a factor of 100 improvement seems possible, if similar improvements of the I_2 stabilized laser are possible. We note that the flicker noise due to the I_2 laser is ~100 Hz (2×10^{-13}), similar to the obtained value for Δv_{beat}. Since the pressure shift in the I_2 cell is ~15 kHz/K, a reservoir temperature stability of 10 μK will be needed. Intensity shifts and gas lens effects are also serious.

This report summarizes an improved Kennedy-Thorndike type experiment based on modern laser metrology. The heterodyne signal between a cavity-locked He-Ne laser and one stabilized on I_2 shows a fractional frequency amplitude at the sidereal frequency of $\Delta v_{beat}/v_c < 2 \times 10^{-13}$ (90% confidence interval). This null result is more accurate by a factor of ≈ 300 than the previous best measurement, by R. J. Kennedy and E. M. Thorndike in 1932! Following the reasoning of Robertson (6), the Lorentz transform of SR can now be based on experimental facts at the 70 ppm level.

We are grateful to J.M. Chartier from the B.I.P.M., Sèvres, France for constructing the reference laser. We acknowledge useful discussions of Fourier techniques with J. Levine, H. Hill, R. Stebbins and K.P. Dinse. We are grateful to M. Winters for developing part of the data acquisition program. We especially would like to thank P.L. Bender for his interest and many helpful discussions on many aspects of the experiment. This work has been supported at JILA in part by the National Institute of Standards and Technology under its program of precise measurement research for possible applications in the basic standards area, and in part by the National Science Foundation, NASA, and the ONR.

References

*   Quantum Physics Division, National Institute of Standards & Technology
1.   A. A. Michelson and E. W. Morley, Am. J. Sci. 34, 333 (1887).
2.   R. J. Kennedy and E. M. Thorndike, Phys. Rev. 42, 400 (1932).
3.   H. E. Ives and G. R. Stillwell, J. Opt. Soc. Am. 28, 215; 31, 369 (1941).
4.   T. S. Jaseja, A. Javan, J. Murray, and C. H. Townes, Phys. Rev. 133, A1221(1964).
5.   A. Brillet and J. L. Hall, Phys. Rev. Lett. 42, 549 (1979).
6.   H. P. Robertson, Rev. Mod. Phys. 21, 378 (1949).
7.   R. M. Mansouri and R. U. Sexl, Gen. Rel. 8, 497 (1977).
8.   R. M. Mansouri and R. U. Sexl, Gen. Rel. 8, 515 (1977).
9.   R. M. Mansouri and R. U. Sexl, Gen. Rel. 8, 809 (1977).
10.   G. F. Smoot, M. V. Gorenstein, and R. A. Muller, Phys. Rev. Lett. 39, 898 (1977).
11.   M. P. Haugen and C. M. Will, Phys. Today 40, 69 (1987).
12.   G. R. Isaak, Phys. Bull. 21, 255 (1970).
13.   M. Kaivola, O. Poulsen, E. Riis, and S. A. Lee, Phys. Rev. Lett. 54, 255 (1985).
14.   E. Riis, L. A. Andersen, N. Bjerre, and O. Poulsen, S. A. Lee and J. L. Hall, Phys. Rev. Lett. 60, 81 (1988).
15.   H. P. Robertson, Rev. Mod. Phys. 29, 173 (1957).
16.   E. F. Taylor, J. A. Wheeler, in *Spacetime Physics*, W.H. Freeman (1963), p. 16.
17.   We note misprints in ref. 8: d^2 should read d^{-2} in Eq.(6.15) and Eq.(6.17). Similarly δ should read -δ in Eq.(6.18).
18.   D. Lynden-Bell, Q. J. Astr.Soc. 27, 319 (1986).
19.   Ch. Salomon, D. Hils and J. L. Hall, J. Opt. Soc. Am. B 5, 1576 (1988).
20.   D. Hils, J. E. Faller, and J. L. Hall, Rev. Sci. Instrum. 57, 2532 (1986).
21.   F. Bayer-Helms, H. Darnedde and G. Exner, Metrologia 21, 49 (1985).

A New Calibration of the Relativistic Doppler Effect in Neon

S. A. Lee, S. J. Sternberg and L.-U. A. Andersen[†]
Department of Physics, Colorado State University
Fort Collins, CO 80523, U.S.A.

We have measured the frequency difference between the neon $1s_5$-$4s_1$"' two photon transition in an rf discharge cell and an I_2 reference transition. In a previous experiment [1], the frequency difference of these two transitions has been measured using a 120kV neon fast beam. In Doppler-free two photon spectroscopy, the first order Doppler shift cancels, leaving the second order Doppler shift to be the dominant term. This shift is a consequence of the time dilation effect in the special theory of relativity. Thus a measurement of the frequency difference between the cell and the fast beam is a direct measure of the time dilation effect. By combining the result of the present experiment with the fast beam measurement, we arrive at a new measurement of the relativistic Doppler effect in neon.

The main improvement of the present experiment over a previous measurement [2] is the use of an intermediate I_2 reference, which decoupled the cell measurement from the fast beam measurement. This allowed us to carry out a careful study of the systematic effects associated with the discharge cell.

The two photon transition in neon was excited with a cw ring dye laser in an rf discharge flow cell. An acousto-optic modulator (AO) acted as an optical isolator, and shifted the laser frequency 40 MHz to the red. The laser was frequency modulated at 30kHz to allow its locking to the center of the neon transition. The cell was placed in a servo-controlled Fabry-Perot cavity to allow a 20X buildup in laser power. The laser intensity in the cavity was typically 1kW/cm^2, corresponding to a circulating power of 1.6W. Stray magnetic fields were compensated by Helmholtz coils to below 30mG. The two photon transition was monitored by observing the fluorescence from the upper level.

A second cw dye laser was tuned to the I_2 reference line. Fig. 1 shows a saturated absorption spectrum of the iodine reference and the neon two photon signal. The laser was frequency modulated and locked to the center of the reference hyperfine component.

The frequency difference between the two lasers was measured by sending ~0.2mW of each laser to an avalanche photodiode. The beat frequency at ~2.8GHz and a level of -65dbm was mixed with a local oscillator at 2.9GHz. The mixer output was amplified to 0dbm and counted with a 100 MHz

---

[†]Present address: Institute of Physics, Aarhus University, DK 8000 Aarhus C, Denmark.

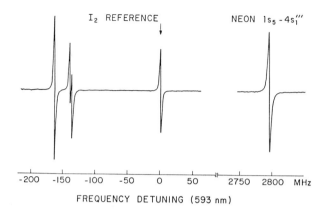

I₂ REFERENCE

NEON 1s₅ - 4s₁'''

FREQUENCY DETUNING (593 nm)

Fig. 1 The saturated absorption spectrum of the I_2 reference, and the neon two photon transition signal at a pressure of 0.2 Torr.

frequency counter. The fluctuation of the beat was ±3kHz with an averaging time of 30 sec.

The main systematic effects associated with this experiment are the pressure effects and the ac Stark effect. The pressure shift results are plotted in Fig. 2a. A least-squares fit gives a zero pressure intercept of 2835.0086(32)MHz, and a pressure shift of -2.90±0.03MHz/Torr. In addition, a pressure broadening of 14.5±1.0MHz/T was obtained. These are in excellent agreement with Ref. 3, but disagrees with Ref. 2. The ac Stark shift is shown in Fig. 2b. A least-squares fit of the data gives an ac Stark shift of -104±4kHz/kW/cm².

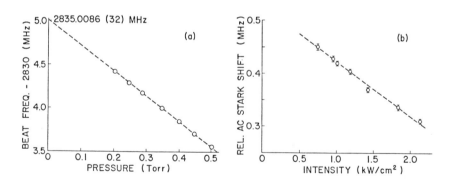

Fig. 2. (a) Pressure shift of the two photon transition. The laser intensity was 1kW/cm². The error bars of the data are smaller than the size of the circles. (b) The relative ac Stark shift of the two photon transition as a function of the laser intensity. The cell pressure was 0.2 Torr.

The result of the experiment is summarized below. Also listed is the result (adjusted for an AO shift) from the fast beam measurement of Ref. 1. The two measurements are combined to yield an experimental value for the relativistic Doppler shift. The theoretical value predicted by special relativity may be calculated from the absolute frequency of the two photon transition and the fast beam velocity, which in turn may be derived from the frequencies of the one photon transitions coupling the lower to intermediate, and the intermediate to upper states [2]. We have measured these frequencies by interferometry. For the $1s_5$-$4s_1$"' two photon transition, we obtained a frequency of 1011921020.8(1.9)MHz, and for the $1s_5$-$2p_4$ transition, a frequency of 504150972.1(1.4)MHz.

```
Cell measurement (this experiment)
    Cell(zero pressure) - Iodine    2835.009 ± .003 MHz
    AO frequency shifter             -40.000 ± .000
    AC Stark effect                      .104 ± .004
    Discharge electric field                ± .009
    Pressure reading                        ± .006
    Combined laser lock                 -.004 ± .014
               Total cell-iodine     2795.109 ± .018 MHz

Fast beam measurement (Ref. 1)
               Total iodine-beam      440.772 ± .006 MHz

Experimental cell - beam            3235.881 ± .019 MHz
Theoretical prediction              3235.865 ± .007 MHz
```

The ±19kHz accuracy of the measurement verifies the relativistic Doppler shift to 6×10^{-6}. This is a 7-fold improvement over our previous experiment [2], and represents the most accurate direct verification of the time dilation effect.

This research was supported by the National Science Foundation. We thank Dr. J. Helmcke for providing the iodine cell, C. Flynn for assistance in part of the experiment, and Drs. J. L. Hall, E. Riis and O. Poulsen for helpful discussions.

References

1. E. Riis, L.-U. A. Andersen, N. Bjerre, O. Poulsen, S. A. Lee and J. L. Hall, Phys. Rev. Lett. 60, 81 (1988).
2. M. Kaivola, O. Poulsen, E. Riis, and S. A. Lee, Phys. Rev. Lett. 54, 255 (1985).
3. G. Grynberg, Thesis, Ecole Normale Superieure, Paris, 1976.

Suppression of a Recoil Component in Ramsey Fringe Spectroscopy

J. Helmcke, F.Riehle, J.Ishikawa[+]

Physikalisch–Technische Bundesanstalt, Bundesallee 100
D – 3300 Braunschweig, Fed. Rep. Germany

[+] National Research Laboratory of Metrology
1–1–4, Umezono, Tsukuba–Shi 305, Japan

In high resolution saturation spectroscopy the recoil splitting /1/ may lead to difficulties to determine the center of the transition, precisely because of extra broadening
and line asymmetry if the recoil doublet is not resolved. On the other hand, if both
components are resolved, the height of the saturation dips will be significantly reduced and their line profiles will be distorted by 2nd order Doppler broadening /2/. We
present a novel method to suppress the low frequency recoil component.

The recoil doublet is a direct consequence of the momentum transfer h/λ of the
photon to the absorber: The saturable absorber belongs to a different velocity group
before and after an absorption/emission of a photon. Saturation dips are detected
when the saturating and the probing laser beams interact with the same velocity
group of absorbers either in the lower state (high frequency recoil component) or in
the higher state (low frequency component). Both components are shifted from the
unperturbed transition energy by the recoil energy $\pm (h \cdot \nu)^2/2mc^2$. Correspondingly,
one of the recoil components can be suppressed either by removing the population
holes in the lower state or the population enhancement in the higher state. Population
decay of the upper state into a variety of other states was applied to transitions of
iodine utilizing enough spatial separation between saturation and probe beams /3/.
However, the natural widths of the I_2 absorption lines did not allow to demonstrate
the suppression.

Here we demonstrate a method suitable for narrow lines which can be used also
in systems with the lower level being the ground state using the calcium intercombi-
nation line $^3P_1 - {}^1S_0$ ($\lambda = 657.46$ nm, $Q \simeq 10^{12}$; recoil splitting $\Delta\nu = 23.1$ kHz) /4/. In
our method we pump the population of the $^3P_1$ state via the shortlived $^3S_1$ state ($\lambda =
612.39$ nm) to the metastable states $^3P_0$ and $^3P_2$.

The experimental setup is shown in Fig. 1 /5/. The Ca atoms of a thermal
beam (T $\cong$ 1000 K) interact with the radiation of a high resolution dye laser spectrometer /6/. Two "cat's-eye" retroreflectors were employed to produce two counterpro-

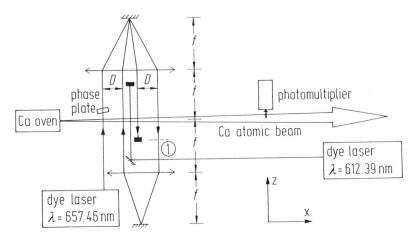

Fig.1: Experimental setup for the suppression of the low frequency recoil component in nonlinear saturation spectroscopy (with the $\lambda = 612$ nm laser beam blocked at position 1) and optical Ramsey fringe spectroscopy in a Ca atomic beam.

pagating equally spaced parallel pairs of traveling waves. Saturated absorption experiments were performed with the same setup by blocking the second cat's eye to

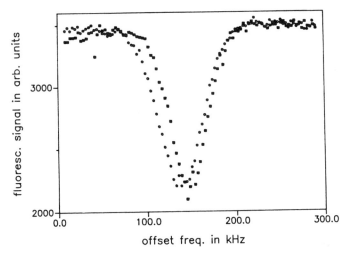

Fig.2: Lamb dip of the saturation resonance of the Ca intercombination line without pump laser beam (circles) and with pump laser beam (squares).

obtain a single pair of counterpropagating waves. For depopulation of the excited $^3P_1$ level, the radiation of a second laser ($\lambda = 612$ nm) interacted with the Ca atoms between the innermost interaction zones of the 657 nm radiation.

The suppression of the low frequency recoil component was first investigated in saturated absorption (Fig. 2). With the pump laser (612 nm) on, the width of the Lamb dip is reduced from about 60 kHz to about 45 kHz and its minimum is shifted to a higher frequency by approximately half of the recoil splitting.

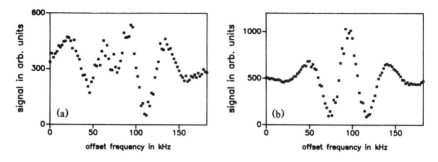

Fig.3:   Ramsey fringes without (a), with (b) suppression of one recoil component

The suppression of one recoil component could be demonstrated directly with Ramsey excitation /7/. The somewhat confusing overlapping structures of the Ramsey interference fringes of both recoil components (Fig. 3 a) could be reduced to that of the high frequency component alone (Fig. 3 b) by removing the $^3P_1$ population after the second interaction. With only one recoil component, the influences of the recoil effect, the second order Doppler effect, and of the Ramsey fringe structure can be separated allowing a precise realization of the line center.

References:

1.   J.L.Hall, Ch.J.Bordé, and K.Uehara: Phys. Rev. Lett. 37, 1339 (1976)
2.   R.L.Barger: Opt. Letters 6, 145 (1981)
3.   A.N.Goncharov, M.N.Skvortsov, and V.P.Chebotaev: Sov. J. Quantum Electron. 13, 1429 (1983) and G.Camy, N.Courtier, J.Helmcke in: *Laser Spectroscopy VIII* (Springer, Berlin–Heidelberg–N.York, 1987) W.Persson, S.Svanberg (eds), p.386
4.   J.C.Bergquist, R.L.Barger, D.J.Glaze in: *Laser Spectroscopy IV* (Springer, Berlin–Heidelberg–N.York, 1979) H.Walther, K.W.Rothe (eds), p.120
5.   F. Riehle, J. Ishikawa, J. Helmcke: Phys. Rev. Lett. 61, 2092 (1988)
6.   J. Helmcke, J.J. Snyder, A. Morinaga, F. Mensing, M. Gläser: Appl. Phys. B 43, 85 (1987)
7.   J.Helmcke, D.Zevgolis, B.Ü.Yen: Appl.Phys. B28, 83 (1982) and Ch.J.Bordé, Ch.Salomon, S.Avrillier, A.van Lerberghe, Ch.Bréant, D.Bassi, G.Scoles: Phys. Rev. A30, 1836 (1984)

Part XI

Molecular Spectroscopy
and Dynamics

Molecular Eigenstate Spectra at the Threshold for Chemical Bond Breaking:
Formaldehyde and Formyl Fluoride

C.B. Moore, Y.S. Choi, D.R. Guyer[a], W.F. Polik[b] and Q.K. Zheng[c]
Department of Chemistry, University of California,
Berkeley, CA 94720, USA

Fully resolved spectra of the vibrational states of formaldehyde and
formyl fluoride at the threshold for dissociation are recorded and
analyzed, Figs. 1 and 2. The spectra of D_2CO show complete randomization
of vibrational energy. The statistical distributions of spectral
intensities and dissociation rate constants provide information on the
mechanism and dynamics of non-radiative decay(1). Preliminary spectra of
HFCO exhibit considerable local mode character in eigenstates even for
energies well above the dissociation threshold(2). The dynamics are
expected to be highly state specific.

I. Stark Level-Crossing Spectra of D_2CO

The vibrational levels of the ground electronic state(S_0) of
formaldehyde are investigated by Stark level-crossing spectroscopy. A
frequency-doubled pulse-amplified cw dye laser provides transform limited
pulses to excite single rovibronic levels of the first excited singlet
state(S_1) of D_2CO in a cooled pulsed molecular beam. At this energy of
28,000 cm^{-1} the vibrational levels of the ground state are broadened, but
not strongly overlapped, as a result of their coupling to the dissociative
continuum, Fig 3. Since the two electronic states have different dipole
moments, the Stark effect scans S_1 levels through resonances with S_0
levels as the voltage of a uniform d.c. electric field is ramped and the
pump laser scanned in coincidence with a single Stark component of the
absorption spectrum. These resonances are observed as decreases in the
non-radiative decay rate (internal conversion of electronic to vibrational
energy) of the S_1 fluorescence. The increase in dissociation rate with
vibrational energy is shown in Fig. 1. When different M states are
probed, spectra are found to scale as the product of M and electric field,
Fig. 4. The increase in the number of lines for the higher fields (low M)
is the result of J not being a good quantum number in the field(1).
Analysis of the resulting resonance lineshapes yields complete distributions of S_0 decay rates (linewidths) and S_1-S_0 coupling matrix elements.
The S_0 decay rates represent the first measurements of unimolecular dissociation rates of a polyatomic molecule at the eigenstate-resolved level.
S_0 decay widths fluctuate between 6.4×10^{-5} and 3.8×10^{-3} cm^{-1}, Fig. 5, and
S_1-S_0 coupling matrix elements vary from 3.5×10^{-7} to 4.7×10^{-5} cm^{-1}, Fig.
6, demonstrating that chemical properties of neighboring eigenstates
fluctuate by over two orders of magnitude. The observed density of S_0
vibrational states is 400 per cm^{-1}, six times greater than an estimate
including first-order anharmonic corrections. The small observed increase of level density with J indicates that K_a is nearly a good quantum
number for J≤4. The barrier height to unimolecular dissociation on the

[a]Present Address: Dept. of Chem., Hope College, Holland, MI 49423, USA
[b]Present Address: Spectra Technology, Inc., 2755 Northup Way, Bellevue,
  WA 98004-1485, USA
[c]Present Address: Dept. of Physics II, Fudan Univ., Shanghai, China

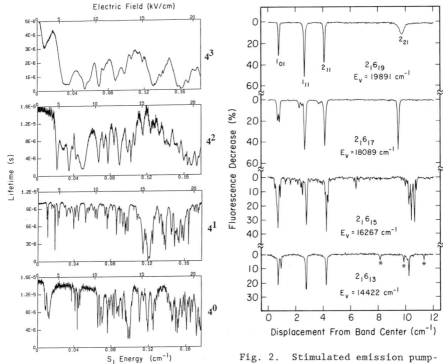

Fig. 1. Stark level-crossing spectra from $1_{10}(1)$ in the four lowest vibronic levels of S_1 D_2CO.

Fig. 2. Stimulated emission pumping spectrum of HFCO. The intermediate level is $6^2 1_{10}$. The pump and dump lasers propagate colinearly and are perpendicularly polarized.

S_0 D_2CO surface is determined to be 80.6 ± 0.8 kcal/mole, corresponding to 79.2 ± 0.8 kcal/mole for H_2CO, in good agreement with ab initio predictions. Quantitative agreement between the magnitude of experimentally determined decay rates and an RRKM rate calculation with all parameters set by ab initio calculation(3) is found.

The spectra are analyzed for evidence of energy level correlations and quantum ergodicity(4). The analysis for short and long range level correlations gives an apparent energy level spacing distribution which lies between the Poisson and GOE limits. However, the Stark level-crossing method diminishes any existing correlations. The true level spacing distribution must be closer to the GOE than to the Poisson limit. The observed spectra are consistent with expectations for levels with GOE distribution statistics at zero field. Complete distributions of S_1-S_0 coupling matrix elements and S_0 decay rates are reported and subjected to statistical tests for ergodicity.

The distribution of S_1-S_0 coupling matrix elements indicates that the dynamics of intramolecular vibrational redistribution of energy (IVR) is very nearly quantum ergodic. The curves shown in Fig. 6 are derived assuming that when a molecular eigenstate is expressed in a basis of eigenstates of a separable Hamiltonian the wavefunction coefficients are random numbers subject only to normalization. Interference of the decay amplitudes through neighboring S_0 states is observed as Fano-like line profiles(5). For molecular eigenstates with random coefficients the shape of the distribution of matrix element magnitudes is given by the average de-

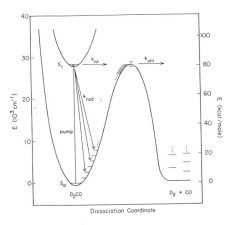

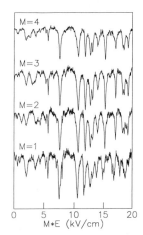

Fig. 3. Potential energy surface diagram for D_2CO. Each S_1 rovibronic state decays both radiatively(k_{rad}) and nonradiatively(k_{nr}).

Fig. 4. Stark level-crossing spectra from $4_{41}(M)$, M=1,2,3,4, in 4^1 D_2CO plotted against M × electric field strength.

viation from a symmetric Lorentzian line profile. This behavior is good evidence for strong coupling and free flow of energy among vibrational degrees of freedom in a molecule above its dissociation threshold(1).

The average S_0 decay rate can be accounted for by RRKM theory with tunneling corrections. The large variation about the average is the quantum statistical fluctuation of rates to be expected for any system with a small number of open decay channels, Fig. 5. When the results are extrapolated to zero electric field, there appear to be about four channels of comparable importance or more with some less important. These channels may be considered as the vibrational levels of the transition state which contribute to the dissociation rate. The results suggest that at least one of the ab initio frequencies for the transition state is too high or is strongly anharmonic.

II. Stimulated Emission Pumping Spectra of HFCO

The potential surfaces of HFCO are similar to those of formaldehyde except that the S_1 state is higher in energy and the barrier is a factor of two lower, near 14,000 cm^{-1}. At this lower dissociation energy,

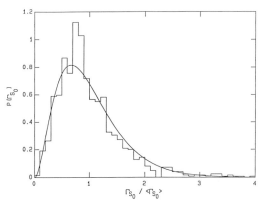

Fig. 5. Distribution of S_0 decay rates for 4^1 S_1 D_2CO J,K,M≠0 Stark level-crossing spectra. $\langle\Gamma_{S_0}\rangle = 6.5\times10^{-4}$ cm^{-1}, or $\langle k_{uni}\rangle = 1.2\times10^8$ s^{-1}. The curved line is a chi-squared distribution with 6.1 degrees of freedom.

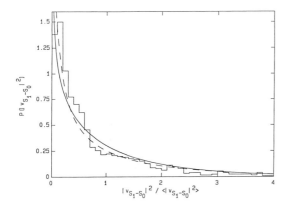

Fig. 6. Distribution of internal conversion matrix elements, $\langle v^2 \rangle = 4.4 \times 10^{-11}$ cm^{-2}. The solid curve is a chi-squared distribution with 1.4 degrees of freedom. The dashed line (Ref. 5) uses no fitting parameters.

vibrational energy is much less likely to be randomized and state specific variations in dynamical properties might be expected. The Franck-Condon factors and energies are perfect for reaching vibrational levels in a wide energy range around the barrier top. Stimulated emission pumping spectra(4) of HFCO in a pulsed jet at 0.05 cm^{-1} resolution in the range of 13,000 to 23,000 cm^{-1} have been recorded, Fig. 2 (2). Above the dissociation threshold, $\approx$14,000 cm^{-1}, single Franck-Condon allowed transitions are spread out over widths varying from 0.05 up to a few cm^{-1} for S_0 vibrational levels within an energy range of a few hundreds of cm^{-1}. The well-resolved level structures within these lines exhibit mixing with roughly 10 levels per cm^{-1}. The spectra in Fig. 2 exhibit increasing mixing as energy increases with addition of the first few out-of-plane bending quanta. Then the spectra sharpen as further quanta are added. High amplitude motion in a single coordinate seems to be particularly stable against energy redistribution(6).

Studies near the dissociation threshold show that there is a strong systematic dependence of dissociation rate on both rotational and vibrational quantum number. Substantial vibrational mode dependence of intramolecular vibrational energy redistribution and/or dissociation rates can be anticipated when SEP spectra and dissociation product state dynamics are fully explored.

Work supported by the National Science Foundation's Chemistry Division.

References

1. D.R. Guyer, W.F. Polik and C.B. Moore, J. Chem. Phys. 84, 6519 (1986). H. Bitto, D.R. Guyer, W.F. Polik and C.B. Moore, Faraday Disc. Chem. Soc. 82, 149 (1986). W.F. Polik, D.R. Guyer and C.B. Moore, SPIE 912, 150 (1988).

2. Y. Choi and C.B. Moore, J. Chem. Phys. 90, 3875 (1989).

3. G.E. Scuseria and H.F. Schaeffer III, J. Chem. Phys., in press.

4. E. Abramson, R.W. Field, D. Imre, K.K. Innes, and J.L.Kinsey, J. Chem. Phys. 80, 2298 (1984); 83, 453 (1985). M. Lombardi, Excited States 7, 163 (1988).

5. W.F. Polik, C.B. Moore, and W.H. Miller, J. Chem. Phys. 89, 3584, (1988).

6. G. Hose and H.S. Taylor, Chem. Phys. 84, 375 (1984).

SPECTROSCOPY ON NO_2: EVIDENCE FOR A SUBSTRUCTURE UNDERLYING THE EXPECTED EIGENSTATES OF A MOLECULE

H.G. Weber

Heinrich-Hertz-Institut and Optisches Institut der Technischen Universität Berlin
Einsteinufer 37, D-1000 Berlin 10, Fed. Rep. Germany

Spectroscopy on the polyatomic molecule NO_2 reveals a substructure underlying the expected eigenstates of this molecule. The substructure is seen in two characteristic times associated with each excited state hyperfine-structure level of this molecule, the radiative decay time $\tau_R \approx 40\mu s$ and a time $\tau_0 \approx 3\mu s$ representing an evolution within the substructure. The time τ_0 is for instance seen in a change of the degree of polarization of fluorescence light.

We report on Hanle effect and optical-radio-frequency double resonance experiments on NO_2. The experimental conditions are similar to measurements on atoms and diatomic molecules, namely: preparation of the molecule into a well defined and isolated fine-structure level and observation of the subsequent radiative decay of this state. This well defined experimental situation is verified by a number of experiments which yield g-factors, angular momentum quantum numbers and radiative lifetimes τ_R in excellent agreement with results from other groups. An example is the very detailed comparison of our spectroscopic data with data obtained by high resolution laser spectroscopy [1]. However, under the same experimental conditions there are also other results, which contradict the expectations associated with an isolated eigenstate of this molecule. In the following we report on these results.

We report on molecular beam experiments and static gas experiments. The arrangement of the molecular beam experiments is as follows. An NO_2 beam is crossed perpendicularly by a single mode cw laser beam, whose light is tuned to molecular transitions either near $\lambda_{ex} \approx 593$ nm or 514 nm. The laser induced molecular fluorescence is detected by two photomultipliers (S 20) placed in opposite directions perpendicularly to the plane defined by the two crossing beams. The fluorescence light passes cutoff filters which transmit light with wavelength $\lambda \geq \lambda_x$, where $\lambda_x > \lambda_{ex}$ in most experiments. Important experimental parameters are: the intensity I of the exciting laser light, the transit time T_L of the molecules through the light beam, the cutoff wavelength λ_x, and the state of polarization of the ligth in excitation and detection with respect to the applied magnetic fields.

The most unusual result is the observation of two characteristic times $\tau_R \approx 40\mu s$ and $\tau_0 \approx 3\mu s$ associated with each hyperfine-structure level (expected eigenstate). Both times appear in several experiments. We consider here Hanle effect measurements, which yield a signal, which is a superposition of two components, the narrow Hanle signal having a width characteristic of τ_R and the broad Hanle signal having a width characteristic of τ_0 (Fig. 1). The width of each component is independent both of the light intensity I and of the transit time T_L. Both parameters were varied by more than a factor 20 [2]. The narrow Hanle signal exhibits an interesting property which was named inversion effect [3]. The narrow Hanle signal (the degree of polarization) changes the sign versus the light intensity I or the transit time T_L. As figure 1 shows the signal amplitude of the narrow component is pointing upward for low I and downward for high I. The broad Hanle signal (degree of polarization) shows no inversion effect. The narrow Hanle signal exhibits also a strong dependence on the cutoff wavelength λ_x, whereas this is not seen for the broad Hanle signal. This dependence on λ_x represents a change of the degree of polarization of fluorescence light and no change of the distribution of fluorescence light [2]. By an appropriate choice of λ_x and T_L it is possible to measure either the narrow or the broad Hanle signal solely. For $\lambda_x \approx \lambda_{ex}$ the signal amplitude of the narrow Hanle signal is at least a factor 10 bigger than the signal amplitude of the broad Hanle signal. Therefore in many experiments the broad component was not seen before.

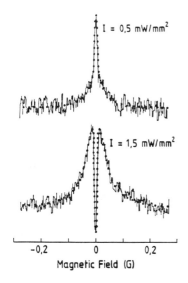

Fig 1: Results of Hanle effect measurements. The dots superimposed on the experimental results represent the sum of two Lorentzian curves which were fitted to the measurements. The two Lorentzian curves are the narrow and the broad Hanle signal respectively.

Another most unusual property of both components of the Hanle signal appears in static gas experiments. Fig. 2 depicts measurements of the width of the narrow Hanle signal versus the NO_2 pressure. These results ($\lambda_{ex} \approx 514$ nm) show that the linear extrapolation to zero pressure is about a factor 7 (for $\lambda_{ex} \approx 593$ nm a factor 2) bigger than the value obtained in molecular beam experiments. Similar results were found for the broad Hanle signal. Properties of the narrow Hanle signal, which are connected with the nonlinear pressure dependence of the width, are firstly a change of the degree of polarization of fluorescence light by nearly a factor 10 similar to the λ_x dependence considered before, and secondly the possibility to increase the degree of polarization of fluorescence light with a weak magnetic field (<30 mG) which is linearly polarized parallel to the exciting laser light (π-excitation) [4].

395

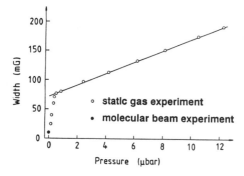

Fig 2: Measurements of the width (HWHM) of the narrow Hanle signal versus the NO_2 pressure in the static gas experiment. The dot represents a measurement under molecular beam conditions.

The Hanle effect measurements were performed with the selection of a single fine-structure level only. However, optical-radio-frequency double resonance experiments, which enable the selection of a single hyperfine-structure level confirm the Hanle effect measurements and show that the described results represent properties of isolated hyperfine-structure levels (hfs-splitting > 10 MHz). The Hanle effect and optical pumping with lasers are complicated phenomena, which are however well understood (see for instance [5]). We examined that the narrow and the broad Hanle signal cannot be described by the well known properties of optical pumping with lasers (see for instance [3] and references given there). It seems that these optical pumping effects are covered by a stronger phenomenon which has obviously no analogue in atomic physics.

The two characteristic times τ_R and τ_0 appear also in several other experiments. Under molecular beam conditions time resolved fluorescence decay measurements yield a single exponential decay with the radiative lifetime τ_R. The time τ_0 is known for a long time from investigations of the integrated absorption coefficient [6]. The same time τ_0 was also determined from the absorption width of laser induced transitions [7]. Finally τ_0 was also obtained in various "magnetic absorption resonances" as a change of the absorption probability versus an applied magnetic field [8]. These investigations show, that it is not possible to associate the time τ_0 with the lifetime of the lower state, with two photon transitions, hyperfine-structure coherence or properties of the light source. It is inevitable to attribute both τ_R and τ_0 to a single isolated hyperfine-structure level.

The experimental results reported here were obtained on all investigated laser induced transitions (more than 10) and seem to represent therefore a general property of NO_2. The existence of two characteristic times τ_R and τ_0, the inversion effect, the nonlinear pressure dependence and the effects associated with this pressure dependence are not in agreement with the excitation of an eigenstate. These results give therefore evidence for a substructure underlying the expected eigenstates.

The following model for this substructure is able to explain (consistently combine) the molecular beam results [9] and presumably also the unusual collision effects [2]. Underlying an expected eigenstate $|e>$ is a substucture represented by the two "states" $|e_1>$ and $|e_2>$. An optical transition populates only $|e_1>$. Being in $|e_1>$ the molecule evolves irreversibly from $|e_1>$ to $|e_2>$ in the time τ_0. To our present knowledge, the only quantity to discriminate between $|e_1 >$ and $|e_2 >$ is the degree of polarization of fluorescence light. This suggests to associate the substructure with different shapes (symmetries) of the molecule. The symmetry determines via P- Q- and R-branch selection rules the degree of polarization of fluorescence light. Evidently, a change of the shape, which results in a change of the degree of polarization of fluorescence light, is possible only on polyatomic molecules. If we consider the shape of a molecule as an effect of the environment [10] we come to the conclusion that the environment adds the substructure to the molecule, and that isolation of the molecule is not possible. Consequently the usual molecular hamiltonian cannot describe this molecule completely. This is presumably the first indication that quantum mechanics may not be a complete theory for polyatomic molecules. We note that to our knowledge neither our experimental results nor the suggested explanation are in disagreement with the whole body of experimental results on NO_2.

References:

[1] F. Bylicki, H.G. Weber, G. Persch, W. Demtröder: J. Chem. Phys. 88, 3532 (1988)

[2] H.G. Weber: Phys. Lett. A 129, 355 (1988) and work to be published

[3] H.G. Weber: Z. Phys. D 6, 73 (1987)

[4] H.G. Weber, F. Bylicki: Z. Phys. D 8, 279 (1988)

[5] C.Cohen-Tannoudji in Atomic Physics 4, 589 (1975) Editors: G.zu Putlitz, E.W.Weber, A.Winnacker, Plenum Press, New York, 1975

[6] V.M. Donnelly, F. Kaufman: J. Chem. Phys. 66, 4100 (1977)

[7] H.G. Weber: Z. Phys. D 1, 403 (1986)

[8] H.G. Weber, F. Bylicki: Chem. Phys. 116, 133 (1987)

[9] H.G. Weber: Phys. Rev. A 31, 1488 (1985)

[10] R.G. Woolley: Chem. Phys. Lett. 125, 200 (1986)

Reaction Dynamics and Spectroscopy in the VUV and XUV

A.H.Kung, E.Cromwell, D.J.Liu, M.J.J.Vrakking and Y.T.Lee
 Chemistry Department, University of California,
 Berkeley, CA 94720
 and
 Materials and Chemical Sciences Division,
 Lawrence Berkeley Laboratory, Berkeley, CA 94720

In the last few years applications of vuv lasers have been expanding. With the development of powerful lasers to generate vuv sources of very high intensity and excellent spectral purity the sensitivity and resolution of experiments have reached new heights. Experiments that require high performance from these vuv sources are becoming feasible. These vuv laser sources are powerful spectroscopic tools for high-resolution photoionization and Rydberg spectroscopy studies. The pulsed nature of these sources makes them highly suitable for time-resolved studies. In chemical kinetics studies, microscopic details of chemical reactions can be elucidated by probing state-to-state distribution of reaction products.

A near-transform-limited vuv-xuv laser source that has a bandwidth of ~250 MHz and is continuously tunable from 70nm to >150nm with typical peak powers of 100 to 1000 watts in a 5 ns pulse has been developed in our laboratory (1). This source has been used to study ultrahigh-resolution spectroscopy of Kr, where the hyperfine splittings and isotope shifts of several Rydberg levels were measured and analyzed (2). A scheme for producing vibrational- and possibly rotational-state selected ions was demonstrated on nitrogen (3). In this report we provide preliminary results of our latest experiment as an example of applying the high resolution vuv laser to study processes of practical interest: mechanism of H_2 elimination and product state distribution of H_2 following uv photolysis of cyclohexadiene (CHDN) at 212 nm. A full account of the experiment will be published elsewhere.

1. Laser improvements

Prior to describing the experiment on CHDN we wish to document the modifications and improvements to our vuv source implemented after the submission of reference 1.

(a) The principal change to the laser system is the replacement of our commercial pulsed dye laser by a pulse-amplified single-mode dye laser as the second laser. This gives the capability of having two independently tunable single frequency systems so that all vuv frequencies generated by any four-wave mixing schemes now have near-transform limited characteristics.

(b) Approximately 5 mj of 210nm to 220nm radiation is now routinely obtained by frequency doubling using KDP and then third harmonic mixing using BBO of our pulse-amplified single-mode DCM laser. This makes possible two-photon resonant four-wave mixing processes to give intense stable (10% peak-peak) vuv radiation. The tradeoff for the improved

amplitude stability is an increase in the vuv bandwidth from
210 MHz to 610 MHz at full power due to Stark broadening of
the two-photon resonance.
(c) A uv achromat is installed to provide good spatial
overlap of uv and visible beams at the focal spot of the lens
used to generate vuv. At least a factor of 5 increase in vuv
power was recorded in non-resonant mixing cases.

2.   Dissociation of Cyclohexadiene

In a molecular beam photofragmentation-translational
spectroscopy study of 1,4-CHDN Zhao et. al. has shown that
the peak of the translational energy distribution for
concerted dissociation in the ground state is determined
mainly by the dynamics of the potential energy release along
the reaction coordinate, and is not sensitive to either the
amount of internal energy or the form of excitation (4). In
that study it was not possible to determine the product
internal energy distribution, which is required to elicit
complete information about the nature of the transition state
of the unimolecular reaction. By using the vuv laser we have
mapped the vibrational- and rotational energy distribution as
well as the Doppler profile of the H_2 product. The results
provide a qualitative but clear understanding of the
mechanism and the dynamics of the reaction.
The reaction under consideration is:

CHDN + h   -- > H_2 + benzene E_A = 43.8 kcal/mol

CHDN in a pulsed molecular beam is excited by 212nm laser
radiation. A second laser, delayed by 10-100 ns relative to
the excitation laser, is used to probed the H_2 product by
vuv-uv resonance-enhanced two-photon ionization (5).
Densities of product states for v" up to 4 are measured. The
Doppler profile of each state is scanned. Both 1,4 and 1,3
isomers of CHDN are studied.
Figure 1 shows the vibrational distribution of product H_2.
About 75% of the H_2 is in the ground vibrational state, 20%
in v=1, ~4% in v=2 and <1% in the higher v states.

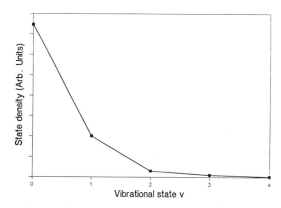

Figure 1. Vibrational energy distribution of product H_2
detected by 1+1 REMPI.

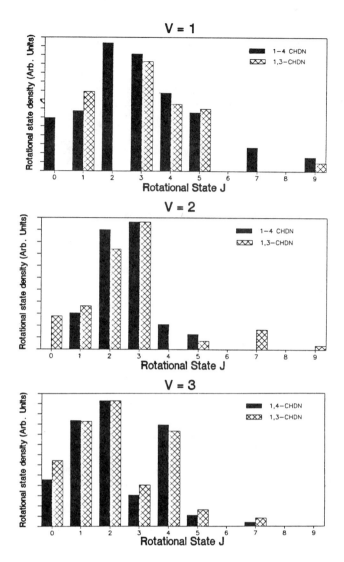

Fig. 2. H_2 rotational state distribution for different vibrational levels as indicated.

The rotational energy of the product is rather low, with state distributions peaking between J=2 to 3 for v=1 to 3 (Fig. 2) and slightly higher for v=0. Furthermore the distributions for 1,3 and 1,4-CHDN are very similar.
    Scanning the time-delay of the probe laser relative to the pump laser gives the time evolution of the H_2 product. It shows that in this collisionless excitation H_2 production is consistent with a concerted elimination mechanism. Several conclusions can also be drawn. The low H_2 vibrational and

rotational energies observed indicate that dissociation is from a highly vibrationally excited ground state of 1,4-CHDN so that the ring is bent sufficiently to bring the H atoms located at each end of the boat configuration to be close to the equilibrium inter-atom separation of the H_2 molecule before elimination occurs. Symmetry conservation forbids 1,3-CHDN from undergoing H_2 elimination. The observed similarity in product state distribution between 1,3- and 1,4-CHDN is strong evidence that 1,3-CHDN first isomerizes to 1,4-CHDN prior to ejecting the H_2 molecule. The Woodward-Hoffman rule is not violated here.

About 50% of the available energy appears in translation and the balance in internal excitation of the product. Our results show that the Doppler profile of every product state has nearly identical halfwidths. This is in good agreement with the observation of Zhao that the translational energy distribution is dominated by the potential energy barrier and is insensitive to the form of excitation.

The results of this experiment point to a transition state where the H_2 molecule is pushed away from the benzene ring as it is being formed, carrying with it little vibrational and rotational energy. It is very likely that there is a strong correlation between the velocity vector and the angular momentum vector of the H_2 product. This can be tested by studying the shape of the Doppler profile. Work is in progress to obtain this information.

Acknowledgement

This work was supported by the Director, Office of Energy Research, Office of Basic Energy Sciences, Chemical Sciences Division of the U.S. Department of Energy under contract No. DE-AC03-76SF00098.

References

1. E. Cromwell, T. Trickl, Y.T. Lee and A.H. Kung, Rev. Sci. Instru. (Aug. 1989).

2. T. Trickl, M.J.J. Vrakking, E. Cromwell, Y.T. Lee and A.H. Kung, Phys. Rev. A 39, 2948 (1989).

3. T. Trickl, E. Cromwell, Y.T. Lee and A.H. Kung, to be published (LBL report 27186).

4. X. Zhao, R.E. Continetti, A. Yokoyama, E.J. Hintza and Y.T. Lee, to be published (LBL report 26333).

5. A.H. Kung, T. Trickl, N.A. Gershenfeld and Y.T. Lee, Chem. Phys. Lett. 144, 427 (1988).

First Observation of Perturbations on the $C^1\Pi_u$ State of Na$_2$ by CW UV Modulated Population Spectroscopy

G.-Y. Yan, B.W. Sterling, and A.L. Schawlow
Department of Physics, Stanford University
Stanford, CA 94305-4060 USA

Modulated population spectroscopy has been used to excite the $C^1\Pi_u$ state of Na$_2$ molecule from its ground state, $X^1\Sigma_g^+$. A strong visible laser depopulates a particular level of the lower state. Rotationally resolved excitation spectra $C^1\Pi_u - X^1\Sigma_g^+$ were obtained by scanning a UV laser, which was generated in a computer-controlled 699-29 dye laser with an intracavity frequency-doubling crystal. The excitation spectra reveal that a few hundred e-levels of the $C^1\Pi_u$ state with $15 \leq v' \leq 26$ are perturbed drastically. We ascribe the perturbations to the $(3)^1\Sigma_u^+$ state. After deperturbation, potentials and Dunham coefficients for $C^1\Pi_u$ and $(3)^1\Sigma_u^+$ states were deduced. Agreement between the indirect observation and a theoretical calculation of the potential for $(3)^1\Sigma_u^+$ state is fairly good. In addition, Λ-type doubling for the $C^1\Pi_u$ state was observed directly.

The experimental setup has been described in detail in a previous paper.[1] A cross oven containing sodium was heated to 400–450° C with Ar buffer gas at a pressure of 1 Torr. After minor modifications, a 699-29 dye laser with intracavity frequency doubling, operating in AutoScan mode, provided CW UV radiation with output power of a few milliwatts. A strong visible beam from a 599-21 dye laser, counterpropagating with respect to the UV beam was directed through the oven. The visible beam was tuned in resonance with a transition of the $A^1\Sigma_u^+ - X^1\Sigma_g^+$ system and was chopped. This modulated the population on the lower level of the transition, $X^1\Sigma_g^+ (v'', J'')$. UV excitation spectra of $C^1\Pi_u (v', J') - X^1\Sigma_g^+ (v'', J'')$, originating on the lower level were recorded when the 699-29 laser scanned. These spectra were simplified by modulated population spectroscopy. A measurement accuracy of energies of observed levels was conservatively estimated to be 0.03 cm^{-1}.

More than 450 vibronic levels with e-parity or f-parity of the $C^1\Pi_u$ state were measured. Vibrational and rotational quantum numbers of the observed levels were in the ranges $11 \leq v' \leq 35$ and $8 \leq J' \leq 70$. The measurements reveal that many e-levels are perturbed. Figure 1 depicts deviations between measured and calculated energies of levels, using Verma's constants,[2] $vs.$ $J'(J'+1)-1$ for $v' = 15$. Obviously, only e-levels around $J' = 39$ or $J' = 62$ are perturbed strongly. This indicates that two rotational series of perturbing levels overlap with one of $v' = 15$ in the $C^1\Pi_u$ state at the two crossing points. The perturbing levels have smaller rotational constants than that of the perturbed level. Like $v' = 15, 18$ crossing points were observed for $15 \leq v' \leq 26$. They are located in the range 31400–32500 cm^{-1}. In addition, the deviations for e-levels and f-levels increase monotonically as v' up to 5.5 cm^{-1} for $v' = 35$. The observations provided a direct measure of Λ-type doubling as a function of J'.

The perturbations are heterogeneous because only levels with e-parity are perturbed. Perturbation matrix elements are proportional to $[J'(J'+1)]^{1/2}$. Thus vibronic levels with lower J' values are less perturbed. These data were used to obtain the rotationless energies for each of vibrational levels with $v' \geq 11$. After depertur-

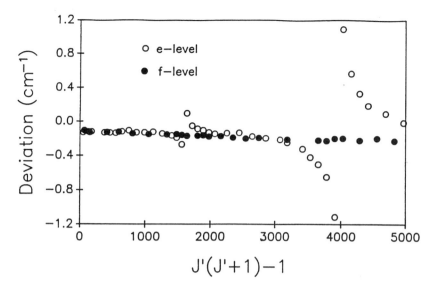

Figure 1. Deviations between measured and calculated energies of rotational levels for $C^1\Pi_u$ with $v' = 15$ vs. $J'(J' + 1) - 1$.

bation, term energies of unperturbed levels for the $C^1\Pi_u$ state were extracted. We could not excite vibronic levels with $v' < 11$ from the ground state due to the limited turning range of our UV laser (305–325 nm). The dangers of fitting based only on data at high vibrational values are well known. Fortunately, some Argon-ion laser and Krypton-ion laser lines are matched with 18 transitions of the $C^1\Pi_u$-$X^1\Sigma_g^+$ system, involving 18 levels of the $C^1\Pi_u$ state with $0 \leq v' \leq 8$ and $12 \leq J' \leq 124$.[3] Half of these levels are of e-parity, the rest f-parity. These were added to our measured data. e-levels and f-levels were separately used for least squares fittings to deduce the Dunham coefficients.[1] Reproductions of term energies of the levels, using the coefficients obtained, were within the measurement uncertainty, except a few levels involved with the transitions of laser lines. $\Delta\nu_{ef}$ is positive. This illustrates that the main interactions are with the low lying $(2)^1\Sigma_u^+$ state. A RKR potential for the $C^1\Pi_u$ state was constructed with the Dunham coefficients, as shown in Fig. 2.

Since only e-levels are perturbed, any perturbing state must be a $^1\Sigma_u^+$ state. The deperturbations indicated that differences between perturbed and perturbing levels are consistently around 0.02 cm^{-1}. According to the differences, tentative lines for rotational energies of the perturbing levels vs. $J(J+1)$, which passed through the crossing points, were plotted. Being vibrational spacing, distances between consecutive lines were around 50-60 cm^{-1}. These implied that all the perturbations could be caused by one $^1\Sigma_u^+$ state. We noticed that vibrational spacings of the $(2)^1\Sigma_u^+$ state over this regime are much less than 50 cm^{-1}.[4] Other $^1\Sigma_u^+$ states are located above the energy regime, except the $(3)^1\Sigma_u^+$ state. Thus, we concluded that all the perturbations are caused by crossing between levels of the $C^1\Pi_u$ state and the $(3)^1\Sigma_u^+$ state.

The crossing points dealt with 18 vibrational levels of the perturbing state. The deperturbations provided energies and fractional J values for each of the crossing

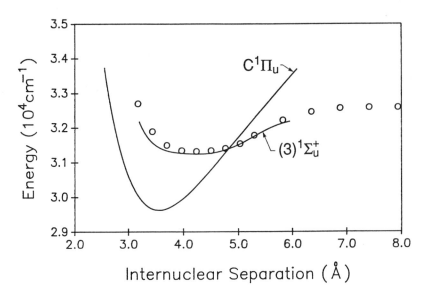

Figure 2. Potentials for the $C^1\Pi_u$ and $(3)^1\Sigma_u^+$ states. Circles represent theoretical calculations for the $(3)^1\Sigma_u^+$ state.

points. Tentative assignments of vibrational quantum numbers of the crossing points were made. 15 of 18 data points were used in a least squares fitting to deduce a set of Dunham coefficients, and then a potential curve for the $(3)^1\Sigma_u^+$ state. A correct assignment was confirmed by a comparison of observed matrix elements of perturbation with calculations of overlap integrals of wavefunctions of the $C^1\Pi_u$ and $(3)^1\Sigma_u^+$ states based on the RKR potentials obtained with its assignment. The experimental potential for the $(3)^1\Sigma_u^+$ state is shown in Fig. 2. The circles in the figure depict theoretical calculations.[5] Both experimental and theoretical potentials have flattened bottoms and local maxima. As a consequence, the vibrational spacing increases at first and then decreases as the vibrational quantum number increases. Comparison of the experimental results and theoretical calculations yields fairly good agreement.

This work was supported in part by the National Science Foundation under grant NSF PHY-86-0441 and in part by the Office of Naval Research under contract ONR N00014-87-K-0265. BWS is an Office of Naval Research predoctoral fellow.

References:

[1] G.-Y. Yan, B.W. Sterling, T. Kalka, and A.L. Schawlow, submitted to J. Opt. Soc. Am. B.

[2] K.K. Verma, T.H. Vu and W.C. Stwally, J. Mol. Spectrosc. **91**, 325-347 (1982).

[3] C. Effantin, J. d'Incan, A.J. Ross, R.F. Barrow and J. Verges, J. Phys. B **17**, 1515-1523 (1984).

[4] D.L. Cooper, R.F. Barrow, J. Verges, C. Effantin and J. d'Incan, Can. J. Phys. **62**, 1543-1562 (1984).

[5] G.H. Jeung, Phys. Rev. A **35** 26-35 (1987).

Analysis of the Two-Photon $D^1\Delta$-$X^1\Sigma^+$ Transition in CO: Perturbations in the (10-0) Band

B.A. Garetz
Department of Chemistry, Polytechnic University
Brooklyn, NY 11201, USA
C. Kittrell
Department of Chemistry, Rice University
Houston, TX 77251, USA

1. Introduction

The $D^1\Delta$ state of CO is the last of the seven strongly bound valence electronic states of carbon monoxide to be correctly characterized, due in part to its inaccessibility via one-photon absorption. We recently reported the analysis of data obtained by two-photon excitation in thirteen vibronic bands of the $D^1\Delta$-$X^1\Sigma^+$ transition in three isotopomers of CO (1). Of particular interest is the extent of perturbations of the D state with nearby triplet states. The resulting states with mixed singlet-triplet character should provide access into the triplet manifold, and should yield useful systems for studying collision-induced intersystem crossing.

We have recently characterized a complex system of perturbations associated with the D-X (10-0) band in $^{12}C^{16}O$ (2). Spin-orbit perturbations among the $D^1\Delta$ (v=10), $d^3\Delta$ (v=14), and $a^3\Pi$ (v=18) states have been observed, and spectra have been analyzed using a nonlinear least-squares fitting method (3). The d~D perturbation is a rare example of an off-diagonal isoconfigurational perturbation. The analysis extends knowledge of the a state up to v=18.

2. Experimental

Experimental details have been described elsewhere (1). Briefly, CO gas was excited with the frequency-doubled output of a narrow-band, tunable dye laser, populating rotational levels of various vibrational states of the $D^1\Delta$ state. The D state is metastable, as the one-photon transition to the ground state is forbidden. Therefore no direct fluorescence is observed. However, at a pressure of ~3 torr, collisional energy transfer populates the nearby $A^1\Pi$ state which fluorescences strongly in the vacuum ultraviolet. The resulting fluorescence is detected with a solar-blind photomultiplier tube, and the signal-averaged photomultiplier output is plotted versus dye-laser frequency. A diagram of the experimental setup is shown in Figure 1.

3. Spectrum

Figure 2 shows the bandhead region of the D-X (10-0) band, and assignments are indicated there. Five rotational branches are observed, as expected by the two-photon selection rule $\Delta J=0,\pm1,\pm2$. In addition, however, there are a number of weaker spectral features which are not assignable to the D-X transition. Many of these features are assignable to the forbidden $d^3\Delta(F2)$-$X^1\Sigma^+$ (14-0) band, whose bandhead falls about 9 cm^{-1} to the high-energy side of the D-X bandhead. Near J'=10 in the D-X band, the intensities of the D-X lines begin to fall off, and some additional weak lines appear in the spectrum. Many of these are assignable to the $a^3\Pi(F3)$-X (18-0) transition. Both the a-X and d-X transitions are borrowing intensity from the two-photon allowed D-X transition. Additional weak lines are assignable to a number of other transitions, including other components of the a-X and d-X transitions, as well as two of the three components of the $a'^3\Sigma^+$-X (19-0) transition.

3. Analysis

An iterative approach is used to fit molecular parameters to this extensively perturbed spectrum. First, parameters are fitted to a subset of less-perturbed lines. Then diagonalizations are extended a few J at a time, with parameters refitted at each step. The initial values of interaction parameters are estimated using theoretical single-configuration expressions given by Field (4,5). In addition, spectral lines from earlier conventional one-photon spectra involving the interacting states are also included in the fit (6).

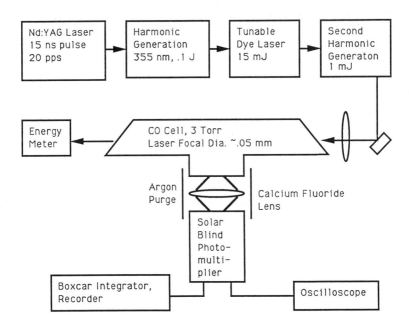

Fig. 1. Apparatus for obtaining the two photon laser induced fluorescence excitation spectrum of carbon monoxide

4. Results and discussion

The analysis of the (10-0) band is nearly completed. It indicates that a multiple crossing occurs at about J=13 which involves the four states $a^3\Pi(F3)$ (v=18), $a'^3\Sigma^+(F3)$ (v=19), $d^3\Delta(F2)$ (v=14) and $D^1\Delta$ (v=10). The fit is sufficiently good to correctly determine thirteen parameters, including the three principal interaction energies, $\alpha(d\sim a)$, $\alpha(d\sim D)$ and $\alpha(a\sim D)$, the origins, T, and rotational constants, B, of the four states a(F3), d(F2,F3), D, and a'. Table 1 lists these parameters.

Table 1. Molecular Parameters for $D^1\Delta$, v=10 and Perturbing Levels[*]

State	D (v=10)	d (v=14)	a (v=18)	a' (v=19)
T_3 (T)	74957.10(10)	75001.84(13)	74903.41(52)	74932.76(03)
T_2	-	74967.26(11)	-	-
T_1	-	-	-	-
B	1.0736(5)	1.0789(4)	1.3080(4)	1.0205(3)
λ	-	-	-	-1.165(47)

Interaction	d~a	d~D	a~D
α	-0.96(30)	1.40(38)	-9.032(67)

[*]For definitions of parameters, see ref. 3.

406

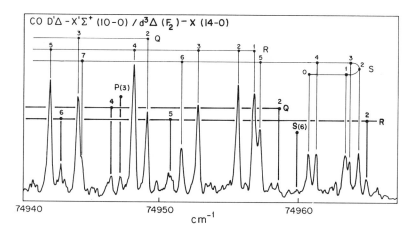

Fig. 2. Two-photon excitation spectrum of the CO D-X (10-0) bandhead. Q, R and S rotational branch assignments are indicated at the top of the figure. Also shown are assignments of some weak d-X (14-0) lines, indicated in the center of the figure.

The d~D interaction is an unusual example of an off-diagonal isoconfigurational perturbation. Because both states belong to the same $\pi^3\pi^*$ configuration, their rotational constants are rather similar, and such pairs of levels generally do not cross in the observed J-range. Thus, their interaction generally gives rise to approximately constant, J-independent energy shifts which are difficult to distinguish from other spin-orbit interactions. In the present case, the proximity of the two states (their origins are less than 10 cm^{-1} apart) coupled with the multiple crossing referred to above allows the determination of the d~D interaction parameter.

5. Acknowledgements

We acknowledge the collaboration of A.C. Le Floch in the analysis of the perturbations discussed above. We also thank R.W. Field for his constant interest and for many useful discussions. The experiments were performed at the M.I.T. Laser Research Center, which is supported by the National Science Foundation, and where B.A. Garetz has been a Visiting Scientist.

References

1. C. Kittrell and B.A. Garetz, Spectrochim. Acta **45A**, 31 (1989).

2. B.A. Garetz, C.Kittrell and A.C. Le Floch, to be published.

3. A.C. Le Floch, F. Launay, J. Rostas, R.W. Field, C.M. Brown and K. Yoshino, J. Molec. Spect. **121**, 337 (1987).

4. H. Lefebvre-Brion and R.W. Field, *Perturbations in the Spectra of Diatomic Molecules*, (Academic Press, New York, 1986), pp. 92-93.

5. R.W. Field and H. Lefebvre-Brion, Acta Phys. Hung. **35**, 51 (1974).

6. G. Herzberg and T.J. Hugo, Can. J. Phys. **33**, 757 (1955).

High Resolution Spectroscopy of Molecules and Small Clusters
in Molecular Beams

Wolfgang E. Ernst and Jörn Kändler*
Institut für Molekülphysik, Freie Universität Berlin,
Arnimallee 14, D-1000 Berlin 33, F.R. Germany

The application of tunable single mode dye lasers to a molecular beam allowed the development of many new spectroscopic techniques involving selective excitation of states, optical pumping and probing and double resonance schemes. The achievable high resolution helps to investigate the dense spectra of radicals and small clusters. A collimated beam of alkaline earth monohalide radicals produced in a high temperature oven interacted with laser and microwave radiation for the study of hyperfine structure in several electronic states (described in section 1). The application of an electric field allowed to measure the permanent electric dipole moment of the molecules in different states (section 2). In addition, zero field forbidden transitions were observed and excited state fine structure splittings could directly be determined from optical spectra. Metal clusters with a large size distribution are produced in a supersonic jet expansion. While in the other molecular beam studies the absorption of microwave or laser radiation was detected by monitoring laser induced fluorescence, resonant two photon ionization with size selective detection of ions has to be employed for the cluster spectroscopy (section 3).

1. Molecular Beam Spectroscopy with Lasers and Microwaves

The hyperfine structure of rotational transitions in the ground state of alkaline earth monohalides has been measured in a laser-microwave double resonance experiment. In principle, the method represents an optical version of a Rabi type apparatus, in which the A- and B-fields are replaced by a pump and a probe laser. In the region between A and B a microwave transition is induced filling the level which was depleted at A and is probed at B by observation of the laser induced fluorescence. In this way microwave transitions were measured with 10 kHz linewidth /1/. In the same experiment optical /1/ and UV spectra /2/ were recorded with 20 MHz linewidth using cw single mode dye lasers with optional intracavity frequency doubling. In this way the hyperfine structure of ground and excited states could be independently determined.

2. High Resolution Stark Spectroscopy

The Stark effect in the electronic ground state of molecules was measured at high precision by applying an electric field in the C region of the molecular beam laser-microwave double resonance experiment. In this way the Stark shift and splitting of microwave transitions was determined, from which the ground state dipole moment of the molecule could be derived /1/.

The dipole moment provides valuable information about the total charge density distribution in a molecule, while the hyperfine structure

*Present Address: VDI-Technologiezentrum, Graf-Recke-Str. 84,
                   D-4000 Düsseldorf 1, F.R. Germany

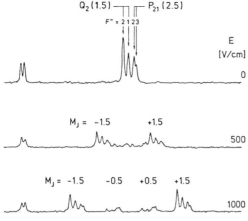

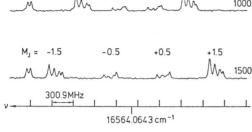

Fig. 1. Stark spectrum of the $Q_2(1.5)$ and $P_{21}(2.5)$ lines in the $A^2\Pi_{3/2}$-$X^2\Sigma^+(0.0)$ band of CaF near 604 nm for different electic field strength E. The hyperfine structure is well resolved.

serves as local probe for intramolecular fields and field gradients. For the understanding of chemical reactions, it is sometimes important to know how the electronic structure and chemical bonding changes, when the molecule is excited into different electronic states. Knowing the Stark effect in the ground state from the experiments above, we performed electric field studies of excited states by crossing the molecular beam with a cw single mode laser beam between two Stark plates. Stark shifts of optical lines were measured for electric field strengths up to 3 kV/cm. As an example Fig. 1 shows a Stark field study of the $A^2\Pi$ -$X^2\Sigma^+(0.0)$ transition of CaF. The narrow optical linewidth of 30 MHz allowed the determination of precise excited state dipole moments. In the depicted spectrum the line shifts and splittings are dominated by the linear Stark effect of the $A^2\Pi$ state. For CaF, excited state dipole moments were measured for the $A^2\Pi$, $C^2\Pi$ /3/ and $B^2\Sigma^+$ /4/ states. The analysis yielded an extraordinarily large dipole moment of the $C^2\Pi$ state at 30200 cm^{-1} with $\mu \approx$ 9D, compared with $\mu \approx$ 2.5D for $A^2\Pi$ (16520 cm^{-1}) and $\mu \approx$ 2D for $B^2\Sigma^+$ (18800 cm^{-1}). This behaviour could be well explained by a change in the orbital structure of the molecule /3/. Our systematic measurements provided the basis for the development of an ionic bonding model for the excited states of alkaline earth monohalides /5/.

The application of an electric field can be very useful for the analysis of an optical transition, if a direct determination of excited state fine structure splittings is needed, but prevented by the zero field selection rules. The zero field trace in Fig. 2 shows two rotational lines in the $C^2\Pi_{3/2}$ -$X^2\Sigma^+(0.0)$ band of CaF. The nomenclature of the transitions is explained in the level diagram on the right hand side of Fig. 2. With the electric field turned on, additional, Stark induced satellite lines appear due to the mixing of states of different parity in the excited state. The lines with a rotational quantum number as high as J = 23.5

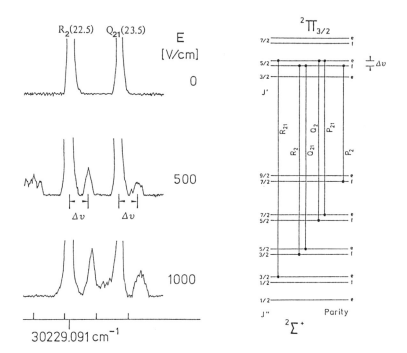

Fig. 2. Portion of the Stark spectrum of the $C^2\Pi_{3/2}$ -$X^2\Sigma^+(0.0)$ band of CaF near 330 nm showing electric field induced satellite lines next to the R_2 and Q_{21} lines. The energy increases to the left. The energy separation $\Delta\nu$ extrapolated to zero field corresponds to the upper state Λ-splitting labeled $\Delta\nu$ in the energy level diagram on the right.

remain nearly unshifted for up to 1 kV/cm and the frequency separation $\Delta\nu$ between main and satellite line is a direct measure of the Λ-type doubling in the $C^2\Pi$ state. Furthermore, the sign of the doubling parameters is given, because the relative positions of the e and f parity levels can also be determined. In the example shown, the information was extremely helpful for the analysis of the spectrum /2/.

3. Mass Selective Double Resonance Spectroscopy

Research on clusters is gaining increasing interest. High resolution spectroscopy is far from being applied to large clusters, but there is a beginning of investigation of small metal clusters with narrowband lasers. Since clusters of all sizes are produced in the commonly used supersonic jet expansion, a cluster size specific detection of spectra is needed. Monitoring laser induced fluorescence is therefore replaced by resonant two photon ionization with size selective detection of ions in a quadrupole mass spectrometer /6/. Even a rather small species like Na_3 exhibits an extremely dense rotational spectrum and Foth and Demtröder developed an optical-optical double resonance scheme with subsequent ion detection /7/. For the investigation of the rotational spectrum and hyperfine structure of alkali trimers we converted our molecular beam laser-microwave double resonance experiment in the way shown in Fig. 3. At A a level is again depleted by a pump laser beam, while the population is probed at B by resonant two photon ionization using a second laser in addition to the probe laser beam. The detection of ions is performed mass selectively.

410

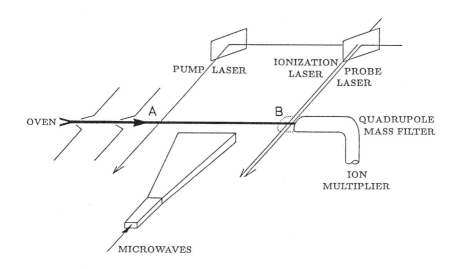

Fig. 3   Molecular beam laser-microwave double resonance spectroscopy observed by resonant two-photon ionization and mass selective detection of ions.

The method will allow a detailed study of the Jahn-Teller effect in the ground state of alkali trimers. Experiments on Na_3 are underway.

The work was financially supported by the Deutsche Forschungsgemeinschaft. One of the authors (W.E.E.) is grateful for the award of a Heisenberg fellowship.

References

1. W.E. Ernst, J. Kändler, and T. Törring, J.Chem.Phys.84, 4769 (1986).
2. W.E. Ernst, J. Kändler, and O. Knüppel, to be published.
3. W.E. Ernst and J. Kändler, Phys.Rev. A 39, 1575 (1989).
4. W.E. Ernst, O. Knüppel, and J. Kändler, to be published.
5. T. Törring, W.E. Ernst, and J. Kändler, J.Chem.Phys. 90, 4927 (1989).
6. A. Herrmann, M. Hofmann, S. Leutwyler, E. Schumacher, and L. Wöste, Chem.Phys.Lett. 62, 216 (1979).
7. H.-J. Foth and W. Demtröder, in "Laser Spectroscopy VIII" (eds. W. Persson and S. Svanberg), Springer, Berlin 1987, p. 248.

Multiphoton Ionization of $(Xe)_n$ and $(NO)_n$ Clusters Using a Picosecond Laser

D. Barton Smith* and John C. Miller
  Chemical Physics Section, Oak Ridge National Laboratory,
  P.O. Box 2008, Oak Ridge, TN 37831-6125, USA

Mass-resolved multiphoton ionization (MPI) spectroscopy is an established technique for detecting and analyzing van der Waals molecules and larger clusters. MPI spectroscopy provides excellent detection sensitivity, moderately high resolution, and selectivity among cluster species. In addition to information provided by the analysis of photoions following MPI, photoelectron spectroscopy can reveal details regarding the structure of ionic states (1). Unfortunately, the technique is limited by its tendency to produce extensive fragmentation. Fragmentation is also a problem with other ionization techniques (e.g., electron impact ionization), but the intense laser beams required for MPI cause additional dissociation channels to become available. These channels include absorption of additional photons by parent ions ("ion ladder" mechanism), absorption of additional photons by fragment ions ("ladder switching" mechanism), and resonances with dissociative states in the neutral manifold. The existence of these dissociation channels can preclude the use of MPI spectroscopy in many situations.

Recently, MPI studies of stable molecules using picosecond lasers (pulse length = $1 - 10$ ps) have indicated that limitations due to fragmentation might be subdued. With picosecond lasers, dissociation mechanisms can be altered and in some cases fragmentation can be eliminated or reduced. Additional photon absorption competes effectively with dissociation channels when a very short laser pulse or, perhaps more importantly, a sufficiently high peak-power is used. In the case where ionic absorption and fragmentation occurs, it has been shown that picosecond MPI might favor the ion ladder mechanism rather than the ladder switching mechanism (2). Larciprete and Stuke (3) have presented the argument that ionic fragmentation can be greatly reduced or even eliminated with the use of short laser pulses. Finally, two-color pump-probe experiments using picosecond lasers can, in principal, provide direct measurements of dissociation rates.

In an effort to extend the application of MPI spectroscopy to the study of weakly bound systems, we have begun a systematic investigation of picosecond MPI in van der Waals molecules and clusters. To our knowledge no previous picosecond MPI studies of weakly bound systems have been reported. We present here results of picosecond MPI of $Xe_n (n = 1 - 20)$ and $(NO)_n (n = 1 - 4)$ clusters. Previous MPI studies using nanosecond lasers have not detected the NO cluster series, presumably because of fast dissociation channels. The use of high peak-power allows resonant and non-resonant photon absorption to the ionization limit to compete effectively with fast dissociative processes.

The apparatus and method used for our picosecond MPI studies is similar to that used for previous MPI experiments in our laboratory (4). In brief, we produce atomic or molecular clusters in a supersonic jet expansion from a pulsed nozzle and ionize clusters with tightly focused laser light. The photoions are detected and analyzed with a time-of-flight (TOF) mass spectrometer. The laser system consists of a dual-operational-mode Nd:YAG laser, an H_2 Raman cell, and a short-cavity dye laser. The Nd:YAG delivers 1.2 J in a 10-ns pulse (Q-switched operation) or

---

*Postdoctoral Research Associate, The University of Tennessee, Knoxville

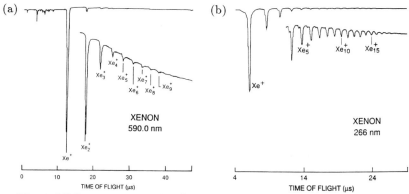

Fig. 1. MPI mass spectra of Xe_n^+ at (a) 590 nm and (b) 266 nm.

75 mJ in a 20-ps pulse (mode-locked operation) at the 1064 nm fundamental wavelength. The second, third or fourth harmonic (532, 355 or 266 nm) of the fundamental and corresponding Raman-shifted wavelengths are available in either operational mode. The dye laser provides 10-ps pulses of red or blue light (580-710 or 440-590 nm), and these wavelength regions can be extended by frequency doubling or frequency mixing.

Electron impact ionization studies of rare-gas clusters have detected large cluster series, but resonant MPI using nanosecond lasers has yielded only rare-gas monomers and dimers. [Nonresonant MPI at 266 nm, however, has yielded an extended series of clusters (5).] In contrast to the results of the nanosecond MPI studies, we have observed $Ar_n^+ (n = 1, 2)$, $Kr_n^+ (n = 1-4)$ and $Xe_n^+ (n = 1-20)$ ions using nonresonant MPI with our picosecond laser. For example, we have observed cluster ions as large as Xe_9^+ or Xe_{20}^+ with picosecond light near 590 nm or 266 nm, respectively, as shown in Fig. 1. At least for uncomplexed Xe atoms, the six-photon process is nonresonant with 590-nm light, the nearest allowed resonance being with the $5d[3/2]_1^0$ state at the five-photon level ($\Delta = 856$ cm^{-1}). Once formed, rare-gas cluster ions are known to be stable, but the relative ease of using high-order nonresonant MPI to observe clusters is rather remarkable since the high laser intensity ($\sim 10^{12}$ W cm^{-2}) might be expected to completely dissociate the ions. (Indeed, the laser intensity was sufficient to produce a small amount of Xe^{+2} in some spectra, an ionization process requiring the absorption of 16 photons at 590 nm). These results belie the conventional wisdom that MPI of "fragile" species must be performed with the least possible number of photons and, if possible, by ionizing the molecule to just above the ionization threshold.

Nitric oxide and $(NO)_n$ van der Waals molecules have been the subjects of numerous spectroscopic studies because NO plays a prominent role in the chemistry of the upper atmosphere. Although NO dimers and rare gas-NO clusters are readily formed in a free jet expansion, nanosecond MPI experiments with these species have failed to detect the $(NO)_2^+$ parent ion (6-7). This failure to detect the dimer ion can be attributed to the presence of dissociative states in the $(NO)_2$ and $(NO)_2^+$ manifolds; the only known excited state of the neutral dimer very rapidly dissociates (8) and the dimer ion is readily photodissociated by visible light (9). Likewise, we do not observe the dimer ion (or larger polymers) when we use visible light (532 nm or 585-605 nm). However, when either nanosecond or picosecond UV light (355 nm, 266 nm, or frequency-doubled 585-605 nm) is used, $(NO)_n^+ (n = 1-5)$ ions are

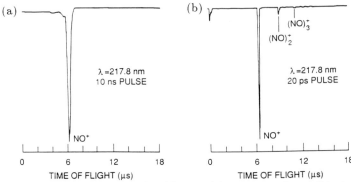

Fig. 2. MPI mass spectra of $(NO)_n^+$ with (a) nanosecond and (b) picosecond pulses.

readily observed. Figures 2a and 2b show TOF spectra following MPI with 217 nm light (the second anti-Stokes Raman-shifted wavelength from the 266-nm output). In this case the two-photon ionization is resonant with the dissociative excited state in the neutral manifold. Only NO^+ ions are observed with nanosecond pulses, while $(NO)_n^+(n = 1 - 4)$ ions are observed with picosecond pulses. Clearly, the higher peak-power available in the picosecond pulses enables MPI to compete effectively with the fast dissociation channel.

In conclusion, resonant or nonresonant MPI with a high peak-power picosecond laser is a versatile and rather general means of detecting and studying cluster distributions. Additional details on the present work are available elsewhere (10). Work in progress incorporates pump-probe techniques and photoelectron spectroscopy with picosecond MPI experiments.

Research sponsored by the Office of Health and Environmental Research, U.S. Department of Energy under contract DE-AC05-84OR21400 with Martin Marietta Energy Systems, Inc.

References

1. R. N. Compton and J. C. Miller, in *Laser Applications in Physical Chemistry*, edited by K. Evans (Dekker, New York, 1989).
2. J. J. Yang, D. A. Gobeli, and M. A. El-Sayed, *J. Phys. Chem.* 89, 3426 (1985); and references therein.
3. R. Larciprete and M. Stuke, *J. Crystal Growth* 77, 235 (1986).
4. J. C. Miller and R. N. Compton, *J. Chem. Phys.* 84, 675 (1986); J. C. Miller, *Anal. Chem.* 58, 1702 (1986).
5. O. Echt, M. C. Cook, and A. W. Castleman, *Chem. Phys. Lett.* 135, 229 (1987); *J. Chem. Phys.* 87, 3276 (1987); and references therein.
6. J. C. Miller and W. C. Cheng, *J. Phys. Chem.* 89, 1647 (1985); J. C. Miller, *J. Chem. Phys.* 86, 3166 (1987); J. C. Miller, *J. Chem. Phys.* 90, 4031 (1989).
7. K. Sato, Y. Achiba, and K. Kimura, *J. Chem. Phys.* 81, 57 (1984); *Chem Phys. Lett.* 126, 306 (1986); K. Sato, Y. Achiba, H. Nakamura, and K. Kimura, ibid. 85, 1418 (1986).
8. J. Billingsly and A. B. Callear, *Trans. Faraday Soc.* 67, 589 (1971).
9. G. P. Smith and L. C. Lee, *J. Chem. Phys.* 64, 5395 (1978).
10. D. B. Smith and J. C. Miller, *J. Chem. Phys.* 90, 5203 (1989); *Trans Faraday Soc. II* (to be published).

GaAs$^+$ Clusters Reaction with NH$_3$

L.H.Wang, L.P.F.Chibante
F.K.Tittel, R.F.Curl, R.E.Smalley

Rice Quantum Institute, Rice University
Houston, Texas 77251-1892

There has been considerable interest in theoretical models of III-V semiconductors, particularly GaAs, motivated by the technological importance of new semiconductor materials and devices. Much of the recent work focuses on the properties of the GaAs surface and its interface with other materials and dopants. Many of the theoretical approaches to describing these surfaces begin with a small cluster model. In this work, the chemical reactivities of real GaAs cluster ions are studied experimentally in order to probe the physical and chemical nature of the GaAs surface on the microscopic scale.

The apparatus used in this work is shown in Fig. 1. The GaAs clusters were produced by pulsed laser vaporization. The second harmonic of Nd:YAG laser (532 nm, 15-20 mJ/pulse) is used to vaporize a pure GaAs disk to generate an atomized plasma[1], which is subsequently quenched in helium carrier gas to promote effective clustering, the resulting clusters seeded in the gas are then cooled by subsequent supersonic free expansion into vacuum[2]. The molecular beam passes through a skimmer resulting in a well collimated beam. Positive residual ions created in the source are then electrostatically extracted and guided via einzel lenses and deflectors to an FT-ICR cell centered in a

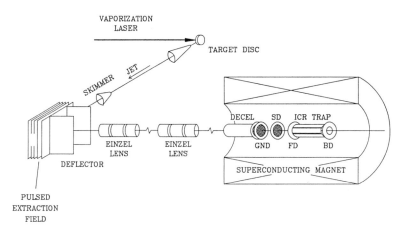

Fig. 1. Schematic of supersonic beam/FT-ICR apparatus. The extracted cluster ions (660 eV) are decelerated by the decelerator (DECEL), then admitted into the cell (ICR TRAP) by momentarily lowering the potential of the screen door (SD). The cluster ions are trapped between front door (FD) and back door (BD) set at 2-10 volts.

6 Tesla superconducting magnet. Ions of interest are mass-selected and decelerated into the cell, where they can remain trapped for minutes. At this point, numerous studies may be done, such as addition of a reaction gas, laser photofragmentation, and collision-induced dissociation.

In studying binary clusters, such as Ga_xAs_y, it is desirable to have high mass resolution in order to differentiate cluster masses corresponding to the various x and y values for a given total number of atoms, as it is expected that for a given size, clusters of varying x to y ratio would have dissimilar chemical and physical properties. In this work, a Fourier Transform-Ion Cyclotron Resonance (FT-ICR) provides the desired mass resolution of about 10^5. Extensive discussion of the design details of this apparatus have appeared elsewhere[2,3].

The FT-ICR mass spectra of $GaAs(9)^+$ clusters before and after reaction with NH_3 are shown in Fig. 2. The Ga_xAs_y ratio for parent/product ions is evaluated by deconvolution of the mass spectra using a least square technique.

To date, reactions of positive $Ga_xAs_y^+$ clusters, with a total atom count of 9 to 13 [9 is shown in Fig. 2 (B)], with NH_3 have been studied in the above mentioned FT-ICR apparatus . In this experiment, once the parent cluster was trapped and thermalized to room temperature, NH_3 (6×10^{-6} torr) was introduced into the cell region for varying exposure times. Parent and product ions were then mass analyzed. The addition chemisorption reaction is observed with up to four NH_3 binding to the cluster ion with no observable decomposition. For any given Ga:As ratio, the rate for total ammonia chemisorption is calculated under pseudo first-order conditions from the loss of a given bare cluster intensity, I_t.

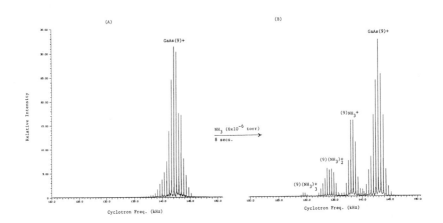

Fig. 2. $GaAs(9)^+$ FT-ICR mass spectrum before (A) and after (B) reaction with NH_3.

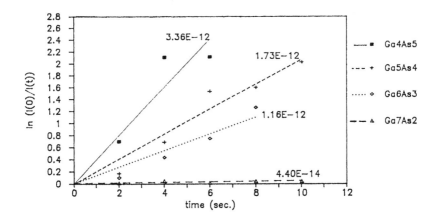

Fig. 3. Concentration decay vs. time (sec.) for GaAs(9)$^+$ reacting with NH_3. Calculated slopes are the corresponding reaction rates in cm^3/sec.

$$\ln(I_0/I_t) = k \, [NH_3] \, t.$$

An exemplary kinetic plot vs. time for various Ga:As ratio in the 9 atom cluster(x+y=9) is shown in Fig. 3. A striking trend of increased reaction rate with increased As in a given size cluster is observed for the cluster sizes 9-13 studied.

By using this FT-ICR device, semiconductor cluster ions can be observed with extremely high mass resolution and the interfacial physics and chemistry can be studied for a wide variety of conditions and reactants. For GaAs, so far no dramatic dependence of reactivity upon cluster size has been observed as in the case of Si[4].

The research described has been supported by the US Army Research Office and the Robert A. Welch Foundation.

1. S.C.O'Brien, Y.Liu, Q.Zhang, J.R.Heath,F.K.Tittel, R.F.Curl, and R.E.Smalley, J.Chem.Phys. **84** (7), 4074(1986).

2. R.E.Smalley, Analytical Instrumentation,**17** (1&2),1-21 (1988).

3. J.M.Alford, P.E.Williams, D.J.Trevor and R.E.Smalley, Int. J. Mass Spec. & Ion Proc. **72**, 33 (1986).

4. J.M.Alford, R.E.Smalley, MRS Conference Proc. **131** (1988).

PHOTODISSOCIATION SPECTROSCOPY OF
COLD MOLECULAR IONS AND CLUSTER IONS

J. A. SYAGE AND J. STEADMAN

Aerophysics Laboratory
The Aerospace Corporation
P. O. Box 92957
Los Angeles, CA 90009

1. Introduction

Recent work in our laboratory has centered on developing methods for (1) recording high resolution spectra of non-fluorescing polyatomic ions, and (2) studying size-specific photodissociation in molecular cluster ions. The first objective requires a means for preparing ions that are rovibrationally cold. These ions are then photodissociated by one-photon resonant absorption to a predissociative state or by multiphoton resonance-enhanced absorption to a dissociative state. The detection of a fragment ion signal as a function of wavelength yields a high resolution vibronic spectrum. The second objective depends on the effectiveness of isolating the properties of a specific cluster ion formed in a distribution of cluster sizes. Both objectives require mass filtering techniques for separating photofragment ions from initially formed fragment ions; however, the cluster ion experiments also require a method for distinguishing the parentage of the photofragments. A simple technique, that is adaptable to conventional single time-of-flight mass spectrometer systems, is described and two applications presented. First, cold ion photodissociation spectroscopy is demonstrated for CH_3I^+ produced by electron-impact ionization in a molecular beam. Second, the resonant photodissociation of size-specific $(CH_3I)_n^+$ cluster ions is reported.

2. Technique

The molecular beam time-of-flight (TOF) mass spectrometer and operating conditions have been described before.[1] The ion optics consist of the conventional three-grid arrangement represented by V0, V1, and V2 in Fig. 1a. The additional items V3 and V4 are part of a pulsed electron impact (EI) ionization source described in Ref. 2.

The discrimination of photofragment ions from primary fragment ions formed in the ionization step was achieved by two methods. In the first method, the photodissociation laser pulse was delayed with respect to the EI pulse. This causes the photofragment ions to arrive at the detector after the primary fragment ions. The second method, illustrated in Fig. 1b, is based on a new but simple concept,[3] whereby the traditional three-grid TOF ion optics assembly is operated as a low mass

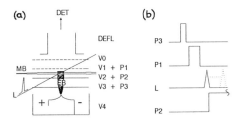

Fig. 1. (a) Crossed electron-laser-molecular beam ion optics assembly. The electron beam (EB) is activated by applying a gate pulse P3 to V3, and the ions are then extracted by applying a pulse P2 to V2. Deflector plates (DEFL) are used to compensate for ion drift resulting from the molecular beam (MB) velocity. (b) Primary ions are first accelerated away from the detector by P1+V1 and then reversed by dropping P1 and activating P2. Lighter ions that have traveled past the V2 grid are not detected.

rejection filter. A reverse bias pulse P1 applied to V1 causes the initially formed primary ions to move away from the detector. The voltage and duration define a cutoff point for light masses which exit the lower V2 grid. A positive bias is then applied by turning on P2. A TOF mass spectrum of only the heavier masses results. By applying the photodissociation laser pulse subsequent to the low mass filter (LMF) period, the photofragment ions appear in the mass spectrum against a zero background.

3. Resonance Ion Dissociation Spectroscopy

In recent years, investigators have turned to molecular beam resonance enhanced multiphoton ionization (REMPI) techniques to produce rovibrationally cold ions for use as precursors for recording high resolution ion photodissociation spectra.[4,5] Our work on the resonance ion dissociation (RID) spectroscopy of cold naphthalene ion is an example of this type of ion spectroscopy.[5] These techniques require two independently tunable lasers, one for ionization and the other for ion photodissociation. We have developed a new approach that makes use of electron impact (EI) ionization of a cold neutral molecular beam, thus enabling high resolution spectroscopy of nonfluorescing ions in a single-laser RID experiment.[6] The EI ionization of jet-cooled molecules produces ions with nearly the same rotational temperature as that of the neutral beam and a vibrational distribution dominated in favorable cases by diagonal Franck-Condon factors. Resonant photodissociation of the cold parent ions and mass selective detection of the photofragments results in highly resolved vibronic spectra.

We demonstrate this method for the $\tilde{A} \leftarrow \tilde{X}(^2\Pi_{1/2})$ transition of CH_3I^+ in Fig. 2. The spectral resolution is comparable to the two-laser method[4] mentioned above. The spectrum in Fig. 2 exhibits narrow rotationally cooled bandshapes as well as a minimal contribution from hot band transitions. It is apparent that the cold rotational temperature of the jet-cooled neutral CH_3I is preserved in the EI ionization process. The near absence of hot band transitions (only the minor presence of the $\nu_2 = 1$ series is apparent) is due to vibrationless preparation of CH_3I in the molecular beam and a favorable (0,0) Franck-Condon factors to the ground electronic state of the ion. Unless the ion geometry is quite different from the neutral geometry (e.g., NH_3), then it is reasonable to expect that the vibrationless transition probability will dominate in molecular beam ionization. Single-photon photoelectron spectra can serve as a guide for choosing favorable molecules on the basis of ion vibrational excitation and electronic state energies. In summary, the crossed molecular-electron-laser beam method yields spectral resolution comparable to the two-laser experiments, but is simpler, more economical, and less restrictive in the choice of precursors for forming cold ions.

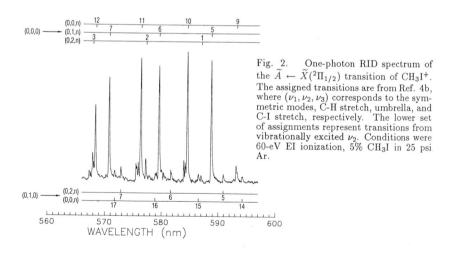

Fig. 2.   One-photon RID spectrum of the $\tilde{A} \leftarrow \tilde{X}(^2\Pi_{1/2})$ transition of CH_3I^+. The assigned transitions are from Ref. 4b, where (ν_1, ν_2, ν_3) corresponds to the symmetric modes, C-H stretch, umbrella, and C-I stretch, respectively. The lower set of assignments represent transitions from vibrationally excited ν_2. Conditions were 60-eV EI ionization, 5% CH_3I in 25 psi Ar.

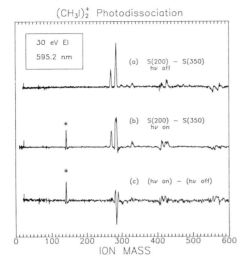

$(CH_3I)_2^+$ Photodissociation

30 eV EI
595.2 nm

(a) $S(200) - S(350)$
$h\nu$ off

(b) $S(200) - S(350)$
$h\nu$ on

(c) $(h\nu$ on$) - (h\nu$ off$)$

ION MASS

Fig. 3. Mass specific resonance photodissociation of $(CH_3I)_2^+$. (a) Window spectrum for dimer only, obtained from the difference of the LMF spectra $S(m_o)$ for $m_o = 200$ and $m_o = 350$ amu. (b) Corresponding difference spectrum showing dimer photodissociation to CH_3I^+ (*) (595.2 nm, 35 mJ, unfocused). (c) Parent (negative signal) and fragment (positive signal) spectrum obtained by subtracting spectrum (a) from (b). Conditions were 30-eV EI ionization, 5% CH_3I in 25 psi Ar.

4. Size-Specific Cluster Ion Photodissociation

The study of molecular cluster properties is motivated to a large extent by the opportunities to investigate the transition from gas phase to bulk phase behavior. In this segment of our work, we are interested in the photochemistry that occurs in cluster ions as a function of cluster size.

Photodissociation mass spectra were recorded for $(CH_3I)_n^+$ clusters produced by 30 eV EI ionization representing the sequence of events; ionization - low mass rejection - photodissociation. A means for studying photodissociation from specific cluster ion sizes is illustrated in Fig. 3 for $(CH_3I)_2^+$. The LMF P1 pulse was first adjusted to remove all masses below the dimer ion. Subsequent photodissociation by resonant 595 nm pulses was observed to give rise to the van der Waals dissociation product CH_3I^+. Differentiating the photofragment parent from among the remaining higher mass clusters is accomplished by filtering out successively larger cluster sizes and observing the disappearance of the corresponding photofragments. The difference spectrum from successive mass spectra consists of a single cluster ion parent and fragments and is illustrated in Fig. 3a and 3b in the absence and presence, respectively of a photodissociation pulse. Subtracting Fig. 3a from Fig. 3b reveals in Fig. 3c a spectrum consisting of the photofragments as positive signals and the specific parent cluster ion as a negative signal. The study of the size-specific photodissociation of $(CH_3I)_n^+$ cluster ions is currently in progress and will be reported in detail at a later date.

References

1. J. A. Syage, *J. Phys. Chem.* **93**, 170 (1989).

2. J. A. Syage, *Chem. Phys. Lett.* **143**, 19 (1988); J. E. Pollard and R. B. Cohen, *Rev. Sci. Instrum.* **58**, 32 (1987).

3. J. A. Syage, to be submitted.

4. A. M. Woodward, S. D. Colson, W. A. Chupka, and M. G. White, *J. Phys. Chem.* **90**, 274 (1986); K. Walter, R. Weinkauf, U. Boesl, and E. W. Schlag, *J. Chem. Phys.* **89**, 1914 (1988).

5. J. A. Syage and J. E. Wessel, *J. Chem. Phys.* **87**, 3313 (1987).

6. J. A. Syage, J. E. Pollard, and J. Steadman, *Chem. Phys. Lett.* submitted.

Excited-state LIF Spectroscopy of Hg and Zn Molecules - The Path to New Excimer Lasers?

J.B. Atkinson, E. Hegazi, W. Kedzierski, L. Krause and J. Supronowicz
Department of Physics, University of Windsor
Windsor, Ontario N9B 3P4, Canada

There has recently been a revival of interest in the spectroscopy of Group 2b excimers because of the possibility of laser action in the blue-green spectral region. The various past attempts to produce laser action in Hg_2 have been unsuccessful because of excited-state absorption from either the upper laser-level or metastable levels in equilibrium with it. The spectroscopy of the Hg_2 excimer has been investigated systematically during recent years using pump and probe methods, and numerous bound excited states and transitions between them have been identified. Analysis of vibrational structures of the fluorescence and excitation bands yielded vibrational frequencies, anharmonicities, energy separations between the electronic states, and relative equilibrium internuclear separations r_e.[1] Recent high-resolution laser-spectroscopic studies of monoisotopic Hg_2 produced partial resolution of the rotational structure as shown in Fig. 1. Some non-linearity in the line-separations is due to the mode structure of the probe-laser output. Analysis of these spectra together with the previously published relative r_e values[1] yielded the absolute r_e values for the various states, which are listed in Table 1 with absolute limits of error that exceed the relative errors.

We have also begun work on laser spectroscopy of the HgZn and Zn_2 excimers about which very little had been known. The HgZn molecule was found to possess a very rich spectrum consisting of bound-continuum and bound-bound fluorescence bands and of excitation bands, all with vibrational structures, which were identified and analysed with the aid of the simultaneously calculated potential energy diagram.[3-5] We find that the spectra can be satisfactorily interpreted on the basis of Hund's case (c) coupling, in terms of transitions between spin-orbit states. A typical HgZn excitation spectrum which is shown in Fig. 2, includes components due to transitions between A0$^+$ and A0$^-$ 'reservoir' states populated by pump-laser excitation, and E1 and F1 states excited with probe laser radiation. We have thus far observed transitions from the A1, A0$^+$ and A0$^-$ 'reservoir' states to the C0$^+$, E1, F1, and E0$^-$ states, with fluorescence emissions from the upper states to the X0$^+$ repulsive ground state and to the bound reservoir and B0$^-$ states. The HgZn excimer continues to yield new spectra some of which exhibit unusual and interesting properties due to predissociation.[6]

Very little laser spectroscopy of the Zn_2 molecule has been reported,[7,8] though several calculations of PE curves for it have been carried out according to Hund's case (a) coupling scheme. Figure 3 shows the $^1\Pi_u \leftarrow {}^3\Pi_g$ excitation spectrum of Zn_2, monitored by observing the structured continuum fluorescence emitted in the $^1\Pi_u \rightarrow X\,^1\Sigma_g^+$ decay. The vibrational assignments in the $^1\Pi_u$ state were obtained from the structure of the continuum band.[10] The assignments of the electronic states are consistent with a theoretical PE diagram[9] and work on further bands is continuing.

Table 1. r_e for Hg_2 states

State	r_e(Å)
A0$_g^+$	2.68±0.10
A0$_g^-$	2.68±0.10
B1$_g$	2.69±0.10
E2$_g$	2.69±0.10
F1$_g$	2.67±0.10
G0$_u^+$	2.93±0.11
H1$_u$	2.78±0.10
I0$_g^+$	2.79±0.10
J1$_u$	2.77±0.10

These experiments aim to identify gaps in the absorption spectra of the excimers, where laser action (and gain) might be obtained. We are also exploring the possibilities of producing laser action in a supersonic jet.[11]

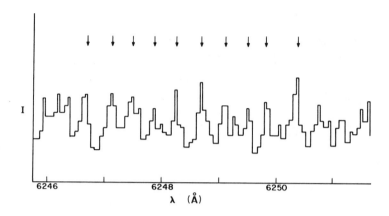

Fig. 1   A section of the $(Hg^{202})_2$ $G0^+_u \leftarrow A0^+_g$ excitation spectrum near 6248Å, containing mainly the v'=1 ← v"=7 vibration-rotation band. The marked peaks are overlapping P and R components in the range J=120-140.

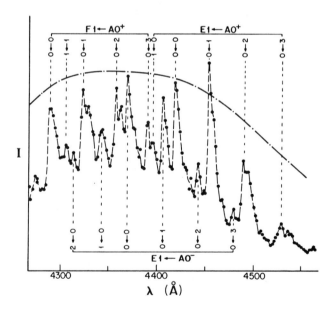

Fig. 2   A HgZn excitation spectrum showing the vibrational structures of the F1 ← A0$^+$, E1 ← A0$^-$, and E1 ← A0$^+$ band systems. The v' ← v" assignments are indicated, the dashed line represents the dye-laser power curve.

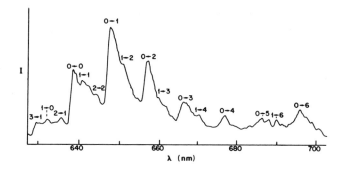

Fig. 3   The $^1\Pi_u \leftarrow {}^3\Pi_g$ excitation spectrum of Zn_2 showing v' ←v" vibrational components.

References

1.   R.J. Niefer, J. Supronowicz, J.B. Atkinson, and L. Krause, Phys. Rev. A 35, 4629 (1987) and references within.

2.   J. Supronowicz, W. Kedzierski, J.B. Atkinson, and L. Krause, J. Phys. B. to be published.

3.   J. Supronowicz, E. Hegazi, G. Chambaud, J.B. Atkinson, W.E. Baylis, and L. Krause, Phys. Rev. A 37, 295 (1988).

4.   J. Supronowicz, E. Hegazi, J.B. Atkinson and L. Krause, Phys. Rev. A 37, 3818 (1988); 39, 4892 (1989).

5.   E. Hegazi, J. Supronowicz, G. Chambaud, J.B. Atkinson, W.E. Baylis, and L. Krause, to be published.

6.   E. Hegazi, J. Supronowicz, J.B. Atkinson, and L. Krause, to be published.

7.   G. Rodriguez and J.G. Eden, Bull. Amer. Phys. Soc. 34, 1361 (1989).

8.   P.L. Zhang, Proc. NICOLS '89.

9.   P.J. Hay, T.H. Dunning, Jr., and R.C. Raffenetti, J. Chem. Phys. 65, 2679 (1976).

10.   W. Kedzierski, J.B. Atkinson, and L. Krause, Optics Lett. in press (1989).

11.   B.P. Stoicheff and T. Efthimiopoulos, Proc. NICOLS '89; Optics Lett., in press (1989). Bull. Amer. Phys. Soc. 34, 1357 (1989).

Improved insight into catalytic reactions by kinetic modelling and LIF detection of intermediate radical species

E. Fridell, B. Hellsing, S. Ljungström, T. Wahnström,
A. Rosén and B. Kasemo
Department of Physics
Chalmers University of Technology and University of Göteborg
S-412 96 Göteborg, Sweden

Heterogeneous catalytic reactions consists of a number of consecutive steps on the surface of the catalyst, starting with the adsorption of reactant molecules and ending with the desorption of product molecules. The intermediate steps involve so called intermediate reaction species which are short-lived and relatively hard to detect. Identification of such species and their properties is, however, a key to understand catalytic reactions. Keeping the catalyst at sufficiently high surface temperature, the *laser induced fluorescence (LIF)* technique is a powerful method to identify and characterize intermediate species desorbed from the catalyst [1]. Particularly challenging is if measured quantities for such intermediates, as the yield, desorption energy as a function of pressure, gas composition, catalyst temperature, can be used as input parameters to simplify the kinetic modelling of the reaction.

Recently [2], the LIF technique has been used for measurements of OH in the oxidation of hydrogen on polycrystalline Pt with a simultaneous kinetic modelling. Fig. 1 shows the desorption yield for OH, I_{OH}, determined with LIF, and production rate of H_2O, determined by the dissipated chemical energy as a function of the hydrogen partial pressure, α ($\alpha = p_{H_2} / (p_{H_2} + p_{O_2})$). The main features of the desorption rate of OH and production rate of H_2O are reproduced by the kinetic model calculations (solid lines) [2,3]. These yields are calculated from the surface coverage of the relevant species, i.e. O, H, OH, H_2O shown in the inset in fig. 1, which are obtained by solving the coupled equilibrium equations by the iterative multidimensional Newton-Rapson method [3].

These works have now been extended to measurements of the temperature dependence of the OH desorption and evaluation of the "apparent" desorption energy, E^a_{OH} in a conventional Arrhenius plot (i.e. $\ln(I_{OH})$ vs. $1/T$). I_{OH} is, however, dependent on the surface coverage of OH, θ_{OH}, which is a non-transparent and complicated function of temperature and α. Knowing the kinetics of the reaction, a detailed comparison between experimentally determined E^a_{OH} and kinetic modelling has been found to give information about the "true" desorption energy, E^d_{OH}, as well as the formation energy on the surface [4]. As can be seen from fig. 2, the apparent desorption energy of OH is dependent on α since the temperature dependence of θ_{OH} is different for different α values [4]. By comparing the experimental E^a_{OH} values with the kinetic model (solid line in fig. 2) we obtain the desorption energy for OH to be $E^d_{OH} = 2.0 \pm 0.15$ eV and the formation energy of H_2O, $E^f_{H_2O}$, to be ≤ 0.1 eV.

References

1. M.C. Lin and G. Ertl, Ann. Rev. Phys. Chem. <u>37</u>, 587 (1986).

2. B. Hellsing, B. Kasemo, S. Ljungström, A. Rosén and T. Wahnström, Surf. Sci. <u>189/190</u>, 851 (1987).

3. B. Hellsing and B. Kasemo, to be published.

4. T. Wahnström, E. Fridell, S. Ljungström, B. Hellsing, B. Kasemo and A. Rosén, to be published.

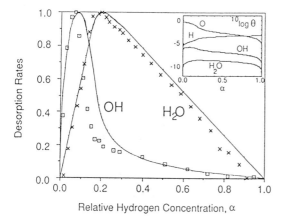

Fig. 1. Experimental data for the OH yield and H_2O production rate vs. α at a catalyst temperature of 1200 K and a total pressure of 100 mTorr. The solid lines represent the result of model calculations. Calculated coverages of O, H, OH and H_2O are presented in the inset.

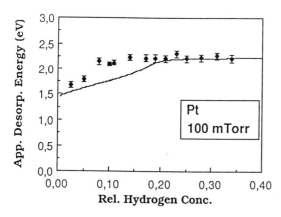

Fig. 2. Experimentally determined apparent desorption energies as a function of the relative hydrogen concentration, α, for a total pressure of 100 mTorr and a catalyst temperature of 1200 K. The solid line represents the result of the kinetic model calculations.

LASER-INDUCED FLUORESCENCE OF EXCITED Zn_2 MOLECULES

Pei-Lin Zhang, Shuo-Yan Zhao and Ding-Ning Xie
Department of Physics, Tsinghua University,
Beijing 100084, China

The molecules of IIB group elements are interested because they exhibit shallow van der Waals ground states and deeply bound excited states, characteristic of excimer dimers. Zn_2 specta were first observed qualitatively in 1931 by Hamada, using hollow cathode discharge at high temperature [1]. Since then, only a few papers about absorption spectra of Zn_2 in matrix of the rare gases [2][3] have been published during last decades for lack of suitable method to produce zinc dimers in the gas phase. Several theoretical papers [4][5] on potential energy curves were published. Interatomic potential for Zn_2 from absorption spectra was obtained in ref.[6]. In this paper we present laser induced fluorescence emission spectra of zinc molecules, especially the emission spectra from triplet states of zinc molecules.

A frequency-doubled Nd:YAG laser was used to pump a dye laser. The dye laser output was again frequency-doubled to produce 307.6nm radiation, corresponding to the $4^3P_1 - 4^1S_0$ intercombination line of zinc atom. The laser pulses were made incident on the zinc vapor (contained in a cross heat-pipe oven with buffer gas He or Ar) and served to pump the atomic zinc states. The fluorescence emission observed at right angle to the pump laser beam was resolved with a mono-chromator followed by a boxcar and their time dependence was registered with an optical multichannel analyzer combined with adjustable delay generator or a boxcar used in scan mode.

Zn_2 molecules were created by association of zinc atoms in the metastable state with other zinc atoms in the ground state or by association of the Zn atoms both in the meta-stable state.

$$Zn(4^1S_0) + h\nu \rightarrow Zn(4^3P_1)$$
$$Zn(4^3P_1) + Zn(4^1S_0) + M \rightarrow Zn_2(^3\Sigma_u^+, \, ^3\Pi_g) + M$$
$$Zn(4^3P_1) + Zn(4^3P_1) + M \rightarrow Zn_2(^3\Sigma_g^+, \, ^3\Pi_g, \, etc.) + M$$

The molecular states can be also populated through other processes: collisional energy transfer after optical excitation, and photoassociation etc.

Fig.1 shows the fluorescence spectrum of zinc molecules between 200-500nm with narrow linewidth excitation. There are continum bands in the range of 224-278nm and 350-480nm with four maxima at 252nm, 359nm, 390nm, and 424nm respectively. In the same wavelength region some sharp discrete lines are also observed. In order to assign the band spectra to transitions between particular Zn_2 states we have done the following experiments: the spectra versus presure of buffer gas and oven temperature, and the spectra versus intensity of the

pump laser light. We found that collisions between the atoms
of buffer gas and zinc vapor play an important role in the
band spectra. At lower pressure of the buffer gas the intensities of the atomic lines are stronger than the peak intensities of the molecular bands, while at higher pressure the
latter are stronger. The intensitiy of the 252nm band
increases rapidly with decreasing the spot size of the pump
laser light. We have also measured the intensities of bands
as a function of time following the pump laser pulse. Fig.2
shows time evolution of 252nm and 424nm bands respectively.
They are quite different so that the two bands can not
originate from the same upper molecular state. From the
theoretical potential energy curves [4][5] we assign 252nm
band with emission from state $^1\Sigma_u^+$ (which tends to $4\,^1P_1 + 4\,^1S_0$
atomic limit) to ground state $X\,^1\Sigma_g^+$, in agreement with the
result of Zn_2 of Hamada [1] and Ault and Andrews [3]. The
390nm band is considered as emission band from state $^3\Sigma_u^+$
($4\,^1S_0 + 4\,^3P_1$) to $X\,^1\Sigma_g^+$. The 359nm and 424nm are probably due to
transitions that originate from excited molecular states
$^3\Sigma_g^+$, $^3\Pi_g$($4\,^3P + 4\,^3P$) to lower molecular state $^3\Sigma_u^+$($4\,^1S + 4\,^3P$).

Spectral scans revealed fluorescence emission lines due to
atomic zinc transitions. There are thirty-one lines in total.
The intensities of these atomic lines increase as the
pressure of the buffer gas decreases. The normal fluorescence
emission lines have been identified to various atomic transi-
tions: $n\,^3D - 4\,^3P_J$, $n=4,5,6,7,8$, $m\,^3S_1 - 4\,^3P_J$, $m=5,6,7$, $J=0,1,2$;
$4\,^3P_1 - 4\,^1S_0$, $4\,^1D_2 - 4\,^1P_1$, and $4\,^1P_1 - 4\,^1S_0$. Seven of them, $4\,^3D - 4\,^3P_J$,
$5\,^3S_1 - 4\,^3P_J$ and $4\,^1D_2 - 4\,^1P_1$, showed stimulated amplification of
spontaneous fluorescence emission, collinear with pump laser
beam. Two additional fluorescence emission lines are from
ionic zinc transitions: ZnII $3d^{10}4p\,^2P_J - 3d^{10}4s\,^2S_{1/2}$.

The experimental and theoretical investigation are now in
progress. This work has been supported by National Natural
Science Foundation of China.

References

1.   H. Hamada, Philos. Mag., 12, 50 (1931)
2.   W.W. Duley, Proc. Phys. Soc. 91, 976 (1967)
3.   B.S. Ault, L. Andrews, J. Mol. Spectrosc. 65, 102 (1977)
4.   P.J. Hay, T.H. Dunning Jr. and R.C. Raffenetti, J. Chem.
     Phys. 65, 2679 (1976)
5.   H. Tatewaki, M. Tomonari and T. Nakamura, J. Chem. Phys.
     82, 5606 (1985)
6.   C.-H. Su, Yu Huang and R.F. Brebrick, J. Phys. B:
     At. Mol. Phys. 18, 3187 (1985)
7.   H. Komine, R.L. Byer, J. Chem. Phys., 67, 2536 (1980)

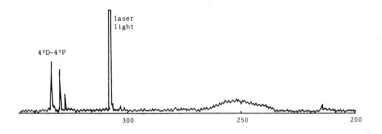

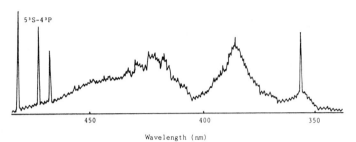

Wavelength (nm)

Fig.1 Fluorescence spectra of Zn$_2$ and ZnI

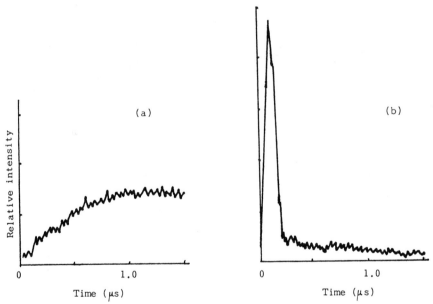

Fig.2 Time evolution of Zn$_2$ 252nm(a) and 424nm(b) bands

Laser Induced Lasing in CS_2 Vapor

Huei Tarng Liou,[*] Peilin Dan, Howard Yan, Ravia W. Joy, Nancy C. Wang
and J. Y. Yuh
 Institute of Atomic and Molecular Sciences, Academia Sinica
P. O. Box 23-166, Taipei, Taiwan

When the CS_2 vaper of 400 mTorr in a 115 cm-long cell was radiated by a 1 mJ pulsed laser (duration 20 ns) at wavelength 343.57 nm and was optically pumped to the R(J) = 28, (ν=0,10,0), a^3A_2 eletronic state, six coherent emissions, wavelengths at 538.4, 514.4, 507.2, 485.2, 465.4, and 447.2 nm were observed bi-coaxially with pumping laser. These emissions possess characteristics of laser but don't need a cavity to gain amplification. Their linewidths have the same order of magnitude as that of the pumping laser.

Figure 1 shows that the peak at 507.2 nm has the highest intensity. Its converting efficiency is about 1.5% with a quartz plate as an entrance window. Each transition has two lines corresponding to P and R branch fine structures (1).

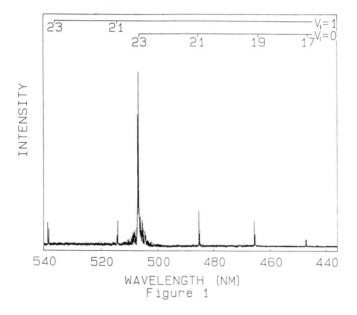

Fig. 1. The dispersive spectrum of laser induced lasing of CS_2 from R(J) =28, (ν = 0, 10, 0), a^3A_2 state with monochromator resolution FWHM= 0.18 nm.

The energy scheme depicted in Figure 2 clearly expresses that the transitions of the laser induced lasing are from a^3A_2 metastable state to high vibrational states of ground electronis state $^1\Sigma_g^+$ which behave as terminal states.

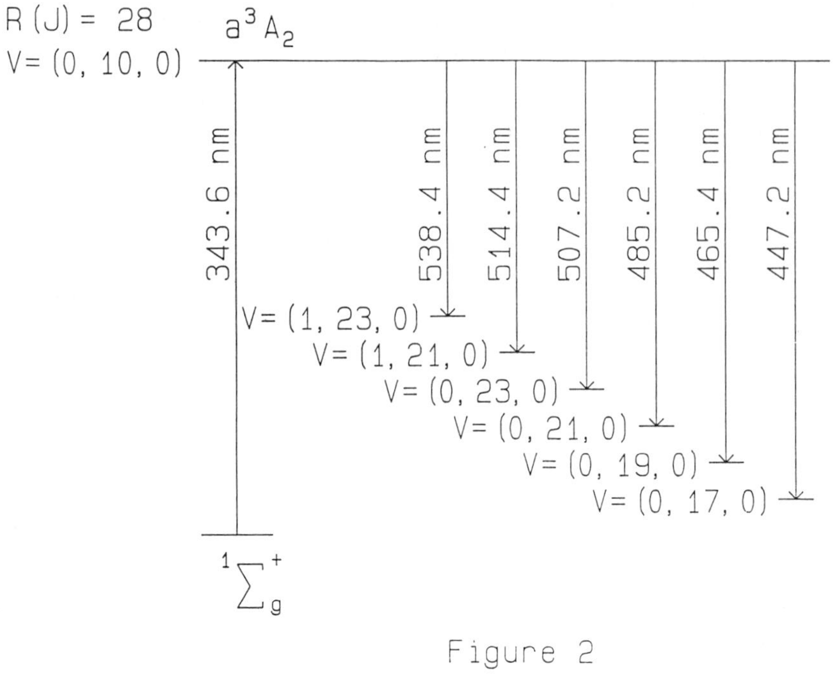

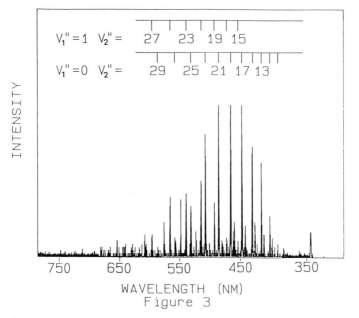

R (J) = 28 a^3A_2
V= (0, 10, 0)

343.6 nm

538.4 nm
514.4 nm
507.2 nm
485.2 nm
465.4 nm
447.2 nm

V= (1, 23, 0)
V= (1, 21, 0)
V= (0, 23, 0)
V= (0, 21, 0)
V= (0, 19, 0)
V= (0, 17, 0)

$^1\Sigma_g^+$

Figure 2

Fig. 2.   The energy scheme of laser induced lasing mechanism.

INTENSITY

$V_1'' = 1$ $V_2'' =$ 27   23   19   15

$V_1'' = 0$ $V_2'' =$ 29   25   21   17 13

750   650   550   450   350

WAVELENGTH (NM)
Figure 3

Fig. 3. The spectra of single vibronic level dispersive fluorescence
was taken with molecular beam configuration. The fluorescence is from R
(J) =4, (V =0,10,0), a^3A_2 state of CS_2.

430

In order to compare the intensity distribution of the laser induced lasing (Fig.2) with that of spontaneous emission, single vibronic level dispersive fluorescence spectrum (upper fluorescence state; R(J) = 4, (ν = 0,10,0)), shown in Figure 3, was taken under a molecular beam configu-ration. It is obviously that two spectra have different intensity dis-tributions. A notable difference is that in the laser induced lasing spectrum the intensity of ν=0,25,0 is absent, that is, ν_1 =0 progressing along ν_2 ends at ν_2=23 whereas ν_1=1 progressing along ν_2 begins to show up. This may be due to transitions to a^3A_2 state are borrowing the intensi-ty from the B_2 state(2) and causing the observed difference between absorption and emission transition probabilities(3). Another possible cause is due to self-absorption in the cell because that the observa-tion of laser induced lasing is coaxial, whereas the observation of dispersive fluorescence is 90°with respect to excitation laser and taken under the molecular beam configuration in which self-absorption can be neglected. Furthermore, it is worth knowing that no significant changes in the intensity distribution of laser induced lasing were seen when we varied the pressure, the length of cell(within 5%), or the polarization direction of pump UV laser; the peak at 507.2nm always do-minates the intensity. So far we can not eliminate the possibilities that the intensity distribution of the laser induced lasing resulting from dynamics of molecule itself such as Fermi resonance, Colioris interaction, collisional induced effects.

Figure 4 shows the relations of time among excitation photon, laser induced lasing and spontaneous emission. The time delay between excitation photon and laser induced lasing decreases as CS_2 pressure increases.

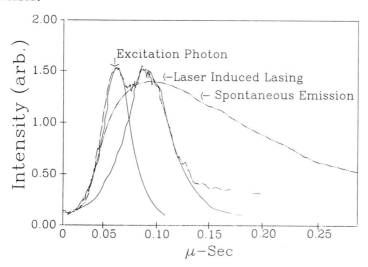

Fig. 4.   Time relation among excitation photon, laser induced lasing and spontaneous emission. Solid curves are Gaussian fitting. The excited state is R(J)=28, (ν=0,10,0), a^3A_2. The final state of laser induced las-ing is ν=(0,23,0), $X^1\Sigma_g^+$. Measurements were taken under 400 mTorr CS_2 pressure condition.

Dicke(4) had pointed out that under highly population inversion, a coope-rative emission such as superradiance can be a coherent emission. and the intensity varies as N^2 (where N is initial population inversion).

What we have observed laser-like emissions possess third order dependence which is shown in Fig.5.The laser induced lasing which possess higher order dependence indicates that it is a new phenomenon instead of super-radiance.

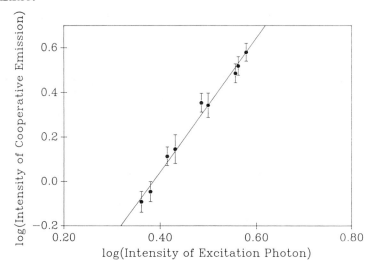

Fig. 5.   Log-Log plot of intensity of laser induced lasing vs. intensity of excitation photon. It gives a third order dependence. The laser induced lasing is from $R(J)=28$, $(V=0,10,0)$, a^3A_2. to $V=(0,23,0)$, $X'\Sigma_g^+$.

So far more experimental works are undergoing in order to understand the phenomenon of laser induced lasing.Our recent observation indicates that high vibrational states of ground electronic states of CS_2 molecule can be terminal states of lasing process.Moreover,when the pumping wave -length changes to pump the molecule to other vibrational states of a^3A_2 , new set of lasers at different wavelengths can be provided. This may be a prospect of laser in the future.

References

1. Stimulated emission pump to $(V=0,23,0)$ and other vibrational levels has carried out in this laboratory,to be published

2. Robin M.Hochstrasser and Douwe A.Wiersma,J.Chem.Phys.,54,4165(1971)

3. P.A.Geldof and P.H.Rettschnick,Chem.Phys.Lett,10,549(1971)

4. R.H.Dicke,Phys.Rev.93,99(1954)

Multi-Quantum-Transition Sampling Spectroscopy for Fast Transient Luminescence Measurement

Yoshihiro Takagi, Tohru Kobayashi, and Keitaro Yoshihara
Institute for Molecular Science, Myodaiji Okazaki,444 Japan

Multi-quantum-excited photoelectron emission in various solids has been studied for high time-resolution sampling spectroscopy of short-lived luminescence.

Previously we proposed [1] a sampling method based upon a photoionization detection due to multi-photon resonant absorption in gases. Here we propose another sampling method simpler in experimental setup and more efficient in measurement than the previous method. A two-photon-induced electron emission [2-4] is detected when a photocathode of a photomultiplier (PMT) is simultaneously irradiated by a pump laser and a luminescence to be examined. If the work function W of the photocathode satisfies an energy relation $2h\nu_1$, $h\nu_2 < W < h(\nu_1+\nu_2)$, where ν_1 and ν_2 are the frequencies of the pump wave and luminescence, respectively, the photoelectron emission can be attributed to the double-quantum excitation induced by ν_1 and ν_2 photons. A competitive process of two-photon transition due to the two ν_1 photons can be excluded by this relation. Photoelectrons are amplified in the same PMT, giving a highly sensitive detection. There is no necessity for an extra nonlinear material such as the optical Kerr shutter and optical frequency mixing. In contrast to the previous method, measurement of time-resolved spectrum does not require a tunable laser but only a monochromator. Highly time-resolved measurements of transient luminescence can thus be realized by a very simple setup.

We investigate suitable photoelectric materials for the two-photon process and observe an autocorrelation profile of

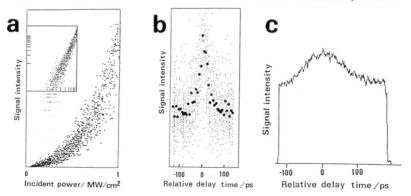

Fig.1(a) Input power dependence of photocurrent with a photo-cathode of Cs_3Sb,(b) autocorrelation signals of L1, and (c) L5.

pump laser pulse. We have investigated commercial PMT's with photocathodes of different spectral responses. The light source used in the experiment was a single pulse of a passively mode-locked Nd:YAG laser (L1) and its second (L2), third (L3), and fourth (L4) harmonics, and a cw mode-locked Nd:YAG laser (L5). Figure 1a shows photocurrent as a function of the input power of L1. The photocathode was Cs_3Sb, which is sensitive in visible and near-UV wavelength ranges. The signal showed a typical quadratic characteristics in the input range up to several megawatts/cm^2. Too high input power produced a saturation of the signal. Figure 1b and 1c show autocorrelation profiles of the input pulse for L1 and L5, respectively, obtained with an ordinary Michelson-type autocorrelator. We also investigated "solar blind" PMT's with photocathodes of Cs_2Te and CsI, and a specially designed PMT with a gold photocathode. Although Cs_2Te and CsI gave a remarkable sensitivity (linear characteristics) in the visible, but no quadratic dependence was obtained in our input range. The gold showed a clear quadratic feature for L3, which is consistent with the known work function of gold (4.8 eV). Nevertheless, a careful autocorrelation measurement showed no peaks at the coincident time of the two input beams and almost flat over the time range of 300 ps. This result is reasonable if we consider the absorption spectra of most metals with no deep band-gap in our spectral range, so that the two-photon transition would take place via a real intermediate state.

In order to find a multi-quantum photoelectric effect for various materials in a more controlable manner, we have designed a vacuum chamber containing a photocathode and an electron multiplier. With a photocathode of CsI film evaporated on a stainless steel with thickness of 2000 Å, we obtained a quadratic characteristics for L3 as shown in Fig. 2a and an autocorrelation signal in Fig. 2b. The width of the correlation profile was determined by the incident pulse width and not by the response time of the photocathode. This is quite reasonable because of the very broad absorption band-gap over the wavelength range 0.2 to 10 μm. An autocorrelation was observed using a photocathode of silver chloride film for L3 with a slightly higher input power. A cross correlation

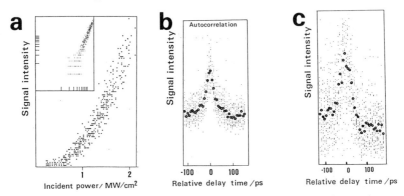

Fig.2 Characteristics of CsI; (a) Input power dependence, (b) autocorrelation for L3 and (c) cross correlation of L2 and L4.

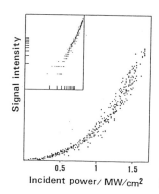 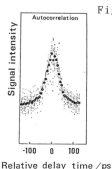

Fig.3 Two-photon-induced photoconductivity in a GaAsP photodiode.

Incident power/ MW/cm² · Relative delay time /ps

signal between L2 and L4 was also detected with CsI as shown in Fig. 2c. The input power of L4 was suppressed so that the two-photon photoelectric effect due to L4 was not detected, whereas the input power of several hundred microjoules was required for L2 to give a cross correlation signal. A background signal in Fig. 2c is due to an illumination of the electron multiplier by the scattered light of L2.

We now try a picosecond sampling of UV fluorescence from a solution of biphenyl using the arrangement of the cross correlation measurement replacing the L4 by the fluorescence beam. Since the sum of the photon energies of L2 and the fluorescence is not well above the threshold of the photoelectron emission of CsI ($W \approx 6.9$ eV), a different fluorescent material would be necessary.

As a photocathode material for the sampling of the deep UV luminescence, alkali-chlorides and alkali-fluorides would be more suitable. A more sophisticated detection system in combination with the photoelectron analyzer such as in the retarding potential method or time-of-flight method will give more information on time-resolved spectrum of short-lived luminescence.

In addition to the two-photon photoelectron emission, we have observed for the first time a quadratic characteristics in the internal photoelectric effect using a commercial photodiode (Hamamatsu Photonics Inc., diffused junction and Schottky type GaAsP photodiodes). Figure 3 shows an autocorrelation profile for L1. The autocorrelation width was nearly equal to that obtained with a Cs_3Sb PMT. Since the photodiode used in the measurement is sensitive in the wavelength range 0.2-0.8 μm, the obtained data strongly indicates the two-photon-induced photoconductivity. An autocorrelator using a photodiode is useful as a simple device for pulse-width measurements and optical alignment in ultrashort time scale.

[1] Y.Takagi and K.Yoshihara, "Ultrashort phenomena", pp.407-409 (Springer-Verlag, 1988).
[2] H.Sonnenberg, H.Heffner, and W.Spicer, Appl.Phys.Lett., 5 95(1964).
[3] E.M.Logothesis, Phys.Rev.Lett., 19 1470(1967).
[4] S.Imamura, F.Shiga, K.Kinoshita, and T.Suzuki, Phys.Rev., 166 322(1968).

High Resolution Stimulated Raman-Brillouin Spectroscopy
in Gases and Solids

William K. Bischel,* Mark J. Dyer, Gregory W. Faris,
A. Peet Hickman, and Leonard E. Jusinski
Molecular Physics Laboratory, SRI International, Menlo Park, CA 94025

1.   Introduction

High resolution, quasi-cw, stimulated Raman-gain spectroscopy has become a standard technique for the study of molecular energy levels. In this paper we apply this technique to the study of two diverse problems: a determination of the ac Stark effect for the vibrational levels of H_2, and a demonstration of the first high resolution stimulated Brillouin gain spectra in glasses and crystals.

2.   Measurement of the ac Stark effect in H_2

The optical Stark effect as applied to molecular transition frequencies has been of considerable interest in recent years.[1] In particular, the ac Stark shift has been found to contribute to the linewidth broadening of the Stokes and anti-Stokes lines generated by stimulated Raman frequency conversion,[2] and influence the ultimate resolution limits and sensitivities of spectroscopic techniques such as stimulated Raman gain spectroscopy and CARS.[3] Also, transient changes in the index of refraction in Raman amplifiers can occur which arise from the difference in the polarizabilities of the v"=0 and v"=1 vibrational state. This degrades the beam quality, and therefore the conversion process, of the pump and Stokes frequencies. By investigating this polarizability difference between v"=1 and v"=0 in H_2 through the ac Stark effect, these changes in the index of refraction may be quantified.

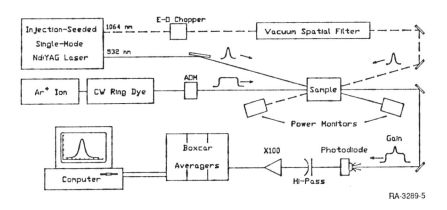

RA-3289-5

Figure 1. Quasi-cw gain/absorption detection diagram for investigation of the ac Stark effect in H_2 and Raman/Brillouin spectroscopy of solids.

We chose to study this effect in molecular hydrogen by constructing the quasi-cw stimulated Raman gain spectrometer[4] illustrated in Fig. 1 using the second harmonic of a single-mode Nd:YAG laser as the fixed frequency pump and a cw ring dye laser as the tunable narrowband probe. By tuning the probe through a Raman transition frequency relative to the pump, a small gain signal was imposed by the pump on the chopped cw probe and monitored by a gated boxcar averager to give lineshape information vs. frequency. A high-intensity, non-resonant optical field provided by the residual IR fundamental of the Nd:YAG was spatially-filtered and applied to the Raman interaction region on alternating pulses of the pump laser as the frequency of the dye laser was tuned through the vibrational Raman transition. This facilitated collection of shifted and unshifted spectral data on the same scan and permitted relative frequency shift measurements accurate to within 10 MHz. Data processed through boxcar averagers were then recorded by computer, normalized to pump pulse energy, and fit to Lorentzian lineshapes for shift determination.

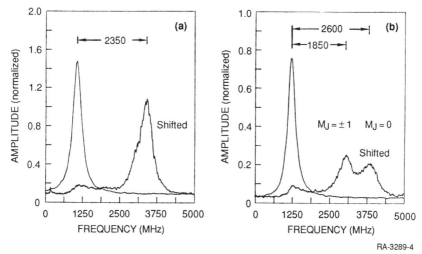

Figure 2. Data for the ac Stark effect on the vibrational Raman transitions in H_2 at a density of 1.81 amagat and an IR laser intensity of ~100 GW/cm^2 for (a) the Q(0) transition and (b) the Q(1) transition.

Typical data is shown in Figure 2. For the Q(0) transition (Fig. 2a), line shifts scaled with IR field intensity and the data yielded a shift coefficient of 23 MHz/GW cm^{-2}. Shifting of the Q(1) transition (Fig. 2b), which split into composite M_J levels under the influence of the optical field, was measured to be 27 MHz/GW cm^{-2} for $M_J=0$ and 19 MHz/GW:cm^{-2} for $M_J=\pm1$. The observed shifts were density independent for both transitions in the absence of four-wave mixing for counterpropagating IR and pump/probe geometries.

Excellent agreement is reached when the measured quantities are compared with the ac Stark shifts and refractive indices predicted from *ab initio* theory by W. Huo.[7]

3.   Brillouin Gain Measurements in Glasses and Crystals

Brillouin scattering can limit the transmission of light through optical materials, ranging from optical fibers for communications to high energy laser fusion devices. With the quasi-cw spectroscopy system described above, high resolution measurements of Brillouin gain coefficients and related acoustic parameters may be performed. The frequency-doubled Nd:YAG laser serves as a pump laser, and the ring dye laser as a probe. While gain spectroscopy of this type has been performed previously in gases[5,6] this is the first use of Brillouin gain spectroscopy in solids to our knowledge. Because of the high peak power of the single mode Nd:YAG laser, large signals can be obtained from bulk samples roughly a centimeter long. And, because the laser frequency is transform-limited (widths < 40 MHz), narrow band measurements may be made.

In solids, both longitudinal (compressional) and transverse (shear) acoustic waves are involved Brillouin scattering. To accurately measure both waves, two pump-probe geometries are used, a counter-propagating geometry and a crossed beam geometry. For the latter, the Nd:YAG pump laser is focused to a line with two cylindrical lenses, and the probe laser passes through the focus at approximately 90°. For the counterpropagating geometry, the crossing angle is about 5 mrad. The spot sizes are relatively large (200 to 600 μm), so line broadening due to beam divergences is negligible. By measuring the percent gain of the probe beam, the peak intensity of the pump beam, and the path length through the sample, absolute gain coefficients are measured. Line widths and line shifts are measured by scanning the probe beam. By fitting a Voigt profile with the proper Gaussian component for the laser, the Lorentzian width due to the acoustic damping time is found. From the measured line shifts, the speed of sound is calculated. For fused silica, the FWHM longitudinal and transverse linewidths were measured to be 162 ± 9 MHz and 43 ± 8 MHz, respectively, and the speed of sound was found to be $5.94 \pm 0.005 \times 10^5$ cm/s.

A measurement of both the longitudinal and transverse gain peaks in fused silica in the crossed beam geometry is shown in Figure 3. This measurement was performed in a right angle turning prism. While the transverse gain peak is due to crossed beam pumping, the longitudinal peak is actually due to a reflection of the edge of the pump beam line focus off the back face giving a counterpropagating-pumped gain. When the pump laser intensity is increased, a central peak due to stimulated Rayleigh scattering is seen. High resolution Brillouin gain spectroscopy should increase the understanding of the Brillouin scattering process and result in more accurate acoustic constants. The use of quasi-cw gain spectroscopy for measurements of ac Stark effects and Brillouin spectroscopy demonstrates the wide applicability and high accuracy of the technique.

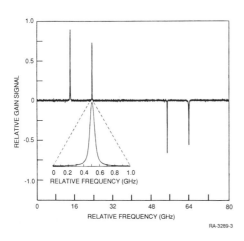

Figure 3. Longitudinal and transverse gain peaks in fused silicon. The two peaks on the left correspond to gain on the pump beam (loss on the probe). The inset shows a Voigt fit (solid line) to a measurement of the transverse gain peak (points).

This research is supported by the Office of Naval Research and Lawrence Livermore National Laboratory

1. L. A. Rahn, R. L. Farrow, M. L. Koszykowski, and P. L. Mattern, Phys. Rev. Lett. **45**, 620 (1980).
2. W. K. Bischel, D. J. Bamford, and M. J. Dyer, SPIE **912**, 191 (1988).
3. M. Pealat, M. Lefebvre, J.-P.E. Taran, and P. L. Kelley, Phys. Rev. A **38**, 1948 (1988).
4. P. Esherick and A Owyoung, *Advances in Infrared and Raman Spectroscopy*, Vol. 9, pp. 130-187 (1983).
5. S. Y. Tang, C. Y. She, and S. A. Lee, Opt. Lett. **12**, 870 (1987).
6. G. C. Herring, M. J. Dyer, and W. K. Bischel, to be published.
7. W. Huo, private communication (1987).

Multiplex CARS Study of Infrared-Multiphoton-Excited OCS

Kuei-Hsien Chen, Cheng-Zai Lü, Eric Mazur and Nicolaas Bloembergen
Division of Applied Sciences, Harvard University, Cambridge, MA 02138

Mary J. Shultz
Department of Chemistry, Tufts University, Medford, MA 02155, and
George R. Harrison Spectroscopy Laboratory, MIT, Cambridge, MA 02139

Coherent anti-Stokes Raman spectroscopy (CARS) is a sensitive and efficient technique to monitor molecular vibrational and rotational distributions. This paper reports on the application of CARS to study the dynamics of OCS after infrared multiphoton excitation. The OCS molecule has three widely separated fundamental modes ($v_1 = 859$ cm^{-1}, $v_2 = 527$ cm^{-1}, $v_3 = 2079$ cm^{-1}). The overtone of the v_2 mode can be excited with CO$_2$ laser frequencies between the $P(10)$ and $P(26)$ lines of the 9.6 μm branch. The experiments were carried out both in bulk samples and in a supersonic molecular jet.

1. Setup

The experimental setup is shown in Fig. 1. The important features of the CARS spectrometer, described previously,[1] are summarized here. The main improvement to the setup is the addition of a supersonic molecular jet described below.

The infrared pulses from a TEA CO$_2$-laser have a 150-ns duration and a maximum energy of 200 mJ. The beam is focused into the interaction region with a 15-cm focal length cylindrical lens down to a beam waist of 110 μm by 18 mm.

Frequency-doubled Nd:YAG laser pulses with a 10-ns duration and 200-mJ average energy are used to generate the CARS signals. The bandwidth of these pulses is reduced with an intracavity line-narrowing etalon to 0.1 cm^{-1}. About 25 mJ of the doubled Nd:YAG laser output is used for each of two beams at $\omega_1 = 532$ nm, while the remainder serves to pump a broad band prism-tuned dye laser at $\omega_2 = 557$ nm. After amplification, 6-ns, 20-mJ pulses of 60-cm^{-1} linewidth are obtained. This linewidth is sufficient to generate multiplex CARS signals[2] over the entire region of interest.

The ω_1 and ω_2 beams are aligned parallel to each other and focused in a folded BOXCARS geometry by a 25-cm focal length lens. The beam waist in the interaction region where the infrared laser beam, the CARS laser beams and the molecular jet cross at right angles is 80μm.

The CARS signal passes through an aperture which spatially rejects the ω_1 and ω_2 beams. The signal is dispersed using an 1-m spectrograph, and the dispersed CARS signal

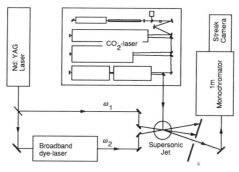

Fig. 1. The experimental setup.

is then recorded on a streak camera system with a detector array. The resolution of the entire system is 0.012 nm (0.46 cm^{-1}). To obtain a good signal-to-noise ratio each spectrum is averaged over 50 shots.

The molecular jet apparatus consists of a nozzle mounted on an XYZ translational manipulator, and a high vacuum chamber pumped by a 700-liter/s diffusion pump. Without molecular jet the chamber pressure was 4×10^{-7} Torr; with a back pressure on the nozzle of about 3 atm, and 1-ms molecular pulses at a 10-Hz repetition rate, the chamber pressure increases to 7×10^{-4} Torr.

The supersonic jet system has several advantages over the static cell: First, it greatly reduces the number of collisions between molecules and thus allows one to study molecules under collisionless conditions. Second, because of the adiabatic expansion the translational, rotational and vibrational temperatures of the sample are drastically reduced. This reduces the total number of populated molecular states and increases the population of individual states, thereby improving the CARS signal which is proportional to the square of the number of molecules in that state.

The density in the supersonic jet was determined by comparing the integrated intensity of ground state peak in the CARS spectrum obtained in the jet with the one in bulk samples at various pressures. For OCS at $x/D = 3$, with x the distance from the nozzle and $D = 1$ mm the diameter of the nozzle aperture, the density in the jet was found to be 5×10^{22} m^{-3}.

The temperature of the supersonic jet was determined using both vibrational and pure rotational[3] CARS. From the integrated line intensities of the CARS spectrum it follows that the vibrational temperature of OCS after expansion is 200 K. Because of rotational cooling, the linewidth of the lines in the vibrational CARS spectrum in the jet is 0.5 cm^{-1}, or about one third of that observed in the bulk. It was not possible to determine the rotational temperature for OCS because the individual rotational lines could not be resolved. For nitrogen, however, the pure rotational spectrum can easily be resolved, showing a rotational temperature of 5 K.

2. Results and Discussion

Results from the measurements carried out in the bulk have been reported previously,[1] and can be summarized as follows. At high excitation, the temperature of the v_2 mode rises up to 2000 K, and hot bands are observed up to the $v_2 = 4$ state. Because of the small anharmonicity and the fast collisional relaxation within the v_2 mode, the bulk measurements always show an equilibrium distribution for the v_2 mode, despite the fact that only the even-numbered states are populated by the CO_2 laser. The time-dependence of the spectra provides information on V-V energy transfer rates. In particular, the measurements put a lower limit of $k_{v_2 \to v_2} = 1$ μs^{-1} Torr^{-1} on the vibrational relaxation rate within the v_2 mode.

Collisions increase the infrared multiphoton excitation because of collisional broadening and rotational hole-filling during the excitation. In the bulk it may be possible to study infrared multiphoton excitation under nearly collisionless conditions by reducing the density of the sample and the duration of the laser pulses. In a supersonic jet it is much easier to obtain collisionless conditions; as the distance from the nozzle, x, is increased, the translational temperature of molecules decreases, and the mean free time between collisions is greatly increased.

Close to the nozzle the spectra obtained in the jet (see Fig. 2) are similar to the ones obtained in the bulk. The right-most peak is the ground state peak for the v_2 mode, and corresponds to the vibrational transition between the $(v_1,v_2,v_3) = (0,0,0)$ and $(1,0,0)$ states. The other transitions, which are shifted because of the 6 cm^{-1} cross-anharmonicity between

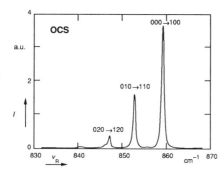

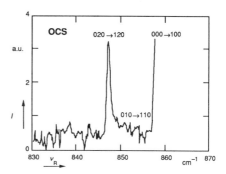

Fig. 2. CARS spectrum of OCS in a supersonic jet 500 ns after 7.2 J/cm² infrared excitation .

Fig. 3. CARS spectrum of OCS in a supersonic jet 30 ns after 7.2 J/cm² infrared excitation, for $x/D = 3.2$.

the v_1 and v_2 modes, are labeled correspondingly in the Figure. As in the bulk, odd-numbered states are populated because of collisions occurring in the jet during the laser pulse. As the distance x to the nozzle is increased, the excitation decreases because of the reduced collision rate. This clearly shows that even in a supersonic expansion collisions contribute to the excitation close to the nozzle. However, at distances $x/D > 3$, the excitation no longer decreases. At this distance, where the temperature reaches its lowest point, about 10% of the molecules are still excited under collisionless conditions. A spectrum obtained in this regime at a delay of 30 ns is shown in Fig. 3. As expected, the spectrum here only shows a single peak corresponding to the excited overtone transition.

At a delay of 60 ns the spectra show that the $v_2=1$ state begins to be populated. At $x/D = 3$ the density determined from CARS measurement is 5×10^{22} m^{-3}. Assuming a translational temperature of 10 K, one obtains a gas-kinetic mean-free-time between collisions of 40 ns for OCS. Therefore, collisional energy transfer occurs within 1.5 gas-kinetic collisions, which corresponds to an energy transfer rate of $k_{v_2 \to v_2} = 5.4 \ \mu s^{-1} Torr^{-1}$ at room temperature.

In the supersonic jet the OCS could only be vibrationally excited by the $P(22)$ and $P(18)$ lines of the 9.6 μm CO$_2$ laser. This is because only a few rotational states are populated and there is virtually no collisional broadening. The Rabi frequency for OCS is given by $\omega_R = 1.3 \times 10^{-5} \sqrt{I}$, with ω_R in cm^{-1} and where I is the peak CO$_2$ laser intensity in W/cm².[4] In this experiment one has $\omega_R = 0.08$ cm^{-1}, and only the $P(22)$ and $P(18)$ CO$_2$ laser lines have a frequency mismatch below the Rabi frequency. These lines pump the P_5 and R_3 transitions of OCS, respectively. This is quite different from the situation in the bulk, where excitation was observed for all CO$_2$ lines between the $P(10)$ to $P(26)$ lines of the 9.6 μm CO$_2$ laser branch. In the bulk, the broader rotational distribution and rotational hole filling by collisions assist the excitation.

This work was funded by the Army Research Office under contracts DAAL03-88-K-0114 and DAAL03-88-G-0078, and the Joint Services Electronics Program under contract N0014-89-J-1023, and by Hamamatsu Photonics K.K.

References

1 K.H. Chen, C.Z. Lü, L. Avilés, E. Mazur, N. Bloembergen, and M.J. Shultz, *J. Chem. Phys.*, to appear in August 1989.

2 W.B. Roh, P.W. Schreiber, J.P. Taran, *Appl. Phys. Lett.* **29**, 174 (1976)

3 D.V. Murphy and R.K. Chang, *Opt. Lett.* **6**, 233 (1981)

4 T.B. Simpson and N. Bloembergen, *Chem. Phys. Lett.* **100**, 325 (1983)

Part XII

Applications in Radiation Forces

Laser cooling and trapping of noble gas atoms

Fujio Shimizu
Department of Applied Physics, University of Tokyo,
Bunkyo-ku, Tokyo 113, Japan
Kazuko Shimizu and Hiroshi Takuma
Institute for Laser Science, University of Electro-
communications,
Chofu-shi, Tokyo 182, Japan.

1. Introduction

The laser cooling of alkali atoms uses the transition between the ground and lowest excited states which ensures the repeated absorption of laser photons through spontaneous decay to the same ground state. The lowest excited state of rare gas is in the vacuum ultraviolet region which cannot be accessed by presently available continuous lasers. However, for rare gases the transition between the $1s_5$ (J=2) metastable state and the $2p_9$ (J=3) state provides an alternative choice. Its wavelength is in the visible to near infrared region accessible by dye or semiconductor lasers. The lifetime of $1s_5$ is long and can be considered as infinite during the cooling process. Most rare gases have both even and odd isotopes. The characteristics of the cooling by a J=2 → 3 transition is similar to that of alkalis which use a F=2 to 3 transition, but has no hyperfine complexity for even isotopes. A high internal energy of the cooling levels is another characteristics of rare gas cooling. Therefore, the laser cooling of rare gases can offer many possibilities not obtainable by alkali atoms. We discuss in this talk our recent results on the metastable Ne cooling (1).

2. Continuous cooling and trapping of ^{20}Ne

Figure 1 shows the experimental setup for the cooling and trapping. The metastable Ne beam traveling along z-axis passed through axially symmetric magnetic field composed of several sections. In the first section the field intensity rose rather sharply to its maximum, then decreased approximately as square root of the traveling distance over 60cm to a lower value of 50 to 100 gauss. In the second section it was kept constant for approximately 20cm to the trap center O, and its direction was reversed in the third section to form a quadrupole field around O. A circularly polarized laser to induce Δ M=+1 transition was sent towards the Ne source (-z direction) to decelerate the atom. In the rising part of the magnetic field the atomic population was pumped to the highest magnetic sublevel M=2. In the falling part of the field the atom stayed in M=2 state and decelerated. It moved towards O keeping approximately constant velocity in the second section. To trap the atom at O, the deceleration laser was reflected back by a mirror placed between the magnet and Ne source to form a standing wave along z axis. Its polarization was reversed using a 1/4 wave plate. The transverse trapping laser consisted of two standing waves along x and y directions, which were generated by three mirrors as shown in Fig. 1 (b) (2). The trap was analyzed by observing the resonance fluorescence, or by an electron multiplier which counts positive ions created at the trap by collisions.

Using the above configuration we observed the trapping of ^{20}Ne up to 4×10^7 atoms with the detuning frequency approximately 0 to 50MHz below the resonance.

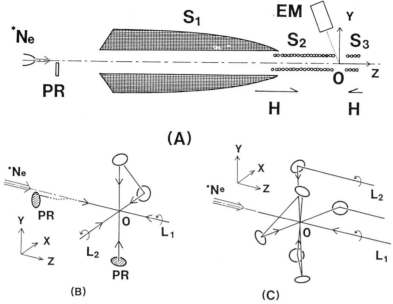

Fugure 1 (A)   Schematic diagram of Ne cooling. Ne*: metastable neon beam source, S_1, S_2, S_3: solenoids, EM: electron multiplier. The laser arrangement to make a trap at O is shown in (B) Raab's type trap and in (C) tetrabeam trap. PR are polarization reversing reflectors.

3. Trapping of isotopes ^{21}Ne and ^{22}Ne.

Neon has three stable isotopes ^{20}Ne, ^{21}Ne and ^{22}Ne with natural abundance 91%, 0.26% and 9%, respectively. Both ^{20}Ne and ^{22}Ne have no nuclear spin and can be cooled and trapped using single frequency. Since ^{21}Ne has hyperfine structure due to the nuclear spin 3/2, multiple frequencies are necessary to avoid atoms spilling from the cooling cycle. Among various possible combinations of transitions the best choice is to pump $\Delta F=+1$ transitions, where the maximum F transition ($F'=7/2 \rightarrow F''=9/2$) works as the main cooling transition. The intervals of $\Delta F=+1$ transitions are 308, 140 and 33MHz for the descending order of F' (3). Using a second laser and an accousto-optic modulator, we superposed the field resonant to $F'=5/2 \rightarrow F''=7/2$ transition and the field in the middle of $F'=3/2 \rightarrow F''=5/2$ and $F'=1/2 \rightarrow F''=3/2$ transitions. Figure 2 shows the electron multiplier current when the first laser frequency is scanned while the other fields are fixed. In addition to the trapping signal of ^{20}Ne and ^{22}Ne, the trapping signal for ^{21}Ne is observed at the frequency of $F'=7/2 \rightarrow F''=9/2$ transition.

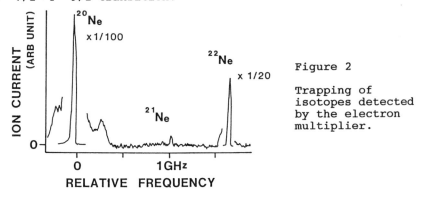

Figure 2

Trapping of isotopes detected by the electron multiplier.

4. The lifetime of the metastable-atom trap

We measured the decay rate of the resonance-fluorescence intensity emitted from the trap after the Ne beam was cut off at various background air pressure. The temporal dependence of the fluorescence intensity was found to fit with the sum of linear and quadratic decays. The quadratic term is due to collisions between two trapped metastable Ne. The pressure dependence of the linear rate is shown in Fig. 3. The value extrapolated to zero pressure is $1/22$ s^{-1}. The rate is affected by the level mixing of $1s_5$ and $2p_9$ states by magnetic and laser fields, population ratio between $1s_5$ and $2p_9$ states and also by the change of the gas composition when the pressure is increased by leaking air from the minimum pressure (6×10^{-8}Pa). A rough estimate on the magnetic and light mixing ratio shows that those effects does not affect the decay time more than 10%. The pressure effect is more difficult to estimate, however, judging from the lifetime of 17s at the minimum pressure, we believe that the observed 22s reflects the lifetime of $1s_5$ state, which is in good agreement with theory (4).

5. Tetrabeam trap

As has been reported (2), the number of trapped atoms is often larger when the laser beam is slightly misaligned. We discuss the trapping efficiency for various laser beam con

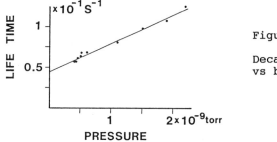

Figure 3

Decay rate of the trap vs background pressure.

figurations with the aid of Monte Carlo simulation based on the rate equation and show that an efficient trap can be constructed using only four laser beams.

The problem on the trap shown in Fig.1 (a) is the existence of copropagating laser beam which induces $\Delta M=-1$ transition and accelerates the atom. In the rising section of the magnetic field, this beam tends to pump the population to the minimum magnetic sublevel, which slips out the deceleration process. Even when the laser frequency is very close to the resonance, the population is equally divided between M=2 and M=-2 levels, thus reducing the number of trapped atoms to one half. More serious damaging effect occurs when the atom approaches the trapping point with the Doppler velocity several times the natural linewidth. In this region the atom in M=2 state starts to interact with the copropagating laser again due to a smaller Zeeman shift, and is populated to lower M levels. Once in the low M level it keep absorbing π_- photons and accelerates to acquire kinetic energy larger than the effective trap potential. The result of the Monte Carlo calculation shows less than 10% of all atoms are captured even at the best condition. A simple method to improve capturing efficiency is to misalign the laser beam by tilting the reflector so that the deceleration and acceleration beams do not overlap. By misaligning also one of the side beam (a race track configuration) by a fraction of the beam diameter, the calculation shows that the capturing efficiency improves drastically.

Since the Zeeman field assisted spontaneous trap works with a large laser-intensity imbalance, various configurations without the copropagating laser along the atomic beam can be designed. We trapped the neon using 4 laser beams with tetrahedral configuration, which is the minimum number to push atoms towards the trap center from all three dimensional directions. Its mirror arrangement is shown in Fig. 1(c). The beam traveling -z direction works as the deceleration laser. The other 3 beams function as transverse trapping beam as well as the counterbalance of the deceleration laser. The polarization of the latter beams are the same as the transverse beams in Fig. 1(b). One might think that atoms leak towards -z direction, since this configuration lacks π_+ laser beam necessary to push back atoms in the negative side of z axis towards the center. However, the field along +z axis is the sum of three transverse beams contains π_0, π_+ as well as π_- components. Therefore, it works as the counterbalance of the deceleration beam. The result of the numerical calculation shows that this configuration can capture essentially all atoms traveling within the 1/e laser beam radius, though the trap size is considerably larger than that of perfectly aligned orthogonal configuration. This was confirmed by experiment. The number of trapped atoms, as well as the lifetime of the trap were approximately the same as the case of orthogonal beams.

References
1. F. Shimizu et al, Phys. Rev. A, 39, 2758 (1989).
2. E.L. Raab et al, Phys. Rev. Lett. 59, 2631 (1987).
3. L. Julien et al, J. Phys. (Paris) Lett. 41, L479 (1980).
4. N.M. Small-Warren, Phys. Rev. A, 11, 1777 (1975).

Laser Cooling of a Ca Atomic Beam and Measurement the $^1P_1$ - $^1D_2$ transition rate

N. Beverini, F.Giammanco, E.Maccioni, F. Strumia, G. Vissani

Dipartimento di Fisica, Università di Pisa, INFN and INFM (Pisa)

Piazza Torricelli 2, 56126 Pisa, Italy

Introduction

The production of an atomic beam of Magnesium (1,2) or Calcium (3) with a subthermal velocity distribution is of great interest for spectroscopy, and particularly for time and frequency metrology (4). Laser cooling can be achieved by irradiating a thermal beam of Magnesium or Calcium with a counterpropagating laser beam in resonance with the $^1S_0$ - $^1P_1$ transition at 285 nm or at 422 nm, respectively. In the present experiment, a single-frequency laser at a fixed frequency was used, and the Doppler effect was compensated by shifting the atomic absorption by Zeeman effect. Owing to the zero total angular momentum of the Mg and Ca ground state and the absence of nuclear spin of the most abundant isotopes, the cooling process occurs in a true two levels system if circularly polarized light is used. The resulting velocity distribution was analyzed by monitoring the fluorescence excited by a second laser beam whose frequency was scanned around the resonance.

In the case of Ca, the presence of the $^1D_2$ level, at an energy lower than the $^1P_1$ one, introduces an additional difficulty (fig.1). As a consequence, the cooling cycle can be interrupted by the spontaneous emission from the $^1P_1$ level towards the $^1D_2$ level, that corresponds to a loss of atoms owing to the very long lifetime of the $^1D_2$ level compared with the duration of the cooling process. In the present research,the influence on the cooling efficiency of the optical pumping in the $^1D_2$ level is investigated. We found that the branching ratio R between the channels of spontaneous decay of the $^1P_1$ level, i.e. towards the ground state and the $^1D_2$ level (R= A[$^1P_1$-$^1S_0$] / A[$^1P_1$-$^1D_2$]) is quite large. As a consequence an efficient Ca cooling can be achieved by using a single laser beam at 422.7 nm regardless of the optical pumping.

Experiment

A collimated thermal beam of Ca atoms is produced by an effusive source at a temperature of about 860 °K, and interacts with a counterpropagating laser beam, resonant with the $^1S_0$ - $^1P_1$ transition. The single frequency CW laser beam is obtained by a Coherent 699/21 ring dye laser with Stilbene 3 dye. At the best conditions, the available power is of several tens of mW at 422.7 nm, in single mode TEM$_{00}$ operation with a r.m.s. line width of $\approx$2 MHz.

The cooling laser beam is tuned at a fixed frequency, within the absorption Doppler profile of the incoming Ca atoms. The modified velocity distribution is analyzed by a second laser beam at 422 nm (obtained by a second dye laser), nearly collinear to the first one, and whose frequency is scanned around the Ca resonance line. The near collinear geometry, where the diameter of the analyzing beam is smaller than the cooling one, and, thus, of the atomic beam diameter , was preferred in order to obtain a sufficiently large analysis signal

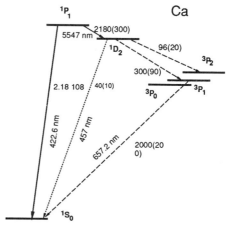

Fig.1 Levels and transition rates of the Ca atom (in s-1) involved in the laser cooling scheme. Continuous line: allowed electr. dipole transition; broken lines: forbidden electr. dipole; dotted lines: allowed electr. quadrupole

and a good resolution in the velocity profile. In fact, the analysis laser beam crosses the center of the cooling beam only in front of the photomultiplier detecting the fluorescence signal. Furthermore, the intensity of the analyzing laser beam is chopped at about 1 kHz, and the fluorescence signal processed by a lock-in amplifier in order to discriminate the fluorescence signal induced by the cooling beam. The axial magnetic field, whose intensity profile fits fairly the theoretical requirements, is generated by an ensemble of three solenoids and properly designed iron poles. The magnetic field intensity B(z) decreases from its maximum to zero in about 25 cm which represents the maximum cooling length with the field on.

Fig.2 displays an example of laser cooling in presence of Zeeman tuning. Curve a) was obtained in absence of magnetic field and with the intensity of the cooling laser beam strongly reduced : the hole in correspondence of v=230 m/s shows the position of the laser frequency in the velocity distribution. Curve b) shows the modified velocity distribution obtained with an appropriate laser intensity and an initial magnetic field B(0)= 0.102 T, corresponding to a tuning of about 600 m/s. The absence of fluorescence signal in 200-800

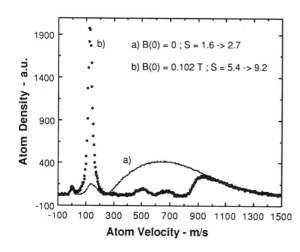

a) B(0) = 0 ; S = 1.6 -> 2.7

b) B(0) = 0.102 T ; S = 5.4 -> 9.2

Fig.2 Evidence of laser cooling in presence of Zeeman tuning. Curve a): velocity distribution with magnetic field off and low intensity of the cooling beam; curve b): deceleration observed with an initial magnetic field B(0)= 0.102 T.

m/s region is the experimental evidence that the atoms lost in the $^1D_2$ state do not return in the ground state, as expected. The peak velocity of the cooled beam correspond to 125 m/s, and the fluorescence at the maximun is five time larger than that of the uncooled beam. In the case of curve b), the atom density in the analysis region was estimate to be reduced by a factor $d\approx0.5$ as an effect of the transverse velocity diffusion. The efficiency of cooling demonstrates that R must be large. Finally, it must be noticed that the velocity distribution of fig. 2, observed at a distance of about 20 cm from the decelerating magnet, demonstrates that the diamagnetic properties of the Ca ground state cooperate to improve the extraction efficiency of the decelerated atoms from the cooling region

A possible way to estimate R is to evaluate the ratio of the integrals of experimental velocity distributions b) and a). This ratio can be computed, as demonstrated in (5), and fitted to the experimental data giving $R=1.0\pm0.6 \times 10^5$. However, the ratio depends critically on d, thus, only a rough estimation of R can be carried out. A more accurate measurement of the branching ratio R between the two spontaneous emissions rates from the P state is obtained by reducing the influence of transverse velocity diffusion, i.e. by increasing the final velocity and excluding the presence of the magnetic field.

Fig.3 shows examples of the experimentally observed velocity density profiles. The observed dependence of the maximum of the fluorescence profile on the saturation parameter, compared with the theoretical behavior (5,6) at different branching ratios R, allows us to carry out the value of R, whereas the variation of the corresponding velocity, quite insensitive to the branching ratio, constitutes a suitable test for the theoretical approach with regard to the involved approximations. Fig.4 shows the calculated dependences of the maximum of the distribution function on the saturation parameter for different branching ratios R. The parameters, entering the equations, were selected according to the experimental conditions. The experimental data, taken from fig.3, represent the observed dependence, in agreement with the theoretical curve, corresponding to

$$R = (1.0\pm0.15) \times 10^5$$

close to the theoretical estimation (4) and the above calculations from the distribution of fig.2. The above value has been obtained by fitting the experimental data with theoretical curves calculated for branching ratio values around 1.0×10^5.

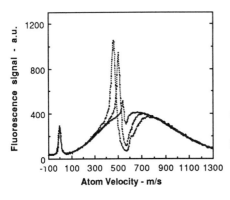 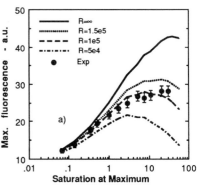

Fig. 3 Velocity profile observed in absence of magnetic field and for three values of the saturation

Fig.4 Maximum of the fluorescence signal versus the saturation parameter S Experimental data are from fig.3

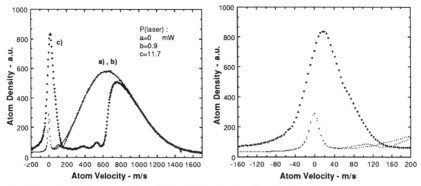

Fig.5 - Example of laser cooling. Left: full velocity distribution; right: expanded view around v=0. The decelerated beam has a velocity at maximum of 15 m/s

The best correlation coefficient (r=.98) has been obtained for the $R=1.0x10^5$ curve and the standard deviation has been estimated by the variation of the r coefficient as a function of the branching ratio R.

Conclusion

We have demonstrated that efficient laser deceleration of Ca atomic beam can be obtained by using a laser tuned on the resonance transition at 422 nm. The branching ratio R between the decay probability, from the $^1P_1$ level to the ground $^1S_0$ state and to the metastable $^1D_2$, has been measured to be large enough that the decay channel, towards the metastable state, slightly affects the cooling efficiency. In conclusion, low velocity beams of Ca atoms, with a velocity as low as 10÷20 m/s or less, corresponding to a thermal source colder than 1 K as shown in fig.5,.can be easily obtained, demonstrating the importance of this element for future atomic frequency standards operating in the submillimeter and optical regions.

References

1- N. Beverini,S. De Pascalis, E. Maccioni, D. Pereira, F. Strumia, G. Vissani, Y.Z. Wang, C. Novero : Opt. Lett., 14, 350 (1989)

2- N. Beverini, E. Maccioni, D. Pereira, F. Strumia, G. Vissani : Ital. Phys. Soc. Conf. Proc., 21, 205 (1989)

3- N. Beverini, E. Maccioni, D. Pereira, F. Strumia, G. Vissani : " Frequency Standards. and Metrology", A. De Marchi ed. , Springer Verlag, pag. 282 (1989)
N. Beverini, E. Maccioni, D. Pereira, F. Strumia, G. Vissani : Proc. IV Conf. Laser Science, 1988, AIP Conf. Proc. 1989, in press

4- F. Strumia ."Application of laser cooling to the atomic frequency standards".in " Laser science and technology " Eds. A.N.Chester and S.Martellucci ,Plenum Press, N.Y. 1988 , pagg. 367-401

5- N.Beverini, F.Giammanco, E.Maccioni, F.Strumia, G.Vissani : subm. J.Opt.Soc.Am.

6- V.G.Minogin, V.S.Letokhov : " Laser light pressure on atoms", Gordon and Breach Sc. Publ. , New York 1987

451

Localization of Atoms with Light-Induced Mirrors and in a Standing Light Wave

V.I. Balykin, V.S. Letokhov, Yu.B. Ovchinnikov, A.I. Sidorov, S.V. Shul'ga

Institute of Spectroscopy, Academy of Sciences of USSR

We have studied laser field configurations allowing the motion of atoms and molecules to be restricted to a region of space free from radiation. Such configurations may be atomic cavities relying for their functioning on the effect of reflection of atoms from a nonuniform field and ring traps based on the channeling of atoms in a standing light wave (Fig. 1). We have implemented experimentally the reflection of atoms from a nonuniform laser field /1/, inverstigated the quantum-state selective reflection of atoms and molecules, and calculated an atomic cavity configuration. The atomic cavity is largely analogous to a optical cavity with high concentration of photons in the modes. One of the main parameters characterizing the field in an optical cavity is its degeneracy. To obtain a degeneracy much in excess of unity is possible only in a laser cavity. It has been demonstrated /2/ that so high a density of atoms can be obtained in the atomic cavity as makes their degeneracy parameter much greater than unity (Fig. 2).

We have studied the channeling of sodium atoms in a standing spherical light wave (Fig. 3) and managed to show them move following the curved front of the wave /3/. The atoms entering the standing wave separate into localized and non-localized groups (Fig. 4). The non-localized atoms emerge from the wave in a direction close to the original propagation direction of the atomic beam, while localized counterparts move along the nodes of the wave, their original direction of motion being changes through an angle $\alpha \approx 2\emptyset$, where $\emptyset$ - angle of atomic beam entry into the standing wave.

This allows us to treat a laser beam with a great radius of curvature as an element of a ring trap. We have also investigated experimentally the channeling and localization of three-level atoms /4/. The optical pumping effect causes a strong enrichment of the atomic beam with slow atoms (effective cooling of the atoms in the standing light wave). The atoms are observed to divide into two ensembles - cold and hot. This effect may prove very important in the three-dimensional cooling of atoms in a standing light wave.

References

/1/ V.I. Balykin, V.S. Letokhov, Yu.B. Ovchinnikov, A.I. Sidorov
 Phys. Rev. Lett 60, 2137 (1988)
/2/ V.I. Balykin, V.S. Letokhov, Appl. Phys. B48, 517 (1989)
/3/ V.I. Balykin, V.S. Letokhov, Yu.B. Ovchinnikov, A.I. Sidorov and
 S.U. Shul'ga, Opt. Lett., 13, 958 (1988)
/4/ V.I. Balykin, V.S. Letokhov, Vu.B. Ouchinnikov, A.i. Sidorov,
 J. Opt. Soc. Am (to be published)
/5/ R. Cook, R. Hill, Opt. Commun., 43, 258 (1982)

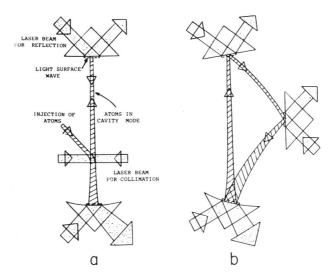

Fig. 1   Schematic diagram of atomic cavities with light induced mirrors. Such mirrors are based on the use of the surface light
wave produced upon total internal reflection of laser beam at
a dielectric-vacuum interface /5/. a) - linear cavity ("atomic
confocal interferometer"), b) - ring atomic configutation. To
inject of atoms in the cavity, use is made of a standing wave.

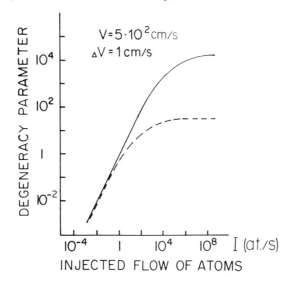

Fig. 2   Atomic degeneracy parameter as a function of injected flow of
atoms in the atomic cavity (dashed curve for linear, solid
for ring atomic trap configuration).

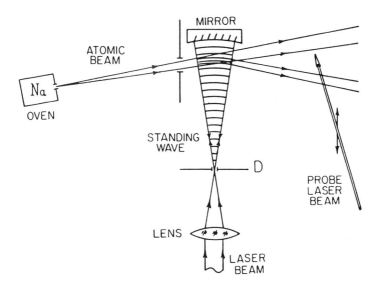

Fig. 3   Localization of atoms in a spherical standing wave. Ø - is
the atom's angle of entry into a standing wave and α - is the
anlge of deflection from the original atomic trajectory. In
the course of localization the atoms are divided into two ensembles: one contains "cold" localized atoms and the other
"hot" nonlocalized ones, followed by their spatial separation.

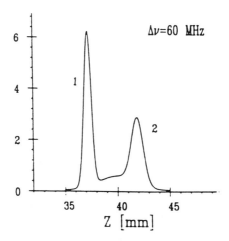

Fig. 4   Atomic beam profile in the registration region after interaction with the standing spherical wave: 1 - atomic beam intensity peak corresponding to localized atoms, 2 - corresponding
to non-localized atoms

Collimation of Atomic Beam Using Ratarted Dipole Force

Y.Z.Wang, Y.D.Cheng, W.Q.Cai, Y.S.Wu, Y.Luo and X.D.Zhang
Quantum Optics (Joint) Laboratory, Shanghai Institute of
Optics and Fine Mechanics, Academia Sinica, Shanghai,
P.R.China

In the recent years the study on mechanical effects of light has acquird great success. The techniques of laser cooling, trapping and controlling of atoms have reached a new high level and have provided a new tool in study on atomic and molecular physics[1-4]. Collimation of moving atoms by radiation pressure is one of the interesting subjects. The first experiment is to collimate a sodium atomic beam by spontanous light pressure[5]. The transverse velocities of the atomic beam are reduced from 5.5×10^2 to 1.6×10^2 cm/s, corresponding to decrease of effective transverse temperature of the moving atoms from 42 to 3.3 mk. A French group collimated a cesium atomic beam by using stimulated optical molasses for the first time[4]. The beam is strongly collimated to a narrow velocity peak of 40 cm/s FWHM, which is about 3 times larger than the Doppler cooling limit. In this paper we report an experimental study on collimation of atomic beam by retarded dipole force and the beam is composed of channeling atoms, we have measured the transverse velocity of the collimate atomic beam to be much lower than the Doppler cooling limit.

The experimental scheme is shown in Fig.1. The transversal velocity of the atomic beam can be seen as slowly moving atoms in the standing light wave. The divergence angle of the atomic beam is 1.5×10^{-3} rad. The most probable velocity is a (a =760 m/s), the most probable transversal velocity of the atoms is $v_t < 0.2$ m/s at the center and $v_t < 1.14$ m/s on the two sides of the atomic beam. The standing wave is 65 cm away from the sodium source. The diameter of the standing wave is 4.5 mm. The time for the atoms flying for across the standing wave is 5.9×10^{-6} sec, which is much greater than the excited state

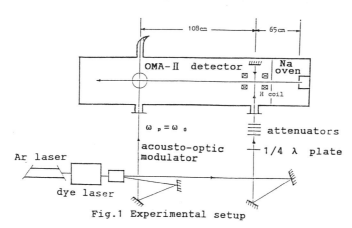

Fig.1 Experimental setup

life-time of the atom $(1.6\times10^{-8}\,\mathrm{sec})$. The atoms after interacting with the standing wave fly for 108 cm length in the drift region and meet the probe laser beam at right angle. The probe laser beam induces resonant fluorescence from the atoms in the space. The intensity of the fluorescence represents the profile of the atomic beam. The spatial distribution of the fluorescence is imaged on the sensitive surface of the detector of an OMA-Ⅱ, and the spatial profile of the atomic beam is directly shown on the screen. The angular resolution of the detection system is $3.0\times10^{-5}\,\mathrm{rad}$. The exposure time is 25 ms, it allows us to collect data in a very short period of time. The laser used is a 301-D dye laser (made in China) pumped with a 360 Ar laser (made in China). After going through an acousto-optic modulator the laser beam is split into two beams, one with frequency ω_1, the other with $\omega_1+\delta$, δ is the modulation frequency. The beam with $\omega_1=\omega_0$ (ω_0-the atomic resonant frequency) is the probe beam and the beam with $\omega_1+\delta$ is the interaction beam.

Fig.2 shows the profile of the collimated atomic beam by the retarded dipole force with positive detuning $\delta = +50\mathrm{MHz}$. The laser power is 110 mw, which leads to a saturation parameter $G=108$ or $P=G/[4(\delta/\Gamma)^2+1]=1.02$, where Γ is the natural line width. The atomic beam is strongly collimated, the transversal velocity decreases from 0.57 m/s to 0.15 m/s and the temperature of transverse motion decreases from 274 μk to 60 μk. The temperature of the transverse motion is lower than the Doppler cooling limit 240 μk. The result is suprising, but it's qualitatively in agreement with the theory of channeling atoms in the standing wave [7,8]. The theory pointed out that,the acting force on atoms in the standing wave consists of two parts: one is the so-called dipole force, which is proportional to the gradient of the light field, the other is the retarded dipole force, which is a function of the atomic velocity. For the atoms with greater υ_t, the dipole force disappears for averaging over the wave length, whereas the retarded dipole force does not disappears and becomes a damping force. When the kinetic energy of the atoms $E=(1/2)m\upsilon_t^2$ is smaller than the potential of the standing wave $U_{(z)}$, the dipole force and the retarded dipole force both act on the atoms and atoms perform damped oscillation in the potential traps. The atoms are forced by the standing wave to

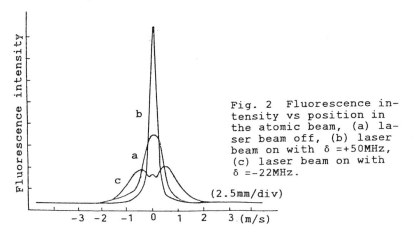

Fig. 2 Fluorescence intensity vs position in the atomic beam, (a) laser beam off, (b) laser beam on with $\delta =+50\mathrm{MHz}$, (c) laser beam on with $\delta =-22\mathrm{MHz}$.

(2.5mm/div)

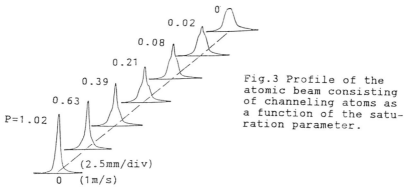

Fig.3 Profile of the
atomic beam consisting
of channeling atoms as
a function of the satu-
ration parameter.

move along the nodes, constituting a set of channeling atomic
beams, which exist not only in the standing wave, but also
out-side it. Fig.3 shows the dependence of the profile of
collimated beam with the intensities of the laser beam. When
the intensity of the standing wave reduces, a "shoulder" of
0.8 mm in width appears at the center of the atomic beam[9],
which is just equal to the diameters of the collimating holes
of the atomic beam. The width of the "shoulder" is unchan-
geable whith the variation of the intensity, where as its
height varies. It gives evidence for channeling effects of
atoms in the atomic beam, only in the case that the atoms are
channeled and collimated, outside the standing wave we could
observe the "shoulder" at the center of the profile of atomic
beam. Fig.4 shows that when the axis of the atomic beam is

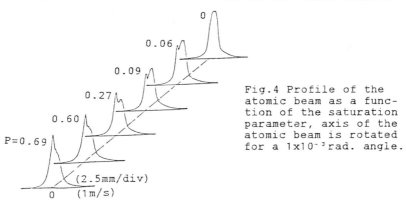

Fig.4 Profile of the
atomic beam as a func-
tion of the saturation
parameter, axis of the
atomic beam is rotated
for a 1×10^{-3} rad. angle.

rotated for an angle of 1×10^{-3} rad, the standing wave makes a
fraction of atoms to enter the channels, the atomic beam
flying along the nodes of the standing wave splits off from
the original beam and forms another narrow peak. It clearly
shows that the atoms are cooled, collimated and trapped in the
channels. The cooled and collimated atoms form a new peak on
the side of the atomic beam. In the above discussion the
momentum diffusion is not taken into consideration. According
to the theory[6.7], the momentum diffusion coefficient D is
spatially inhomogeneous, if we take the minimum value of D
near the node of the standing wave, $D=(hk)^2 \Gamma /(\Omega_0/\delta)^2/2$ [7],
hk is the photon momentum, Ω_0 is the Rabi-frequency, the

width of the atomic momentum diffusion $\sqrt{\langle(\Delta v_t)^2\rangle}=\sqrt{2Dt}$, t is the acting time of the atom with the standing wave. In our case the velocity diffusion $\sqrt{\langle(\Delta v_t)^2\rangle}\approx 0.35$ m/s is more than the measured 0.15 m/s. It seems that the diffusion coefficient is smaller than the theoretical predication, a new theory to explain the effects is needed.

In conclusion here we report the experimental results of collimation of moving atoms by retarded dipole force. The collimated atomic beam is composed of a set of atomic beams arranged at the intervals of $\lambda /2$. It has a very high degree of collimation and the temperature of the transverse movement of the atoms may be less than 60 µk. The channeling atoms can be used in high resolution spectroscopy study. (A paper about the spectroscopy of the channeling atoms in an absorption cell will be published elsewhere.) Moreover this method may be used in the study of controlling of atomic motion, imaging and focusing of atoms [5.8].

References:

[1] S.Chu,J.E.Bjorkholm, A.Ashkin and A.Cable, Phys. Rev. Lett. 57,314(1986).
[2] P.D.Lett, R.N.Watts, C.I.Wesbrook, W.D.Phillips, P.L.Gould and H.J.Metcalf, Phys. Rev. Lett. 61, 169(1988).
[3] A.Aspect, J.Dalibar, A.Heidmann, C.Salomon and C.Cohen-Tannoudji, Phys. Rev. Lett. 12, 876(1986).
[4] V.I.Balykin, V.S.Letokhov, Yu.B.Ovchinnikov, A.I.Sidorv and S.V.Shnl'ga, Optics Lett. 13, 958(1988).
[5] V.I.Balykin, V.S.Letokhov, V.G.Minogin, Yu.V.Rozhdestvensky, and A.I.Sidorov, J.Opt. Soc. Am.B, 2 1776(1985).
[6] A.P.Kazantsev, V.S.Smirnov, G.I.Surdutovich D.O.Chudesnikov and V.P.Yakovlev, J.Opt. Soc. Am.B, 2 1731(1985).
[7] A.P.Kazantsev, G.I.Surdutovich, and V.P.Yakovlev, Opti.Commu. 68 103(1988).
[8] Y.Z.Wang, W.Q.Cai, Y.D.Cheng and L.Liu, Proceeding of the Topical Meeting on Laser Materials and Laser Spetroscopy, Ed.Z.J.Wang and Z.M.zhang, World Scientific,P 351 (1988).
[9] Y.Z.Wang and et al, Opt. Comm.,70 462 (1989). In this paper we thought that the "shoulder" is due to the maximum point of the force, it's not correct.
[10] V.I.Balykin, V.S.Lekohov and V.G.Minogin Phys. Scripta 22, 119(1988).

Part XIII

Highly Excited States
and Dynamics

Tribute to Sylvain Liberman

P. Jacquinot
    Laboratoire Aimé Cotton, CNRS
    91400 Orsay Cedex, France

Sylvain Liberman should have been with us today. He attended many Laser Spectroscopy Conferences and he was a member of the steering Committee of this one. Instead, this session is dedicated to his memory. I have been asked to talk a few minutes about him and his work. Like all the friends of Sylvain, I greatly appreciate this idea of the organisers of the Conference and I am grateful to them.

Our friend died suddenly, of a heart attack on August 5 last year. He was 54. For all those who knew him that was a dreadful shock, and that was a great loss for our community of atomic physics and laser spectroscopy.

Those who met Sylvain Liberman in this series of Conferences keep the memory of a communicative and friendly man with a warm and radiant smile. It was a real pleasure to talk with him and he gave very good and lively papers, with a very peculiar sense of humor.

In his scientific work, he was a very keen researcher. He liked to work deeply on specific problems, but he also had broad views on fields adjacent to his own one. He was mainly an experimentalist but with good theoretical capabilities.

A great part of his research was made in collaboration with several coworkers. This was partly due to the sophistication of the experiments often covering different fields, for instance atomic physics and nuclear physics. But this was also due to his ability to work with others and to his qualities of leader. His most constant coworker since the beginning of his work with tunable lasers was Jacques Pinard whom many of you know : they were like two brothers.

Sylvain devoted also much of his time and activity, and may be of his life, to the scientific community. He was a member of many - too many - committees and he never refused the duties that unavoidably fall upon those who are able to do them. Since 1981, he was director of Laboratoire Aimé Cotton and he threw himself heart and soul in that enterprise.

For those who knew him more intimately, he was a very modest and sensitive man, a great-hearted man.

The history of the scientific work of Liberman closely follows the history of laser spectroscopy. It starts with the very beginning of high resolution spectroscopy using lasers. That was around 1966, at a time when tunable lasers were still unknown. The only means of taking advantage of the wonderful sharpness of the lines emitted by a laser was then to study these lines themselves by tuning the cavity. The laser acted then both as the source and as the spectrometer. That was rather tricky and indirect, and, of course, the number of lines one could study was small. That was, so to speak the prehistory of laser spectroscopy. The thesis of Sylvain was devoted to hyperfine structure and isotope shift of infrared lines of rare gases. He not only measured a

number of lines but also gave good theoretical interpreta-
tions of his results.

Then came the time of tunable lasers and it became possible to make **absorption** spectroscopy of almost anything. At
Laboratoire Aimé Cotton it was chosen to work with atomic
beams : first, thermal beams with perpendicular excitation,
and later on fast beams with collinear excitation.

The first thing was to look for the most suitable methods of detection ot optical resonances induced by the
laser light in atomic beams. Different methods were explored : fluorescence, deflexion of atomic beams, photoionisa-
tion or ionisation by an electric field from the upper state
of the transition. Later on another method was proposed by
H.T. Duong and J.L. Vialle at Laboratoire Aimé Cotton : it
uses the focalisation by a magnetic field of the atoms which
have undergone an optical pumping by the laser light. It is
by this method that most of the results on radioactive atoms
have been obtained.

The next period was devoted to the study of Rydberg
states. First that was purely spectroscopy, that means
measurements of level positions, fine and hyperfine structures and perturbations that arise when a Rydberg series
crosses a valence state, for instance. Then came the effects
of electric and magnetic fields which are of particular interest when the effects of external fields become comparable
or larger than those of internal fields. This type of work
is still going on at Laboratoire Aimé Cotton with the effects of simultaneous electric and magnetic fields on hydrogen Rydberg atoms.

In 1975, starts a long period of work on long chains of
radioactive isotopes. It was tempting to make use of the
high resolution, and essentially of the high **sensitivity**, of
laser spectroscopy on atomic beams to study long series of
rare and short-lived isotopes produced by nuclear reactions
in accelerators. The aim was to follow the variation of some
nuclear properties when neutrons are added to a nucleus. Hyperfine structures give nuclear moments, and isotopes shifts
give nuclear radii. And it may happen that in a long series
of isotopes these quantities exhibit non trivial variations
essentially around a magic number of neutrons. The first
results were obtained at Orsay in 1975 on sodium 21 to 30,
in collaboration with the laboratory of nuclear physics of
Robert Klapisch. This was very encouraging and this type of
work was continued with the ISOLDE facility at CERN. A great
number of results were obtained on very long series of K,
Rb, Cs. I only want to show here two transparencies that
Sylvain was very fond of and that he presented several times. The first one (fig. 1) is relative to Cesium and summa-
rises hyperfine structures and isotope shifts of 36 isotopes
or isomeres. On this figure, one can see at a glance interesting variations of nuclear moments and of nuclear radii.
These radii are reflected in the centers of gravity of the
structures, represented by dots on the figure. The other
transparency (fig. 2) shows the same kind of results for
Francium. An interesting point is that, despite the fact
that Francium was discovered in 1939, not a single line of
its optical spectrum had ever been observed before these

461

experiments at CERN by the Orsay group in which Sylvain
played an eminent part. And on this figure on can see high
resolution spectroscopy of 16 of its isotopes obtained in
1979. Sylvain was very proud of that ! This type of work is
still going on other atoms, Au, Pt, etc... and Sylvain was
active in the preparation of a vast project on the Bohr-
Weisskopf effect or hyperfine structure anomaly which reflects the distribution of magnetism in nuclei. A poster has
been presented on this subject in this Conference by H.
Stroke.

Very briefly, I also mention the work on superradiance
and subradiance that was initiated in 1974 when Sylvain made
a one-year stay at MIT with M. Feld who was then obtaining
experimental evidence of superradiance on the molecule HF.
This cooperative effect of interfering atoms on spontaneous
emission can be constructive or destructive, giving rise to
enhanced pulses - superradiance - or to short periods of
inhibition - subradiance. Sylvain worked with Anne Crubel-
lier on both effects : in particular they were able to give
in 1983 the first experimental evidence of subradiance.

The next project should have started for Sylvain in one
or two years at CERN on optical transitions of the atom p $\bar{\text{p}}$
with the low energy antiproton ring LEAR.

Such was the scientific itinerary of our friend, so
prematurely interrupted. It was in complete harmony with his
human personality.

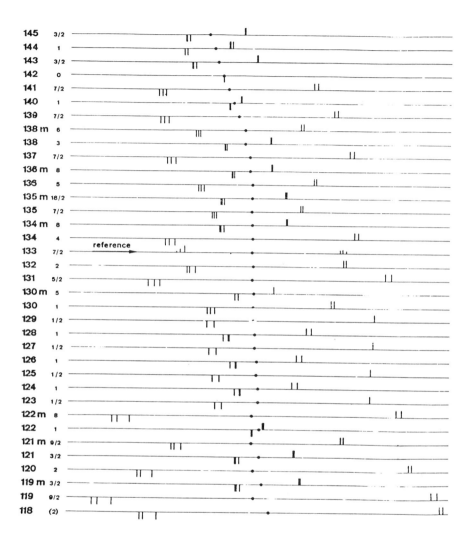

Fig. 1

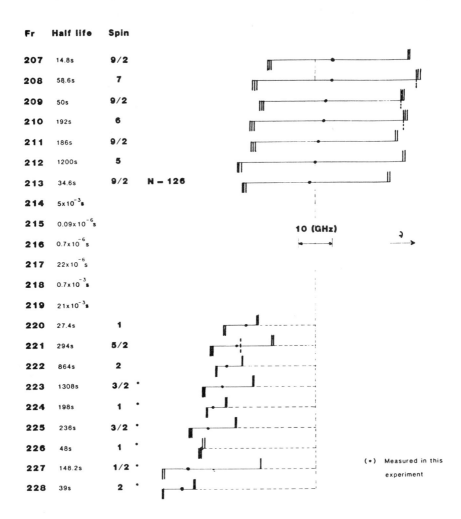

Fig. 2

ELLIPTIC ATOMIC STATES

Jean-Claude GAY and Dominique DELANDE
Laboratoire de Spectroscopie Hertzienne * de l'E.N.S.
Tour 12 - E01 - 4 place Jussieu - 75252 PARIS Cedex 05 - FRANCE

How classical is a Rydberg electron ? In spite of evidence, this is one of the most challenging question in Quantum Mechanics. A quantum picture of the classical Kepler motion is still to settle. No receipe for building minima wavepackets, computer - runned or laser - built is presently available [1], and the question of their stability is elusive.

We will address this question by building a new class of atomic states - the elliptic states - showing perfect, within quantum mechanical constraints, localization on a Kepler trajectory [2]. They are eigenstates of the hydrogen atom in the n shell, obtained by coherent mixing of the n^2 spherical states [3]. Various applications are briefly discussed. Next, the problem of localized wavepackets is addressed from a new point of view : the one of squeezing of the atomic structure. The relevance will be established in the special case of squeezed circular states.

1 - The elliptic eigenstates of the atom.

The geometrical localization of a quantum eigenfunction on a classical orbit can be achieved in various ways. But the set for which fluctuations are minimum is unique and we name it "elliptic". It can be built from the concept of coherent states for the angular momentum in 3 dimensions. If $\vec{j}$ points in the $\vec{u}$ direction, taking its maximum value j, then the fluctuations $\Delta j = (< \vec{j}^2 > - < \vec{j} >^2)^{1/2}$ are $\sqrt{j}$ and minimum.

A classical Kepler orbit is entirely defined given the angular momentum $\vec{\ell}$ and the Lenz vector $\vec{a}$. The operator $\vec{\mathcal{L}} (\vec{\ell} , \vec{a})$ is a 4 dim. angular momentum which expresses the rotational symmetry of the Coulomb interaction in a four dimensional space [4]. Hence, quantum states with semi-classical properties are coherent states of SO(4). They are deduced from some minimum uncertainty atomic state, through a 4 dim. rotation.

The simplest choice for minimum uncertainty atomic states is obviously the circular state $| n \ell \ell_z >$ which is an eigenvector of ℓ^2 and ℓ_z with $\ell = \ell_z = n - 1$ while the other components of $\vec{\mathcal{L}}$ average to zero. The fluctuations are $\Delta \ell_x = \Delta a_x = \Delta \ell_y = \Delta a_y = ((n - 1)/2)^{1/2}$ and $\Delta a_z = \Delta \ell_z = 0$. Hence $< \vec{\mathcal{L}} > = (n - 1) \vec{z}$ and $\Delta \mathcal{L} = [2 (n - 1)]^{1/2}$ which expresses the minimum-fluctuation properties in SO(4).

The most general minimum uncertainty state is thus :

$$| n \alpha > = e^{i(\vec{\ell} \cdot \vec{m} + \vec{a} \cdot \vec{n})} | n \ell = \ell_z = n - 1 > \tag{1}$$

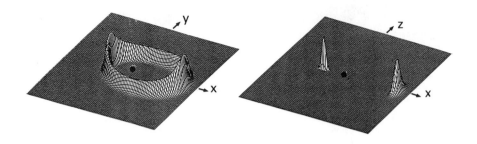

Fig.1. Density plots of the elliptic state $(n = 100; \alpha = \pi/4)$ in position representation.

which by dropping the unessential geometrical ($\overrightarrow{\ell}$) rotation and phase factors, and choosing a convenient frame can be written :

$$| n \; \alpha > = \; e^{-i\alpha a_y} \; | n \; \ell = \ell_z = n - 1 > \tag{2}$$

The transformation $e^{-i\alpha a_y}$ is actually a rotation, of non-geometrical type, which transforms a circular state into an elliptical one, with angular momentum (n-1) $\cos \alpha$ and eccentricity $\sin \alpha$. The $| n \; \alpha >$ state minimizes the Heisenberg relations as :

$$\Delta a_x.\Delta a_y \;\; = \;\; \Delta \ell_x.\Delta \ell_y \;\; = \;\; \frac{1}{2} \; |< \ell_z >|$$
$$\Delta a_z.\Delta \ell_y \;\; = \;\; \Delta a_y.\Delta \ell_z \;\; = \;\; \frac{1}{2} \; |< a_x >|$$

The electronic density plots are shown on Fig. (1). They confirm the localization on a classical ellipse in position representation. The peaking at aphelion is expected from semiclassical arguments. The one at perihelion comes from the focusing effect of the nucleus in the z=0 plane. In momentum representation, the localization is on a circle as expected from the classical Hodograph.

This builds the unique set of stationary states of the Hydrogen atom having the best geometrical localization it is possible to achieve, on a Kepler trajectory. The hyperbolic set of the continuous spectrum can be simply deduced through complexification of the Lenz vector.

466

2 - Properties and experimental production.

Explicit expansions of the elliptic states can be deduced simply by noticing that the operator $\vec{\lambda}\,(a_x, a_y, \ell_z)$ builds a 3-dim. angular momentum operator [2]. Hence the elliptic state $\mid n\ \alpha >$ is a coherent state of $\vec{\lambda}$ which yields immediately the expansion :

$$\mid n\ \alpha > = \sum_m \left[\frac{2\lambda!}{(\lambda - m)!(\lambda + m)!}\right]^{1/2} \left(\sin\frac{\alpha}{2}\right)^{\lambda - m} \left(\cos\frac{\alpha}{2}\right)^{\lambda + m} \mid \lambda = n - 1 \ \ \lambda_z = m >$$

bearing on *(2n-1)* values of λ_z and with $\lambda = n - 1$. Further expansion on the spherical states $\mid n\ \ell\ m>$ leads to :

$$\mid n\ \alpha > = \sum_{\ell m} C_{n\ell m} \left(\sin\frac{\alpha}{2}\right)^{n - m - 1} \left(\cos\frac{\alpha}{2}\right)^{n + m - 1} \mid n\ \ell\ m > \qquad (3)$$

where $C_{n\ell m}$ does not depend on the eccentricity.

The elliptic state is thus a coherent superposition of the n^2 spherical states [2]. This is shown on Fig. (2). It seems hardly possible to build such a distribution by optical laser techniques.

But elliptic states can be generated simply, from a crossed electric and magnetic field arrangement and laser excitation of Rydberg atoms, making full use of the dynamical properties of the atom [5]. Among the set $\{\mid n\ \alpha >\}$ are circular states ($\alpha = 0$) and the parabolic Stark states ($\alpha = \frac{\pi}{2}$: classically a straight line along $\hat{x}$ axis). The latter being $\ell_x = 0$ states can be laser excited easily. The problem of the production thus amounts to finding an experimental way of realizing the (4.dim!) rotation $e^{-i\alpha a_y}$.

The crossed fields hamiltonian at 1st order is :

$$V = \omega_s.\,a_x + \omega_L.\ell_z$$

where ω_s, ω_L are the Stark and Larmor frequencies. This can be diagonalized through the rotation $e^{-i\beta a_y}$ (along the $\vec{E} \times \vec{B}$ axis) with angle $\beta = tg^{-1}(\omega_s/\omega_L)$:

$$V = (\omega_s^2 + \omega_L^2)^{1/2}\, e^{-i\beta a_y}\, \ell_z\, e^{i\beta a_y} \qquad (4)$$

The eigenfunctions are $e^{-i\beta a_y} \mid n \, \ell\ell_z >$ and for $\ell = \ell_z = n-1$ coincide with elliptical states. Starting from a laser excited Stark state with ($\ell_x = 0$; $a_x = n-1$) elongated along the x axis, one builds an elliptic state by adiabatically tuning the field strengths to the desired eccentricity $\sin \beta$ [5].

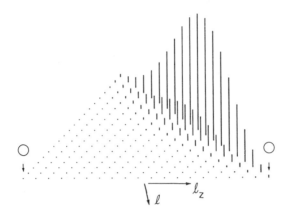

Fig.2. Squared amplitudes of the elliptic state ($n = 20$; $\alpha = 1$) on the spherical basis.

The feasability of these dynamical manipulations of the atom have been experimentally demonstrated in the transformation of linear Stark states into circular states [6][7]. Also direct static measurements have been done, years ago [8][9] on elliptical states for investigating the crossed-fields energy diagram.

Among various applications, those centered on spectroscopy of the dynamical properties of the atom seems important. The $\overrightarrow{\lambda}$ type symmetry of these states complies with the one of most Rydberg systems in external fields or in interaction, and with doubly excited systems. The tunable ℓ-mixed character of the states is a way of overcoming the low ℓ values restriction arising from ℓ selection rules in optical excitation. Hence, combining elliptic states and stepwise excitation should lead to completely new experimental views on autoionization and double excitation. Such a "dynamical spectroscopy" of the atom looks promising.

3 - Minimum wavepackets and squeezed atomic states.

To further localize the electron on the "quantum ellipse", one should coherently mix several n values, using e.g. pulsed laser excitation. The elliptic state so produced is no longer stationary but becomes a wavepacket. In order to respect various criteria such as maximum localization or/and semi-classical behaviour which, though equivalent in the harmonic oscillator, are not necessarily complying in the atom, the n mixing should obeye rules as definite as the ones for building elliptical states.

The general solution are squeezed states of a 4-dim. oscillator under constraint. This follows from the equivalence of the dynamical properties of such an oscillator and of the Coulomb problem. This will be discussed elsewhere.

468

We focus here on a simple situation which allows to illustrate the relevance of the notion of squeezing for dealing with atomic wavepackets. We consider here squeezed circular states. They are defined from the elliptical states for $\alpha=0$ by applying the dilatation operator $U_y = (\ \vec{r} \cdot \vec{p} \ + \ \vec{p} \cdot \vec{r} \ )/2$ as :

$$| \, n \, \theta \, \alpha = 0 > \ = \ e^{-i\theta u_y} \ | \, n \, \alpha = 0 > \qquad (5)$$

Such a state builds a wavepacket which will radially evolve exhibiting collapse and revival. The angular variables are frozen to $\ell = \ell_z = n - 1$ as the components of $\vec{\ell}$ commute with the dilatation. This leads to the picture of a pulsating circle and obviously the electron is not localized on the circle.

At time t=0, such a squeezed circular state is a minimum uncertainty state of the hydrogen atom which can be established by noting that the action of the dilatation in position representation $| \ \vec{r} >$ is $| \, e^{\theta} \, \vec{r} >$ and in momentum representation $| \ \vec{p} >$ leads to $| \, e^{-\theta} \, \vec{p} >$. The squeezed circular state is similar to the n circular state as concerns its shape, but the values of the parameters do no coherently fit with each other. This mismatching of the boundary conditions in the radial potentiel well leads to subsequent time evolution.

The amplitudes on the hydrogenic states with $N \geq n$ can be deduced in a way completely similar to squeezing of the harmonic oscillator, although the values are obviously different. In particular, in contrast with the oscillator, the effective squeezing angle for the N shell is $\theta_N = \theta + Log \, \frac{N}{n}$. Next, time evolution is likely to introduce a phase dispersion depending on N, which means that, at time t, the expression of the wavepacket can no longer match with (5). For harmonic systems, Eq. (5) would still hold true (to within a uniform rotation of the axis of squeezing) which is the key point in the stability of squeezed states.

Collapse and revival of such squeezed atomic states can be studied <u>analytically</u> while keeping complete parallelism with the completely understood case of the harmonic oscillator. This is likely to afford important clarifications on this controversial matter of wavepackets, and advances in the understanding of how to build those with maximum stability.

It is worth noticing that squeezed atomic states are not academic monsters. For example, the squeezed circular state $| \, n \, \theta \, \alpha = 0 >$ is the circular (stationary) eigenstate for $\ell = \ell_z = n - 1$ of the hydrogenic system with charge $Z = e^{-\theta}$. Hence a process somewhat equivalent to those involved in the generation of two-photons coherent states of the E.M. field, would be, in atomic physics, the spontaneous disintegration of tritium into helium. But much more realistic situations are likely to exist in atoms or excitons, in which squeezed atomic states are involved.

* Laboratoire Associé à l'Université P. et M. Curie, à l'Ecole Normale Supérieure et au Centre National de la Recherche Scientifique (UA 18).

REFERENCES

(1) J. PARKER and C.R. STROUD Phys. Rev. Lett. 56, 716 (1986)
and references therein

(2) J.C. GAY, D. DELANDE and A. BOMMIER, Phys. Rev. A39, 6587 (1989)

(3) A. BOMMIER, D. DELANDE and J.C. GAY in "Atoms in strong Fields" C.A. Nicolaïdes et al. Ed. (Plenum 1989)

(4) V. FOCK, Z. Phys. 98, 145 (1935)

(5) D. DELANDE and J.C. GAY Europhys. Letters 5, 4 (1988)

(6) J. HARE, M. GROSS and P. GOY, Phys. Rev. Lett. 61, 1938 (1988)

(7) G. SPIESS, M. CARRE, L. ROUSSEL, M. GROSS and J. HARE, Europhys. Letters 9 (1989) 236.

(8) F. PENENT, D. DELANDE, F. BIRABEN and J.C. GAY, Optics Comm. 49, 184 (1984)

(9) F. PENENT, D. DELANDE and J.C. GAY, Phys. Rev. A37, 4707 (1988).

Tests of Basic Molecular Physics Using Laser Spectroscopy of H_2

E.E. Eyler
Department of Physics, Yale University
New Haven, CT 06511

The hydrogen molecule has long been used as a proto-type molecular system for which theoretical and experimental methods can be compared in detail. The first accurate quantum mechanical calculation of its binding energy by James and Coolidge in 1933 was a major milestone in the early history of quantum chemistry.[1] Since then, both experimental and theoretical results have improved by several orders of magnitude, although in recent years the accuracy of experiments on the ionization potential and dissociation energy has lagged somewhat behind the rapidly increasing accuracy of theory. The theoretical energies are now accurate to about 0.03 cm^{-1}, sufficiently good that new experimental measurements can sensitively test the relativistic and quantum electrodynamic corrections in this simple molecular system. Results are reported from recently completed experiments in which the ionization potential has been measured to 0.015 cm^{-1}, and near-threshold dissociation has been investigated with very high resolution near the $H(1s) + H(2\ell)$ limit.

1. The Ionization Potential

As in the classic experiment by Herzberg and Jungen,[2] and the more recent laser experiment by Glab and Hessler,[3] our group has determined the ionization potential by extrapolating the singlet Rydberg np series. Our new measurement is a two-part effort: first two-photon intervals from the ground state to the $E,F(2s\sigma)$ $^1\Sigma_g^+$ state near 100,000 cm^{-1} were determined, then transitions from the E,F state to the high Rydberg states were measured in a double resonance experiment. For the first part, cw laser radiation near 420 nm was amplified into 3 mJ, 10 nsec pulses in an excimer laser-pumped amplifier chain. After frequency doubling, the pulsed light was intersected at right angles with a collimated supersonic beam of vibrationally excited H_2, yielding Doppler free spectra.[4] The absorption spectrum of tellurium was used for wavelength calibration. A total of 22 transitions in the (0,2), (1,2) and (0,1) bands were measured, with a typical uncertainty of 0.01 cm^{-1}.

For the second part, a pulsed dye laser was used to populate the E,F state and a pulse-amplified, frequency-doubled cw laser was operated near 397 nm to excite the Rydberg states. Four rotational branches to states with n=42-89 were studied. One- and two-channel quantum defect theory analyses describe the results within experimental uncertainties. When combined with the E,F state measurements and calculated rotational intervals for H_2^+, these results yield four consistent values for the ionization potential. The average of 124417.524 ± 0.015 cm^{-1} agrees with the theoretical value of 124417.503 cm^{-1} and is 7 times more accurate than previous experiments. Alternatively, the results for the ionization potentials of ortho- and parahydrogen can be regarded as independent measurements, and the difference of 60.257 ± 0.009 cm^{-1} between the ionization potentials from N=0 and N=1 can be regarded as a confirmation of the calculated interval of 60.256 cm^{-1}. The close agreement of experiment and theory is a remarkable confirmation of the calculations, including the calculated radiative correction of 0.378 cm^{-1} to the ionization potential. A more detailed account of these results has very recently been published elsewhere.[5] Similar measurements are now underway for the D_2 and HD isotopes, and are expected to be somewhat more accurate than the current result for H_2.

2. Spectroscopy Near the Dissociation Limit

As shown in Fig. 1, a similar double resonance scheme is used to study the region near the second dissociation limit of H_2, converging to H(1s) + H(2s or 2p). The high resolution laser can be used to excite molecules from the E,F state either to the highest bound vibrational levels just below this limit or to the vibrational continuum just above it. While most of our work to date has been concentrated on the bound levels, preliminary observations of the adjoining continuum have also been made.

The bound spectrum just below the dissociation limit consists mainly of transitions to very high vibrational levels of the B $^1\Sigma_u^+$ state, which has a broad potential curve with a $1/R^3$ dependence at long range. Using simple arguments based on the form of the long-range potential, together with experimental data from Herzberg's work, Stwalley predicted the spacing of these high vibrational levels some time ago.[6] We have studied this region with a resolution of about 0.005 cm^{-1} using multiphoton ionization as a detection scheme. Signals can be observed by searching for H^+, H^-, or H_2^+ ions, indicating that the ionization dynamics from these long-range states are very complex. The observed energy levels are not in good agreement with the predicted spectra in the region within 10 cm^{-1} of threshold, as can be seen in the stick spectrum shown in Fig. 2. In this figure, showing transitions from the v=6, N=1 level of the E,F state, extra lines are observed near v=37 and v=38 of the B state, and the levels near threshold is irregularly spaced. The situation for N=0 is is also complicated, with far more observed transitions near the threshold than can easily be accounted for. A likely explanation is the non-adiabatic coupling of singlet and triplet states by the spin-orbit coupling, which has not been taken into account in existing calculations. A more complete theoretical picture is obviously required. Several of these levels are bound by only a fraction of an inverse centimeter, so the situation is a very unusual one in which two nearly separated atoms interact at extremely long range.

To observe the continuum itself, Lyman α fluorescence from excited hydrogen atoms is observed with a solar blind photomultiplier tube. Normally only the H(2p) atoms are observed, but by applying a small field to quench the metastable H(2s) atoms the total dissociation signal can be measured. Preliminary observations with a rather poor signal to noise ratio indicate that the continuum edge corresponds to a dissociation energy of 36118.1 ± 0.2 cm^{-1}. Although much less accurate than the new measurement of the ionization potential described above, this result is in agreement with theory and is about twice as accurate as previous measurements. Preliminary results from Stoicheff's group using vacuum uv laser excitation directly from the ground state yield a similar value.[7] Somewhat surprisingly, we find that nearly all of the excited atoms are produced in the H(2s) state in the first several cm^{-1} above threshold. Since both theory and the results of our below-threshold measurements suggest that strong non-adiabatic couplings occur in this near-theshold region, the absence of atoms in the 2p state is not yet understood.

Clearly, much remains to be explained about the behavior even of this simplest of molecules very near the dissociation limit. In the near future, this study will be extended by using the same techniques to investigate the thresholds leading to H(1s) + H(3ℓ) and to the ion pair, H$^+$ + H$^-$.

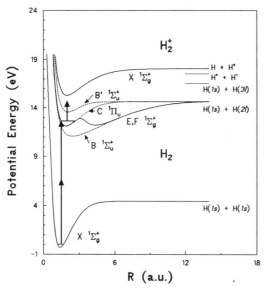

Fig. 1. Excitation scheme for investigating the second dissociation limit

473

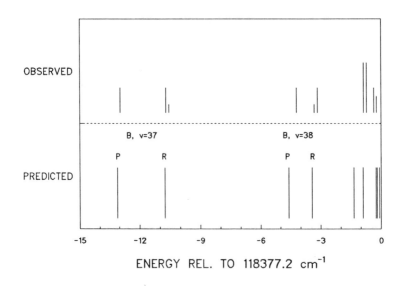

Fig. 2. Predicted and observed transitions from
v=6, N=1 of the E,F state

These experiments were conducted with graduate stu-
dents Elizabeth McCormack, Andre Nussenzweig, and Jonathan
Gilligan, and a postdoctoral fellow, Christian Cornaggia.
They were supported by the National Science Foundation.

References

[1] H.M. James and A.S. Coolidge, J. Chem. Phys. 1, 825
(1933).
[2] G. Herzberg and Ch. Jungen, J. Mol. Spec. 41, 425 (1972).
[3] W.L. Glab and J.P. Hessler, Phys. Rev. A 35, 2102 (1987).
[4] E.E. Eyler, J. Gilligan, E. McCormack, A. Nussenzweig,
and E. Pollack, Phys. Rev. A 36, 3486 (1987).
[5] E. McCormack, J.M. Gilligan, C.Cornaggia, and E.E. Eyler,
Phys. Rev. A 39, 2260 (1989).
[6] W.C. Stwalley, Chem. Phys. Lett. 6, 241 (1970).
[7] B. Stoicheff, private communication.

Spatial Distributions of Ions in Short Pulse High Intensity Multiphoton Ionization

R.R. Freeman and L. D. Van Woerkom
 AT&T Bell Laboratories, Murray Hill, NJ 07974
W.E. Cooke
Dept. of Physics, University of Southern California,
Los Angeles, CA 90089
T.J. McIlrath
Institute for Physical Science and Technology,
U. of Maryland, College Park, Md 20742

At laser intensities of about 10^{13} W/cm^2 and greater, most atoms and molecules become unstable against multiphoton ionization (MPI), regardless of the wavelength of the light. In these circumstances the usual weak field MPI processes become dominated by higher order processes. The most prominent of these is above threshold ionization (ATI), where an electron absorbs more than the minimum number of photons required to ionize. The clear manifestation of this effect is found in the photoelectron kinetic energy spectrum[1]. Here, multiple electron peaks are observed, separated by the photon energy. This phenomenon has been found to be quite general with respect to both target atom and laser wavelength[2].

In an intense electromagnetic field an electron has a ponderomotive energy of $U_p(I) = [2\pi e^2 I]/[m_e c\omega^2]$. Since high intensities usually require tight focusing, within the interaction volume the laser beam contains photons with the same energy but different vector momenta. Conservation of energy considerations show that for laser pulses of sufficiently long duration, a free electron is able to exchange momentum with the laser field, altering its trajectory but not its total energy.[2] It has been the general conclusion of several year's experiments that long pulse ATI measurements can yield only modest information about the role of the target atom upon the photoemission process, largely due to the masking effects of these ponderomotive effects. Recent experiments[4], however, involving short pulses of laser light, have uncovered a rich unexpected photoelectron spectrum that depends completely on the atomic structure.

In an intense optical field the energy levels of an atom or molecule shift by approximately U_p due to the AC Stark effect.[6] For a given multiphoton absorption event, this means that excited states may be shifted into resonance with the mth harmonic of the laser frequency when the appropriate intensity, I_R is reached. Such a resonance would increase the ionization rate and hence the total number of emitted electrons with kinetic energy $E_R = s\hbar\omega - E_b$ where E_b is the binding energy of the excited state which has shifted into resonance, and s is the number of "extra" photons absorbed necessary to give a positive E_R. These separate peaks arising from the atomic states show up in the photoemission spectrum for short pulse excitation only, being masked by ponderomotive forces of the intense light on the electrons for long pulse.[5]

The AC Stark shift at any position (r, z, ϕ) and time t within the interaction volume is directly proportional to the local intensity, $I(r, z, \phi, t)$. Consider a gaussian pulse in space and time, at the temporal peak, t=0. If the resonant intensity I_R is less than the peak intensity, I_o, then electrons with kinetic energy E_R will be produced by those atoms located in a narrow spatial region centered at $r_R = r_o[\ln(I_o/I_R)]^{1/2}$. Electrons from the same resonance, however, with the same energy, E_R will also be produced at all $r < r_R$. The resonance intensity is still *exactly* the same for these atoms, but these resonances occur twice, once when the intensity passes through I_R on the rising edge, and once on the falling edge of the laser pulse.

We use a model based on Landau-Zener level crossings to model this kind of ionization process.[7,8] Consider two states: state 1 is the excited state of the atom and state 2 is the ground state dressed by m photons (state 2 is assumed to lie above state 1). Before the optical field turns on, the two states are separated in energy by $\hbar\omega_R$. As the field increases state 1 is shifted upward according to $\omega = \omega_R(1-I/I_R)$. Figure 1a shows the energy levels as a function of time for a position within the interaction volume where the local peak intensity is equal to the intensity required to shift the state into resonance. In this case the time for resonance is relatively long due to the flat slope. Figure 1b shows the case for positions where the local peak intensity, $I_L > I_R$. Here the transition occurs on the rising and falling edges of the pulse, and the resonance time is considerably shorter than in Fig. 1b.

The first and second derivatives of the energy separation are used to parametrize the level separation with time. For a gaussian distribution in time and $I = I_R$, $\left|\partial\omega/\partial t\right| = (2\omega_R/\tau)[\ln(I_L/I_R)]^{1/2}$. If $I = I_R$, so that the resonance occurs at the temporal peak where $\partial\omega/\partial t = 0$, then $\left|\partial^2\omega/\partial t^2\right| = 2\omega_R/\tau$.

Normal Landau-Zener theory for a curve crossing linear in time gives a probability of $q = 2\pi V_R^2 \tau_c^2$ in the limit of small probability. In this expression $\hbar V_R$ is the interaction strength (proportional to $(I_R)^m$) and $\tau_c^2 = \left|\partial\omega/\partial t\right|^{-1}$. When $I_L = I_R$, τ_c is determined by the second derivative and can be shown to be $\tau_c^2 = \alpha^2 \left|\partial^2\omega/\partial t^2\right|^{-2/3}$, where $\alpha^2 = (1/2\pi)(4/3)^{1/3}(\Gamma(1/3))^2$ and $\Gamma(x)$ is the gamma function. A total characteristic time is found by adding the inverse of the linear and quadratic times in quadrature. This still yields a low probability of $q = 2\pi V_R^2 \tau_c^2$, with

$$\tau_c^2 = \tau_c^2(r,z,\omega_R) = (\tau/\omega_R)\frac{\epsilon}{1 + \epsilon[\ln(I_L/I_R)]^{1/2}}$$

where $\epsilon = \Gamma^2(1/3)(\omega_R\tau/3)^{1/3}/2\pi = 0.79(\omega_R\tau)^{1/3}$. In this expression $I_L = I_L(r,z) = I_o(r_o/r(z))^2\exp(-(r/r(z))^2)$ which is the local peak intensity at $t=0$ at position $(r,z=0)$ in the laser beam. For $z=0$, I_L and r are related to I_o, the peak intensity at $r=0$, and r_o, the spot size at the focus by $r = r_o[\ln(I_o/I_L)]^{1/2}$.

As the probability grows large and approaches unity, saturation of the transition must be taken into account. There are two types of saturation. First, the probability of a transition for a single resonance may become large. The second type of saturation deals with the depletion of the ground state atoms. The presence of strong transitions and/or many resonances will cause the number of ground state atoms to decrease. The total number of electrons created at position (r,z) is proportional to $N(r,z)P(r,z)$, where $N(r,z)$ is the number density of ground state atoms at (r,z) available for ionization. If q is small, then $N(r,z) \approx N_o$, the uniform density of the neutral atoms. As the probability gets larger, $N(r,z)$ will decrease and become a function of position due to the spatial variation in the ionization probability.

Consider a single resonance and an atom located at an r such that $I_L > I_R$. Only the temporal edges of the pulse can contribute and the rising edge causes the first transition. The number of electrons created in a volume $dV = 2\pi r dr dz$ is $S_{rising}(r,z)$, and is given by $S_{rising}(r,z) = N_o P_{rising}(r,z)$. The resonance is traversed again on the falling edge. Now, however, the ground state density, N_o has been decreased due to the rising edge ionization. For the falling edge ionization the correct ground state density is given by $N_o - S_{rising}(r,z) = N(r,z) = N_o \exp(-q_{edge})$ where $q_{edge} = q_{rising} = q_{falling} = q/2$. This new density is used to calculate the number of electrons created in the small volume due to the falling edge, $S_{falling}(r,z)$, and yields $S_{falling} = N_o \exp(-q_{edge}) P_{falling}(r,z)$. The total number of electrons created in the volume dV is just the sum of the edges, $S_{total}(r,z) = S_{rising} + S_{falling}$.

In the general case where there is more than one resonance each resonance is coupled to all of the others through the ground state depletion. The ionization probability for a given position is computed by starting with the resonance with the lowest I_R (since this one will be crossed first in time). The number of electrons created is used to calculate the new ground state density seen by the next resonance. This process is continued sequentially up the rising edge through all resonances with $I_R < I_L$, the spatially local peak intensity).

To illustrate the saturation effects described above, we have calculated the spatial distribution of ionization for a case close to recent experimental results.[4] Our case study uses the $5p^5(^2P^o_{3/2})nf$ states of Xenon for n=4-10. The following assumptions are made: (1) 2 eV (620 nm) photons are used, creating a 6 photon resonance for the f state series; (2) The states all shift uniformly by the ponderomotive energy. This approximation appears to be well justified by experiment, although deviations from ponderomotive shifts for other states in xenon with lower values of l are being currently studied; (3) The interaction strength is $V_R = \beta(I_R)^6(n_R^*)^{-3}$, where n_R^* is the effective quantum number of the resonance state, and β is the coupling strength which depends on the laser wavelength, detunings from resonance, and atomic cross section. All of these states will shift into 6 photon resonance within 1 eV of ponderomotive potential ($\approx 3 \times 10^{13} W/cm^2$). Note that once a specific target atom and wavelength are chosen, the only adjustable parameter is the laser pulse duration. Increasing the laser intensity will only serve to move the location in space of the ionization of each atomic level. *Each atomic state will undergo resonance at its characteristic I_R, regardless of how high the peak intensity of the pulse is increased.* However, there will be a small effect on the total ionization yield due to this change in the spatial location of the ionization. That is, higher intensities will push a resonance further out in r and z, thus increasing the volume over which that resonance will contribute. For spatial distributions, the number of electrons produced at (r,z) will be proportional to $S_{total}(r,z) dV = S_{total}(r,z) 2\pi r dr dz$. All of the spatial calculations performed here are plotted as $r S_{total}(r,z)$ versus r. Figure 2 shows the ion spatial and electron energy distributions for different pulse durations. For each of the three pulse widths, the solid line is the total ionization signal, the dashed curve is rising edge contribution, and the dashed-dotted curve is the falling edge signal. Figure 2a shows the case for $\tau = 100$ fsec. The coupling parameter, β is chosen such that the largest transition (the 4f state, due to it having the largest I_R) shows a weak hint of saturation. The 5f transition is not saturated at all since both the rising and falling edge ionization contribution signals are the same.

The dramatic discontinuity in each resonance is due to the inherently narrow atomic states. The rapid decrease in resonance height for increasing quantum number arises from the highly nonlinear dependence of V_R on I_R.

The spatial distribution for $\tau = 250$ fsec is plotted in Fig. 2b. The 4f resonance is now more saturated, although still not heavily depleted. The falling edge is dropping relative to the rising edge for the 4f, while the 5f and above still show no evidence of depletion. Finally, Fig. 2c illustrates the heavily saturated regime for which $\tau = 1$ psec. The 4f resonance is extremely saturated, with the rising edge ionization depleting almost all of the ground state. Furthermore, the transition itself is saturated and the small r side of the resonance is due entirely to the linear r factor in the volume element. The 5f is still unsaturated, although some depletion may be evident.

In this paper we have outlined how saturation of short pulse multiphoton resonance transitions leads to complex structure in the spatial distribution of ions. We have modeled the ionization on both the rising and falling edges of the laser pulse while accounting for substantial ground state depletion. We found that the effect of the laser intensity is only to set the spatial location of the ionization through the spatial location of the various resonance intensities, I_R. The coupling strength is determined by the laser wavelength and target species, and is thus bound within fairly strict limits. This means that the laser pulse duration becomes the only useful parameter with which to study saturation.

References

1. P. Agostini, F. Fabre, G. Mainfray, G. Petite, and N.K. Rahman, Phys. Rev. Lett. *42*, 1127 (1979).

2. T.J. McIlrath, P.H. Bucksbaum, R.R. Freeman, and M. Bashkansky, Phys. Rev. A. *35*, 4611 (1987).

3. R.R. Freeman, P.H. Bucksbaum, and T.J. McIlrath, IEEE JQE *24*, 1461 (1988).

4. R.R. Freeman, P.H. Bucksbaum, H. Milchberg, S. Darack, D. Schumacher, and M.E. Geusic, Phys. Rev. Lett. *59*, 1092 (1987).

5. H.G. Muller, A. Tip and M.J. van der Wiel, J. Phys. B: At. Mol. Phys. *16*, L679 (1983); T.W.B. Kibble, Phys. Rev. *150*, 1060 (1966).

6. L. Pan, L. Armstrong and J.H. Eberly, J. Opt. Soc. Am. B *3*, 1319 (1986); P. Avan, C. Cohen-Tannoudji, J. Dupont-Roc and C. Fabre, J. Phys. (Paris) *37*, 993 (1976); M.D. Perry, A. Szoke, O.L. Landen and E.M. Campbell, Phys. Rev. Lett. *60*, 1270 (1988).

7. W.E. Cooke, R.R. Freeman, T.J. McIlrath, and L.D. Van Woerkom, to be published.

8. T.J. McIlrath, R.R. Freeman, W.E. Cooke, and L.D. Van Woerkom, Phys. Rev. A (to be published).

Figure Captions

1. Schematic of the level crossing model with the ground state dressed by m photons labeled state 2, and the excited state subjected to the AC Stark shift labeled state 1. (a) position in space where $I_L = I_R$ and quadratic time dependence is needed. (b) position at smaller (r,z) where I_R is passed

on the rising and falling edges, requiring only the linear time dependence.

2. Calculated spatial structures for various laser pulse durations.

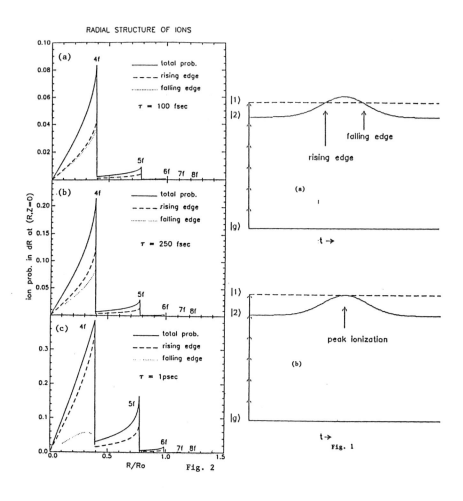

Fig. 1

Fig. 2

POLARIZATION EFFECTS IN ATI

P. H. Bucksbaum
AT&T Bell Laboratories,
Murray Hill, NJ 07974 USA

ABSTRACT

Above-threshold ionization is strongly affected by the state of the laser polarization. Sub-nanosecond circular and elliptical polarization experiments in the rare gases have provided the best information on the wiggling electron continuum that leads to ATI. Now, new above-threshold ionization experiments with sub-picosecond 616nm circularly polarized light suggest that this form of ionization is truly nonresonant, in contrast to linear polarization, where bound state resonant photoionization dominates the ATI cross section.

1. INTRODUCTION

Above-threshold ionization,[1-3] where atoms absorb more than the minimum number of photons during ionization in strong laser fields, is still not well understood. Part of the difficulty is comparing calculations, which are generally carried out using steady-state, plane-wave, and electric dipole approximations, to experiments using tightly focused ultra-short laser pulses, where electrons are subjected to strong ponderomotive forces.[4] These forces are substantially reduced in sub-picosecond laser pulses;[5] however, the spectra from these experiments show that bound state resonances facilitate the ionization.[5]

These resonances complicate ATI calculations. For this reason, we began to study ATI with circularly polarized light. Angular momentum considerations restrict the excited states that may contribute to the process. Those that may have very weak multiphoton matrix elements to the ground state, so that we might expect direct photoionization to the continuum to dominate the total rate. Recent measurements confirm this,[6] and point the way to a new technique to study the final states in the ATI process which are quantum-mechanical wiggling electrons imbedded in the coulomb continuum.

2. PONDEROMOTIVE FORCES

In most ATI spectra such as figure 1, the regular arrangement of peaks follows an n-photon photoelectric formula, $E_{electron} = nh\nu - E_0$, where E_0 is the ionization potential in the absence of any A.C. Stark shift. Both the atoms and the electrons are actually shifted in energy by several eV at these intensities, but the shifts nearly cancel.

The free electron "ponderomotive" shift[7] $U_P(x,t) = (e^2 I(x,t))/(2\pi m_e c \nu^2)$, which equals 1 eV for an electron in a Nd:YAG laser field (hν=1.165eV) with an intensity of about 10^{13}W/cm^2, gives rise to forces along the negative gradient of the intensity energy. Ponderomotive effects can scatter and accelerate electrons to several eV in the fields necessary for ATI, disrupting our ability to obtain valid electron spectra.[8]

Since a focused laser beam contains a continuous intensity distribution, we might expect the ponderomotive accelerations to produce a continuous spectrum of photoelectron energies in an ATI experiment. The reason that this does not happen is because the atomic levels undergo A.C. Stark shifts as well, which closely follow the free electron energy shifts. In other words, the Stark effect and the ponderomotive potential effectively hide each other.

3. CALCULATING ATI

The exact solution to a hydrogenic atom in the presence of a laser vector potential has not been found. In the present case, the laser-atom interaction cannot be treated as a small perturbation. However, other approximate methods have been applied to the problem, which have had some success.[9-13]

3.1 Volkov States: The Free Electron in a Laser Field

If the continuum is dominated by the laser, so that coulomb field of the ion may be neglected Schroedinger's equation can be written (in the **A·p** gauge)

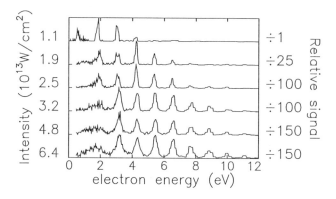

Figure 1: Photoelectron spectrum showing above-threshold ionization in xenon by 1.165 eV photons. The laser was linearly polarized and 100 psec in duration, focused to a 12 μm gaussian waist, with peak intensity shown on the left axis. The electrons were detected along the laser polarization direction, and energies were analyzed by time-of-flight. (From reference 3.)

$$\frac{[P - \frac{e}{c} A(x,t)]^2}{2m_e} \psi(x,t) = i\hbar \frac{\partial \psi(x,t)}{\partial t} \qquad (1)$$

where $A(x,t)$ is the classical (non-quantized) vector potential, which can be approximated as $A(x,t) \approx A(t) = \hat{\epsilon} A_0 \sin\omega t$ in the electric dipole approximation. The solutions, called Volkov states, are the quantum-mechanical versions of free wiggling electrons.[10] Like plane waves, the solutions are eigenstates of the *canonical* momentum with eigenvalue **p**. The extra time-dependent phases are periodic energy fluctuations caused by the oscillating field. They correspond to the energy fluctuations in a classical wiggling electron, which modulate the energy at frequencies ω and 2ω. However, while classical wiggling electrons cover a continuous but finite distribution of energies, the Volkov state is an infinite superposition of plane wave energy eigenstates that are separated by integer multiples of $\hbar\omega$:

$$\psi_{Volkov}(x,t) = e^{\frac{i}{\hbar}(p \cdot x - \frac{p^2 t}{2m_e} - U_p t)}$$

$$\times \sum_{n=-\infty}^{\infty} \sum_{m=-\infty}^{\infty} J_m(\frac{e^2 A_0^2}{8\hbar\omega m_e c^2}) J_{n-2m}(|\hat{\epsilon} \cdot p| \frac{eA_0}{\hbar\omega m_e c}) e^{in\omega t} \qquad (2)$$

Here J_m's are cylindrical Bessel functions, whose arguments are the coefficients of the oscillating terms in the Volkov phase.

This infinite series of eigenstates shows that Volkov states have an energy structure similar to the ATI final states of figure 1: they both consist of a series of energy peaks separated by $\hbar\omega$.

3.2 Circular Polarization

Volkov states can be incorporated into a scattering theory of photoionization originally put forward by Keldysh.[11] These have not been particularly successful. However, in 1986, H. Reiss suggested that certain Keldysh-type theories[12] proposed by him and by F. Faisal could accurately reproduce the circular polarized ATI spectra, with laser intensity as the only adjustable parameter.[13]

4. ATI WITH SUB-PICOSECOND PULSES

As explained in the introduction, A.C. Stark shifts can be seen directly if very short laser pulses are used.[5] For pulses shorter than 1 psec, the ponderomotive force acts for too short a time to significantly alter the electron's energy. The top part of figure 3 shows spectra using 140 fsec linearly polarized

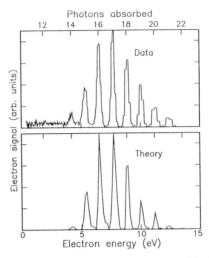

Figure 2. Top: ATI spectrum for circularly polarized 1064 nm light in xenon (from reference 14). Bottom: A computer simulation of the experiment, using ionization rates from the Keldysh theory of H. Reiss (reference 13), integrated over the measured temporal and spatial profile of the laser pulse.

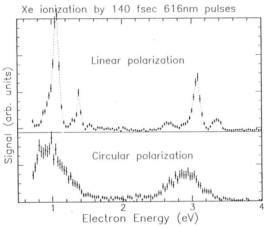

Figure 3. Top: Photoelectron spectrum in xenon for ionization by 140 fsec linearly polarized 616 nm pulses. The peaks correspond to the expected positions of six-photon resonant enhancements, provided that the resonant intermediate states shift by the full ponderomotive potential of a free electron in the laser field. (From reference 6.) Bottom: Same conditions as the top spectrum, only now the laser is circularly polarized.

616nm pulses, which reveal that each ATI peak really consists of a "fine structure" of electrons with different energies, ionized when the laser intensity was at different values. This fine structure has been associated with excited atomic states that are temporarily Stark shifted into resonance by the intense radiation.

The dominance of intermediate resonances complicates the business of calculating ATI. We would like know if systems exist for which high order multiphoton ionization is truly nonresonant. It now appears that at least one such system has been found: sub-picosecond photoionization of rare gases with *circularly polarized light.*

The bottom trace in figure 3 the xenon ati spectrum for spectrum for 140 fsec 616 nm pulses with circular polarization. The main difference is the absence of resonances in the circularly polarized case. Why have the resonances gone away? We speculate that the situation is essentially the same as for long pulses with circular polarization. There we found that the centrifugal barrier was an effective means of reducing the overlap, and hence the transition rate, between the ground state and continuum states of high angular momentum but low energy. A similar argument may be made for the bound states that come into resonance during ionization. Essentially, these are high lying states that are relatively circular, i.e. have their angular momentum L approaching their principal quantum number n. Thus they are quite far from the core. Evidently, transitions to continuum states is favored over transitions to these bound states, since even though high lying continuum states must have absorbed more photons and thus carry still more angular momentum, they can penetrate close to the ground state more easily because of their high energy. This argument is explained in more detail in reference 14, and gives very similar results to the KFR calculations for circular polarization, as in reference 13.

In any case, the absence of sharp resonances in the 140 fsec electron spectra is very good evidence that bound states play a minimal role in ATI with circular polarization. We may thus proceed to use these data as a test of theories such as the KFR model, which treat the ATI process as scattering between the ground state and Volkov states in the continuum.

For a pulse of known spatial and temporal shape, these data may be converted into photoelectron yields vs. intensity. Figure 4 shows such yield curves, along with KFR calculations for ATI in xenon.

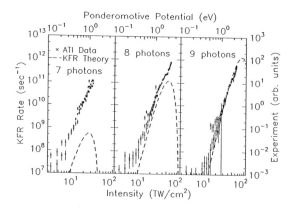

Figure 4 Experimentally determined yield curves for ATI in xenon by circularly polarized 616nm light.

The main advantage of these yield curves over the calculations done previously to compare to the sub-nanosecond data, is that the shorter pulse spectra remove all ambiguity about the laser intensity. This is because the electron energy is tagged to the intensity where it was born in the ionization event.

5. CONCLUSIONS

These experiments have shown that above-threshold ionization is strongly affected by the state of the laser polarization. Previous experiments employing 0.1 nsec circular and elliptical polarization were useful in testing different approximate theories of the ionization process, but are ultimately limited by the inability to know the intensity of the ionizing region accurately. New above-threshold ionization experiments with sub-picosecond 616nm circularly polarized light overcome this problem. The new results show (1) ATI with circularly polarized light is truly nonresonant, unlike ATI using linear polarization; and (2) ionization spectra can be used to generate yield vs. intensity for direct comparison to theory.

All of the experimental work described here has been the result of collaborations at AT&T Bell Laboratories, involving at various times the efforts of M. Bashkansky, L. DiMauro, R.R. Freeman, M. Geusic, T.J. McIlrath, H. Milchberg, D.W. Schumacher, L. Van Woerkom, in addition to myself. I particularly thank Van Woerkom, Freeman, and Schumacher, for permitting me to discuss our most recent work on sub-picosecond circularly polarized ATI.

REFERENCES

1. P. Agostini, F. Fabre, G. Mainfray, G. Petite, and N. Rahman, Phys. Rev. Lett. **42**, 1127 (1979).

2. P. Kruit, J. Kimman, H. G. Muller, and J. J. van der Wiel, Phys. Rev. **A28**, 248 (1983).

3. T. J. McIlrath, P. H. Bucksbaum, R. R. Freeman, and M. Bashkansky, Phys. Rev. **A35**, 4611 (1987).

4. P.H. Bucksbaum, R.R. Freeman, M. Bashkansky, and T.J. McIlrath, Jour. Opt. Soc. Am. B **4**, 760 (1987).

5. R.R. Freeman, P.H. Bucksbaum, H. Milchberg, S. Darack, D. Schumacher, and M.E. Geusic, Phys. Rev. Letters **59**, 1092 (1987).

6. P.H. Bucksbaum, L. Van Woerkom, R.R. Freeman, and D.W. Schumacher, to be published.

7. L. S. Brown and T. W. B. Kibble, Phys. Rev. **133**, A705 (1965);

8. P. H. Bucksbaum, M. Bashkansky, and T. J. McIlrath, Phys. Rev. Lett. **58**, 349 (1987).

9. K.C. Kulander, Phys. Rev. A **38**, 778 (1988).

10. D. M. Volkov, Zeit. fur Physik **94**, 250 (1935).

11. L. V. Keldysh, Sov. Phys. JETP **20**, 1307 (1965).

12. F. H. M. Faisal, J. Phys. B **6**, L89 (1973); H. R. Reiss, Phys. Rev. A **22**, 1786 (1980).

13. H. R. Reiss, J. Phys. B **20**, L79 (1987).

14. P. H. Bucksbaum, M. Bashkansky, R. R. Freeman, T. J. McIlrath, and L. F. DiMauro, Phys. Rev. Lett. **56**, 2590 (1986).

15. M. Bashkansky, P.H. Bucksbaum, and D.W. Schumacher, Phys. Rev. Lett. **60**, 2458 (1988).

16. P. Lambropoulos and X. Tang, Phys. Rev. Lett **61**, 2506 (1988); H.G. Muller, G. Petite, and P. Agostini, Phys. Rev. Lett. **61**, 2507 (1988).

17. S. Basile, G. Ferrante, and F. Trombetta, Phys. Rev. Lett. **61**, 2435 (1988); J. Phys. B **21**, L539 (1988); J. Phys. B **21**, L377 (1988).

Bond Localization of Molecular Vibration

T. Shimizu, T. Kuga, K. Nakagawa, and Y. Matsuo
Department of Physics, University of Tokyo,
Hongo, bunkyo-ku, Tokyo 113, Japan

Overtone transitions of $\Delta v=2,4,5$ and 6 of NH_3 are systematically measured by high resolution laser spectrometers. Deviation from the nomal mode representation becomes appreciable even in the v=3 state and the local mode picture is remarkable in the v=5 state.

1. Introduction

It may be interesting to know whether the vibrational energy temporarily locates in some proper bond or always distributes among all bonds in a molecule. It is known that the local mode character is prominent in the molecule with a heavy atom-light atom bond such as -C-H or -O-H.[1] Several vibrational quanta tend to localize in a bond and each bond oscillates independently in higher vibrational states. This report describes high resolution spectroscopic studies of NH_3 molecule. Stationary and dynamic properties of vibrational states from v=0 up to v=6 are systematically investigated with infrared diode lasers and ring dye lasers. A process of localization of vibrational energy is traced as a function of vibrational quantum number.

Highly excited vibrational states of molecules have been attracting a great interest in recent years. Their properties are directly concerned with mechanisms of multiphoton dissociation, chemical reaction, relaxation in lasing media, and atomic and molecular processes

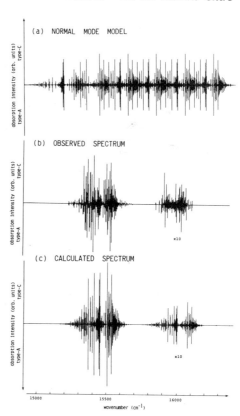

(a) NORMAL MODE MODEL

(b) OBSERVED SPECTRUM

(c) CALCULATED SPECTRUM

Fig.1 Calculated and observed spectra for the v=5←0 transitions

in interstellar space and in planetary atomosphere. In the 1930's absorption spectra of H_2O and NH_3 in the visible wavelength region due to changes of several vibrational quanta investigated by Mecke et al[2] and Badger[3], respectively. The observation was made at the high pressure, so that the rotational structure was rarely resolved. Rotational analysis of the 645 nm band of NH_3 was atempted by McBride and Nicholls.[4]

Recent developments in tunable dye lasers and diode lasers have allowed us to observe detailed rotational structure in a very weak spectrum at high resolution. The spectrum of H_2O was studied by an optoacoustic technique in the frequency range of 16750-17030 cm^{-1}.[5] The 645 nm band of NH_3 was also studied with dye lasers.[6] Only a few works with diode lasers have been reorted.

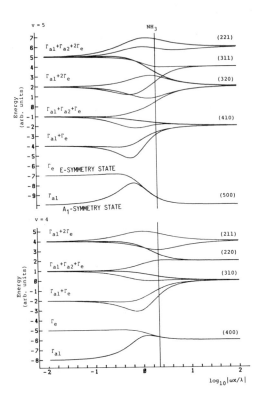

Fig.2 Energy levels as a function of $(\omega)_x/\lambda$)

In the normal mode representation, NH_3 has four vibrational modes. Since the frequencies, ν_1, ν_3, $2\nu_4$, and $4\nu_2$ are nearly degenerate, many vibrational states overlap with each other in higher energy region. For example in the wavelength region of a DCM dye laser (v=5) nearly 60 ($5\nu_1$, $5\nu_3$, $4\nu_1+\nu_3$, ...) vibrationa bands may be expected (Fig.1a). However, an actually observed spectrum is quite simple (Fig.1b). Only a single (but degenerate) band is prominent. The next band is roughly 1/30 times weaker than the first band.[7]

The calculation based on the localization of vibrational energy reproduces the observed spectrum very well (Fig. 1c). The strongest band may be represented by (500) and the much weaker (second strongest) band by (4,1,0), where the

Rotational constants

V=5 Band origin		B	B-C
E,s	15451.248	9.15831	3.0878
A_1,s	15449.89	9.14260	3.0314
E,a	15448.489	9.14297	3.07451
A,a	15447.24	9.11527	3.0210

Table I. Molecular constants [cm^{-1}] in the v=5 state

figure means the number of vibrational quanta in each bond among three idential N-H bonds in NH_3. Not only global spectral pattern, but also precise transition frequencies the theory can predict.

The theory[8] starts with three identical independent unharmonic oscillations of

$$E(n_1,n_2,n_3)/hc = (v+\tfrac{3}{2})\omega - \sum_{i=1}^{3}(n_i+\tfrac{1}{2})^2 \omega_x$$

in the Morse potential, where ω is the angular frequency of vibration and ω_x represents an unharmonicity. Now the coupling between the bonds

$$H' = \lambda \sum [(n_i+1)n_j]^{1/2} \, \delta_{n_i' n_i+1} \, \delta_{n_j' n_j-1} \, \delta_{n_k' n_k}$$

is introduced so that the total vibrational quanta $v=n_1+n_2+n_3$ is conserved. The energy is calculated as a function of ω_x/λ (Fig.2). The limit of $\omega_x/\lambda \to \infty$ represents that pure bond mode, while that of $\omega_x/\lambda \to 0$ the pure normal mode. The calculation well fits to the observation at $\omega_x/\lambda = 2.1$ (the vertical line in Fig.2). This means that the vibration is almost localized.

As seen in Fig.2, the A_1- and E-symmetry levels are degererate. Actually the observation gives nearly equal band origins and the rotational constants for both states (Table I).

The rotational constants for various vibrational levels are accurately determined. These show appreciable deviation from those predicted by the normal mode theory (Fig.3).

Schematic representations of the wavefunctions are shown in Fig.4. In lower vibrational level the wave function is a well-defined function of the normal coordinate (say Q_1), while in higher vibrational level the probability density distributes in the direction of a linear combination of the normal coordinates Q_1 and Q_2. The localization becomes appreciable even in the v=3 state.

Although vibrational characteristics in the high energy level are quite different from those in the ground state, the collisional dynamics in well defined rotational levels in both states is very similar (Fig.5). The energy level structure of the v=5 state is confirmed by the pulsed mm-wave-optical double resonance methods.[10]

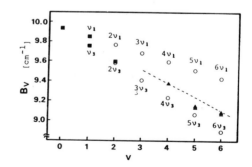

Fig.3 Rotational Constant as a function of v

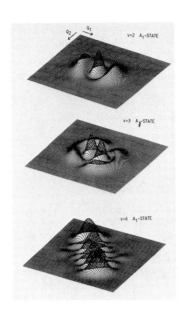

Fig.4. Vibrational wave
function in low and
high vibrational
states

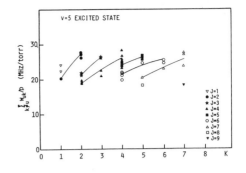

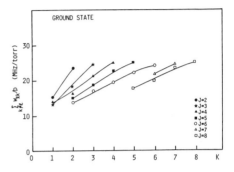

Fig.5 Rotational dependences
of presure broadening
parameters as funcitons
of rotational quantum
numbers, J and K

References

1) M.S. Child and R.T. Lawton ; Faraday Disc. 71 (1981) 273
2) R. Mecke et al. ; Z. Phys. 81 (1933) 313, 445, 465
3) R.M. Badger ; Phys. Rev. 35 (1930) 1038
4) J.O. McBride and R.W. Nicholls ; Can. J. Phys. 50 (1972) 93
5) A.B. Antipovetal ; J. Mol. Spectrosc. 89 (1981) 449
6) K.K. Lehmann and S.L. Coy ; J. Chem. Phys. 81 (1984) 3744, 84 1986 5239
7) T. Kuga, T. Shimizu, and Y. Ueda ; Jpn. J. Appl. Phys. 24 (1985) L147, 25 (1986) L1084
8) L. Halonen and S. Child ; J. Chem. Phys. 79 (1983) 4355, Mol. Phys. 46 (1982) 239
9) Y. Matsuo, K. Nakagawa, T. Kuga, and T. Shimizu ; J. Chem. Phys. 86 (1987) 1878
10) Y. Matsuo, Y. Endo, e. Hirota, and T. Shimizu ; J. Chem. Phys. 87 (1987) 4395, 88 (1988) 2852

Second-order Doppler free Spectroscopy

S.N.Bagayev,   V.P.Chebotayev,   A.K.Dmitriyev,
A.E.Om,   Yu.V.Nekrasov,   B.N.Skvortsov
Institute of Thermophysics, Siberian Branch of the USSR
Academy of Sciences, Novosibirsk-90, SU-630090

1. It is well known that methods of Doppler-free laser spectroscopy allow one to obtain resonances with a homogeneous linewidth 2Γ with maxima coinciding with the transition frequency ω_0. When obtaining very narrow resonances with a relative width less than 10^{-11} the influence of the second-order Doppler effect (SODE) becomes essential. A particle moving with a transverse velocity u radiates (absorbs) at the frequency shifted relative ω_0 by $\Delta = -1/2(u/c)^2\omega_0$. If $\Delta_0 \ll \Gamma$ (Δ_0 – the shift for the particle with an average thermal velocity v_0), we have inhomogeneous broadening of the saturated absorption resonance due to SODE. The width and the shift of the resonance will have a value about Δ_0. At least, there are two ways to eliminate the SODE influence. The first one is connected with a deep radiative cooling of particles [1]. The second one is connected with an optical selection of cold particles in a gas [2]. As it has been experimentally shown [2], the dispersion resonance width in transit-time conditions when $\Gamma\tau_0 \ll 1$ ($\tau_0 = a/v_0$ is a transit time of the particle flying with an average thermal velocity through a light beam with radius a), can be determined by homogeneous width of 2Γ. In this case, the effective temperature of particles that gives the main contribution to the saturation resonance is equal to $T_{ef} = (\Gamma\tau_0)^2 T_0$, T_0 is a gas temperature. Correspondingly, due to SODE the resonance shift has a magnitude $\delta = (\Gamma\tau_0)^2 \Delta_0$ which is small at $\Gamma\tau_0 \ll 1$. Here we report on the observation of second-order Doppler free optical resonance with the halfwidth of about 50 Hz. The experiments have been carried out by using the specially developed laser spectrometer, supposed to obtain resonance with a relative width of $10^{13} \stackrel{.}{-} 10^{14}$. The SODE influence on a form of nonlinear optical resonance in transit-time conditions have been analysed.

2.   The influence of SODE on the resonance shape in transit-time conditions for the case $\Gamma \gg \Delta_0$ has been studied in [3,4]. Results [3,4] can be also applied to the case $\Gamma < \Delta_0$. The expression for the resonance shape in the transit-time conditions with allowance for SODE has a complicated form [3]. Therefore here we will use a simple approximate expression:

$$\alpha_s = \alpha_0 \{1 - \varkappa(\Gamma\tau_0)^2 (\Gamma^2/4) \int_0^\infty \frac{W(u)\,du}{[\Omega + (u/c)^2\omega_0/2]^2 + \gamma^2} \}, \qquad (1)$$

where $\Omega = \omega - \omega_0$, $W(u)$ is the Maxwell distribution of particles over transverse velocities u, $\varkappa = (2dE/\hbar\Gamma)^2$ is the saturation parameter for cold particles, 2E is a field amplitude, d is the dipole matrix element of the absorbing transition, α_0 and α_s are an unsaturated and saturated absorption coefficients. The physical sense of (1) is obvious: the resulting resonance form consists of a combination of resonances, located at the frequencies $\omega = \omega_0 - (u/c)^2 \omega_0 / 2$ and having the halfwidth of $\gamma = \Gamma + u/a$. The additional numerical analysis of (1) has shown that the resonance form with $\Gamma\tau_0 < 1$ and $\Gamma < \Delta_0$ has been well enough described by this formula qualitatively as well as quantitatively. The resonance form for the homogeneous case ($\Gamma \gg \tau_0^{-1}$) with $\Gamma/\Delta_0 = 0.1$ is shown by Fig.1a (curve 1). The resonance is seen to have a strongly asimmetrical form which could be divided into two characteristic parts. The first one corresponds to $\Omega > 0$ and has a homogeneous halfwidth Γ. The second low-frequency resonance part ($|\Omega| \gg \Gamma$) is determined by its inhomogeneous broadening and a shift caused by SODE and has a value of about Δ_0. In a space-limited beam when transit effects are considerable ($\Gamma < \tau_0^{-1}$) the width of separate resonances for particles with different velocities u is also different ($\gamma = \Gamma + u/a$). Consequently, a resulting resonance form is altered greatly (see curve 2).

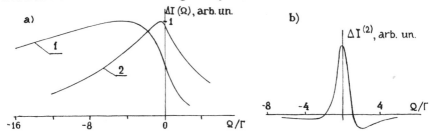

Fig.1. Calculated shape of the CH$_4$ resonance according to (2).

The first resonance part ($\Omega > 0$) is broadened, but the second one narrowed due to contribution of the transit effects. Hence, the role of SODE inhomogeneuos broadening decreases. Fig. 1b shows the form of second derivative of the saturated absorption line at $\Gamma\tau_0 = 0.1$. As it has been already noted in [2], this resonance width is determined by homogeneous width, because the main contributionin it is made by slow particles with the velocities u < aΓ. So, SODE influence on a resonance broadening and shift can be effectively eliminated at $\Gamma \ll \tau_0^{-1}$. Its width is .6Γ and its maximum shift is of $\delta \approx .01\,\Delta_0$.

3. To obtain the resonances with a halfwidth of $\Gamma \ll \Delta$ it is necessary to use an absorbing gas at very low pressure of about 10^{-6} Torr. This circumstance, as well as a small part of cold particles in a gas and saturation power, leads to the sharp resonance intensity decrease. To overcome these difficulties, a special laser spectrometer with intracavity telescopic beam expander (TBE) has been developed. First experiments have been carried out on $F_2^{(2)}$ P(7) ν_3 -absorption line ($\lambda = 3.39\mu m$) of methane [5]. The spectrometer consists of three lasers: a frequency stable He-Ne/CH$_4$ laser with a narrow radiation line of about 1 Hz, an auxiliary He-Ne laser and investigated He-Ne/CH$_4$ laser with TBE. The telescopic laser cavity was formed by the system of six mirrors (see [5]). The absorption cell length and the light beam diameter in it were of 850 cm and 30 cm, respectively. The absorption cell was protected from the magnetic field. We could operate

490

at $T_0 = 300K$ and $T_0 = 77K$ gas temperature in the cell. A narrow radiation line of the telescopic laser was provided by its phase locking to the stable laser. The resonances were recorded by synchronous detection of the second harmonic signal in laser power with modulation of its frequency.

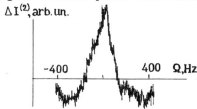

$\Delta I^{(2)}$, arb. un.

-400   400 Ω, Hz

Fig. 2. Record of the single recoil component in methane on the MHFS transition 7 -> 6 of $F^{(2)}P(7)\nu_3$ – absorption line ($T_0 = 300K$, a = 15cm, P = 5 μTorr, modulation frequency - f = 60Hz, deviation amplitude – $\Delta f = 100Hz$, recording time - 20min).

Fig. 2 shows the record of the second derivative of CH_4-resonance. The resonance halfwidth is about 50 Hz. The resonance maximum shift from the transition center of recoil component has a magnitude less than 1 Hz due to SODE.

It was difficult to obtain smaller width due to low-frequency fluctuations of laser intensity because of angular detuning of the cavity. The magnitude of these fluctuations was higher by several orders. Elimination of this amplitude noise by automatic angular cavity mirrors tuning will enable us to obtain resonances with the absolute width about of 10 Hz.

Fig. 3 shows the possible spectral resolution made by He-Ne/CH_4 laser with TBE. The resonance halfwidth is about 4Hz (relative halfwidth is about $4\cdot10^{-14}$).

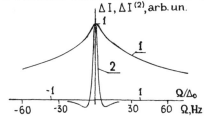

$\Delta I, \Delta I^{(2)}$, arb. un.

-60   -30   30   60 Ω, Hz
-1   1 Ω/Δ_0

Fig. 3. The resonance form (curve 1) and its second derivative (curve 2) in CH_4 on $F_2^{(2)}$ absorption line for one recoil component (the calculation experiment results): a = 15 cm, CH_4 - pressure - .2 μTorr, T = 77K, $\Gamma\tau_0 = .01$, T = 8mK, δ = .006 Hz.

Thus, we have an every reason to claim a new field in laser spectroscopy - "Second-order Doppler free spectroscopy".

Acknowlegements. The authors are greatful to Prof. E.A.Titov and Dr. V.M.Semibalamut for valuable discussions, to Dr. V.G.Goldort for working out the stable power supply.

References
1. V.G.Minogin, V.S.Letokhov. Laser radiation pressure on atoms, Moscow, Nauka, 1986.
2. S.N.Bagayev, A.E.Baklanov, V.P.Chebotayev, A.S.Dychkov, P.V.Pokasov: Pis'ma Zh.Eksp.Teor.Fiz. 45, 371 (1987); Appl.Phys. B48. 31 (1989).
3. E.V.Baklanov, B.Ya. Dubetsky: Kvant.Elektr.2,2041 (1975).
4. V.M.Semibalamut, E.A.Titov, V.A.Ulybin: In Optical Time and Frequency Standards, ed. by V.P.Chebotayev (Novosibirsk, 1985) p.98.
5. First results were reported at the IVth Frequency Standards and Metrology Symposium (Italy, September, 1988).

Part XIV

Laser Spectroscopy for
Biomedicine

Prospects of Laser Spectroscopy of Biomolecules with Nanometer
Spatial Resolution

V.S.Letokhov
Institute of Spectroscopy, USSR Academy of Sciences, 142092
Troitzk, Moscow Region, USSR

1. Introduction

Many papers read at the eight preceding Conferences on Laser
Spectroscopy (1973-1987) considered various ways to improve
drastically the main characteristics of optical spectroscopy,
namely,

$1°$. Spectral resolution, to be capable of covering the sub-
Doppler range.

$2°$. Sensitivity, to be capable of detecting single atoms
and molecules.

$3°$. Temporal resolution, to be capable of covering the femtosecond range.

$4°$. Detection selectivity, to be capable of detecting trace
atoms, isotopes, and molecules.

The present paper discusses the possibility of improving radically, with the aid of laser radiation, one more spectroscopic characteristic -

$5°$. Spatial resolution, to be capable of covering the subwavelength range, down to atomic-molecular resolution.

At present, a number of different approaches to the solution
of this problem are being lively discussed, and the first attempts at realizing them have already been made. Let us mention them briefly.

(1) Near-Field Optical-Scanning Microscopy based on the use
of a small aperture of subwavelength radius located and scanned
very close to the sample. Light passing through such an aperture diaphragm has a very small spot in the near field, the size of which is less than λ . This method was proposed and demonstrated in the microwave range [1] and has recently been extended to the optical range with a resolution of $\lambda/20$ [2-5].

(2) Scanning Tunneling Microscopy (STM) [6,7] and Spectroscopy (STS) with laser excitation of the imaging molecules adsorbed on a surface. The first experiments have been performed
on laser-frequency mixing using STM [8], but the STM/STS capabilities of differentiating between the molecules over which
the needle tip is scanning can apparently be materially extended on account of resonant laser excitation of the molecular
structures under study.

(3) Laser Projective Photoion Microscopy [9-11] based on ultrafast resonant laser photodetachment of molecular photoions
from chromophores of biomolecules adsorbed on the tip of an ion

projector needle. It is exactly this approach that is discussed in more detail in this paper.

(4) Atomic Scanning Microscopy [12] based on the possibility of deep focusing of an atomic beam into the Å-range by means of the gradient force of a laser beam with an appropriate configuration.

2. Spatial Resolution of Photoion Microscope

The femtosecond laser fragmentation of molecules involving the formation of charged particles (photoions) offers a new possibility for realization of an Å -high resolution in the study of the structure of large molecules. This possibility is based on the following two key factors: (1) the photoionization fragmentation site (photoion or photoelectron detachment site) can be localized with an accuracy much better than the photoionization laser wavelength and (2) the ion projective microscopy technique makes it possible to observe the ejection site of a charged particle with a high magnification. This forms the basis for an interesting possibility of detecting molecules and chromophores with a subwavelength resolution, i.e., for the development of femtosecond photoionization microscopy.

The idea of spatial photoionization localization of molecular bonds or chromophores of a large molecule, based on a combination of the spectrally selective multiphoton photoionization (MPI) of the chosen bond or chromophore with ion projective microscopy, is illustrated in Fig. 1. A hemispherical needle tip with a radius of curvature of r is used as a cathode and a confocal screen with a radius of curvature of R, as an anode. A molecule is adsorbed on the cathode surface. Under the action of a sequence of ultrashort (femtosecond) light pulse the chosen chromophore undergoes MPI, i.e., the ejection of the molecular ion M^+. In the presence of a strong electric field near the tip, the photoions move along radial lines toward the screen. The electric field serves only to transfer the photoions to the screen and not to ionize the adsorbed molecule, and so the field strength can be as low as 0.5 V/Å , a value at which no field-induced nonselective and destructive molecular evaporation and decomposition can take place. Here

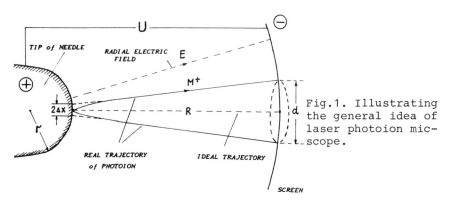

Fig.1. Illustrating the general idea of laser photoion micscope.

495

is the difference between the Müller field-ionization ion microscope [13] and this laser-induced photoionization ion microscope.

The ultimate spatial resolution of the laser photoion microscopy is limited by the nonzero tangential velocity of the ejected photoions, which makes them move not along strictly radial lines of force, but within the confines of some quasi-cone. As a result, the photoion ejection site is displayed on the microscope screen as a spot of finite diameter d (Fig. 1). That the tangential photoion velocity is nonzero is due to three fundamental reasons: (a) the Heisenberg uncertainty principle, (b) the nonzero initial kinetic energy of photoions, and (c) the Coulomb interaction between neighboring photoions (space charge effect). Simple estimates show that the laser photoionization microscopy can have a spatial resolution of a few Ångström units, which may be sufficient to visualize large molecules.

For example, at r = 1000 Å, a field strength of $E = 0.5$ V/Å near the tip surface, and a photoion mass of $M = 10^2$ at.un., the limit on the spatial resolution of the technique due to the uncertainty principle is as low as 1 Å. The limitation due to the nonzero initial photoion velocity is much more serious. With the kinetic energy of the tangential photoion movement being as high as $E_{kin}^t = 0.001$ eV and the values of E and r remaining the same as above, the spatial resolution of the technique is $2\Delta x \simeq 3-5$ Å. This means that the needle tip must be cooled and the final state in MPI must be very close to the photoion detachment limit. The same applies also to the energy of the Coulomb interaction between photoions spaced a distance of a apart ($E_{Coul} = e^2/2a$): it must not exceed 0.001 eV, i.e., no more than one photoion must be allowed to be detached from a tip with r = 1000 Å during a single femtosecond laser pulse. The complete picture of a molecular structure can be obtained in the course many laser pulses by accumulating information about the photoion ejection sites. For this reason, the large biomolecule must not be subject to desorption or destruction during the MPI of its chromophore.

3. Femtosecond MPI Spectrometry of Biomolecules on Surface

To develop a photoion microscope, it is necessary to learn how to detach with a laser pulse a chromophore from a large molecule adsorbed on the projector tip surface. The formation of photoions upon irradiation of a surface with powerful UV laser pulses was revealed in [14-17] where it was demostrated that the production of photoions resulted from the MPI of neutral molecules undergoing laser-induced thermal desorption. To avoid this effect, the laser pulse duration, i.e., the time of deposition of the required energy in the chromophore, τ_{depos}, must satisfy the following obvious condition:

$$\tau_{depos} \ll \tau_{detach} \simeq \langle a \rangle / \langle v \rangle \ll \tau_{transf}$$

where τ_{detach} is the time it takes for a photoion with an average size of $\langle a \rangle \simeq 1$ Å to move with an average velocity of $\langle v \rangle \simeq 10^4$ cm/s for a distance of the order of $\langle a \rangle$ from the projector tip surface and τ_{transf} is the time of excitation transfer from the chromophore to the other portions of the molecule. Since $\tau_{detach} \simeq 1$ ps and $\tau_{transf} \simeq 0.1 - 1$ ps, it is obvious that to effect the MPI of molecular chromophores directly on the surface necessitates the use of femtosecond laser pulse.

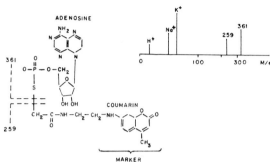

Fig. 2. Chemical structure of coumarin-labeled adenosine absorbing in the blue region and its femtosecond laser-induced mass spectrum (λ = 308 nm, τ_n = 300 fs, Φ = 0.006 J/cm^2) with distinct lines of coumarin (M = 259 at. un.) and adenosine (M = 361 at. un.) fragments (from [19]).

The first experiments [18] on the femtosecond desorption of ions showed the mechanism of production and detachment of ions in this case to be substantially different from the one indicated above. These experiments were conducted with tryptophan and a tryptophan-containing tripeptide and revealed some specific features (a drastic reduction in the threshold energy fluence for the appearance of photoions in particular) which were especially evident from comparison between the mass spectra resulting from irradiation with femtosecond and nanosecond UV laser pulses.

To avoid too fast an energy transfer and enhance the selectivity of laser excitation of specific sites in large biomolecules, it seems very promising to use their chemical labelling. This is especially important in the mapping of the sequences of DNA nucleotide bases having very close spectral properties. The first successful experiments [19] were performed at the Institute of Spectroscopy, USSR Academy of Sciences, with the adenosine-coumarin complex synthesized at the Institute of Physical Chemistry, Heidelberg University. Figure 2 shows the chemical structure of this compound and its femtosecond laser pulse-induced mass spectrum. The distinct mass lines of the coumarin chromophore and adenosine fragments can be resolved – fairly easily.

4. Towards the Laser Photoion Microscopy

The next natural steps are the study of the mechanisms of photodetachment of dye chromophore ions from molecules adsorbed

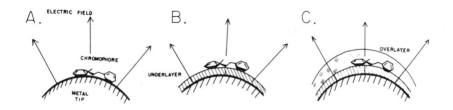

Fig. 3. Three possible locations of a molecule on the tip
of a photoion projector needle.

on metal or dielectric surfaces and the search for conditions
necessary to effect preferable MPI of labeled chromophores
while causing no laser-induced detachment of intact molecules.
It is necessary to find conditions under which data can be accumulated into a photoion image from many (a few hundred) laser shots at a low MPI yield allowing one to avoid the space-
change effects for femtosecond laser pulses that can reduce
spatial resolution.

We have at least three possibilities here (Fig. 3). First,
a biomolecule is adsorbed on a metal surface. In that case,
the surface binding energy of the molecule is high, and so is
its excitation quenching rate, which hinders the MPI and detachment of chromophore photoions. Second, the metal tip surface can be coated with a thin dielectric layer for the biomolecule to be adsorbed on. In that case, the surface binding
energy of the molecule and its excitation quenching rate can
be materially reduced. And finally, the biomolecule can be coated with a thin layer of a dielectric which will emit photoions on account of charge transfer from the molecular chromophore being excited.

The first experiments have already been conducted [20,21]
on the laser-induced MPI of molecules on the projector tip of
a photoion microscope in the presence of a strong pulsed electric field. The next steps will be systematic experiments with
nucleotide-dye compounds adsorbed on the tip surface under various conditions (Fig. 3) in the presence of a strong field and
the search for conditions conducive to a highly selective photodetachment under the joint action of laser and electric field
pulses.

The objective of this research is to develop a new method
for direct investigations into the spatial structure of large
molecules with a resolution of a few Angström units. An immedi-
ate practical result of the work will be the possibility of reading the primary genetic information encoded in the base sequences of short (about a thousand bases) DNA fragments. The
achievement of the ultimate spatial resolution (of a few Angström units) is a very difficult goal, but to reach a spatial
resolution of several nanometers, i.e., hundreds of times better than the laser wavelength, seems quite realistic.

References

1. E.A.Ash, G.Nichols. Nature $\underline{237}$, 510 (1972).

2. A.Lewis, M.Isaakson, A.Murray, A.Harootunian. Biophys.J. $\underline{41}$, 405a (1983).

3. D.W.Pohl, W.Denk, M.Lonz. Appl.Phys.Lett. $\underline{44}$ (7),651(1984).

4. A.Harootunian, E.Betzig, M.Isaakson, A.Lewis. Appl.Phys. Lett. $\underline{49}$ (11), 674 (1986).

5. U.Dürig, D.W.Pohl, F.Rohrer. J.Appl.Phys. $\underline{59}$ (10),3318(1986).

6. G.Binning, H.Rohre, Ch.Gerber, E.Weibel. Phys.Rev.Lett.$\underline{49}$, 57 (1982).

7. G.Binning, H.Rohrer. Nobel Lecture, Stockholm, December 8, 1986.

8. L.Arnold, W.Krieger, H.Walther. Appl.Phys.Lett. $\underline{51}$, 786 (1987); J.Vac.Sci.Technol. $\underline{16}$ (2), 466 (1988).

9. V.S.Letokhov. Kvantovaya Elektron. $\underline{2}$, 930 (1985)(in Russ.); Phys.Lett. $\underline{A51}$, 231 (1975).

10. V.S.Letokhov. Comm.Atom.Molec.Phys. $\underline{11}$, 1 (1981).

11. V.S.Letokhov. Laser Photoionization Spectroscopy (Academic Press, Orlando, 1987), Chap.12.

12. V.I.Balykin, V.S.Letokhov. Optics Comm. $\underline{64}$, 151 (1987); Zh.Eksp.Teor.Fiz.$\underline{94}$, 140 (1988).

13. E.W.Müller, T.T.Tsong. Field Ion Microscopy and Field Evaporation. In: Progress in Surface Sciences (Pergamon Press, Oxford, 1973), vol.4.

14. V.S.Antonov, V.S.Letokhov, A.N.Shibanov. JETP Lett. $\underline{21}$,471 (1980); Appl.Phys. $\underline{25}$, 71 (1981).

15. V.S.Antonov, V.S.Letokhov, Yu.A.Matveetz, A.N.Shibanov. Laser Chemistry $\underline{1}$, 37 (1982).

16. S.E.Egorov, V.S.Letokhov, A.N.Shibanov. In: Surface Studies with Lasers, ed. by F.Ausseneg, A.Leither, M.E.Lippitsch Springer-Verlag, Berlin, 1983), p.156.

17. S.E.Egorov, V.S.Letokhov, A.N.Shibanov. Sov.J.Quant.Electr. $\underline{121}$, 1393 (1984).

18. S.V.Chekalin, V.V.Golovlev, A.A.Kozlov, Yu.A.Matveetz, A. P.Yartzev, V.S.Letokhov. J.Phys.Chem. $\underline{92}$, 6855 (1988).

19. S.V.Chekalin, V.V.Golovlev, K.-O.Greulich et al. to be published.

20. V.S.Antonov, V.S.Letokhov. In: Multiphoton Processes, ed. by P.Lambropoulos and S.J.Smith (Springer-Verlag, Berlin, 1984), p.182.

21. S.E.Egorov, V.S.Letokhov, E.V.Moskovets. Appl.Phys. $\underline{B45}$, 53 (1988).

Medical Applications of Laser Spectroscopy

S. Andersson-Engels, R. Berg, J. Johansson, K. Svanberg[*] and S. Svanberg

Department of Physics, Lund Institute of Technology, S-221 00 Lund, Sweden

[*]Department of Oncology, Lund University Hospital, S-221 85 Lund, Sweden

Lasers are finding an increasing number of applications in medicine and biology. Thermal interactions have dominated the field but spectroscopic aspects are playing a more important part in several new medical laser methods, which are briefly covered in this paper. Medical analytical laser spectroscopy provides sensitive analysis of body fluids and tissue samples. Photodynamic tumor therapy, using tumor seeking agents in combination with laser light, is a potent method of laser-induced chemistry in the body. Laser-induced fluorescence provides new possibilities for diagnosis of malignant tumors and atherosclerotic plaques. The potential of these techniques can be enhanced by performing time-resolved studies. Spectral studies of the propagation of picosecond optical pulses in tissue can be utilized for the assessment of tissue oxygenation and tissue drug concentrations. Enhanced contrast in transillumination can also be demonstrated by using temporal discrimination. Finally, we consider multicolor fluorescence imaging as a powerful technique for the visualization of malignancies. Some examples from work performed by our group in Lund are given. For more detailed reviews and for references to most of the relevant literature we refer the reader to (1-5).

1. Medical Analytical Laser Spectroscopy

Several powerful laser spectroscopic techniques now complement conventional atomic absorption and emission spectroscopy techniques which are routinely used for the analysis of constituents in body fluids. Elements of interest may be alkali ions or heavy metals (toxicology). Laser-enhanced ionization (LEI) spectroscopy, resonance ionization spectroscopy (RIS) and resonance ionization mass spectroscopy (RIMS) provide unprecedented sensitivity and accuracy in atomic analysis. Liquid chromatography and capillary zone electrophoresis with fluorescence detection have, in the same way, increased the sensitivity and selectivity in molecular analysis. Fluorescence techniques are also widely used in biology and medicine for immunoassay and DNA sequencing using fluorescent tags. Similar techniques are used in cytofluorometry and automatic cell sorting.

Another technique, which provides real time diagnostics is laser-induced breakdown spectroscopy (LIBS). A beam from a powerful pulsed laser focussed onto tissue gives rise to a hot plasma emitting atomic and ionic lines. Such techniques have been used to characterize the vessel wall in connection with atherosclerotic plaque ablation and for the characterization of gall- and kidney stones in connection with stone lithotripsy.

2. Photodynamic Tumor Therapy

For some time, tumor-seeking agents such as hematoporphyrin derivative (HPD) have been used in combination with laser radiation to localize and treat malignant tumors. Progress in this field is described, e.g. in Refs. 6 and 4. The two aspects of the technique are illustrated in Fig. 1. HPD is intravenously injected at low concentration into the body where it spreads. It is subsequently cleared out through natural processes. How-

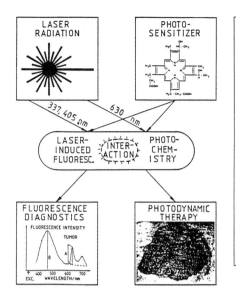

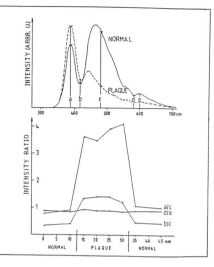

Fig. 1. Tumour fluorescence detection
and treatment using HPD and laser (4)

Fig. 2. LIF spectra and results from
scan through an atherosclerotic
plaque region (7).

ever, for still partially unknown reasons, the HPD molecules are selectively retained in tumor tissue. The major absorption peak of HPD occurs
in the Soret band peaking at 405 nm. Fluorescence follows with a characteristic, dual-peaked distribution in the red spectral region. In tissue
the HPD fluorescence is superimposed on the natural fluorescence spectrum
of the tissue (autofluorescence) as shown in the figure. The spectral
fingerprint of HPD identifies the tumor.

The excited HPD molecules can, alternatively, transfer their acquired
energy to oxygen molecules that end up in a singlet state. Singlet oxygen
is a strong toxic agent which violently oxidizes the surrounding (tumor)
tissue leading to necrosis. This photodynamic therapy (PDT) is performed
at 630 nm where HPD still absorbs and where the body is more "transparent"
than at shorter wavelengths. The result of such a treatment is shown in
the figure for a human basalioma 2 days after uniform irradiation of the
lesion and surrounding tissue.

Presently, dihematoporphyrin ether/ester (DHE) is generally used in
clinical applications. Much work is now being invested in the development
and testing of more efficient sensitizing drugs. Among interesting drugs,
tetrasulphonated phthalocyanine, mono-L-aspartyl chlorine e6 (MACE) and
bensoporphyrin mono acid (BPMA) can be mentioned. The last one has particularly attractive features including strong PDT action and fluorescence
emission, absorption extended towards larger (less absorbed) wavelengths
(690 nm), and fast metabolization.

3. Fluorescence Diagnostics of Tissue

As has been mentioned in the previous section, HPD fluorescence is
useful in identifying malignant tumors, which may otherwise be difficult

to observe. Various point monitoring systems for tissue analysis have been designed. We have shown, that improved demarcation of tumors can be achieved by combining the specific red drug fluorescence with the blueish autofluorescence, which has been observed to be reduced in tumors of different kinds. By forming the ratio between the increasing red signal and the decreasing blue signal, contrast enhancement can be achieved. Such a ratio is a dimensionless quantity, which has the attractive properties of being immune to distance variations, surface topography and changes in excitation and detection efficiency. Recently, we have performed clinical studies of pure tissue autofluorescence using a mobile fluorosensor. Brain tissue was shown to provide a great deal of spectral information and tumors in the oral cavity were shown to accumulate endogeneous porphyrins.

A particularly interesting task is to try to spectroscopically distinguish atherosclerotic plaque from normal artery wall to guide laser plaque ablation. N2-laser excited spectra from plaque and normal wall are shown in Fig. 2. In scans through a plaque region certain intensity ratios are shown to demarcate the plaque well, whereas other ratios exhibit no contrast. Studies of plaque fluorescence have been performed by several groups (See e.g. (8,7). Reabsorption of fluorescence light by blood changes the shape of the spectra substantially and for clinical application it is important to eliminate the blood interference.

4. Picosecond Laser-induced Fluorescence Studies

Since fluorescence spectra from tissue do not exhibit much structure it is important to utilize the temporal characteristics of the fluorescence (9,10). We have performed measurements with a picosecond dye laser system used in conjunction with delayed coincidence detection. We have observed that DHE fluorescence at 630 nm is considerably more long-lived than the autofluorescence background at the same wavelength when excited at 325 nm (11). Thus, by using delayed detection, the DHE-related fluorescence can be lifted off from the spectra. Time-resolved fluorescence spectroscopy also offers improved demarcation of plaque, since it has been found that plaque fluorescence at 400 nm is considerably more long-lived than that from normal vessel wall (12). The fluorescence is long-lived enough to allow the use of a short-pulse N2 laser and a dual-channel boxcar integrator to discriminate between "late" and "early" fluorescence. A recording of the ratio between the two signals is shown in Fig. 3. Since only one detection wavelength is used immunity to blood reabsorption is also obtained.

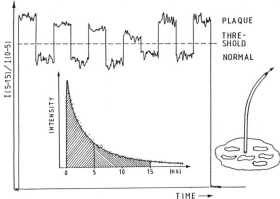

Fig. 3. Demarcation of plaque using time-resolved LIF spectroscopy (7).

5. Propagation of Picosecond Pulses in Tissue

The optical properties of tissue are of considerable interest, e.g. for light dosimetry in connection with PDT and for tissue transillumination (diaphanography). Time-resolved measurements on red picosecond pulses offer new possibilities to study the complex interplay between multiple scattering and absorption. Several groups have recently performed studies in this area, for example to assess tissue oxygenation and concentration of injected drugs (13). We have explored another aspect of time-resolved propagation in tissue to achieve enhanced contrast in diaphanography. The principle and some results are shown in Fig. 4. Due to multiple scattering, no sharp shadows of, for instance, bones are obtained in transillumination. A very broad distribution of arrival times of transmitted photons is obtained, as shown in the figure for the case of a 4 mm diameter metal rod in a strongly scattering liquid. However, by gating on the photons that arrive first at the detector enhanced viewing is obtained, since photons traveling in a straighter path are utilized (14). Thus, an enhanced shadow effect is obtained in a scan across the rod as illustrated in the figure.

6. Multi-color Fluorescence Imaging

Imaging techniques for laser-induced tissue fluorescence are of considerable diagnostic interest. Such systems have been used to visualize bronchogenic and bladder tumors (15) after injection of HPD. We have applied our previously developed concepts for contrast enhancement to multi-color fluorescence imaging, where spatial, as well as spectral, resolution is obtained simultaneously (16). Fluorescence light from an extended object is collected by a Cassegranian telescope with the main mirror divided into 4 individually adjustable segments. Light falling on each segment is filtered through interference filters and the resulting four fluorescence images are adjusted to fall as quadrants on a CCD detector preceded by an

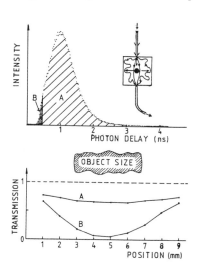

Fig. 4. Illustration of enhanced contrast in time-resolved diaphanography.

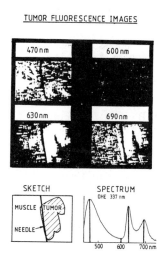

Fig. 5. Multi-color LIF imaging of a rat malignant tumor in a muscle environment.

image intensifier tube (11). In Fig. 5 an example of multi-color fluorescence imaging is shown for the case of a tumor of a rat that had previously been injected with DHE. A tumor fluorescence spectrum is shown with selected fluorescence bands indicated. From the individual images a contrast enhanced image can be calculated. Since the intensifier can be gated down to 5 ns, further contrast enhancement can be obtained by combining information from "early" and "late" images.

7. Conclusions

As has been exemplified above, laser spectroscopy in the wavelength and the time domain adds several new aspects to the use of lasers in medicine. Apart from basic studies on the molecular and cell levels, an important impact on the clinical management of cancer and circulating system diseases can be foreseen. The field presents a major challenge on laser spectroscopists in collaboration with physicians.

This work was supported by the Swedish Cancer Society (RmC), the Swedish Medical Research Council (MFR), the Swedish Board for Technical Development and the Knut and Alice Wallenberg Foundation.

1. J.L. Boulnois, Lasers Med. Sci. **1**, 47 (1986).

2. K. Greulich and J. Wolfrum (Eds.), Lasers in the Life Sciences, Ber. Bunsenges. Phys. Chem. **93**, No. 3, 233-415 (1989), special issue.

3. S. Svanberg, Physica Scr. **T26,** 90 (1989).

4. K. Svanberg, The Interaction of Laser Light with Tissue. PhD Thesis (Lund University Hospital, Lund 1989)

5. Special Issue on Lasers in Biology and Medicine, IEEE J. Quant. Electr. **QE-23,** 1701-1852 (1987)

6. T.J. Dougherty, in CRC Critical Reviews in Oncology/Hematology, edited by S. Davis (CRC, Boca Raton 1984).

7. S. Andersson-Engels, J. Johansson, U. Stenram, K. Svanberg and S. Svanberg, Lasers Med. Sci, in press, and to appear.

8. R. Richards-Kortum *et al.*, Spectrochim. Acta **45A**, 87 (1989) and Refs. therein.

9. D.B. Tata, M. Foresti, J. Cordero, P. Thomachefsky, M.A. Alfano, and R.R. Alfano, Biophys. J. **50**, 463 (1986)

10. J.J. Baraga *et al.*, Spectrochim. Acta **45A**, 95 (1989).

11. S. Andersson-Engels, J. Johansson and S. Svanberg, to appear.

12. S. Andersson-Engels *et al.*, J. Photochem. Photobiol., in press

13. B. Chance (ed.), *Photon Migration in Tissue* (deGruyter, Berlin 1989)

14. R. Berg, Diploma Paper (Lund Inst. of Techn., Lund 1989) to appear.

15. A.E. Profio, SPIE Proc. **905**, Contr. 11 (1988)

16. P.S. Andersson, S. Montán and S. Svanberg, IEEE J. **QE-23**, 1798 (1987)